J.

Methods
in
Animal Physiology

Editors
Zdeněk Deyl
Associate Professor
Head
Department of Extracellular Physiology
Institute of Physiology
Czechoslovakia Academy of Sciences
Prague, Czechoslovakia

Josef Zicha
Physiologist, Research Worker
Department of Developmental Physiology
Institute of Physiology
Czechoslovakia Academy of Sciences
Prague, Czechoslovakia

CRC Press, Inc.
Boca Raton, Florida

Library of Congress Cataloging-in-Publication Data

Methods in animal physiology / editors. Zdeněk Deyl, Josef Zicha.
   p. cm.
  Includes bibliographies and index.
  ISBN 0-8493-6965-7
  1. Physiology, Experimental—Methodology. 2. Pathology.
Experimental—Methodology. 3. Diseases—Animal models. I. Deyl,
Zdeněk. II. Zicha, Josef.
QP43.M47 1989
619—dc19

88-21005
CIP

Direct all inquiries to CRC Press, Inc., 2000 Corporate Blvd., N.W., Boca Raton, Florida, 33431.

© 1989 by CRC Press, Inc.

International Standard Book Number 0-8493-6965-7

Library of Congress Card Number 88-21005
Printed in the United States

INTRODUCTION

Nobody is going to deny that the spectrum of methods exploited in physiology is large ranging from behavioral studies, chronical or acute tests in whole animals, experiments with isolated organs to ultrastructural studies and purely biochemical analyses. This diversity sometimes makes the communication between physiologists rather difficult. Naturally a particular category of problems can be solved by a limited number of techniques. The choice of methods is based first on the available literary data; the personal experience, the overall scientific milieu, and frequently also the availability of the specific equipment play the important role.

The aim of the present volume was to give an overview over different available methodological approaches. The specialists may, perhaps, object that in their particular field the level of information is superficial. However, let them look at other chapters in which different approaches are discussed and which, surely, will appear less superficial from the more general point of view. We hope, at least, that crucial references can be traced throughout the book that would enable the readers to go in more detail when desired. It was really one of our ideas to draw the survey of possibilities available. If this can stimulate the readers to use other methods than those they are routinely using the goals will be met.

In order to keep the volume within reasonable limits we had to cut the individual topics somehow. First, this book does not deal with neuroscience methodology as such. In our eyes, this is a specialized area in which monographs are available.[1-8] We also did not attempt to review biochemical, separational, or immunological techniques and in this respect we limited ourselves to quoting more specialized monographs or reviews.

At the beginning we had the feeling which later turned to be fact, that the search for methodical approaches within the physiological literature is rather difficult. Therefore we prepared this set of reviews. Because of the diversity mentioned above, we are aware of the fact our presentation may not be complete or may suffer from other disadvantages. Therefore we would appreciate any comments and hints of the potential readers.

Zdeněk Deyl
Josef Zicha

REFERENCES

1. **Brown, C. C.,** *Methods in Psychophysiology,* Williams & Wilkins, Baltimore, 1967.
2. **Bureš, J., Petráň, M., and Zachar, J.,** *Electrophysiological Methods in Biological Research,* 3rd ed., Academic Press, New York, 1967.
3. **Goldstein, N. N. and Free, M. J.,** *Foundations of Physiological Instrumentation,* Charles C Thomas, Springfield, 1979.
4. **Myers, R. D.,** *Methods in Psychobiology,* Vol. 1 and 2, Academic Press, New York, 1971—1972.
5. **Skinner, J. E.,** *Neuroscience, A Laboratory Manual,* W. B. Saunders, Philadelphia, 1971.
6. **Thompson, R. F. and Patterson, M. M., Eds.,** *Methods in Physiological Psychology,* Vol. 1, Part A—C, Academic Press, New York, 1973—1974.
7. **Justice, J. B., Jr.,** *Voltammetry in the Neurosciences, Principles, Methods and Applications,* Humana Press and Wiley, Chichester, 1987.
8. **Tuffery, A. A.,** *Laboratory Animals, An Introduction for New Experimenters,* Wiley, Chichester, 1987.

THE EDITORS

Zdeněk Deyl, Ph.D., D.S. is an Associate Professor of Analytical Medicinal Chemistry and Head of the Department of Extracellular Physiology of the Institute of Physiology. Czechoslovakia Academy of Sciences, Prague. Dr. Deyl received his education in chemistry at Prague Institute of Technology and in physiology at Charles University in Prague. He received a Ph.D. degree in chemistry in 1960 and in biology in 1968. In 1974 he was awarded a D.S. degree in nutrition science from Prague Institute of Technology. In 1966 he joined the Institute of Physiology. Czechoslovakia Academy of Sciences and moved through the ranks to his present position. Dr. Deyl's education was enhanced by spending 2 years as a visiting scientist at the Biology Department, MIT, and Biochemistry Department, Boston University Medical School. Since 1967 he is one of the acting co-editors of the *Journal of Chromatography* and since 1985 editor of the Special Volumes of the *Journal of Chromatography*, Biomedical Applications, and since 1974 an editorial board member of *Mechanisms of Aging and Development*. His main interests are devoted to the physiology and biochemistry of the extracellular matrix and to the separation of complex biochemical mixtures. Courses which he regularly teaches deal with the last subject. He is the author or editor of two monographs on chromatography and electrophoresis and two additional ones dealing with connective tissue physiology. In 1979 he was awarded the Czechoslovak State Prize for eminence in science.

Josef Zicha, M.D., Ph.D. is a research worker in the Department of Circulatory Homeostasis of the Institute of Physiology, Czechoslovakia Academy of Sciences in Prague. He joined this institute in 1974 after his graduation as M.D. on the Faculty of Paediatric Medicine at Charles University in Prague. Dr. Zicha was engaged in the study of developmental aspects of renal and cardiovascular physiology. In 1983 he received his Ph.D. degree in physiology and pathophysiology for his study of age-dependent development of salt hypertension.

His main scientific interest is devoted to the role of environmental factors in the development of induced and genetic forms of experimental hypertension. In 1984 he was awarded the Young Investigator's Award of International Society of Hypertension. The last 2 years he spent in the laboratory of Professor J. Duhm (Institute of Physiology, University of Munich) as a research fellow of the Alexander von Humboldt Foundation by studying red blood cell ion transport in experimental hypertension.

CONTRIBUTORS

Milan Adam
Professor and Chief
Department of Connective Tissue and
 Research
Rheumatism Research Institute
Prague, Czechoslovakia

Marie Baudyšová
Department of Physiology
Institute of Physiology
Czechoslovakia Academy of Sciences
Prague Czechoslovakia

Jan Černý
Associate Professor of Surgery
Head
Department of Cardiac and Transplant
 Surgery
J. E. Purkyne University
Brno, Czechoslovakia

Zdeněk Deyl
Associate Professor
Head
Department of Extracellular Physiology
Institute of Physiology
Czechoslovakia Academy of Sciences
Prague, Czechoslovakia

Stanislav Ďoubal
Assistant Professor
Department of Physical Chemistry
Charles University
Hradec Kralové, Czechoslovakia

Ivan Dvořák
Chief, Bioengineering, Applied
Mathematics and Information Department
Psychiatric Research Institute
Prague, Czechoslovakia

Tomaš Havránek
Doctor
Institute of Computer Science
Czechoslovakia Academy of Sciences
Prague, Czechoslovakia

Jiří Heller
Institute of Clinical and Experimental
 Medicine
Prague, Czechoslovakia

J. Hökl
Institute of Clinical and Experimental
 Medicine
Prague, Czechoslovakia

Karel Janáček
RNDr.
Institute of Microbiology
Czechoslovakia Academy of Sciences
Prague, Czechoslovakia

Jiří Jelínek
Head
Department of Circulatory Homeostasis
Czechoslovakia Academy of Sciences
Prague, Czechoslovakia

Daniela Ježová
Senior Researcher
Institute of Experimental Endocrinology
Slovak Academy of Science
Bratislava, Czechoslovakia

Ludmila Kazdová
Experimental Laboratory
Institute of Clinical and Experimental
 Medicine
Prague, Czechoslovakia

Ali A. Khraibi
Research Fellow
Department of Physiology and Biophysics
Mayo Clinic and Foundation
Rochester, Minnesota

Petr Klemera
Department of Physical Chemistry
Charles University
Hradec Kralové, Czechoslovakia

Pavel Klír
V.M.D.
Department of Biological Experimental
  Models
Czechoslovakia Academy of Sciences
Prague, Czechoslovakia

Jan Knopp
Institute of Experimental Endocrinology
Slovak Academy of Sciences
Bratislava, Czechoslovakia

Vladimír Kočandrle
Director Professor of Surgery
Institute of Clinical and Experimental
  Medicine
Prague, Czechoslovakia

Frantisek Kolář
Physiologist, Research Worker
Institute of Physiology
Czechoslovakia Academy of Sciences
Prague, Czechoslovakia

Jaroslav Kuneš
Physiologist, Research Worker
Institute of Physiology
Czechoslovakia Academy of Sciences
Prague, Czechoslovakia

Věra Nováková
Department of Ethology
Institute of Physiology
Czechoslovakia Academy of Sciences
Prague, Czechoslovakia

Marie Nožičková
Research Biologist
Experimental Laboratory
Institute of Clinical and Experimental
  Medicine
Prague, Czechoslovakia

Bohuslav Ošťadal
Physiologist, Research Worker
Institute of Physiology
Czechoslovakia Academy of Sciences
Prague, Czechoslovakia

František Paleček
Chairman Institute of Pathophysiology
Faculty of Pediatrics
Charles University
Prague, Czechoslovakia

Petr Pavel
Research Associate
Department of Surgery
Institute of Clinical and Experimental
  Medicine
Prague, Czechoslovakia

Rudolf Poledne
Senior Scientist
Department of Metabolic Studies
Institute of Clinical and Experimental
  Medicine
Prague, Czechoslovakia

Michal Pravenec
Dipl. Ing.
Department of Experimental Models
Institute of Physiology
Czechoslovakia Academy of Sciences
Prague, Czechoslovakia

R. Rybová
Institute of Microbiology
Czechoslovakia Academy of Sciences
Prague, Czechoslovakia

Jiří Samohýl
Department of Plastic Surgery
KUNZ
Brno, Czechoslovakia

Pavel Šebesta
Research Associate
Deparment of Surgery
Institute of Clinical and Experimental
  Medicine
Prague, Czechoslovakia

Thomas L. Smith,
Research Assistant Professor
Department of Physiology and
  Pharmacology
Bowman Gray School of Medicine
Wake Forest University
Winston Salem, North Carolina

David Štepán
Department of Plastic Surgery
KUNZ
Brno, Czechoslovakia

Vladimír Štrbák
Head
Department of Developmental
  Endocrinology
Institute of Experimental Endocrinology
Slovak Academy of Sciences
Bratislava, Czechoslovakia

P. Tatár
Fellow Researcher
Institute of Experimental Endocrinology
Slovak Academy of Sciences
Bratislava, Czechoslovakia

Jiří Vaněček
Physiologist, Research Worker
Institute of Physiology
Czechoslovakia Academy of Sciences
Prague, Czechoslovakia

Antonín Vrána
Senior Scientist
Experimental Laboratory
Institute of Clinical and Experimental
  Medicine
Prague, Czechoslovakia

Josef Zicha
Physiologist, Research Worker
Department of Developmental Physiology
Institute of Physiology
Czechoslovakia Academy of Sciences
Prague, Czechoslovakia

TABLE OF CONTENTS

Biological Models Availability

Chapter 1

MODELS IN EXPERIMENTAL MEDICINE

P. Klír and M. Pravenec

TABLE OF CONTENTS

I. INTRODUCTION

In addition to new methods and methological aspects, a dynamic development of biomedical research during the last 50 years has given rise to a new multi- and interdisciplinary fields of science. The overlapping of these fields naturally resulted in changes of scientific nomenclature.

During the last 15 years "model" and "modeling" have often appeared in biomedical research. MEDLINE/MEDLARS system data covering the period of 1972 to 1982 showed a remarkably stable frequency of papers dealing with modeling (2 to 2.5%) in medical literature. This is thought to be an evidence that medical science, even in the period when computers are increasingly utilized, remains at the level of empiricism.[1] On the other hand, modeling seems to be used in biomedical research more extensively than at the low frequency given above, provided "model" and "modeling" are understood in a broader sense than a mere mathematical modeling.

II. BASIC MODEL CLASSIFICATION

The terms of model and modeling are ambiguous. Philosophically, a model formulates a way, a method, and a means of cognition consisting of a reflection of reproduction of the studied phenomenon utilizing an artificially formed system. A similarity exists between the model and the original; the model substitutes for the object studied in the process of scientific cognition, and the investigation of the model enables us to obtain information on the original. The limits of modeling are those of analogy between the model and the object.

Models can be classified from various viewpoints. If a basic philosophical classification of models (material, ideal) is adopted, they can be employed in biomedical research even if it involves some specific features. Any real object, either natural or artificial, must be considered a material model, whereas any imaginary object represents an ideal model. Statistical models are most often used as the ideal ones in biomedical research.[2] Logically, a biological model, an animal model, or an animal disease model belong to the material models that are natural and exist in reality. In order that the concepts can be defined, their mutual relationships must be specified.

The first of the above-mentioned models was defined as the animal disease model.[3] It represents a living organism with an inborn, acquired, or induced pathological abnormality, similar in one or more aspects to the same phenomenon in man. This definition was adopted even by *ILAR News*.[4] National Research Council introduced a simple classification of animal models in 1979:[5] (1) spontaneous disease models, (2) experimentally provoked disease models, and (3) hitherto unavailable models necessary for research.

Both Wessler's definition and the above-mentioned classification are concepts in a narrower sense aimed mostly at pathological processes. As the biomedical research comprises more aspects than only pathological ones, the animal model has a broader reach. In 1981, the Institute of Laboratory Animal Resources (ILAR) adopted a modified Wessler's definition[6] that could be considered a general definition of the animal model: "An animal model is a living organism in which normative biology or behavior can be studied, or in which the phenomenon in one or more respects resembles the same phenomenon in humans or other species of animals."[11] This new definition specifies not only the animal disease model, but also, more generally, the animal model itself used to study normal (physiological) and abnormal (pathological) processes. Therefore, from the point of view of concept hierarchy, the animal model is superior to that of animal disease model. In both cases it represents a living organism (individual) characterized by a set of physical features and life functions, e.g., metabolism, capability of growing and development, reproduction, irritability (sensibility), heredity, and nourishment, that ensure the organism's homeostasis and enable it to exist independently in its environment. However, not only living organisms (animals) are employed as models in biomedical research. Animal and plant models that use only body organs and tissues (at the cellular and subcellular level) have also been developed, as well as models employing populations or organisms. All these models can be generally called the biological models (biomodels).

III. CONCEPT DEFINITIONS

A. Biological Model

A biological model is defined as a living system which, if studied according to fixed rules, allows reproduction and analogical derivation of the behavior and properties of the original system (i.e., of another living system studied by using a model analysis).[7] The definition suggests that any living system in nature, both in vivo and in vitro, consisting of a set of constituents exhibiting mutual relationships and effects can become the model.

B. Animal Model

An animal model can be defined as a living organism which, if studied according to fixed rules, allows reproduction and analogical derivation of the behavior and properties of the original object (mostly man) studied by a model analysis. The definition includes models evolved to study normal and abnormal processes and regulating principles in living organisms. A problem is that the boundaries among different types of models defined below are not clear; sometimes the individual categories overlap. It is true especially for the boundaries among models used for the study of normal (physiological) and abnormal (pathological) processes. This is caused by vaguely defined biological standards and by the fact that concepts of illness and health are not distinct enough. Another reason is that often only a functional defect helps in understanding a normal physiological mechanism. Figure 1 shows a simplified scheme of classification of animal models that is generally valid for the animal models developed to study normal and abnormal processes, i.e., the animal disease models.

C. Animal Model of Disease

An animal disease model is represented by a living organism with an inborn, natural, or artificially acquired defect, suffering from a pathological process or having a disposition to the disease, the examination of which according to defined rules allows reproduction and analogical derivation of a pathological behavior and properties of the original object studied by the method of model analysis (i.e., etiology, pathogenesis, and therapy of a chosen nosological unit).

The concept of the animal disease model is inferior to that of the animal model and superior to those of spontaneous disease and evoked disease in animals.

1. Spontaneous Animal Model of Disease

A spontaneous animal disease model is represented by a living organism with an inborn or natural defect, suffering from a pathological process or having a disposition to the disease, the examination of which according to defined rules allows reproduction and analogical derivation of a pathological behavior and properties of the original object studied by the method of model analysis (i.e., etiology, pathogenesis, and therapy of a chosen nosological unit).

The spontaneous animal disease model is not identical to a genetic animal disease model; the latter is included in the former model.

a. Genetic Model of Disease

A genetic animal disease model is a living organism with a genetic defect, suffering from a pathological process, or having a disposition to such defects or processes.

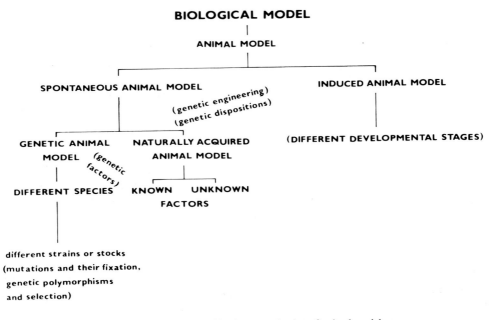

FIGURE 1.   Scheme of basic categorization of animal models.

b. Naturally Acquired Animal Model of Disease

A naturally acquired animal disease model is a living organism with a naturally obtained defect or suffering from a pathological process.

2. Artifically Evoked Animal Model of Disease

An artificially evoked animal disease model is a living organism with an artificially evoked defect, suffering from a pathological process, or having dispositions to the disease, the examination of which according to defined rules allows reproduction and analogical derivation of a pathological behavior or properties of the original object studied by the method of model analysis (i.e., etiology, pathogenesis, and therapy of a chosen nosological unit).

Clear boundaries, however, do not exist between the spontaneous and evoked animal disease models. It is well known that if we select animals exhibiting a strong response to a certain stimulus, the result may be a decrease of the sensitivity threshold leading to the observation of the responses without corresponding stimuli, which means that the induced model becomes a spontaneous one. Dahl rats sensitive to salt in diet[8] or rats sensitive to cholesterol in diet[5] are thought to be typical examples of such a situation. As dispositions are considered to be genetically coded properties of the animal, it is impossible to define in the case of some diseases a clear boundary between the induced and spontaneous models as well as between the induced and genetic animal disease models. Consequently, these models can be classified either as spontaneous or genetic animal disease models. With an increasing ability to define the models used, the extrinsic factors will be better characterized, which will enable us to reclassify many spontaneous models as the induced ones, and, thus, the genetic models will be unambiguously separated from the rest of the spontaneous ones.

The models used to study normal (physiological) processes can also be divided into spontaneous and induced models in a similar way as the animal disease models.

To sum up the classification of animal models employed in biomedical research, we must stress that:

1.   The concept of biomodel covers all living systems
2.   According to the classification of animal models, that of plant models can be made out

3. It is possible to derive analogically the classification of models at the level of organs, tissues, cells, and even at a subcellular level with some limitations

4. Clear boundaries cannot be defined between various categories of animal models, especially, if normal (physiological) and abnormal (pathological) processes and regulating principles are studied; therefore, the classification presented is not enumerative and cannot be considered exhaustive; however, it may be useful in helping to unify the nomenclature.

IV. GENETIC ANIMAL MODEL OF DISEASE

Genetics determine both similar and different properties of the organism including the inter- and intraspecifically conditioned specificity of the phenomenon studied as well as the inter- and intraspecifically bound sensitivity and reactivity to a chosen stimulus. In many cases the selection of a suitable species or strain may be very important for finding a proper solution. Comparative medicine includes the identification, the study, and the use of animal diseases similar to human ones. It has been defined as follows: "Comparative medicine is the study of the nature, cause and cure of abnormal structure and function in people, animals, and plants for the eventual application to and benefit of, all living things."[10]

The selection of a suitable animal species is determined by many factors. The most important criterion should be as good an extrapolation of the results obtained as possible. Traditions and economy also must be taken into account. It is hardly understandable from the biological viewpoint why so many experiments have been conducted with albinic mice and rats. Spontaneous animal disease models (more exactly called genetic animal disease models) are thought to be the most important in experimental medicine. It is because the models obtained by natural infection only rarely allow reproduction and standardization of the model.

In looking for a genetic model, two processes must be distinguished, i.e., identification and formation of the model. During the formation of a model and its further genetic manipulation, the nature of genetic determination of the pathological process must be taken into account. Most present day diseases of unclear etiology are multifactorial. This makes it difficult to formulate a suitable model. Therefore, it is necessary to make use of the data of quantitative genetics. Quantitative genetics, however, is incapable of determining the effect of individual genes on the character studied and, consequently, cannot always predict what the response to the selection will be. Classical genetic models (Mendel, Galton) were derived for a formal study of various phenotypes and only secondarily employed to investigate diseases. The problem is, however, that the diseases are often defined inadequately and artificially.

In formulating a genetic disease model, two cases can be encountered in general:

1. Fixation of mutations (in case of the properties determined by one gene); further genetic manipulation allows preparation of inbred, congenenic, and recombinant-inbred strains; other genetic manipulations are possible by using the methods of molecular biology; fixation of mutations would be facilitated by a possibility of detecting heterozygotes, which often, however, may be done only at the level of DNA

2. Screening for a quantitative property (in case of the properties determined by several genes); such a property is gradually fixed during a strict selective inbreeding or is fixed on the genetic background of an outbred stock (usually during a strong inbreeding depression); the positive alleles are gradually fixed in the model, while the negative alleles are fixed in the control or vice versa (in case of the additive type of heritability); genetic manipulation can produce recombinant-inbred strains from an inbred model and from a second, highly inbred, contrastive strain

The advantages of genetic animal models include spontaneity, high reproducibility, high quality of background data, standard properties of the animals, and predictability. On the other hand, narrow genetic spectrum and inadequate controls are unfavorable features of genetic animal models.

V. EXTRAPOLATION OF EXPERIMENTAL DATA

A biological model must be capable of simulating basic features of the system that will be studied by using the model. Such an analysis can be made before the experiment is conducted, the differences defined, and a decision made whether they can be considered in extrapolation or after some experiments have been carried out. Subsequently, it is necessary to resolve whether it is worth obtaining additional information about the model (in order to define it better) before it is used routinely in research.

If biological models are studied with the aim of obtaining information only about themselves, the word "model" should be avoided in this case. Neither should it be employed on condition that the model was found inadequate. No model, however, can be expected to yield adequate responses to a great many questions. Every model usually helps only partially in understanding the physiological and pathological processes studied. Consequently, it is necessary to keep looking for new models all the time.

Attempts have been made to define general principles that could be used for the extrapolation of the data obtained. For example, "The most striking guiding principle employed for interspecies prediction deals with the recognition of scaling factors, whereby numerous biological parameters are a mathematical function of body weight and these relationships are quite constant over a broad range of species."[11] Each extrapolation beyond such a general framework, however, brings forward many problems because it is limited by inadequate knowledge of basic biological processes. Therefore, it is necessary to be very careful and take into account a detailed knowledge of interspecific differences concerning physiology of the system studied. Nevertheless, extrapolation of the data obtained with diseased animals (animal disease models) to the human organism is even more difficult.

The question is what an adequate disease model should look like. As to a genetic model, it ought to be a living organism whose disease was caused by a gene product(s) changed identically to the one(s) in the human organism.[12]

The following possibilities may occur as a result of an extrapolation (Figure 2).

As far as monofactorially determined diseases are concerned, it is sometimes possible to find homology between animal and human mutations. On the other hand, it is most likely that a comparative study of pathological genetic defects will not suffice for bringing evidence for homology of the two mutated genes. Similar mutations in homologous loci, however, may be expressed in a different way. It is known that many human genetic defects that have long been believed to be simply determined are rather heterogenous, actually. Therefore, it is not possible to verify the adequacy of a model only on the basis of similarity of phenotypic expression of two mutated genes. Such evidence could be obtained by finding a comparable defect at the molecular level, which did not count yet for most mutations. Homology of two mutated genes could be proven by finding out that both genes belong to a conserved linkage group.[13]

In case of multifactorially determined diseases, it is much more difficult to assess the adequacy of animal models. One animal model cannot involve all inner (genetic) and outer (environmental) aspects of the human disease. Extrapolation of the pathological data is further obstructed by the fact that some animal species are not fully defined and our knowledge of their physiological and biochemical processes is fragmentary. Another problem stems from a heterogeneity of many multifactorially determined diseases, when the interaction of inner factors (genetic predisposition) with specific outer stimuli (noxes) is often important for the development of the disease. This interaction can be explained only if such outer and inner factors are well defined.

ANIMAL MODEL   HUMAN DISEASE

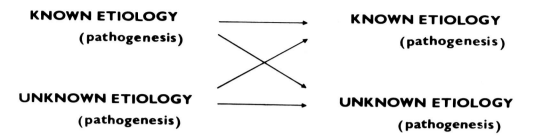

FIGURE 2.   Possibilities of extrapolation of results obtained to human disease.

Consequently, as far as the latter diseases are concerned, even a well-defined and most suitable animal model (adequate animal model) is able to provide only limited information about the human disease. An animal model involving all the aspects of the disease would be a monster.[14] It is not necessary to look for such a model. Models should increase our understanding of basic mechanisms of pathological physiology.

A case, however, may occur in which the model is adequate, even if mutations in homologous loci need not be a condition of the corresponding defect because pathogenetic mechanisms can be relevant. The existence of various models suggesting an inter- and intraspecific variability of symptoms of the disease shows that different human forms of the disease can be considered a part of the biological disease spectrum. Consequently, in the case of monofactorially determined diseases, it seems to be possible (with some limitations) to look up models with homologous mutations. On the other hand, many outer and inner factors do not allow such precision in the case of multifactorially determined diseases. In this situation, models only offer some important possibilities as to how to obtain general information on the disease and thus provide important circumstances for understanding fragmentary data of direct studies of the disease by using the human organism. Such a parallel research employing animal models and the human organism is necessary because, though it seems to be paradoxical, an animal model can be perfect only to the extent the human disease is understood.[15]

Animal extrapolation is a science. It involves a rational process in which the basis of differences and similarities is dissected and evaluated. This process of analysis is only as good as the research base from which it draws its assessments. There is no magic to it.[11]

REFERENCES

1. **Giberg, B. K.,** New computer technologies and their potential for expanded vistas in biomedicine, *Physiologist,* 25, 2, 1982.
2. **Murphy, E. A.,** The quantitative genetics of disease, ambiguities, *Am. J. Med. Genet.,* 7, 103, 1980.
3. **Wessler, Y.,** Introduction, what is a model?, in *Animal Models of Thrombosis and Hemorrhagic Diseases,* Vol. 11, National Institutes of Health, Bethesda, Md., 1976.
4. What is an animal model?, *ILAR News,* 22, 31, 1978.
5. Laboratory animal medicine, guidelines for education and training, *ILAR News,* 22, M1, 1979.
6. National Research Council Committee on animal models for research on aging, in *Mammalian Models for Research on Aging,* National Academy Press, Washington, D.C., 1981.

7. **Klír, P.,** What is a biological model?, *Acta Univ. Carol. Med.,* 1987, in press.

8. **Rapp, J. P. and Dene, H.,** Development and characteristics of inbred strains of Dahl salt-sensitive and salt-resistant rats, *Hypertension,* 7, 340, 1985.

9. **Imai, J. and Matsumara, H.,** Genetic studies on induced and spontaneous hypercholesterolemia in rats, *Atherosclerosis,* 18, 59, 1973.

10. **Bustad, L. K., Gorham, J. R., Hegreberg, G. A., and Padgeth, G. A.,** Comparative medicine. Progress and prospects, *J. Am. Vet. Med. Assoc.,* 169, 90, 1976.

11. **Calabrese, E. J.,** *Principles of Animal Extrapolation,* Neltcalf, R. K. and Stermm, W., Eds., John Wiley & Sons, New York, 1983.

12. **Harris, J. B. and Slater, C. R.,** Animal models. What is their relevance to the pathogenesis of human muscular dystrophy?, *Br. Med. Bull.,* 36, 193, 1980.

13. **Francke, M.,** Cytogenetic approaches to mouse models of human genetic diseases, *Am. J. Pathol.,* 101, 41, 1980.

14. **Herberg, L.,** Spontaneously hyperglycemic animals—models of human diabetes?, *Z. Versuchstierk d.,* 24, 3, 1982.

15. **Reid, Z. M.,** Needs for animal models of human diseases of the respiratory system, *Am. J. Pathol.,* 101, 89, 1980.

Chapter 2

QUALITY OF LABORATORY ANIMALS

P. Klír and M. Pravenec

TABLE OF CONTENTS

I. INTRODUCTION

Increasing the body of knowledge increases the demands that must be met by animal models. The best example is growing requirements concerning the quality of laboratory animals, which have resulted in establishing conventional colonies of laboratory animals that develop into modern barrier breeding units protecting animals against undesirable microorganisms. Consequently, the laboratory animal can be singled out from the group of animals used for experiments. Any animal obtained by chance, caught, or bought, including true laboratory animals, can be considered to be "an experimental animal". Thus, a true laboratory animal clearly differs by its properties. A laboratory animal is a living organism with more or less known genetic properties, bred only for scientific reasons. Consequently, the laboratory animal has both general properties of the experimental animal and the following additional properties: (1) it is bred to be used for laboratory research, (2) it is bred under defined conditions according to strict rules applied for its reproduction, (3) its genetic properties and health are defined, (4) its defined properties and living conditions are standardized during time, and (5) it possesses properties suitable for a laboratory project.

A set of properties is what is meant by the "quality" of a laboratory animal. This is determined by outer and inner factors acting on the breeding population.

A laboratory animal is defined by determining and subsequently meeting the demands of inner and outer factors that act on the breeding population. Therefore, both the properties of the animal itself and the external conditions under which it has been kept before or during experiment must be specified. The extent to which the animal is defined and, also, the fact whether the outlaid properties have been accomplished are the parameters that are normally checked out. The defined properties of the animal and the external conditions must be standardized in time so that the reproducibility of the results obtained with the laboratory animal can be ensured. Standard properties of animals are guaranteed by repeated checking on established requirements concerning the animal and its external life conditions. Consequently, a standard, defined animal can be discussed only if a repeated check on the properties of the bred animal population and its external life conditions is established. The capability of defining and standardizing the animal results not only from the requirements established but also from the level of methods used and the reliability of the control.

II. EXOGENOUS FACTORS AFFECTING QUALITY OF LABORATORY ANIMALS

Attention is paid to the factors that affect the quality of a laboratory animal from the point of view of classification, influence on physiological values of the animal, possibility to ensure suitable life conditions, breeding technology, and the control of the bred animal.[1-7]

Exogenous factors include all that constitutes the animal environment (physical, chemical, and biological factors).

A. Physical Factors

Climatic factors (temperature, humidity, atmospheric pressure, and air flow) are the most important among the physical factors that constitute the environment. The temperature and relative humidity are maintained within the recommended limits,[6] and good air flow is ensured by ventilation.[1] Light represents a very important physical factor. The length of time during which the organism is exposed to light synchronizes cyclic diurnal and seasonal changes of tissues, physiological functions, and behavior.[5,8] The intensity of light affects animal reproduction. The photoperiod and light intensity are relevant to the function of the eyes, especially in albinic animals.[6,8] The photoperiod of small laboratory rodents (rat, mouse) that do not require daylight is controlled artificially by using a 12L:12D schedule. The light color also seems to be important. Sometimes, it can affect maturation and animal reproduction.[8]

Other physical factors include sound (intensity, frequency); electromagnetic signals and waves (wavelength, frequency); ionization; and radiation. These factors must be minimized. From the practical viewpoint, it is essential especially with respect to sound, since the background in a laboratory animal farm should not exceed 60 dBA.[6]

The properties of materials used in farms (construction materials, cages, bedding, physical character of nutrition, dust, particles) may represent important physical factors. The room available to the animal, as given by the size of the cage or run, can be another example of these factors. The sharing of the room by several or many individuals results in intraspecies interactions (territoriality, hierarchy, aggressiveness) that affect neurohumoral regulation of physiological processes.[9] Each kind of laboratory animal has not only specific demands concerning the room necessary for its existence (cage or room size) but also a typical behavior pattern that must be respected both in breeding and during the experiment.[2,9,10]

B. Chemical Factors

Chemical factors are represented by the natural composition of the inhaled air (gas proportion), of food (nutritional components), and of water (salts present), and also by undesirable contaminants (e.g., heavy metals, toxins, carcinogens, and other harmful compounds). The effects of bedding, disinfectants, and other chemicals or drugs also cannot be neglected.[7]

Nutrition of laboratory animals represents an environmental factor of which optimalization and standardization are considered very important.[11-14] This is quite understandable as the food represents the animal's intimate contact with the environment. Three aspects seem to be essential:

1. The amount of food (feeding regimen); small laboratory rodents are usually fed *ad libitum* and bigger animals, e.g., dogs, minipigs, and monkeys, according to a feeding program
2. Good diet formulation that involves demands of the individual kinds of animals as far as the diet composition is concerned (amino acids, fatty acids, carbohydrates, fiber, minerals, vitamins); these questions are being intensively studied; literary data on the amount of the nutrient components needed were shown to differ as much as 10- to 100-fold;[11,13,14] as the content of various diet components in food can not only significantly influence the animal growth, weight gains, reproduction, and life span but also markedly affect the experiment itself,[15,16] much attention must be paid to these questions as well
3. The third aspect is tightly bound with ecological problems; undesirable compounds contaminate food and, by taking part in various interactions within the organism, may distort conclusions of the experiment; a source of these contaminants is natural food, e.g., cereals, alfalfa, fish flour, meat and bone flour, etc.; therefore, commercial diets must be duly checked

C. Biological Factors

Exogenous biological factors affecting the quality of a laboratory animal can be divided into macro- and microbiological ones.[7] Macrobiological factors are represented by the human factor (staff, researcher), other animal kinds (wild rodents), and animals of the same species with which the individual animal is in defined social relationships.

From the point of view of the relationship between man and animal, handling is also important. Small animals, such as rats, mice, guinea pigs, birds, hamsters, and the like, usually do not pose any particular handling problems to trained personnel.[17]

The individual experience of the animal gained by handling favorably affects the organism by getting it ready for the experiment from the psychological point of view. All animals, however, must be prepared by handling in a standard and similar way.

Microbiological factors of the environment include all microorganisms present—viruses, bacteria, yeasts, molds, protozoa, and arthropoda. Microorganisms in the environment take part in long-lived interactions with the animal. The result of these interactions depends on the amount and the properties (virulence, pathogenicity, toxicity) of the microorganisms, on the one hand, and on the properties of the macroorganism (mainly its resistence), on the other hand. If a microorganism invades the macroorganism and colonizes it, the latter becomes a host and the former acts as a significant endogenous factor.

III. ENDOGENOUS FACTORS AFFECTING QUALITY OF LABORATORY ANIMALS

Endogenous factors always have biological character and are for the most part responsible for the quality of the animal. Two groups of factors must be mentioned in this connection that are important for laboratory animal breeding, i.e., genetic factors and microbial colonization of the organism. On the basis of these factors, animals are classified into groups and categories.

A. Genetic Factors
From the genetic point of view, laboratory animals can be divided as follows:

1. Isogenic, i.e., genetically identical (inbred and derived strains and F_1 hybrids resulting from mating of two inbred strains)
2. Nonisogenic, i.e., genetically nonidentical (outbred and random-bred stock)

A number of articles were published dealing with inbred strains, their formation, reproduction, control, and use.[19,28,29] As mostly outbred stocks of rats and mice are used in physiological research, we will mention some disadvantages resulting from lower genetic defining of outbred stocks.

An outbred stock is a population reared as a closed colony, without selection, and in a way ensuring the rise of the inbreeding coefficient does not exceed 1% per generation. The aim is to keep the genotypic spectrum constant. The level of heterozygoticity is not determined for an outbred stock. The defining is in the applied way of mating of animals, and its genetic consequences depend on the genetic history of the stock.[20] It is impossible in practice to keep the genotypic spectrum and characteristics of the stock constant (a fundamental requirement when trying to define and standardize laboratory animals).

In outbred stocks gametes transferring genes to the next generation do not contain all genes of parental generation, contrary to an inbred strain, but only a sample of genes. When this sample is not sufficiently large, i.e., in small populations where breeding animals are not counted in hundreds, changes in gene frequencies will necessarily occur. These random changes in gene frequencies, so-called genetic drift, cause changes of population characteristics. Thus, it is, in principle, impossible to maintain the genetic stability of an outbred stock.

Narrowing of a population is the main cause of the increase of inbreeding and of genetic drift. The reproduction that follows does not influence the inborn inbreeding, but only reduces the rise of new inbreeding. Thus, the size of a population relates to the actual speed of inbreeding accumulation only. Nowadays, Wistar, Spraque-Dawley, and Long-Evans outbred stocks are mostly used. Each individual stock is unique though they all have the same name, e.g., Wistar. This is caused by genetic drift, and so stock standard, background data, and good reproducibility of results do not exist. All outbred stocks were several times strongly reduced in the past.[23]

Festing[24] estimated the inbreeding coefficient of outbred stocks used in Great Britain to correspond to four or more generations of mating brother × sister (above 0.594). Such animals are more akin than siblings, i.e., their relatively degree is that between identical twins and siblings. It cannot be surprising because it is known that, for example, the stock Long-Evans is derived from a single male captured in nature and several Wistar females,[25] and how many albino rats have founded Wistar stocks?

Moreover, outbred stocks are not a good model of an outbred human population. They are extrapolated mostly from rats to humans and not from one population to another. Even if population-to-population extrapolation is the case, such a model cannot be adequate as the human population is much more outbred.

Due to the aforementioned disadvantages it is better to recommend production of synthetic outbred stocks by mating several inbred strains[26] or, as proposed by Festing,[27] to use several inbred strains.

B. Microbial Colonization and Contamination of an Organism

As mentioned before, microbial colonization of an organism results from an interaction of a living organism with outer space where microorganisms are found. Microbial colonization markedly influences the physiology of an animal, above all that of the GI tract whose proper function derives from saprophytic microflora as well.[30-33,38-42] Saprophytic microflora yielding essential metabolites also increase resistance of an organism to pathogenic and facultatively pathogenic microorganisms.[38,39] The composition of an intestinal microbial population can thus influence the biological response of the host[40,41] and its metabolism of tested compounds as far as number and kind are concerned. An optimal situation occurs when animals are maintained so that the invasion of unwanted microorganisms can be prevented and a high degree of microbial stability throughout the life span can be provided.[45] This degree cannot be achieved in conventional breeding units and experimental facilities that are not protected from invading pathogenic microflora.[43,44] The infectious germs can be transferred in air, water, food, insects, contaminated animals or humans, or, in experimental stations, transplanted organs.[43,48,49]

It was possible to decrease the danger of epizootical infections of laboratory animals by means of excluding the horizontal way of infection invasion using barrier systems (specific pathogen-free units) or isolators preventing penetration of undesirable microflora. Moreover, a simple system of flexible isolators[18] enabled use of gnotobiology and, except for transplacentally transferable diseases, preparation of young animals for sterile isolators and so interruption of vertical transfer of infection from mother to progeny.[43] The cessation of both vertical and horizontal infection transfer helped maintain populations of laboratory animals, above all small laboratory rodents.

It was possible to define and standardize laboratory animals, above all small rodents, to a higher level, to increase health of breeding colonies, and to initiate classification of animals based on microbial status.[35-37,45-47] The method of hysterectomy, characteristics, and breeding of young animals in germ-free isolators followed by transfer to transbarrier space enabled simultaneous directed and standard association with saprophytic microflora and investigation of GI colonization resistance.[30-32]

From the point of view of microbial colonization and the depth of control, and resulting from the technology of breeding, basic groups of animals were established. Conventional animals generally are reared under open animal conditions and with usually uncontrolled microbial status or controlled only as far as germs transferable to humans are concerned; specific barrier-free animals or barrier-maintained animals are defined as microbially associated animals removed from an isolator and placed in a barrier and checked for the presence of deliberately or accidentally acquired microorganisms.[45] Gnotobiotic animals are obtained by way of hysterectomy and bred in an isolator using germ-free techniques, free from all forms of associated life (germ-free or axenic animals) or intentionally associated with one, two, or more microbial strains (mono-, di-, or polyxenic animals).[50] Further terms were established, such as "monitored animals", "disease-free animals", etc.[43,45] This simple classification soon proved to suffice for basic orientation in ways of breeding, its technologies, and ways of origin of animals. It is limited only as far as control and presence of pathogens are concerned. Due to these reasons complete catalogues of potential animal pathogens were compiled[46,47] or different ways of grouping animals to quality categories were chosen.[36] Based on the technology of breeding and depth of control, selected pathogenic bacteria, viruses, and parasites were classified to five[36,37] or six[35] categories.

The aim of both given approaches is to yield a first-rate, defined, and standard laboratory animal. It ought to be of interest to a researcher not only to collect data on depth and results of a control, but also, in the case of a long-term experiment, to continue in the control and find out whether the animal is suitable for a designed experiment. For example, animals with some viral diseases in a latent or subchronic form are unsuitable for immunological experiments,[34] and animals contaminated with *Pseudomonas aeruginosa* influence the response of an animal during extreme tests and after irradiation.[51] The quality of animals when testing compounds and drugs is defined in The Good Laboratory Practice Regulations of the Food and Drug Administration as follows: "At the initiation of a nonclinical laboratory study animals shall be free of any disease or any condition that might interfere with the purpose or conduct of the study."[52]

It is clear from the above that a researcher is supposed to have deep knowledge about exogenous and endogenous factors affecting the quality of a laboratory animal. In accord with this is a call of The Working Committee for the Biological Characterization of Laboratory Animals (GV-SOLAS) for more accurate characterization and specification of used animals when presenting results.[53]

REFERENCES

1. **McSheehy, T., Ed.,** Control of the animal house environment, *Laboratory Animal Handbooks,* Vol. 7, Laboratory Animals, London, 1976.
2. Universities Federation for Animal Welfare, *The UFAW Handbook on the Care and Management of Laboratory Animals,* 5th ed., Churchill Livingstone, Edinburgh, 1976.
3. ILAR — Institute of Laboratory Animal Resources, Guide for the Care and Use of Laboratory Animals, U.S. Department of Health, Education and Welfare, NIH Publ. No. 80-23, Animal Resources Program, Bethesda, Md., 1980.
4. Recommendations of the Committee for the Standardisation of Methods in Laboratory Animal Husbandry of the Gesellschaft für Versuchstierkunde, Rossbach, W., Ed., Society for Laboratory Animals Science (GV-SOLAS), Basal, 1980.
5. **Clough, G.,** Environmental effects on animals used in biomedical research, *Biol. Rev.,* 57, 487, 1982.
6. **Clough, G.,** Environmental factors in relation to the comfort and well-being of laboratory rats and mice, in *Proc. Standards in Laboratory Animal Management,* Universities Federation for Animal Welfare, London, 1984, 7.
7. **Heine, W.,** Importance of quality standards and quality control in small laboratory animals for toxicological research, in *Animals for Toxicological Research,* Bartošek, I., Guaitani, A., and Pacei, E., Eds., Raven Pess, New York, 1982, 1.
8. **Weihe, W. H.,** The effect of light on animals, in *Laboratory Animal Handbooks,* Vol. 7, McSheehy, T., Ed., Laboratory Animals, London, 1976, 63.
9. **Price, E. O.,** The laboratory animal and its environment, in *Laboratory Animal Handbooks,* Vol. 7, McSheehy, T., Ed., Laboratory Animals, London, 1976, 7.
10. **Klír, P., Bondy, R., Lachout, J., Jelen, P., and Haniš, T.,** The effect of the number of animals in a cage on the body and organ weight and haematological values in rats, in *Proc. Standards in Laboratory Animal Management,* Universities Federation for Animal Welfare, London, 1984, 93.
11. National Research Council, Nutrient Requirements of Laboratory Animals, No. 10, 2nd ed., 1972.
12. **Edney, A. T. B., Ed.,** *Dog and Cat Nutrition,* Pergamon Press, Oxford, 1982.
13. Laboratory Animals Centre Diets Advisory Committee, Dietary Standards for Laboratory Animals, Medical Research Council, Laboratory Animals Centre, Carshalton, 1977.

14. National Research Council, Nutrient Requirements of Rabbits, 2nd ed., National Academy of Sciences, Washington, D.C., 1977.
15. **Fortmeyer, H. P.**, The influence of exogenous factors such as maintenance and nutrition on the course and results of animal experiments, in *Animals in Toxicological Research*, Bartošek, I., Quaitoni, A., and Pacei, E., Eds., Raven Press, New York, 1982.
16. **Tucker, M. J.**, Nutrition — an important factor, in *Proc. Standards in Laboratory Animal Management*, Universities Federation for Animal Welfare, London, 1984, 51.
17. **Mitruka, B. M.**, Introduction, in *Animals for Medical Research, Models for the Study of Human Disease*, Mitruka, B. M., Rawnsley, M. M., and Vadehra, D. V., Eds., John Wiley & Sons, New York, 1976, 7.
18. **Trexler, P. C. and Reynolds, L. I.**, Flexible film apparatus for the rearing and use of germ free animals, *Appl. Microbiol.*, 5, 406, 1957.
19. **Festing, M. F. W.**, *Inbred Strains in Biomedical Research*, Macmillan, New York, 1979.
20. **Festing, M. F. W., Kondo, K., Loosli, R., Poiley, S. M., and Spiegel, A.**, International standardised nomenclature for outbred stocks of laboratory animals, *ILAR News*, 15, 6, 1972.
21. **Yamada, J., Nikaido, H., and Matsumoto, S.**, Genetic variability within and between outbred Wistar strains of rats, *Exp. Anim.*, 23, 259, 1979.
22. **Lovell, D. P. and Festing, M. F. W.**, Relationships among colonies of the laboratory rat, *J. Hered.*, 73, 81, 1982.
23. Annual Report, Zentralinstitut für Versuchstierzucht, Hannover, 1982, 78.
24. **Festing, M. F. W.**, Phenotypic variability of inbred and outbred mice, *Nature*, 263, 230, 1976.
25. **Robinson, R.**, *Genetics of the Norway Rat*, Vol. 24, Pergamon Press, Oxford, 1965, 449.
26. **Hansen, C.**, Development of the National Institutes of Health genetically heterogenous rat stock, *Alkoholism Clin. Exp. Res.*, 8, 477, 1984.
27. **Festing, M. F. W.**, A case for using inbred strains of laboratory animals in evaluating the safety of drugs, *Food Cosmet. Toxicol.*, 13, 369, 1975.
28. **Festing, M. F. W.**, Properties of inbred strains and outbred stocks, with special reference to toxicity testing, *J. Toxicol. Environ. Health*, 5, 53, 1979.
29. **Festing, M. F. W.**, Inbred strains and the factorial experiment design in toxicological screening, in *Proc. 7th ICLAS Symp.*, Spiegel, A., Erichsen, S., and Solleveld, H. A., Eds., Gustav Fischer Verlag, Stuttgart, 1980, 59.
30. **Koopman, J. P., Janssen, F. G. J., and Van Druten, J. A. M.**, The relation between the intestinal microflora and intestinal parameters in mice, *Z. Versuchstierkd.*, 19, 54, 1977.
31. **Koopman, J. P., and Kennis, M. M.**, Characterization of anaerobic caecal bacteria in mice, *Z. Versuchstierkd.*, 21, 185, 1979.
32. **Koopman, J. P. and Welling, G. W.**, Converting germ-free mice to the normal state with defined anaerobic bacteria, in *Proc. 7th ICLAS Symp.*, Spiegel, A., Erichsen, S., and Solleveld, H. A., Eds., Gustav Fischer Verlag, Stuttgart, 1980, 193.
33. **Veninga, T. S., Dankert, J., and Beenaker, K.**, Variations in the microbial gut flora of conventional mouse colony maintained under controlled conditions, in *Proc. 7th ICLAS Symp.*, Spiegel, A., Erichsen, S., and Solleveld, H. A., Eds., Gustav Fischer Verlag, Stuttgart, 1980, 223.
34. **Hsu, C. K., New, A. E., and Mayo, J. G.**, Quality assurance of rodent models, in *Proc. 7th ICLAS Symp.*, Spiegel, A., Erichsen, S., and Solleveld, H. A., Eds., Gustav Fischer Verlag, Stuttgart, 1980, 17.
35. **Klír, P., Bondy, R., Přibylová, M., Jelen, P., Svoboda, T., Pospíšil, M., Pravenec, M., Boubelík, M., and Čapková, J.**, Categorization of laboratory rats and mice on the basis of microbiological status, presented at 22nd Int. Symp. on Biological Models, Hrubá Skála, April 24 to 28, 1984, 39.
36. The Accreditation and Recognition Schemes for Suppliers of Laboratory Animals, Manual Ser. No. 1, Medical Research Council, Laboratory Animals Centre, Carshalton, 1974.
37. Standardised Laboratory Animals, Manual Ser. No. 2, Medical Research Council, Laboratory Animals Centre, Carshalton, 1977.
38. **Dubos, R. J. and Schaedler, R. W.**, The effect of the intestinal flora on the growth rate of mice and on their susceptibility to experimental infection, *J. Exp. Med.*, 111, 407, 1960.
39. **Dubos, R. J. and Schaedler, R. W.**, Effect of diet on the fecal bacterial flora of mice and on their resistance to infection, *J. Exp. Med.*, 111, 1161, 1960.
40. **Dubos, R. J.**, Lasting biological effects of early influences, *Perspect. Biol. Med.*, 12, 479, 1969.
41. **Dubos, R. J., Schaedler, R. W., Costello, R., and Hoet, P.**, Indigenous, normal and autochthoncus flora of the gastrointestinal tract, *J. Exp. Med.*, 122, 67, 1965.
42. **Williams, R. T.**, Toxicological implications of biotransformations by intestinal microflora, *Toxicol. Appl. Pharmacol.*, 23, 769, 1972.
43. **Myers, D. D.**, Control of microbial and parasitic contamination in the production of laboratory rodents, *Lab. Anim. Sci.*, 30, 330, 1980.
44. **Fujiwara, K.**, Health assessment for laboratory rodents production colonies, *Lab. Anim. Sci.*, 30, 298, 1980.

45. Long-term holding of laboratory rodents, report of the Committee on Long-Term Holding of Laboratory Rodents, Institute of Laboratory Animal Resources, *ILAR News,* 19(4), LI, 1976.

46. Annual Report, Zentralinstitut für Versuchstierzucht, Hannover, 1985, 15.

47. Liste von Erregern zur Specifizierrung bei SPF Versuchstieren, Gesellschaft für Versuchstierkunde (SOLAS), No. 2, Basel, 1977.

48. **Thunert, A.,** The development of bacteriological and virological standards for quality control of laboratory animals, in *Animals in Toxicological Research,* Bartošek, I., Quaitani, A., and Pacei, E., Eds., Raven Press, New York, 1982, 91.

49. **Thibert, P.,** Control of microbial contamination in the use of laboratory rodents, *Lab. Anim. Sci.,* 30, 339, 1980.

50. **Müller, G. and Keissig, R.,** Einführung in die Versuchstierkunde, in *Allgemeine Versuchstierkunde,* VEB Gustav Fischer Verlag, Jena, 1977, 97.

51. **Flynn, R. J.,** *Pseudomonas aeruginosa* infection and radiobiological research at the Argonne National Laboratory: effects, diagnosis, epizootiology and control, *Lab. Anim. Care,* 13, 25, 1963.

52. Department of Health, Education and Welfare, Food and Drug Administration: nonclinical laboratory studies, good laboratory practice regulations, *Fed. Regist.,* 43, 59986, 1978.

53. Working committee for the Biological Characterization of Laboratory Animals/GV-SOLAS, Quidelines for specification of animals and husbandry methods when reporting the results of animal experiments, *Lab. Anim.,* 19, 106, 1985.

Chapter 3

SOME ANIMAL MODELS FOR THE STUDY OF HUMAN DISEASES

M. Pravenec and P. Klír

TABLE OF CONTENTS

I. INTRODUCTION

Systematic classification of animal models for the study of etiology, pathogenesis, and therapy of human diseases was markedly focused on in the literature. A number of monographs and reviews[1-8] can be recommended for further study of the problem. A bibliography of the use of animal models in biomedical research is regularly published in *ILAR News*.

The interest of the public in animal models is usually related to research trends oriented to serious human diseases (hypertension, artherosclerosis, diabetes mellitus, and cancer). Besides classical animal models bred as laboratory animals, more and more frequently domestic and wild animals and those kept in zoos are used as animal models.[9-11] This is enabled by the growing knowledge of veterinary and comparative medicine, i.e., knowledge of physiology and pathophysiology of these animals.

Although the requirements on defining and standardizing wild and zoo animals cannot be applied to such an extent as described in the previous chapter, the influence of exogenous and psychogenic factors (life in capture, fixation during an experiment) on data obtained using these animal models should not be neglected.

The following tabular classification (Tables 1 through 5) of animal models for the study of diseases of human organ systems reflects our choice of models. Further information can be found in corresponding monographs.[4-7,12]

In the tables, the animal kind or strain, disease, and reference are always given. The names of animals do not correspond to the current zoological classification (impossible in many cases); these were taken from authors of corresponding references. Similarly, under the term "disease" in most cases the defined nosologic terms are not included, but these represent morphological or functional changes of the corresponding organ or system as described by the author.

The following tables cannot be exhaustive. They emphasize the vast array of existing animal models and literary sources for the study of serious medical problems.

Table 1
MUSCULAR SYSTEM

Species/strain	Disease	Ref.
Mice		
129/ReJ-dy	Muscular dystrophy	13
C57BL/6J-dy	Muscular dystrophy	14
C57BL/6J-dy[RJ]	Muscular dystrophy	15
Poultry		
Chicken	Muscular dystrophy	16
Duck	Muscular dystrophy	17
Chicken	Muscular dystrophy	18
Turkey	Muscular dystrophy	19
Dogs		
Scottish terrier	Stiff-man syndrome	20
Labrador retriever	Myopathy with type II skeletal muscle fiber deficiency	21
Dog	Malignant hyperthermia	22
Jack Russell terrier	Myasthenia gravis	23
Fox terrier	Myasthenia gravis	24
Dog	Localized myositis ossifficans	25
Cats		
	Myasthenia gravis	26
	Localized myositis	27
Pigs	Myofibrillar hypoplasia	28
	Myofibrillar hypoplasia	29
	Myofibrillar hypoplasia	30
	Malignant hyperthermia	31
	Malignant hyperthermia	32
	Myositis ossificans	33
Other species		
Mink	Muscular dystrophy	34
Cattle	Myopathy associated with hydrocephalus	35

Table 2
CARDIOVASCULAR SYSTEM

Species/strain	Disease	Ref.
Rats		
ALR	Atherosclerosis	36
RICO	Atherosclerosis	37
Koletsky	Atherosclerosis	38
Zucker	Atherosclerosis	39
PHH	Atherosclerosis	40
ExHC	Atherosclerosis	41
SHC, SLC	Atherosclerosis	42
NAR	Atherosclerosis	43
Rat	Atherosclerosis	44
	Atherosclerosis	45
SHR	Systemic hypertension	46
SHRSP	Systemic hypertension	47
GH	Systemic hypertension	48
MHS	Systemic hypertension	49
S/JR, R/JR	Systemic hypertension	50
Münster	Systemic hypertension	51
RI (SHR × BN.1x)	Systemic hypertension	52
BHR	Systemic hypertension	53
LH, LN, LL	Systemic hypertension	54

Table 2 (continued)
CARDIOVASCULAR SYSTEM

Species/strain	Disease	Ref.
Sabra strain	Systemic hypertension	55
MIR	Myocardial ischemia	56
SHR/N-cp	Cardiomyopathy	57
Rat	Cardiomyopathy	58
	Cerebral infarction	59
STR	Spontaneous thrombogenic disease	60
Mice		
NZB	Atherosclerosis	61
C 57 BL/cdJ	Atherosclerosis	62
Mouse	Systemic hypertension	63
	Aortic aneurysm	64
BALB/c, C3H, DBA/2,C57BL/6	Dystrophic cardiac calcinosis	65
Bar Harbor Re129	Cardiomyopathy	66,67
Dogs		
Dog	Ventricular tachyarrhythmia	68
	Pulmonary hypertension	69
	Pulmonary hypertension	70
	Cyanotic heart disease	71
	Portacaval shunt syndrome	72
	Portacaval shunt syndrome	73
	Systemic hypertension	74
	Cardiomyopathy	75
	Mitral incompetence	76
	Neonatal endocardial fibroelastosis	77
Collie	Pericardial diaphragmatic hernia	78
Great dane	Mitral incompetence	79
	Endocardial fibroelastosis	80
Dog	Mitral valve insufficiency	81
	Tricuspid insufficiency	85
	Patent dictus arteriosus	83
Newfoundland	Subaortic stenosis	84
Keeshound dog	Tetralogy of Fallot	85
Dog	Coarctation of the aorta	86,87
Cats		
Cat	Aortic embolism	88
	Aortic thrombosis	89
	Cardiomyopathy	90
	Displasia of tricuspid valve	91
	Pulmonary stenosis	92
	Tetralogy of Fallot	93
Rabbits		
Rabbit	Systemic hypertension	94
	Systemic hypertension	95
	Heart failure	96
	Cardiomyopathy	97
	Atherosclerosis	98
	Atherosclerosis	99
Watanabe	Atherosclerosis	100
Pig	Nutritional cardiomyopathy	101
	Atherosclerosis	102
	Atherosclerosis	103
	Atherosclerosis	104
	Atherosclerosis	105
	Atherosclerosis	106
	Myocardial ischemia	107
Yucatan miniature	Atherosclerosis	108

Table 2 (continued)
CARDIOVASCULAR SYSTEM

Species/strain	Disease	Ref.
Poultry		
Turkey	Systemic hypertension	109
Chicken	Systemic hypertension	110
Turkey	Cardiomyopathy	111
	Cardiomyopathy	112
	Ventricular septal defect	113
Fowl	Ventricular septal defect	114
Birds	Atherosclerosis	115
Pigeon	Atherosclerosis	116
	Atherosclerosis	117—119
Japanese quail	Atherosclerosis	120
Nonhuman primates		
Cynomogulus monkey	Atherosclerosis	121
Rhesus monkey	Atherosclerosis	122
	Atherosclerosis	123
	Atherosclerosis	124
Macaca mulatta	Atherosclerosis	125
Squirrel monkey	Atherosclerosis	126
M. irus	Atherosclerosis	127
M. fasciularis	Atherosclerosis	128
African green monkey	Atherosclerosis	129
Kenya baboon	Atherosclerosis	130
M. nemestrina	Atherosclerosis	131
M. speciosa	Atherosclerosis	132
M. irus	Atherosclerosis	133
Baboon	Atherosclerosis	134
Papio anubis	Frequency-induced hypertension	135
Baboon	Chronic cerebral vasospasm	136
Other species		
Cattle	Atherosclerosis	137
Horse	Systemic hypertension	138
Cattle	Pulmonary hypertension	139
	Pulmonary hypertension	140
	Patent ductus arteriosus	141
Horse	Patent ductus arteriosus	142
Cattle	Tetralogy of fallot	143
Horse	Tetralogy of fallot	144
	Pulmonary stenosis (bicuspid pulmonary valve)	145
Cattle	Ventricular septal defect	146
Hybrid hare	Atherosclerosis	147

Table 3
ENDOCRINE SYSTEM (DIABETES MELLITUS)

Species/strain	Disease	Ref.
Mice		
C57BL/6J-ob	Diabetes mellitus	148
Mouse obese (ob)	Diabetes mellitus	149
C57BL/KsJ-db diabetes (db, db^{2J}, db^{3J}, db^{ad})	Diabetes mellitus	150
Yellow (A^y, A^{vy}, A^{iy})	Diabetes mellitus	151
NZO	Diabetes mellitus	152
KK	Diabetes mellitus	153

Table 3 (continued)
ENDOCRINE SYSTEM (DIABETES MELLITUS)

Species/strain	Disease	Ref.
Rats		
Zucker (fa)	Diabetes mellitus	154
BB	Diabetes mellitus	155
WBN/Kob	Diabetes mellitus	156
Dog	Diabetes mellitus	157
	Diabetes mellitus	158
	Diabetes mellitus	159
	Diabetes mellitus	160
	Diabetes mellitus	161
Keeshond dog	Diabetes mellitus	162
Cat	Diabetes mellitus	163
	Diabetes mellitus	164
	Diabetes mellitus	165
	Diabetes mellitus	166
	Diabetes Mellitus	167
Nonhuman primates		
Squirrel monkey	Diabetes mellitus	168
Macaca nigra	Diabetes mellitus	169
M. cyclopis	Diabetes mellitus	170
Mandrillus leucophaeus	Diabetes mellitus	170
Rhesus monkey	Diabetes mellitus	171
Urogale everetti	Diabetes mellitus	172
Nonhuman primates	Diabetes mellitus	173
	Diabetes mellitus	174
	Diabetes mellitus	175
Other species		
NZW rabbit	Diabetes mellitus	176
Chinese hamster	Diabetes mellitus	177
Mastromys albicaudatus	Diabetes mellitus	178
M. albicaudatus	Diabetes mellitus	179
Guinea pig	Diabetes mellitus	180
	Diabetes mellitus	181
Yucatan miniature swine	Diabetes mellitus	182
Sand rat	Diabetes mellitus	183

Table 4
RESPIRATORY SYSTEM

Species/strain	Disease	Ref.
Rats		
	Bronchiectasis	184
	Emphysema	185
	Asthma	186
Dogs		
	Chronic bronchitis	187
	Asthma	188
	Bronchiectasis	189
	Pulmonary eosinophilia	190
Cattle and horses		
Cattle	Bronchiectasis	191
Horse	Emphysema	192
Cattle	Fibrosing alveolytis	193
	Hypersensitivity	194
	Pneumonitis	195
Horse	Pneumonitis	195
	Hyaline membrane disease	196

Table 4 (continued)
RESPIRATORY SYSTEM

Species/strain	Disease	Ref.
Other species		
Sheep	Pulmonary adenomatosis	197
Blotchy mouse	Emphysema	198
Rabbit	Emphysema	199

Table 5
URINARY SYSTEM

Species/strain	Disease	Ref.
Mice		
Mouse	Renal agenesis	200
CFW$_w$	Cystic disorder	201
CBA/CaH-kd	Nephronophthisis	202
Mouse	Renal papillary necrosis	203
NZ	Glomerulonephritis	204
Mouse (kdkd)	Interstitial nephritis	205
DDD	Hydronephrosis	206
Rats		
MRC/H	Hereditary hydronephrosis	207
Rat	Cystic disorder	208
Gunn rat	Renal papillary necrosis	209
Rat	Glomerulonephrosis	210
	Nephrotic syndrome	211
Dog	Renal agenesis	212
Beagle	Unilateral renal agenesis	213
Dog	Renal hypoplasia	214
	Glomerulonephritis	215
	Glomerulonephritis	216
	Glomerulonephritis	217
Samoyed dog	Hereditary nephritis	218
Dog	Hereditary chronic nephritis	219
	Ectopic ureter	220
	Vesicoureteral reflux	221
	Vesicoureteral reflux	222
	Renal glycosuria	223
	Cystinuria	224
	Cystinuria	225
	Fanconi syndrome	226
	Urolithiasis	227
	Urolithiasis	228
	Urolithiasis	229
	Urolithiasis	230
Cats	Glomerulonephritis	231
	Glomerulonephritis	232
	Cystic disorder	233
	Ectopic ureter	234
Other species		
Syrian hamster	Glomerulonephritis	235
Sheep	Glomerulonephritis	236
	Glomerulonephritis	237
Horse	Glomerulonephritis	238
Macaca mulatta	Glomerulonephritis	239
M. irus	Glomerulonephritis	240

Table 5 (continued)
URINARY SYSTEM

Species/strain	Disease	Ref.
Mink	Glomerulonephritis	241
Mastomys natalensis	Glomerulonephritis	242
Guinea pig	Glomerulonephritis	243
Rabbit	Cystic disorder	244
Hamster	Cystic disorder	245
Goldfish	Cystic disorder	246
Horse	Ectopic ureter	247
Cattle	Ectopic ureter	248
Pig	Ectopic ureter	249
Monkey	Vesicouretal reflux	250
Mink	Urolithiasis	251

REFERENCES

1. **Mitruka, B. J., Rawnsley, H. M., and Vadehra, D. V.**, *Animals for Medical Research, Models for the Study of Human Disease*, John Wiley & Sons, New York, 1976.
2. **Hegreberg, G. and Leathers, C., Eds.**, *Bibliography of Induced Animal Models of Human Disease*, Student Book Corp., Pullman, Wash., 1982.
3. **Hegreberg, G. and Leathers, C., Eds.**, *Bibliography of Naturally Occurring Animal Models of Human Disease*, Student Book Corp., Pullman, Wash., 1982.
4. **Hay, J., Ed.**, *Animal Models of Immunological Processes*, Academic Press, New York, 1982.
5. National Research Council Committee on Animal Models for Research on Aging, Mammalian Models for Research on Aging, National Academy Press, Washington D.C., 1981.
6. Animals Models of Thrombosis and Hemorrhagic Diseases, National Institutes of Health, Bethesda, Md., 1976.
7. **Sclafani, A.**, Animal models of obesity: classification and characterization, *Int. J. Obesity*, 8, 491, 1984.
8. **Andrews, E. J., Ward, B. C., and Altmann, N. H., Eds.**, *Spontaneous Animal Models of Human Disease*, Academic Press, New York, 1979.
9. **Fine, J., Quimby, F. W., and Greenhouse, D. D.**, Annotated bibliography on uncommonly used laboratory animals: mammals, *ILAR News*, 29, 1A, 1986.
10. **Montali, R. J. and Migaki, G., Eds.**, *The Comparative Pathology of Zoo Animals*, Smithsonian Institution Press, Washington D.C., 1980.
11. **Montali, R. J. and Bush, M.**, A search for animal models at zoos, *ILAR News*, 26, 11, 1982.
12. **Sokol, H. W. and Valtin, H.**, The Brattleboro rat, *Ann. N.Y. Acad. Sci.*, 394, 1982.
13. **Michelson, A. M., Russell, E. S., and Harman, P. J.**, Dystrophia muscularis: a hereditary primary myopathy in the house mouse, *Proc. Natl. Acad. Sci. U.S.A.*, 41, 1079, 1955.
14. **Russell, E. S., Silvers, W. K., Loosli, R., Wolfe, H. G., and Southard, J. L.**, New genetically homogenous background for dystrophic mice and their normal counterparts, *Science*, 135, 1061, 1962.
15. **Meier, H. and Southard, J. L.**, Muscular dystrophy in the mouse caused by an allele at the dy-locus, *Life Sci.*, 9, 137, 1970.
16. **Asmundson, V. S., Kratzer, F. H., and Julian, L. M.**, Inherited myopathy in the chicken, *Ann. N.Y. Acad. Sci.*, 138, 49, 1966.
17. **Rigdon, R. H.**, Hereditary myopathy in the white Peking duck, *Ann. N.Y. Acad. Sci.*, 138, 28, 1966.
18. **Wagner, W. D. and Peterson, R. A.**, Muscular dystrophy syndrome in the cornish chicken, *Am. J. Vet. Res.*, 31, 331, 1970.
19. **Harper, J. A. and Parker, J. E.**, Hereditary muscular dystrophy in the domestic turkey, *J. Hered.*, 58, 189, 1967.
20. **Meyers, K. M., Lund, J. E., and Padgett, G. A.**, Hyperkinetic episodes in Scottish terrier dogs, *J. Am. Vet. Med. Assoc.*, 155, 124, 1969.
21. **Cardinet, G. H., Wallace, L. J., Fedde, M. R., Guffy, M. M., and Bardens, J. W.**, Developmental myopathy in the canine with type II muscle fiber hypotrophy, *Arch. Neurol. (Chicago)*, 21, 620, 1969.

22. **Short, C. E.**, Malignant hyperthermia in the dog, *Anesthesiology*, 39, 462, 1973.
23. **Fraser, D. C., Palmer, A. C., Senior, J. E. B., Parkes, J. D., and Yealland, M. F. T.**, Myasthenia gravis in the dog, *J. Neurol. Neurosurg. Phychiatr.*, 33, 431, 1970.
24. **Jenkins, W. L., Van Dyk, E., and McDonald, C. B.**, Myasthenia gravis in a fox terrier litter, *J. S. Afr. Vet. Assoc.*, 47, 59, 1976.
25. **Liu, S. K. and Dorfman, H. D.**, A condition resembling human localized myositis ossificans in two dogs, *J. Small Anim. Pract.*, 17, 371, 1976.
26. **Mason, K. V.**, A case of myasthenia gravis in a cat, *J. Small Anim. Pract.*, 17, 467, 1976.
27. **Liu, S. K., Dorfman, H. D., and Patnaik, A. K.**, Primary and secondary bone tumors in the cat, *J. Small Anim. Pract.*, 15, 141, 1974.
28. **Deutsch, K. and Done, J. T.**, Congenital myofibrillar hypoplasia of piglets: ultrastructure of affected fibres, *Res. Vet. Sci.*, 12, 176, 1971.
29. **Thurley, D. C., Gilbert, F. R., and Done, J. T.**, Congenital splayleg of piglets: myofibrillar hypoplasia, *Vet. Rec.*, 80, 302, 1967.
30. **Ward, P. S.**, The splayleg syndrome in newborn pigs — a review, *Vet. Bull.*, 48, 279, 1978.
31. **Campion, D. R. and Topel, D. G.**, A review of the role of swine skeletal muscle in malignant hyperthemia, *J. Anim. Sci.*, 41, 779, 1975.
32. **Hall, L. W., Trim, C. M., and Woolf, N.**, Further studies of porcine malignant hyperthermia, *Br. Med. J.*, 2, 145, 1972.
33. **Seibold, H. R. and Davis, C. L.**, Generalized myositis ossificans (familial) in pigs, *Pathol. Vet.*, 4, 79, 1967.
34. **Hamilton, M. J., Hegreberg, G. A., and Gorham, J. R.**, Histochemical muscle fiber typing in inherited muscular dystrophy of mink, *Am. J. Vet. Res.*, 35, 1321, 1974.
35. **Urman, H. K. and Grace, O. D.**, Hereditary encephalomyopathy, a hydrocephalus syndrome in newborn calves, *Cornell Vet.*, 54, 229, 1964.
36. **Yamori, Y.**, A selection of arteriolipidosis-prone rats (ALR), *Jpn. Heart J.*, 18, 602, 1977.
37. **Müller, K. R., Dinh, D. M., and Subbiah, M. T. R.**, The characteristics and metabolism of a genetically hypercholesterolemic strain of rats (RICO), *Biochim. Biophys. Acta*, 574, 334, 1979.
38. **Koletsky, S.**, Obese spontaneously hypertensive rats—a model for study of atherosclerosis, *Exp. Mol. Pathol.*, 19, 53, 1973.
39. **Bray, G. A.**, The Zucker-fatty rat, a review, *Fed. Proc., Fed. Am. Soc. Exp. Biol. Med.*, 36, 148, 1977.
40. **Poledne, R.**, Effect of diet on cholesterol metabolism in the Prague hereditary hypercholesterolemic rat, in *Nutritional Effects of Cholesterol Metabolism*, Beynen, A. C., Ed., Voorthuizen-Holland, 1986, 112.
41. **Imai, Y. and Matsumura, H.**, Genetic studies on induced and spontaneous hypercholesterolemia in rats, *Atherosclerosis*, 18, 59, 1973.
42. **Boissel, J. P., Crouset, B., Bourdillon, M. C., and Blaes, N.**, Selection of a strain of rats with spontaneous high cholesterolemia, *Atherosclerosis*, 39, 11, 1981.
43. **Yamori, Y. and Horie, R.**, Development of normotensive atherogenic rats, in *Prophylactic Approach to Hypertensive Diseases*, Yamori, Y., et al., Eds., Raven Press, New York, 1979.
44. **Gustafsson, K. and Kiessling, H.**, Studies on serum lipoproteins of rats developing spontaneous hyperlipidemia, *Artery*, 9, 456, 1981.
45. **Wexler, B. C. and McMurtry, J. P.**, Genetically mediated resistance to naturally occurring aortic sclerosis in spontaneously hypertensive as against Sprague-Dawley and Wistar-Kyoto breeder rats, *Br. J. Exp. Pathol.*, 63, 66, 1982.
46. **Okamoto, K.**, Spontaneous hypertension in rats, *Int. Rev. Exp. Pathol.*, 7, 227, 1969.
47. **Yamori, Y., Horie, R., Akiguchi, J., Kihara, M., and Nara, J.**, Symptomological classification in the development of stroke in stroke-prone spontaneously hypertensive rats, *Jpn. Circ. J.*, 46, 274, 1982.
48. **Phelan, E. L.**, The New Zealand strain of rats with genetic hypertension, *N. Z. Med. J.*, 67, 334, 1968.
49. **Bianchi, G., Fox, U., and Imbasciati, E.**, The development of a new strain of spontaneously hypertensive rats, *Life Sci.*, 14, 339, 1968.
50. **Rapp, J. P. and Dene, H.**, Development and characteristics of inbred strains of Dahl salt-sensitive and salt-resistant rats, *Hypertension*, 7, 340, 1985.
51. **Samizadeh, A., Losse, H., and Wessels, F.**, Einfluss von Kochsalz und β-Sympathykolytika auf den Blutdruckverlauf der erblichen spontanen Hypertonie der Ratte, *Med. Welt.*, 28, 2050, 1977.
52. **Pravenec, M.**, Recombinant inbred strains—a model for the study of spontaneous hypertension in the rat, *Acta Univ. Carol. Med.*, in press.
53. **Lawler, J. E. and Cox, R. H.**, The borderline hypertensive rat (BHR), a new model for the study of environmental factors in the development of hypertension, *Pavlov J. Biol. Sci.*, 20, 101, 1985.
54. **Vincent, M. and Sassard, J.**, Les souches lyonnaises de rats génétiquement hypertendus (LH), normotendus (LN) et normotendus-bas (LL), *Paroi Artérielle/Arterial Wall*, 6, 139, 1980.
55. **Ben-Ishai, M.**, Exaggerated response to isotonic saline loading in genetically hypertension-prone rats, *J. Lab. Clin. Med.*, 82, 597, 1973.

56. **Yamori, Y.,** Myocardial-ischemic rats (MIR). Coronary vascular alteration induced by a lipid-rich diet, *Atherosclerosis,* 42, 15, 1982.

57. **Ruben, Z.,** A potential model for a human disease; spontaneous cardiomyopathy-congestive heart failure in SHR/N-cp rats, *Hum. Pathol.,* 15, 902, 1984.

58. **Fein, F. S., Capasso, J. M., Aronson, R. S., Cho, S., and Nordin, Ch.,** Combined renovascular hypertension and diabetes in rats, a new preparation of congestive cardiomyopathy, *Circulation,* 70, 318, 1984.

59. **Kaneko, D., Nakamura, N., and Ogawa, T.,** Cerebral infarction in rats using homologous blood emboli; development of a new experimental model, *Stroke,* 16, 76, 1985.

60. **Yamori, Y., Horie, R., Ohtaka, M., Nara, Y., and Ikeda, K.,** Genetic and environmental modification of spontaneous hypertension, *Jpn. Circ. J.,* 42, 1151, 1978.

61. **Svendson, U. G.,** Spontaneous hypertension and hypertensive vascular disease in the NZB strain of mice, *Acta Pathol. Microbiol. Scand. Sect. A,* 85, 548, 1977.

62. **Breckenridge, W. C., Roberts, A., and Kuksis, A.,** Lipoprotein levels in genetically selected mice with increased susceptibility to atherosclerosis, *Arteriosclerosis,* 5, 256, 1985.

63. **Schlager, G. and Weiburst, R. S.,** Genetic control of blood pressure in mice, *Genetics,* 55, 497, 1967.

64. **Andrews, E. J., White, W. J., and Bullock, L. P.,** Spontaneous aortic aneurysms in blotchy mice, *Am. J. Pathol.,* 78, 199, 1975.

65. **Eaton, G. J., Custer, R. P., Johnson, F. N., and Stabenov, K.,** Dystrophic cardiac calcinosis in mice, genetic, hormonal and dietary influences, *Am. J. Pathol.,* 90, 173, 1978.

66. **Forbes, M. S. and Sperelakis, N.,** Ultrastructure of cardiac muscle from dystrophic mice, *Am. J. Anat.,* 134, 271, 1972.

67. **Jasmin, G. and Bajusz, E.,** Myocardial lesions in strain 129 dystrophic mice, *Nature,* 193, 181, 1962.

68. **Michelson, E. L.,** Canine models for ventricular tachyarrhythmia *Ann. Intern. Med.,* 95, 648, 1981.

69. **Shelub, I., Van Grondelle, A., McCullough, R., Hofmeister, S., and Reeves, J. T.,** A model of embolic chronic pulmonary hypertension in the dog, *J. Appl. Physiol.,* 56, 810, 1984.

70. **Vogel, J. A., Genovese, R. L., Powell, T. L., Bishop, G. W., Bucci, T. J., and Harris, C. W.,** Cardiac size and pulmonary hypertension in dogs exposed to high altitude, *Am. J. Vet. Res.,* 32, 2059, 1971.

71. **Kohler, J., Silverman, N. A., Levitski, S., Pavel, D. G., Eckner, F. A., and Fang, R. B.,** A model of cyanotic heart disease: functional, pathological and metabolic sequelae in the immature canine heart, *J. Surg. Res.,* 37, 309, 1984.

72. **Barrett, R. E., de Lahunta, A., Roenick, W. J., Hoffer, R. E., and Coons, F. H.,** Four cases of congenital portacaval shunt in the dog, *J. Small Anim. Pract.,* 17, 71, 1976.

73. **Bohn, F. K., Patterson, D. F., and Pyle, R. L.,** Atrial fibrillation in dogs, *Br. Vet. J.,* 127, 485, 1971.

74. **Cox, R. H., Peterson, L. H., and Detweiler, D. K.,** Comparison of arterial hemodynamics in the mongrel dog and the racing greyhound, *Am. J. Physiol.,* 230, 211, 1976.

75. **Tilley, L. P. and Liu, S. K.,** Cardiomyopathy in the dog, *Recent Adv. Stud. Card. Struct. Metab.,* 10, 641, 1975.

76. **Dear, M. G.,** Mitral incompetence in dogs 0—5 years of age, *J. Small Anim. Pract.,* 12, 1, 1971.

77. **Eliot, T. S., Eliot, F. P., Lushbaugh, C. C., and Slager, U. T.,** First report on the occurrence of neonatal endocardial fibroelastosis in cats and dogs, *J. Am. Vet. Med. Assoc.,* 133, 271, 1958.

78. **Eyster, G. E., Evans, A. T., Blanchard, G. L., Krahwinkel, D. J., Chaffe, A., Deyoung, D., Karr, D. R., and Handley, P. O.,** Congenital pericardialdiaphragmatic hernia and multiple defects in a litter of collies, *J. Am. Vet. Med. Assoc.,* 170, 616, 1977.

79. **Hamlin, R. L. and Harris, S. G.,** Mitral incompetence in great dane pups, *J. Am. Vet. Med. Assoc.,* 154, 790, 1969.

80. **Krahwinkel, D. J. and Coogan, P. S.,** Endocardial fibroelastosis in a great dane pup, *J. Am. Vet. Med. Assoc.,* 159, 327, 1971.

81. **Lord, P. F., Wood, A., Liu, S. K., and Tilley, L. P.,** Left ventricular angiocardiography in congenital mitral valve insufficiency of the dog, *J. Am. Vet. Med. Assoc.,* 166, 1069, 1975.

82. **Weirich, W. E., Blevins, W. E., Conrad, C. R., Ruth, G. R., and Gallina, A. M.,** Congenital tricuspid insufficiency in a dog, *J. Am. Vet. Med. Assoc.,* 164, 1026, 1974.

83. **Patterson, D. F., Pyle, R. L., Buchanan, J. W., Trautvetter, E., and Abt, D. A.,** Hereditary patent ductus arteriosus and its sequelae in the dog, *Circ. Res.,* 29, 1, 1971.

84. **Pyle, R. L., Patterson, D. F., and Chacko, S.,** The genetics and pathology of discrete subaortic stenosis in the Newfoundland dog, *Am. Heart J.,* 92, 324, 1976.

85. **Satyanarayana, R., Anderson, R. C., and Edwards, J. E.,** Anatomic variations in the tetralogy of Fallot, *Am. Heart J.,* 81, 361, 1971.

86. **Eyster, G. E., Carrig, C. B., Baker, B., Handley, P. O., and Eberling, G.,** Coarctation of the aorta in a dog, *J. Am. Vet. Med. Assoc.,* 169, 426, 1976.

87. **Parker, G. W., Jackson, W. F., and Patterson, D. F.,** Coarctation of the aorta in a canine, *J. Am. Anim. Hosp. Assoc.,* 7, 353, 1971.

88. **Buchanan, J. W., Baker, C. J., and Hill, J. D.,** Aortic embolism in cats. Prevalence, surgical treatment and electrocardiography, *Vet. Rec.,* 79, 496, 1966.
89. **Holzworth, J., Simpson, R., and Wino, A.,** Aortic thrombosis with posterior paralysis in the cat, *Cornell Vet.,* 45, 468, 1955.
90. **Liu, S. K., Tilley, L. P., and Lord, P. F.,** Feline cardiomyopathy, *Recent Adv. Stud. Card. Struct. Metab.,* 10, 627, 1975.
91. **Liu, S. K. and Tilley, L. P.,** Displasia of the tricuspid valve in the dog and cat, *J. Am. Vet. Med. Assoc.,* 169, 624, 1976.
92. **Will, J. A.,** Subvalvular pulmonary stenosis and aorticopulmonary septal defect in the cat, *J. Am. Vet. Med. Assoc.,* 154, 913, 1969.
93. **Bolton, G. R., Ettinger, S. J., and Liu, S. K.,** Tetralogy of Fallot in three cats, *J. Am. Vet. Med. Assoc.,* 160, 1622, 1972.
94. **Alexander, N., Hinshaw, L. B., and Drury, D. R.,** Development of a strain of spontaneously hypertensive rabbits, *Proc. Soc. Exp. Biol. Med.,* 86, 855, 1954.
95. **Fox, R. R., Schlager, G., and Laird, C.,** Blood pressure in thirteen strains of rabbits, *J. Hered.,* 60, 312, 1969.
96. **Hatt, P. Y., Berjal, G., Moravec, J., and Swynghedauw, B.,** Heart failure: an electron microscopic study of the left ventricular papillary muscle in aortic insufficiency in the rabbit, *J. Mol. Cell. Cardiol.,* 1, 235, 1970.
97. **Weber, H. W. and Van der Walt, J. J.,** Cardiomyopathy in crowded rabbits, *S. Afr. Med. J.,* 47, 1591, 1973.
98. **Minick, C. R. and Murphy, G. E.,** Experimental induction of atheroarteriosclerosis by the synergy of allergic injury to arteries and lipid rich diet. II. Effect of repeatedly injected foreign protein in rabbits fed a lipid rich cholesterol poor diet, *Am. J. Pathol.,* 73, 265, 1973.
99. **Schenk, E. A., Gaman, E., and Feigenbaum, A. S.,** Spontaneous aortic lesions in rabbits. I. Morphologic characteristics, *Circ. Res.,* 19, 80, 1966.
100. **Tilton, G. D., Buja, L. M., Bilheimer, D. W., April, P., Ashton, J., Nicnatt, J., Kita, T., and Willerson, J. T.,** Failure of a slow channel calcium antagonist, verapamil, to retard atherosclerosis in the Watanabe heritable hyperlipidemic rabbit; an animal model of familial hypercholesterolemia, *J. Am. Coll. Cardiol.,* 6, 141, 1985.
101. **Van Vleet, J. F., Ferrans, V. J., and Ruth, G. R.,** Ultrastructural alterations in nutritional cardiomyopathy of selenium-vitamin E deficient swine. I. Fiber lesions, *Lab. Invest.,* 37, 188, 1977.
102. **Daoud, A. S., Jones, R., and Scott, R. F.,** Dietary-induced atherosclerosis in miniature swine. II. Electron microscopy observations. Characteristics of endothelial and smooth muscle cells in the proliferative lesions and elsewhere in the aorta, *Exp. Mol. Pathol.,* 8, 263, 1968.
103. **Getty, R.,** The gross and microscopic occurrence and distribution of spontaneous atherosclerosis in the arteries of the swine, in *Comparative Atherosclerosis,* Roberts, J. C., Jr. and Straus, R., Eds., Harper & Row, New York, 1965.
104. **Gottlieb, H. and Lalich, J.,** The occurrence of atherosclerosis in the aorta of swine, *Am. J. Pathol.,* 30, 851, 1954.
105. **Lee, K. T., Jarmolych, J., Kim, D. M., Grant, C., Krasney, J. A., Thomas, W. A., and Bruno, A. M.,** Production of advanced coronary atherosclerosis, myocardial infarction and sudden death in swine, *Exp. Mol. Pathol.,* 15, 170, 1971.
106. **St. Clair, R. W., Bullock, B. C., Lehner, N. D. M., and Lofland, H. B.,** Long-term effects of dietary sucrose and starch on serum lipids and atherosclerosis in miniature swine, *Exp. Mol. Pathol.,* 15, 21, 1971.
107. **Verdouw, P. D., Wolffenbuttel, B. H. R., and Van Der Giessen, W. J.,** Domestic pigs in the study of myocardial ischemia, *Am. Heart J.,* 4 (Suppl. 6), 61, 1983.
108. **Reitman, J. S., Mahley, R. W., and Fly, D. L.,** Yucatan miniature swine as a model for diet-induced artheriosclerosis, *Atherosclerosis,* 43, 119, 1982.
109. **Krista, L. M., Waibel, P. E., Shoffner, R. N., and Sautter, J. H.,** Natural dissecting aneurysm (aortic rupture) and blood pressure in the turkey, *Nature,* 214, 1162, 1967.
110. **Sturkie, P. D., Weiss, H. S., Ringer, R. K., and Sheahan, M. M.,** Heritability of blood pressure in chickens, *Poult. Sci.,* 38, 333, 1959.
111. **Einzig, S., Jankus, E. F., and Moller, J. H.,** Round heart disease in turkeys: a hemodynamic study, *Am. J. Vet. Res.,* 33, 557, 1972.
112. **Noren, G. R., Staley, N. A., Jankus, E. F., and Stevenson, J. E.,** Myocarditis in round heart disease of turkeys. A light and electron microscope study, *Virchows Arch. A,* 352, 285, 1971.
113. **Einzig, S., Jankus, E. S., and Moller, J. H.,** Ventricular septal defects in turkeys, *Am. J. Vet. Res.,* 33, 563, 1972.
114. **Siller, W. G.,** Ventricular septal defects in the fowl, *J. Pathol.,* 76, 431, 1958.
115. **Clarkson, T. B., Middleton, C. C., Prichard, R. W., and Lofland, H. B.,** Naturally occurring atherosclerosis in birds, *Ann. N.Y. Acad. Sci.,* 127, 685, 1965.

116. **Prichard, R. W., Clarkson, T. B., Goodman, H. O., and Lofland, H. B.,** Aortic atherosclerosis in pigeons and its complications, *Arch. Pathol.,* 77, 244, 1964.

117. **Wagner, W. D. and Clarkson, T. B.,** Mechanisms of the genetic control of plasma cholesterol in selected lines of Show Racer pigeons, *Proc. Soc. Exp. Biol. Med.,* 145, 1050, 1974.

118. **Wagner, W. D., Clarkson, T. B., Feldner, M. A., and Prichard, R. W.,** The development of pigeon strains with selected atherosclerosis characteristics, *Exp. Mol. Pathol.,* 19, 304, 1973.

119. **Wagner, W. D. and Nohlgren, S. R.,** Aortic glysosaminoglycans in genetically selected WC-2 pigeons with increased atherosclerosis susceptibility, *Arteriosclerosis,* 1, 192, 1981.

120. **Shih, J. C., Pullman, E. P., and Kao, K. J.,** Genetic selection, general characterization, and histology of atherosclerosis-susceptible and -resistant Japanese quail, *Atherosclerosis,* 49, 41, 1983.

121. **Bullock, B. C. and Moossy, J.,** Cerebral infarction with atherosclerosis occlusion of the middle cerebral artery in a squirrel monkey *(Saimiri sciureus)* fed an atherogenic diet, *J. Neuropathol. Exp. Neurol.,* 31, 177, 1972.

122. **Armstrong, M. L.,** Atherosclerosis in rhesus and cynomolgus monkeys, *Primates Med.,* 9, 16, 1976.

123. **Bond, M. G., Bullock, B. C., Bellinger, D. A., and Hamm, T. E.,** Myocardial infarction in a large colony of nonhuman primates with coronary artery atherosclerosis, *Am. J. Pathol.,* 101, 675, 1980.

124. **Chawla, K. K., Murthy, C. D. S., Chakravarti, R. N., and Chuttani, P. N.,** Arteriosclerosis and thrombosis in wild rhesus monkeys, *Am. Heart J.,* 73, 85, 1967.

125. **Chakravarti, R. N., Mohan, A. P., and Komal, H. S.,** Atherosclerosis in *Macaca mullata:* histopathological morphometric, and histochemical studies in aorta and coronary arteries of spontaneous and induced atherosclerosis, *Exp. Mol. Pathol.,* 25, 390, 1976.

126. **Clarkson, T. B., Lehner, N. D. M., Bullock, B. C., Lofland, H. B., and Wagner, W. D.,** Atherosclerosis in New World monkeys, *Primate Med.,* 9, 90, 1976.

127. **Kramsch, D. M. and Hollander, W.,** Occlusive atherosclerotic disease of the coronary arteries in monkeys *(Macaca irus)* induced by diet, *Exp. Mol. Pathol.,* 9, 1, 1968.

128. **Kramsch, D. M., Hollander, W., and Renaud, S.,** Induction of fibrous plaques versus foam cell lesions in *Macaca fasciularis* by varying the composition of dietary fats, *Circulation,* 48, 41, 1973.

129. **Kritchevski, D., Davidson, L. M., Krendel, D. A., and Kim, H. K., and Malhotra, S.,** Influence of semipurified diet on atherosclerosis in African green monkeys, *Exp. Mol. Pathol.,* 26, 28, 1977.

130. **McGill, H. C., Strong, J. P., Holman, R. L., and Werthessen, N. T.,** Arterial lesions in the Kenya baboon, *Circ. Res.,* 8, 670, 1960.

131. **McMahan, M. R., Rhyne, A. L., Lofland, H. B., and Sackett, G. P.,** Effects of sex, age, and dietary modification on plasma lipids and lipoproteins of *Macaca nemestrina, Proc. Soc. Exp. Biol. Med.,* 164, 27, 1980.

132. **Pick, R., Johnson, P. J., and Glick, G.,** Deleterious effects of hypertension on the development of aortic and coronary atherosclerosis in stumptail macaques *(Macaca speciosa)* on an atherogenic diet, *Circ. Res.,* 35, 472, 1974.

133. **Prathap, K.,** Spontaneous aortic lesions in wild adult Malaysian long-tailed monkeys *(Macaca irus),* J. *Pathol.,* 110, 135, 1973.

134. **Strong, J. P. and McGill, H. C., Jr.,** Diet and experimental atherosclerosis in baboons, *Am. J. Pathol.,* 50, 669, 1967.

135. **Cavangh, D., Papineni, M. D., Rao, S., Knuppel, R. A., Desai, U., and Balis, J. V.,** Pregnancy-induced hypertension, development of a model in the pregnant primate *(Papio anubis), Am. J. Obstet. Gynecol.,* 1, 151, 1985.

136. **Oppie, L. H., Bruyneel, K. J., and Lubbe, W. F.,** What has the baboon to offer as a model of experimental ischemia?, *Eur. Heart J.,* 4 (Suppl. C), 55, 1983.

137. **Alibasoglu, M., Dunne, H. W., and Guss, S. B.,** Naturally occurring arteriosclerosis in cattle infected with Johne's disease, *Am. J. Vet. Res.,* 23, 49, 1962.

138. **Garner, H. E., Coffman, J. R., Hahn, A. W., Ackerman, N., and Johnson, J. H.,** Equine laminitis and associated hypertension. A review, *J. Am. Vet. Med. Assoc.,* 166, 56, 1975.

139. **Alexander, A. F. and Jensen, R. W.,** Pulmonary vascular pathology of high altitude-induced pulmonary hypertension in cattle, *Am. J. Vet. Res.,* 24, 1112, 1963.

140. **Hecht, H. H., Kuida, H., Lange, R. L., Thorne, J. L., and Brown, A. M.,** Brisket disease. II. Clinical features and hemodynamic observations in altitude dependent right heart failure of cattle, *Am. J. Med.,* 32, 171, 1962.

141. **Wiseman, A. and Murray, M.,** A case of uncomplicated patent ductus arteriosus in a calf, *Vet. Res.,* 94, 16, 1974.

142. **Carmichael, J. A., Buergelt, C. D., Lord, P. F., and Tashjian, R. J.,** Diagnosis of patent ductus arteriosus in a horse, *J. Am. Vet. Med. Assoc.,* 158, 767, 1971.

143. **Dear, M. G. and Price, E. K.,** Complex congenital anomaly of the bovine heart, *Vet. Rec.,* 86, 219, 1970.

144. **Prickett, M. E., Reeves, J. T., and Zent, W. W.,** Tetralogy of Fallot in a thoroughbred foal, *J. Am. Vet. Med. Assoc.,* 162, 552, 1973.

145. **Critchley, K. L.,** An interventricular septal defect, pulmonary stenosis and bicuspid pulmonary valve in a Welsh pony foal, *Equine Vet. J.,* 8, 176, 1976.
146. **Belling, T. H.,** Ventricular septal defect in the bovine heart. Report of three cases, *J. Am. Vet. Med. Assoc.,* 138, 595, 1961.
147. **Pearson, T. A.,** Clonal characteristics of experimentally induced "atherosclerotic" lesions in the hybrid hare, *Science,* 206, 1423, 1979.
148. **Ingalls, A. M., Dickie, M. M., and Snell, G. D.,** Obese. A new mutation in the mouse, *J. Hered.,* 41, 317, 1950.
149. **Ashwell, M., Meade, C. J., Medawar, P., and Sowter, C.,** Adipose tissue: contribution of nature and nurture to the obesity of an obese mutant mouse (ob/ob), *Proc. R. Soc. London, Ser. B.,* 195, 343, 1977.
150. **Coleman, D. L. and Hummel, K. P.,** Studies with the mutation, diabetes, in the mouse, *Diabetologia,* 3, 238, 1967.
151. **Wolff, G. L.,** Composition and coat color correlation in different phenotypes of "Viable Yellow" mice, *Science,* 147, 1146, 1960.
152. **Bielschowsky, M. and Bielschowsky, F.,** The New Zealand strains of obese mice. Their response to stilbesterol and to insulin, *Aust. J. Exp. Biol. Med.,* 34, 181, 1956.
153. **Camarini-Davalos, R. H., Opperman, N., Mittrell, R., and Ehrenreich, T.,** Studies of vascular and other lesions in KK mice, *Diabetologia,* 6, 324, 1970.
154. **Zucker, L. M. and Zucker, T. F.,** Fatty, a new mutation in the rat, *J. Hered.,* 52, 275, 1961.
155. **Nakhooda, A. F., Like, A. A., Chappel, C. I., Wei, C. N., and Marliss, E. B.,** The spontaneously diabetic Wistar rat (the BB rat). Studies prior to and during development of the overt syndrome, *Diabetologia,* 14, 199, 1978.
156. **Tsuchitani, M., Saegusa, T., Narama, J., Nishikawa, T., and Gonda, T.,** A new diabetic strain of rat (WBN/Kob), *Lab. Anim.,* 19, 200, 1985.
157. **Berkow, J. W. and Ricketts, R. L.,** Spontaneous diabetes mellitus in dogs, *J. Am. Vet. Med. Assoc.,* 146, 1101, 1965.
158. **Cotton, R. B., Cornelius, L. M., and Theran, P.,** Diabetes mellitus in the dog. A clinicopathologic study, *J. Am. Vet. Med. Assoc.,* 159, 863, 1971.
159. **Dixon, J. P. and Sanford, J.,** Pathological features of spontaneous canine diabetes mellitus, *J. Comp. Pathol.,* 72, 153, 1962.
160. **Ling, G. V., Lowenstine, L. J., Pulley, L. T., and Kaneko, J. J.,** Diabetes mellitus in dogs: a review of initial evaluation, immediate and long-term management, and outcome, *J. Am. Vet. Med. Assoc.,* 170, 521, 1977.
161. **Engerman, R. L. and Kramer, J. W.,** Dogs with induced or spontaneous diabetes as models for the study of human diabetes mellitus, *Diabetes,* 31 (Suppl. 1) 2, 1982.
162. **Kramer, J.,** Animal model of human disease: inherited early-onset, insulin-requiring diabetes mellitus in Keeshond dogs, *Am. J. Pathol.,* 105, 194, 1981.
163. **Finn, J. P., Martin, C. L., and Manns, J. G.,** Feline pancreatic islet cell hyalinosis associated with diabetes mellitus and lowered serum-insulin concentrations, *J. Small Anim. Pract.,* 11, 607, 1970.
164. **Gembardt, C. and Loppnow, H.,** Zur Pathogenese des spontanen Diabetes mellitus der Katze, *Vet. Pathol.,* 11, 461, 1974.
165. **Gepts, W. and Toussaint, D.,** Spontaneous diabetes in dogs and cats. A pathological study, *Diabetologia,* 3, 249, 1967.
166. **Holzworth, J. and Coffin, D. L.,** Pancreatic insufficiency and diabetes mellitus in a cat, *Cornell Vet.,* 43, 502, 1953.
167. **Schaer, M.,** A clinical survey of thirty cats with diabetes mellitus, *J. Am. Anim. Hosp. Assoc.,* 13, 23, 1977.
168. **Davidson, I. W. F., Lang, C. M., and Blackwell, W. L.,** Impairment of carbohydrate metabolism of the squirrel monkey, *Diabetes,* 16, 395, 1976.
169. **Howard, C. F., Jr.,** Spontaneous diabetes in *Macaca nigra, Diabetes,* 21, 1077, 1972.
170. **Howard, C. F., Jr. and Palotay, J. L.,** Spontaneous diabetes mellitus in *Macaca cyclopis* and *Mandrillus leucophaeus,* case reports, *Lab. Anim. Sci.,* 25, 191, 1975.
171. **Kirk, J. H., Casey, H. W., and Harwell, J. F., Jr.,** Diabetes mellitus in two rhesus monkeys, *Lab. Anim. Sci.,* 22, 245, 1972.
172. **Rabb, G. B., Getty, R. E., Willamson, W. M., and Lombard, L. S.,** Spontaneous diabetes mellitus in tree shrews, *Urogale everetti, Diabetes,* 15, 327, 1966.
173. **Howard, C. F., Jr.,** Nonhuman primates as models for the study of human diabetes mellitus, *Diabetes,* 31 (Suppl. 1), 37, 1982.
174. **Howard, C. F., Jr.,** Diabetes mellitus: relationships of nonhuman and other animal models to human forms of diabetes, *Adv. Vet. Sci. Comp. Med.,* 28, 115, 1984.
175. **Jones, S. M.,** Spontaneous diabetes in monkeys, *Lab. Anim.,* 8, 161, 1974.
176. **Roth, S. I. and Conaway, H. H.,** Animal model of human disease. Spontaneous diabetes mellitus in the New Zeland white rabbit, *Am. J. Pathol.,* 109, 359, 1982.

177. **Eto, M., Watanabe, K., Iwashima, Y., Morikawa, A., Takebe, T., and Ishii, K.,** Elevation of plasma high density lipoprotein-cholesterol in spontaneously diabetic Chinese hamsters, *Tohuku J. Exp. Med.,* 144, 281, 1984.

178. **Goecken, J. A., Packer, J. T., Rose, S. D., and Stuhlman, R. A.,** Structure of the islets of Langerhans; pathological studies in normal and diabetic *Mystromys albicaudatus, Arch. Pathol.,* 93, 123, 1972.

179. **Packer, J. T., Kraner, K. L., Rose, S. D., Stuhlman, R. A., and Nelson, L. R.,** Diabetes mellitus in *Mystromys albicaudatus, Arch. Pathol.,* 89, 410, 1970.

180. **Lang, C. M. and Munger, B. L.,** Diabetes mellitus in the guinea pig, *Diabetes,* 25, 434, 1976.

181. **Lang, C. M., Munger, B. L., and Rapp, F.,** The guinea pig *(Cavia porcellus)* as an animal model of diabetes mellitus, *Lab. Anim. Sci.,* 27, 789, 1978.

182. **Philips, R. W., Panepinto, L. N., and Westmoreland, N.,** Yucatan miniature swine as a model for the study of human diabetes mellitus, *Diabetes,* 31 (Suppl. 1), 30, 1982.

183. **Rice, M. G. and Robertson, R. P.,** Reevaluation of the sand rat as a model for diabetes mellitus, *Am. J. Physiol.,* 239, E340, 1980.

184. **Lindsey, J. R., Baker, H. J., Overcash, R. G., Cassel, G. H., and Hunt, C. E.,** Murine chronic respiratory disease, *Am. J. Pathol.,* 64, 675, 1971.

185. **Paleček, F. and Holuša, R.,** Spontaneous occurrence of lung emphysema in laboratory rats. A quantitative functional and morphological study, *Physiol. Bohemoslov.,* 20, 325, 1971.

186. **Holme, G. and Piechuta, H.,** The derivation of inbred line of rats which develop asthma-like symptoms following challenge with aerosolized antigen, *Immunology,* 42, 19, 1981.

187. **Pirie, H. M. and Wheeldon, E. B.,** Chronic bronchitis in the dog, *Adv. Vet. Sci. Comp. Med.,* 20, 253, 1976.

188. **Patterson, R.,** Investigations of spontaneous hypersensitivity of the dog, *J. Allergy,* 31, 351, 1960.

189. **Wheeldon, E. B., Pirie, H. M., Fisher, E. W., and Lee, R.,** Chronic respiratory disease in the dog, *J. Small Anim. Pract.,* 18, 229, 1977.

190. **Eikmeier, H. and Manz, D.,** Untersuchungen zur Eosinophilie des Hundes. II. Vorkommen der Eosinophilie bei verschiedenen Erkrankungen mit Ausnahme der Verdauugsstörungen, *Berl. Muench. Tieraerztl. Wochenschr.,* 79, 84, 1966.

191. **Jensen, R., Pierson, R. E., Braddy, P. M., Saari, D. A., Lauerman, L. H., Benitez, A., Christie, R. M., Horton, D. P., and McChesney, A. E.,** Bronchiec̆·͟͞˞ in yearling feedlot cattle, *J. Am. Vet. Med. Assoc.,* 169, 511, 1976.

192. **Gillespie, J. R. and Tyler, W. S.,** Chronic alveolar emphysema in the h̤ᴏᴿˢᵉ. *Adv. Vet. Sci. Comp. Med.,* 13, 59, 1969.

193. **Pirie, H. M. and Selman, I. E.,** A bovine disease resembling diffuse fibrosing alveolitis, *Proc. R. Soc. Med.,* 65, 987, 1972.

194. **Breeze, R. G., Pirie, H. M., Dawson, C. O., Selman, I. E., and Wiseman, A.,** The pathology of respiratory diseases of adult cattle in Britain, *Folia Vet. Lat.,* 5, 95, 1975.

195. **Mansmann, R. A., Osburn, B. I., Wheat, J. D., and Frick, O.,** Chicken hypersensitivity pneumonitis in horses, *J. Am. Vet. Med. Assoc.,* 166, 673, 1975.

196. **Mahaffey, L. W. and Rossdale, P. D.,** Convulsive and allied syndromes in new-born foals, *Vet. Rec.,* 69, 1277, 1957.

197. **Martin, W. B., Scott, F. M. M., Sharp, J. M., and Angus, K. W.,** Experimental production of sheep pulmonary adenomatosis (jaagsiekte), *Nature,* 264, 183, 1976.

198. **Fisk, D. E. and Kuhn, C.,** Emphysema-like changes in the lung of the blotchy mouse, *Am. Rev. Respir. Dis.,* 113, 787, 1976.

199. **Strawbridge, H. T. G.,** Chronic pulmonary emphysema (an experimental study). II. Spontaneous pulmonary emphysema in rabbits, *Am. J. Pathol.,* 37, 309, 1960.

200. **Gleucksohn-Waelsch, S. and Rota, T.,** Development in organ tissue culture of kidney rudiments from mutant mouse embryos, *Dev. Biol.,* 7, 432, 1963.

201. **Werder, A. A., Cuppage, F. E., and Nielsen, A. H.,** Naturally occurring polycystic renal disease in CFWw mice, in Proc. 9th Annu. Meet. Am. Soc. Nephrol., 1976, 68.

202. **Lyon, M. F. and Hulse, E. V.,** An inherited kidney disease of mice resembling human nephronophthisis, *J. Med. Genet.,* 8, 41, 1971.

203. **Cornelius, E. A.,** Amyloidosis and renal papillary necrosis in male hybrid mice, *Am. J. Pathol.,* 59, 317, 1970.

204. **Dixon, F. J., Oldstone, J. B., and Tonietti, G.,** Pathogenesis of immune complex glomerulonephritis of New Zealand mice, *J. Exp. Med.,* 134, 65s, 1971.

205. **Neilson, E. G., McCafferty, E., and Feldman, A.,** Spontaneous interstitial nephritis in kdkd mice. I. An experimental model of autoimmune renal disease, *J. Immunol.,* 133, 2560, 1984.

206. **Nakajima, Y., Imamura, K., Onodera, T., Motoi, X., and Goto, N.,** Hydronephrosis in the inbred mouse strain DDD, *Lab. Anim.,* 17, 143, 1983.

207. **Lozzio, B. B., Chernoff, A. I., Machado, E. A., and Lozzio, C. B.,** Hereditary renal disease in a mutant strain of rats, *Science,* 156, 1742, 1967.

208. **Solomon, S.,** Inherited renal cysts in rats, *Science*, 181, 451, 1973.

209. **Axelsen, R. A. and Burry, A. F.,** Bilirubin-associated renal papillary necrosis in the homozygous Gunn rat, Light and electronmicroscopic observations, *J. Pathol.*, 120, 165, 1976.

210. **Elema, J. D., Koudstaal, J., Lamberts, H. B., and Arends, A.,** Spontaneous glomerulosclerosis in the rat, *Arch. Pathol.*, 91, 418, 1971.

211. **Abramowsky, C. R., Aikawa, M., Swinehart, G. L., and Snajdar, R. M.,** Spontaneous nephrotic syndrome in a genetic rat model, *Am. J. Pathol.*, 117, 400, 1984.

212. **Murti, G. S.,** Agenesis and dysgenesis of the canine kidneys, *J. Am. Vet. Med. Assoc.*, 146, 1120, 1965.

213. **Robbins, G. R.,** Unilateral renal agenesis in the beagle, *Vet. Rec.*, 77, 1345, 1965.

214. **Bruyere, P., Posada, G. A., and Gouffaux, M.,** Hypoplasia du cortex renal chez le chien, *Ann. Med. Vet.*, 119, 23, 1975.

215. **Kurtz, J. M., Russell, S. W., Lee, J. C., Slauson, D. O., and Schechter, R. D.,** Naturally occurring canine glomerulonephritis, *Am. J. Pathol.*, 67, 471, 1972.

216. **Asheim, A.,** Comparative pathophysiological aspects of the glomerulonephritis associated with pyometra in dogs, *Acta Vet. Scand.*, 5, 188, 1964.

217. **Muller-Peddinghaus, R. and Trautwein, G.,** Spontaneous glomerulonephritis in dogs. II. Correlation of glomerulonephritis with age, chronic interstitial nephritis and extrarenal lesions, *Vet. Pathol.*, 14, 121, 1977.

218. **Jansen, B., Thorner, P. S., Singh, A., Patterson, J. M., and Lumsden, J. H.,** Animal model of human disease: hereditary nephritis in Samoyed dogs, *Am. J. Pathol.*, 116, 175, 1984.

219. **Finco, D. R.,** Congenital and inherited renal disease, *J. Am. Anim. Hosp. Assoc.*, 9, 301, 1973.

220. **Hayes, H. M.,** Ectopic ureter in dogs: epidemiologic features, *Teratology*, 10, 129, 1974.

221. **King, L. R. and Idress, F. S.,** The effect of vesicoureteral reflux on renal function in dogs, *Invest. Urol.*, 4, 419, 1967.

222. **King, L. R. and Sellards, H. F.,** The effect of vesicoureteral reflux on renal growth and development in puppies, *Invest. Urol.*, 9, 95, 1971.

223. **Bovee, K. G., Reynolds, R., Yost, B., and Segal, S.,** Renal glycosuria in dogs, in Proc. 9th Annu. Meet. Am. Soc. Nephrol., 1976, 92.

224. **Bovee, K. C., Thier, S. O., Claire, R., and Segal, S.,** Renal clearance of amino acids in canine cystinuria, *Metab. Clin. Exp.*, 23, 1974.

225. **Tsan, M. E., Jones, T. C., and Thornton, G. W.,** Canine cystinuria. Its urinary aminoacid pattern and genetic analysis, *Am. J. Vet. Res.*, 33, 2455, 1972.

226. **Bovee, K. G., Joyce, T., Reynolds, R., and Segal, S.,** Spontaneous Fanconi syndrome in the dog, *Metab. Clin. Exp.*, 27, 45, 1978.

227. **Brown, N. O., Parks, J. L., and Greene, R. W.,** Canine urolithiasis—retrospective analysis of 438 cases, *J. Am. Vet. Med. Assoc.*, 170, 414, 1977.

228. **Finco, D. R.,** Current status of canine urolithiasis, *J. Am. Vet. Med. Assoc.*, 158, 327, 1971.

229. **Porter, P.,** Urinary calculi in the dog. II. Urate stones and purine metabolism, *J. Comp. Pathol. Ther.*, 73, 119, 1963.

230. **White, E. G.,** Symposium on urolothiasis in the dog, *J. Small Anim. Pract.*, 7, 529, 1966.

231. **Anderson, L. J. and Jarrett, W. F. H.,** Membranous glomerulonephritis associated with leukemia in cats, *Res. Vet. Sci.*, 12, 179, 1971.

232. **Brown, P.,** A case of feline membraneous glomerulonephritis, *Vet. Rec.*, 89, 557, 1971.

233. **Battershell, D. and Garcia, J. P.,** Polycystic kidney in a cat, *J. Am. Vet. Med. Assoc.*, 154, 665, 1969.

234. **Benko, R. L., Prier, J. E., and Biery, D. N.,** Ectopic ureters in a male cat, *J. Am. Vet. Med. Assoc.*, 171, 738, 1977.

235. **Maguire, S., Hamilton, J. M., and Fulker, M. F.,** Spontaneous glomerular lesions in Syrian hamster *(Mesocricetus auratus)* Br. J. Pathol., 55, 562, 1974.

236. **Angus, K. W., Sykes, A. R., Gardiner, A. C., Morgan, K. T., and Tomson, D.,** Mesangiocapillary glomerulonephritis in lambs. I. Clinical and biochemical findings in a Finnish landrace flock, *J. Comp. Pathol.*, 84, 309, 1974.

237. **Lerner, K. A., Dixon, F. J., and Lee, S.,** Spontaneous glomerulonephritis in sheep. II. Studies on natural history, occurrence in other species, and pathogenesis, *Am. J. Pathol.*, 53, 501, 1968.

238. **Banks, K. L. and Henson, J. B.,** Immunologically mediated glomerulitis of horses. II. Antiglomerular basement membrane antibody and other mechanisms in spontaneous disease, *Lab. Invest.*, 26, 708, 1972.

239. **Feldman, D. B. and Bree, M. M.,** The nephrotic syndrome associated with glomerulonephritis in a rhesus monkey *(Macaca mulatta)*, J. Am. Vet. Med. Assoc., 155, 1249, 1969.

240. **Poskitt, T. R., Fortwengler, H. P., Jr., Bobrow, J. C., and Roth, G. J.,** Naturally occurring immune-complex glomerulonephritis in monkeys *(Macaca irus)*, Am. J. Pathol., 76, 145, 1974.

241. **Henson, J. B., Gorham, J. R., Padgett, G. A., and Davis, W. C.,** Pathogenesis of the glomerular lesions in Aleutian disease of mink. Immunofluorescent studies, *Arch. Pathol.*, 87, 21, 1969.

242. **Van Pelt, F. C. and Blandwater, M. J.,** Immunological and clinical-chemical investigations on blood, urine and kidneys in aging praomys *(Mastomys natalensis)* with spontaneous glomerulonephritis, *Gerontologia,* 18, 200, 1972.

243. **Stablay, R. W. and Rudofsky, U.,** Spontaneous renal lesions and glomerular deposits of IgG and complement in guinea pigs, *J. Immunol.,* 107, 1192, 1971.

244. **Fox, R. R., Krinsky, W. L., and Crary, D. D.,** Hereditary cortical renal cysts in the rabbit, *J. Hered.,* 62, 105, 1971.

245. **Gleiser, C. A., Van Hoosier, G. L., and Sheldon, W. C.,** A polycystic disease of hamsters in a closed colony, *Lab. Anim. Care,* 20, 923, 1970.

246. **Schlumberger, H. G.,** Polycystic kidney (mesonephros) in the goldfish, *Arch. Pathol.,* 50, 400, 1950.

247. **Ordidge, R. M.,** Urinary incontinence due to unilateral ectopia in a foal, *Vet. Rec.,* 98, 384, 1976.

248. **Pearson, H. and Gibbs, C.,** Urogenital abnormalities in two calves, *Vet. Rec.,* 92, 463, 1973.

249. **Benko, L.,** Cases of bilateral and unilateral duplication of ureters in the pig, *Vet. Rec.,* 84, 139, 1969.

250. **Roberts, J. A.,** Experimental pyelonephritis in the monkey. IV. Vesicoureteral reflux and bacteria, *Invest. Urol.,* 14, 198, 1976.

251. **Nielsen, I. M.,** Urolithiasis in mink—pathology, bacteriology and experimental production, *J. Urol.,* 75, 692, 1956.

Chapter 4

ACUTE AND CHRONIC SURGICALLY INDUCED MODELS

J. Zicha

TABLE OF CONTENTS

I. INTRODUCTION

The study of various physiological functions often requires the use of laboratory animals in which different surgical interventions have been made. These are typically the implantation of sensors, vascular catheterizations, or organ preparations done to improve access for the collection of data (see Chapter 12). In addition, a wide spectrum of different pathophysiological states could be induced using various surgical approaches (see Chapters 19, 20, and 26).

The appropriate choice of the laboratory animal (species, strain, sex, and age), adequate anesthesia, good surgical technique, and sufficient animal care prior, during, and after the surgery are necessary prerequisites of the rational use of these models. It is also important to consider the choice of animal and/or surgical techniques with regards to the length of the experiment, the type of information required, and the demands on the quality and quantity of the data. General ethical principles of the work with laboratory animals should be respected (Guiding Principles for Research Involving Animals and Human Beings, American Physiological Society, 1980).

Since the rat is the most commonly used laboratory animal, this chapter will describe preferentially rat models. Though surgery in rats can be complicated by their small size, this is compensated by their easy availability, low breeding costs, good standardization, excellent tolerance of surgical treatments, and by the possibility of working on a large number of animals. Dogs and cats are frequently used if more detailed physiological information is required. They are most aptly suited to long-term experiments in which repeated measurements of several physiological parameters are carried out simultaneously. The detailed description of such models is, however, beyond the scope of this chapter.

II. ANIMAL CARE

Though the rat does not require special preoperative care, some points should not be neglected. The stress of long transport, a new environment, the sudden change of the dietary regimen, long fasting or water deprivation, etc. decrease the tolerance of the animals to severe surgical interventions. It is, therefore, desirable to adjust the animals to the new conditions for about 1 week. If postoperative restraint is scheduled, the rats should also be conditioned to this procedure before the operation to minimize the harmful effects of postoperative restraint stress.[1] Longer preoperative fasting is not necessary except in the case of surgery on the GI tract.

A. Anesthesia

The detailed description of methods and agents (including their dosage) used for animal anesthesia can be found in various monographs or review articles.[2-6] Anesthetic agents are usually administered by inhalation or by injection (i.p., i.v., or i.m.) (Table 1).

1. Inhalation Anesthesia

At the present time ether is still the most widely used anesthetic agent that is relatively safe for the animal, easy to use, and inexpensive.[7] In the last years new volatile agents (halothane, enflurane, isoflurane, etc.) also have been preferred because they are less hazardous for the laboratory staff and they influence the hemodynamics of the animals[8] less than ether.[9] Inhalation anesthesia can be induced with a bell jar, nose cone, or endotracheal tube.[6,7] The last method is used especially whe open-chest surgery is desired. The principal advantages of inhalation anesthesia are its rapid induction, the possibility to maintain it for 1 to 2 hr, and the ability to change its depth, as well as the fast recovery of animals after withdrawal of the anesthetic agents. However, it is reasonable if another person controls the depth of anesthesia and monitors the breathing of the rat. Using nose cone ether sup-

Table 1
THE DOSAGE OF DRUGS COMMONLY USED FOR THE ANESTHESIA OF THE RAT

Inhalation anesthesia	Dosage (vol %)	Injection anesthesia	Dosage (mg/kg)
Ether	3—10	Pentobarbital	30—50
Isoflurane	1—1.5	Inactin	80—100
Halothane	0.5—1.5	Ketamine	60—120
Enflurane	2—2.5	α-Chloralose	80—120
		Urethane	1000
		Urethane + chloralose	500 + 100

plementation, the rat can sometimes wake quickly or stop breathing suddenly. Pulmonary irritation and excessive salivation predisposing the rat to subsequent respiratory infections are the main disadvantages of ether-induced anesthesia.

2. Injection Anesthesia

Barbiturates (mainly pentobarbital and Inactin), ketamine, chloralose, and urethane are the most frequently used drugs for injection anesthesia. These drugs are usually administered i.p. (or i.v. in larger animals), but they can also be injected i.m. Pentobarbital can be given by intragastric gavage as well. I.v. barbiturate injection induces anesthesia rapidly, but there are often signs of acute barbiturate overdosage. On the other hand, i.p. administration is relatively safe, but care should be paid to avoid the possible injection of the drug into the GI tract. The reabsorption of sodium pentobarbital from the GI tract is slow and could be used for sustaining of the anesthesia induced by i.v. or i.p. drug administration. The barbiturate anesthesia can also be prolonged by ether supplementation, additional barbiturate injection (a quarter of the initial dose is given at the onset of the recovery), or by the use of anesthetic agents with protracted effects, e.g., Inactin. Though a single dose of these drugs is usually sufficient to cause lasting and deep sleep, some disadvantages of barbiturate anesthesia should be also mentioned. The full anesthetic dose is close to the lethal dose, the analgesic effects of pentobarbital is rather poor, and there are severe cardiovascular alterations.

3. Anesthesia Choice

The choice of the appropriate drug, the dosage, and the route of its administration depend on the type of surgical intervention as well as on the further use of operated-upon animals. Inhalation anesthesia is used not only for minor or short operations, but it is also preferred in those cases where rapid postoperative revival is desired. Longer deep sleep of the animals after surgery can sometimes contribute to further deterioration of their physiological state (hypothermia, hypercapnia). This might result in protracted recovery periods and/or an increased mortality potential in the operated-upon animals.

Anesthesia choice is especially important in those experiments in which anesthetized animals are objects of physiological studies. It is important to note that there is no ideal anesthetic agent at the present time. Various anesthetics interfere with circulation, metabolism, neurohumoral regulation, etc. Moreover, their effects might be quite different according to the drug used. Most of the anesthetics lower blood pressure. In ether-anesthetized rats this is due to the decrease of systemic resistance, in contrast to pentobarbital-treated rats whose cardiac output is suppressed.[9,10] The effects of anesthetic agents also depend on the state of the organism. The circulation of ketamine-anesthetized rats seems to be close to the awake state, but this is true only in the normovolemic state. Isoflurane seems to be better for hypovolemic animals.[8] The acute pronounced decrease of plasma and extracellular fluid volume (due to its redistribution) is a characteristic finding in animals subjected to a major surgery.[11] Even minor manipulations with the anesthetized animal could cause profound changes of its cardiovascular and renal functions.[12] The opening of the abdominal cavity is always followed by a dramatic blood pressure fall. Sufficient attention should also be paid to the interpretation of the data obtained in animals recently recovered from anesthesia and surgery.[12]

B. Surgery and Postoperative Care

The outcome of the surgery does not depend only on the used surgical technique. Prolonged deep anesthesia threatens the organism by exposure to potentially severe hypoxia and hypercapnia due to a breathing depression. Failing thermoregulation and progressive loss of body fluids should be opposed in order to improve the survival of the operated-upon animals. This is very important in longer acute experiments (lasting several hours) in which body temperature of the anesthetized rat should be monitored (e.g., by a rectal thermometer) and sustained at 37°C by an external feedback-controlled heating system (e.g., IR). Body fluid loss can be controlled by the slow i.v. infusion of a hypotonic Ringer solution that matches diuresis and perspiration.

Though rats and other rodents (but not dogs or cats) do not require strictly aseptic sterile methods, the results are better if an aseptic approach is used.[6,13] The outcome of complicated surgeries could be improved if they are carried out in two or more steps. Subtotal reduction of renal mass is a typical example. It is more appropriate to damage extensively one kidney and then to remove the opposite one after 1 to 2 weeks than to do the whole surgery on the same day.

The postoperative animals should be isolated from other rats until they recover fully from the anesthesia. Even then, these handicapped rats should be protected from the attacks of healthy animals. After an operation rats must be provided with easily accessible food and enough water to drink. Moreover, after some endocrine gland extirpation, glucose (hypophysectomy) or saline (adrenalectomy) should be given in the drinking fluid to aid the recovery and/or to lower the postoperative mortality. The recovery of the rats also will be delayed by any kind of stress or load imposed on these animals. If possible, the onset of any stressful experimental regimen (including special dietary treatments) should be postponed until after the end of the acute postoperative phase.

If the surgery is carried out in rats younger than 3 to 4 weeks that are still kept with their mother, special care must be paid to her behavior because some mothers kill their wounded young. This danger can be decreased considerably by the tranquilization of the mother or, more effectively, by concomitant ether anesthetization followed by the slight contamination of the mother's hair with some adhesive material (e.g., collodium). This will draw her attention away from her young to self-care.

If catheters and/or measuring probes are implanted, the animals must be housed individually to prevent the destruction of exteriorized tubing ends or wires by other animals. Nevertheless, some types of surgery require postoperative restraint of animals (e.g., Bollman cage) to avoid self-mutilation.[1,6] However, all forms of restraint represent severe forms of stress for the operated-upon animals.[1] The restraint should, therefore, be avoided (whenever it is possible) or minimized by preconditioning of rats to this procedure.

Implanted wires and cannulas must be exteriorized at regions that are relatively safe from self-destruction. The most commonly used method is to pass the cannula though a s.c. tunnel to the back of the neck and to exteriorize it at a midpoint between the ears or clavicles. The other possibility is to fit a special holder on the rat's skull and to pass the wires and tubing through this cranial pedestal.[1,6,14]

C. Long-Term Maintenance

The long-term survival of surgically induced models can be threatened by several factors. Besides the potential for an early death caused by inappropriate anesthesia or surgery, the animals are endangered by various infections (namely, by the acceleration of respiratory diseases) and by increased intravascular blood coagulation. The death of operated-upon animals that have already completed the recovery period can be caused by the progressive deterioration of some important physiological functions resulting from the nature of the surgical intervention, e.g., renal, cardiovascular, or hepatic failure. Other frequent complications result from implanted materials (probes, tubing, etc.) that can cause obliteration or constriction of important blood vessels, mechanical damage of organs, or abnormal reactive proliferation of the interstitium. Some of the above-mentioned complications can be minimized by the careful designing and performing of the surgical procedures as well as by the appropriate choice of the extent of intervention. The possibility to oppose the initially increased coagulation is rather limited. Even mild heparinization of operated-upon animals causes frequently severe (namely, internal) bleeding. Troubles resulting from infection can be diminished by (1) the use of healthy animals (specific pathogen-free), (2) rather aseptic surgery, and (3) preventive and especially postoperative antibiotic treatment. Both local (neomycin powder or ointment) and systemic antibiotic application (procaine penicillin 100,000 U daily) are used.

Long-term care should be paid especially to the patency of implanted cannulas. Arterial and other vascular catheters must be initially filled with isotonic saline or 50% dextrose containing 100 to 500 IU heparin per milliliter. Routine daily flushing of cannulas with saline or dextrose (0 to 100 IU heparin per milliliter) is highly recommended. After careful aspiration of the fluid contained in the cannula (for removing possible thrombi), the cannula is flushed by a fresh solution (about 1.5 volume of the cannula) before resealing of its exteriorized end.[1] Another more complicated method is to maintain the respective catheter under permanent pulse flushing (20 $\mu\ell$)[15] or continuous slow infusion (0.5 mℓ/hr).[16]

III. SURGICAL INTERVENTIONS

A. Organ Surgery

Useful introductions to basic surgical techniques in the rat were assembled by Singh and Avery[17] and Waynforth.[6,18] Waynforth's monograph[6] on experimental techniques in the rat contains a large section about the principal methods necessary for the extirpation of particular endocrine glands, visceral surgery, and cannula implantation, as well as methods for kidney and skin transplantation (Table 2). Detailed description of each technique is accompanied by very instructive illustrations. Techniques for hepatectomy and bile collection were also reviewed by Cocchetto and Bjornsson.[1] Techniques for organ transplantation are presented in detail in Chapter 19.

In newborn rats (younger than 3 to 5 days of age) certain operations can be carried out under the condition that the rats are anesthetized and immobilized by hypothermia (20 min at 5°C until breathing and other movements cease). In this manner pineal gland,[19] thymus,[20] testes,[21] and ovaries[22] can be simply removed in neonatal rats.

B. Catheterization and Vascular Surgery

Principal methods of vascular access in the laboratory rat have been extensively reviewed by Cocchetto and Bjornsson.[1] This excellent article summarizes available information on cannula materials and constructions, catheterization of blood vessels, fixing and exteriorizing of in-dwelling cannulas, and methods of maintaining the patency of catheters as well as numerous methods for arterial and venous access in the rat. An interesting new construction of an aortic catheter that can be implanted via renal artery stump was recently reported.[23]

In the last 5 to 10 years a lot of work has been done to develop methods of long-term physiological measurements in conscious, unrestrained animals with chronically implanted catheters. Some of the most advanced techniques used in cardiovascular research are described in more detail in Chapter 12. However, it is necessary to mention here briefly some

Table 2
SPECIAL SURGICAL TECHNIQUES IN ADULT RATS

Endocrine Gland Extirpation

Hypophysectomy (both parapharyngeal and intraaural approaches)
Adrenalectomy (and adrenal demedullation)
Thyroidectomy (including parathyroidectomy)
Orchidectomy
Ovariectomy
Pinealectomy
(Thymectomy)

Visceral Surgery

Nephrectomy (unilateral and subtotal)
Hepatectomy (partial and total)
Pancreatectomy (partial, subtotal, and total)
Bile duct ligation
Gastric fistula
Splenectomy
Thoracic duct catheterization
Lymphadenectomy
Hysterectomy
Cesarian section

From Waynforth, H. B., *Experimental and Surgical Technique in the Rat,*
Academic Press, London, 1980. With permission.

other systems that were recently developed to measure (simultaneously) different physiological functions in conscious rats.

Gellai and Valtin[24] introduced the measurement of renal functions in conscious rats over a period of weeks. Preconditioned, unanesthetized animals with cannulas (in the urinary bladder, descending aorta, and superior vena cava) and with implanted cuffs for vessel constrictions represent ideal objects for long-term serial studies of renal and cardiovascular functions.[12,25]

The considerable progress in the construction of swivel joints[1,26,27] that can be equipped with several fluid channel and electrical circuits facilitated studies monitoring various physiological functions for a longer time period. The combination of long-term blood pressure recording with chronic salt and water loading through chronic infusions[28] or with salt and water balance studies in metabolic cages[16] contributes substantially to the actual research of pathophysiological abnormalities in experimental hypertension. Chronic cannulation of the descending aorta and the jugular vein enables continuous recording of plasma sodium concentration and circulating blood volume in awake rats.[29,30] The implantation of a hydraulic occluder on the pulmonary artery[31] or the insertion of a balloon-tipped catheter to the right atrium[32] made it possible to determine mean circulatory filling pressure, venous capacitance, and venous compliance in conscious rats. The implantation of chronic arterial and venous catheters is a necessary prerequisite for repeated dye dilution measurements of cardiac output, systemic resistance, central blood volume, arterial compliance, etc.[33]

Besides infusions or indicator administration, chronic vascular cannulas can be used for obtaining blood samples. The relationship of different circulating vasoactive factors and blood pressure regulation was studied in conscious, mobile, and undisturbed rats using a new system of chronic arterial and venous catheterization of tail vessels.[34]

If small volumes of fluid are infused for several days or weeks, the external infusion system can be successfully replaced by osmotic minipumps (e.g., Alzet) that deliver desired amounts of drugs, hormones, or inhibitors s.c., i.v., or intracerebroventricularly. A remarkable example of this method is the study in which chronic administration of competitive inhibitors of particular vasopressin effects revealed their participation in the pathogenesis of deoxycorticosterone-salt hypertension.[35]

C. Implantation of Sensors

The rat is rather small for the implantation of larger measuring devices, but there are several methods for the determination of blood flow (thermodilution, electromagnetic, or ultrasonic flowmetry) that require the surgical positioning of the respective probes. For the thermodilution determination of cardiac output it is necessary to place the thermistor-tipped catheter into the aortic arch.[9,36,37]

Since the time of Ledingham's and Pelling's[38] pioneer experiments with the implantation of electromagnetic flowmeters around the ascending aorta in the rat, considerable progress has occurred in this field. Though these flow probes were initially used mainly in anesthetized, open-chest rats,[39] subsequent miniaturization of these probes resulted in their more frequent use in conscious animals.[40-42] The present use of Doppler flow probes for the simultaneous determination of blood flow in several vascular beds (e.g., renal, mesenteric, and femoral) of the conscious rat is an even more sophisticated hemodynamic technique.[43,44] Macro- and microcirculatory measurements could be performed in the same animal which has implanted electromagnetic flow probes and a dorsal microcirculatory chamber.[45] The diameter of greater vessels can also be monitored in conscious rats using an implanted electrolytic strain gauge.[46]

The combination of available hemodynamic methods with on-line computers facilitates the evaluation of blood pressure and other hemodynamic parameters simultaneously in a larger series of animals monitored continuously for many hours or days.[28,47-49] Moreover, all derived parameters (stroke volume, systemic resistance, heart work, etc.) can be calculated in a beat-to-beat manner. For more details of these progressive methods (see Chapter 12).

IV. ACUTE AND CHRONIC MODELS

A. Anesthetized vs. Conscious Animals

There is a long tradition of the use of anesthetized animals for various acute physiological studies due to many apparent advantages of such models. The anesthetized animals do not interfere with the experimental procedure by their activity, there is minimal uncontrolled impact of environmental factors, the laboratory staff is not exposed to increased hazard during experiments with toxic or radioactive agents, the animals can be fixed in desirable positions, etc. There are certain research fields in which anesthesia is a necessary prerequisite of the successful experiment, e.g., micropuncture study of nephron functions.

However, it is evident that anesthetics may modify many physiological functions.[50] Such changes could be detected from the CNS up to cardiovascular or renal functions, and they are dependent on the anesthetic agent used. The present trend is to use conscious and unrestrained animals whenever it is possible. This approach enables the study of animals whose regulations are within physiological limits. Moreover, long-term serial measurements can be performed in the same animals. The principal advantage of such models is that the rats are well adapted to the experimental procedure. Thus, they are mostly free of the stress that is common during various experimental manipulations.

Of course, the use of unanesthetized animals also means some complications. Besides the above-mentioned factors, it requires routine daily care to check implanted tubings, cannulas, electrodes, wires of probes, etc. Additional troubles can result from the insufficient recovery of animals after stressful operations or their inadequate adjustment to the measuring conditions.[12] If these problems are solved, the value of obtained data is much greater than in anesthetized animals. The data from conscious rats also reveal the extent of their natural spontaneous fluctuations within physiological limits.

Table 3
METHODS FOR CHRONIC COLLECTION OF DIFFERENT BODY FLUIDS IN THE RAT

Body fluid	Site of collection	Ref.
Blood	Arterial (aorta, carotid artery)	26, 51—53
	Venous (vena cava, jugular vein)	51, 54—56
	Venous (retroorbital plexus)	57
	Mixed (tail tip)	—
Lymph	Intestinal lymphatics	58, 59
	Thoracic duct	58, 60
Cerebrospinal fluid	Cisterna magna	61
Saliva	Parotid salivary duct	62
Bile	Common bile duct	63—66
Urine	Metabolic cage	67
	Cystostomy	68
	Chronic urinary bladder cannula	24

B. Long-Term Sampling Techniques

These methods include repeated collection of blood and other body fluids, metabolic and bilance studies, as well as continuous or repeated measurements of different physiological functions. Using some of the above-mentioned approaches, the data can be gathered from the same individuals over a long period in the course of which some physiological (development, aging, pregnancy, etc.) or pathophysiological events occur.

Methods for the collection of blood, lymph, cerebrospinal fluid, bile, saliva, and urine were extensively reviewed elsewhere.[1] Though some of these methods are appropriate only for the acute experiments, the others can be used for chronic sampling (Table 3). Various kinds of long-term bilance studies require individual housing of animals in metabolic cages in which quantitative determination of food and fluid intake, urine and feces excretion, and carbon dioxide expiration, as well as the sampling of different body fluids are possible.[1]

The metabolic cages for long-term monitoring of some physiological functions should be constructed so as to provide enough air flow for the rat and to minimize its discomfort. Good access to food and water supplies is a necessary condition of such experiments. Basic principles of the design and construction of metabolic cages for different animal species were outlined earlier.[69] At the present time, more plastic materials are used for the building of these cages.[67] Some cages are constructed so as to be easily disassembled for cleaning or radioisotope decontamination.[70] Many experiments require the collection of perfectly separated urine and feces.[71] A method for safe, low-temperature urine collection has been also designed.[72] There is potential for occasional problems with the accurate quantification of food and fluid intake in metabolic cages. However, the construction of spill-proof food containers[67,73] and methods for precise fluid intake measurements[74-76] were already reported.

If animals with implanted cannulas are studied in metabolic cages, special care must be paid to cage construction in order to facilitate manipulation of the exteriorized ends of catheters or wires.[24,77] This problem is especially important in long-term cardiovascular studies.[15,16,28,40,43,45,47-49]

C. Acute vs. Chronic Experiments

The decision to study any physiological phenomenon in an acute or chronic experiment depends primarily on the aim of the experiment. However, many experimental questions cannot be answered adequately without the use of chronic, surgically induced models among which two main categories of animals can be distinguished. There are animals in which the physiological state was chronically altered by organ or vascular surgery. Other animals were surgically prepared for chronic physiological measurements (implantation of sampling catheters, infusion tubings, stimulation and recording electrodes, measuring probes, etc.). Of course, the approaches could be combined.

In acute experiments, the relationship of certain physiological parameters is studied in an organism that remains otherwise unchanged in its basic properties. On the other hand, chronic experiments, during which the organism develops a pathological state, reacts to environmental influences, or changes some of its properties due to its maturation or aging, reveal the dynamics of the studied alterations of physiological relationships. Though there might be some problems connected with the long-term use of chronic models, the validity of the results is usually much greater.

REFERENCES

1. **Cocchetto, D. M. and Bjornsson, T. D.,** Methods for vascular access and collection of body fluids from the laboratory rat, *J. Pharm. Sci.,* 72, 465, 1983.
2. **Dripps, R. D., Engle, H. M., and Dunner, E.,** Anaesthesia in laboratory animals, *Fed. Proc., Fed. Am. Soc. Exp. Biol.,* 28, 1373, 1969.
3. **Barnes, C. D. and Eltherington, L. C.,** *Drug Dosages in Laboratory Animals, A Handbook,* University of California Press, Berkeley, 1973.
4. **Green, J. C.,** Animal anaesthesia, *Laboratory Animal Handbook 8,* Spottiswoode Ballantyne, Colchester, 1979.
5. **Baker, H. J., Lindsey, J. R., and Weisbroth, S. H.,** *The Laboratory Rat,* Vol. 2, Academic Press, New York, 1980.
6. **Waynforth, H. B.,** *Experimental and Surgical Technique in the Rat,* Academic Press, London, 1980.
7. **Ben, M., Dixon, R. L., and Adamson, R. H.,** Anesthesia in the rat, *Fed. Proc., Fed. Am. Soc. Exp. Biol.,* 28, 1522, 1969.
8. **Seyde, W. C. and Longnecker, D. E.,** Anesthetic influences on regional hemodynamics in normal and hemorrhaged rats, *Anaesthesiology,* 61, 686, 1984.
9. **Salgado, M. C. O. and Krieger, E. M.,** Cardiac output in unrestrained conscious rats, *Clin. Exp. Pharmacol. Physiol.,* Suppl. 3, 165, 1976.
10. **Smith, T. L. and Hutchins, P. M.,** Anesthetic effects on hemodynamics of spontaneously hypertensive and Wistar-Kyoto rats, *Am. J. Physiol.,* 238, H539, 1980.
11. **Maddox, D. A., Price, D. C., and Rector, F. C.,** Effects of surgery on plasma volume and salt and water excretion in rats, *Am. J. Physiol.,* 233, F600, 1977.
12. **Walker, L. A., Buscemi-Bergin, M., and Gellai, M.,** Renal hemodynamics in conscious rats: effects of anaesthesia, surgery and recovery, *Am. J. Physiol.,* 245, F67, 1983.
13. **Popp, M. B. and Brennan, M. F.,** Long-term vascular access in the rat: importance of asepsis, *Am. J. Physiol.,* 241, H606, 1981.
14. **Steffens, A. B.,** A method for frequent sampling of blood and continuous infusion of fluids in the rat without disturbing the animal, *Physiol. Behav.,* 4, 833, 1969.
15. **Garthoff, B. and Towart, R.,** A new system for the continuous direct recording of blood pressure and heart rate in the conscious rat, *J. Pharmacol. Method.,* 5, 275, 1981.
16. **Gill, A. and Beyer, K. H.,** Sustained concurrent blood pressure and salt/water balance in unrestrained rats, *Pharmacology,* 26, 303, 1983.
17. **Singh, D. and Avery, D. D.,** *Physiological Techniques in Behavioral Research,* Brooks/Cole, Monterey, Calif., 1975.
18. **Waynforth, H. B.,** Animal operative techniques (in mouse, rat, guinea-pig and rabbit), in *Techniques in Protein Biosynthesis,* Campbell, P. N. and Sargent, J. R., Eds., Vol. 2, Academic Press, New York, 1969, 209.
19. **Kincl, F. A. and Benagiano, G.,** Failure of pineal gland removal in neonatal animals to influence reproduction, *Acta Endocrinol.,* 54, 189, 1967.
20. **Hard, C. C.,** Thymectomy in the neonatal rat, *Lab. Anim.,* 9, 105, 1975.
21. **Křeček, J.,** Pineal gland and development of salt intake pattern in male rats, *Dev. Psychobiol.,* 9, 181, 1976.
22. **Křeček, J.,** Effect of ovarectomy of females and estrogen administration to males during neonatal critical period on salt intake in adulthood in rats, *Physiol. Bohemoslov.,* 27, 1, 1978.

23. **Dworkin, B. R., Filewich, R. J., DaCosta, J., Eissenberg, E., and Miller, N. E.,** A chronic arterial catheter and low compliance system for recording blood pressure and heart rate from the rat, *Am. J. Physiol.,* 239, H137, 1980.

24. **Gellai, M. and Valtin, H.,** Chronic vascular constrictions and measurements of renal function in conscious rats, *Kidney Int.,* 15, 419, 1979.

25. **Conrad, K. P.,** Renal hemodynamics during pregnancy in chronically catheterized, conscious rats, *Kidney Int.,* 26, 24, 1984.

26. **Burt, M. E., Arbeit, J., and Brennan, M. F.,** Chronic arterial and venous access in the unrestrained rat, *Am. J. Physiol.,* 238, H599, 1980.

27. **Panol, G. R., Carvalho, A. A., and Carvalho, J. S.,** High-pressure swivel for infusions into rats, *Physiol. Behav.,* 30, 317, 1983.

28. **Norman, R. A., Enobakhare, J. A., DeClue, J. W., Douglas, B. H., and Guyton, A. C.,** Arterial pressure-urinary output relationship in hypertensive rats, *Am. J. Physiol.,* 234, R98, 1978.

29. **Tanaka, Y., Morimoto, T., Miki, K., Nose, H., and Miyazaki, M.,** On-line control of circulating blood volume, *Jpn. J. Physiol.,* 31, 427, 1981.

30. **Nose, H., Sugimoto, E., Morimoto, T., Usui, S., and Aomi, T.,** Continuous recording of plasma sodium concentration and blood volume in awake rats, *Jpn. J. Physiol.,* 36, 607, 1986.

31. **Samar, R. E. and Coleman, T. G.,** Measurement of mean circulatory filling pressure and vascular capacitance in the rat, *Am. J. Physiol.,* 234, H94, 1978.

32. **Yamamoto, J., Trippodo, N. C., Ishise, S., and Frohlich, E. D.,** Total vascular pressure-volume relationship in the conscious rat, *Am. J. Physiol.,* 238, H823, 1980.

33. **Zicha, J., Karen, P., Krpata, V., Dlouhá, H., and Křeček, J.,** Hemodynamics of conscious Brattleboro rats, *Ann. N.Y. Acad. Sci.,* 394, 409, 1982.

34. **Fejes-Tóth, G., Náray-Fejes-Tóth, A., Ratge, D., and Frölich, J. C.,** Chronic arterial and venous catheterization of conscious, unrestrained rats, *Hypertension,* 6, 926, 1984.

35. **Hofbauer, K. G., Mah, S. C., Baum, H. P., Hänni, H., Wood, J. M., and Kraetz, J.,** Endocrine control of salt and water excretion: the role of vasopressin in DOCA-salt hypertension, *J. Cardiovasc. Pharmacol.,* 6 (Suppl. 1), S 184, 1984.

36. **Richardson, A. W., Cooper, T., and Pinakatt, T.,** Thermodilution method for measuring cardiac output of rats using a transistor bridge, *Science,* 135, 317, 1962.

37. **Muller, B. and Mannesmann, G.,** Measurement of cardiac output by the thermodilution method in rats. II. Simultaneous measurement of cardiac output and blood pressure in conscious rats, *J. Pharmacol. Method.,* 5, 29, 1981.

38. **Ledingham, J. M. and Pelling, D.,** Cardiac output and peripheral resistance in experimental renal hypertension, *Circ. Res.,* 20, 21 (Suppl. 2), 187, 1967.

39. **Pfeffer, M. A. and Frohlich, E. D.,** Electromagnetic flowmetry in anesthetized rats, *J. Appl. Physiol.,* 33, 137, 1972.

40. **Smith, T. L. and Hutchins, P. M.,** Central hemodynamics in the developmental stage of spontaneous hypertension in the unanesthetized rats, *Hypertension,* 1, 508, 1979.

41. **Smits, J. F. M., Coleman, T. G., Smith, T. L., Kasbergen, C. M., Van Essen, H., and Struyker-Boudier, H. A. J.,** Antihypertensive effect of propranolol in conscious spontaneously hypertensive rats: central hemodynamics, plasma volume, and renal function during beta-blockade with propranolol, *J. Cardiovasc. Pharmacol.,* 4, 903, 1982.

42. **Kawaue, X. Y. and Iriuchijima, J.,** Cardiac output and peripheral blood flows on pentobarbital anaesthesia in the rat, *Jpn. J. Physiol.,* 34, 283, 1984.

43. **Haywood, J. R., Shaffer, R. A., Fastenow, C., Fink, G. D., and Brody, M. J.,** Regional blood flow measurement in the conscious rat with pulsed Doppler flowmeter, *Am. J. Physiol.,* 241, H273, 1981.

44. **Van Orden, D. E., Farley, D. B., Fastenow, C., and Brody, M. J.,** A technique for monitoring blood flow changes with miniaturized Doppler flow probes, *Am. J. Physiol.,* 247, H1005, 1984.

45. **Smith, T. L., Osborne, S. W., and Hutchins, P. M.,** Long-term micro- and macrocirculatory measurements in conscious rats, *Microvasc. Res.,* 29, 360, 1985.

46. **Compagno, L. T., Leite, J. V. P., and Krieger, E. M.,** Continuous measurement of aortic caliber in conscious rats. Effect of acute hypertension, *Mayo Clin. Proc.,* 52, 433, 1977.

47. **Laffan, R. J., Peterson, A., Hitch, S. W., and Jeunelot, C.,** A technique for prolonged, continuous recording of blood pressure of unrestrained rats, *Cardiovasc. Res.,* 6, 319, 1972.

48. **Fink, G. D., Bryan, W. J., Mann, M., Osborn, J., and Werber, A.,** Continuous blood pressure measurement in rats with aortic baroreceptor deafferentation, *Am. J. Physiol.,* 241, H268, 1981.

49. **Su, D. F., Cerutti, C., Barrès, C., Paultre, C. Z., Vincent, M., and Sassard, J.,** Computer analysis of cardiovascular activity in conscious unrestrained hypertensive rats of the Lyon strain, *Clin. Exp. Hypertens. A,* 7, 413, 1985.

50. **Buelke-Sam, J., Holson, J. F., Bazare, J. J., and Young, J. F.,** Comparative stability of physiological parameters during sustained anaesthesia in rats, *Lab. Anim. Sci.,* 28, 157, 1978.

51. **Weeks, J. R.,** Method for administration of prolonged intravenous infusion of prostacyclin (PG I₂) to unanesthetized rats, *Prostaglandins,* 17, 495, 1979.

52. **Carvalho, J. S., Shapiro, R., Hooper, P., and Page, L. B.,** Methods for serial study of renin-angiotensin system in unanesthetized rat, *Am. J. Physiol.,* 228, 369, 1975.

53. **Popovic, V. and Popovic, P.,** Permanent cannulation of aorta and vena cava in rats and ground squirrels, *J. Appl. Physiol.,* 15, 727, 1960.

54. **Kaufman, S.,** Chronic, non-occlusive and maintenance-free central venous cannula in the rat, *Am. J. Physiol.,* 239, R123, 1980.

55. **Brown, R. J. and Breckenridge, C. B.,** Technique for long-term blood sampling or intravenous infusion in freely moving rat, *Biochem. Med.,* 13, 280, 1975.

56. **Popovic, V., Kent, K. M., and Popovic, P.,** Technique of permanent cannulation of the right ventricle in rats and ground squirrels, *Proc. Soc. Exp. Biol. Med.,* 113, 599, 1963.

57. **Salem, H., Grossman, M. H., and Bilbey, D. L. J.,** Micromethod for intravenous injection and blood sampling, *J. Pharm. Sci.,* 52, 794, 1963.

58. **Bollman, J., Cain, J. C., and Grindlay, J. H.,** Techniques for the collection of lymph from the liver, small intestine or thoracic duct of the rat, *J. Lab. Clin. Med.,* 33, 1349, 1948.

59. **Tasker, R. R.,** The collection of intestinal lymph from normally active rats, *J. Physiol. London,* 115, 292, 1951.

60. **Gallo-Torres, H. E. and Miller, O. N.,** A modified Bollmans technique for cannulation of rats thoracic duct — lymph flow standardization, *Proc. Soc. Exp. Biol. Med.,* 130, 552, 1969.

61. **Franklin, G. M., Dudzinski, D. S., and Cutler, R. W. P.,** Amino acid transport into cerebrospinal fluid of the rat, *J. Neurochem.,* 24, 367, 1975.

62. **Piraino, A. J., DiGregorio, G. J., and Ruch, E. K.,** Small animal model utilizing salivary drug excretion for pharmacokinetic determinations, *J. Pharmacol. Method.,* 3, 1, 1980.

63. **Fisher, B. and Vars, H. M.,** A method of collecting bile in rats: normal values on rat bile, *Am. J. Med. Sci.,* 222, 116, 1951.

64. **Johnson, P. and Rising, P. A.,** Techniques for assessment of biliary excretion and enterohepatic circulation in rat, *Xenobiotica,* 8, 27, 1978.

65. **Balabaud, C., Saric, J., Gonzales, P., and Delphy, C.,** Bile collection in free moving rats, *Lab. Anim. Sci.,* 31, 273, 1981.

66. **Klauda, H. C., McGovern, R. F., and Quackenbush, F. W.,** Use of bile duct T-cannula as a new technique for studying bile acid turnover in rats, *Lipids,* 8, 459, 193.

57. **Weigelt, O.,** Ein Rattenstoffwechselkäfig im Baukastenprinzip, *Z. Versuchstierk.,* 12, 68, 1970.

68. **Hoy, P. A. and Adolph, E. F.,** Diuresis in response to hypoxia and epinephrine in infant rats, *Am. J. Physiol.,* 187, 32, 1956.

69. **Lazarow, A.,** Design and construction of metabolic cages, *Methods Med. Res.,* 6, 216, 1954.

70. **Lambooy, J. P.,** An inexpensive and efficient small animal metabolism cage, *Lab. Anim. Care,* 17, 351, 1967.

71. **Lesthwood, P. D. and Plummer, D. T.,** Enzymes in rat urine. I. A metabolism cage for complete separation of urine and feces, *Enzymologia,* 37, 240, 1969.

72. **Lartigue, C. W., Driscoll, T. B., and Johnson, P. C.,** Low-temperature urine collection apparatus for laboratory rodents, *Lab. Anim. Sci.,* 28, 594, 1978.

73. **Guest, G. M., Brodsky, W. A., and Nelson, N.,** Metabolism cage for rats, with feeding device that minimizes food scattering, *Metabolism,* 1, 89, 1952.

74. **Lazarow, A.,** Methods for quantitative measurement of water intake, *Methods Med. Res.,* 6, 225, 1954.

75. **Krpata, V. and Křeček, J.,** Circadian rhythm of water intake in rats with hereditary diabetes insipidus as measured by automatic apparatus, *Physiol. Bohemoslov.,* 24, 449, 1975.

76. **Robbins, R. J.,** Accurate, inexpensive, calibrated drinking tube, *Lab. Anim. Sci.,* 27, 1038, 1977.

77. **Toon, S. and Towland, M.,** A simple restraining device for chronic pharmacokinetic and metabolism studies in rats, *J. Pharmacol. Method.,* 5, 321, 1981.

Chapter 5

ISOLATED ORGANS AND EXPLANTED TISSUE

M. Baudyšová

TABLE OF CONTENTS

I. INTRODUCTION

Beginnings of the in vitro culture go back to the first years of this century. It was Harrison[1] who tried for the first time to keep a tissue dissected from metazoan organism in in vitro conditions. At that time, nobody could have expected that in such a simple system the basis of important intercellular relations might be studied and elucidated. During the last 2 decades, however, the results achieved by many scientists stressed that culture in vitro represents a suitable model for studying fundamental processes in animals and human beings occurring at cellular level as well as for studying the panoply of factors governing these processes. Thus, a curious experiment started an era of culture techniques widely applied in contemporary physiology, cytogenetics, oncology, immunology, biochemistry, virology, and other disciplines. The results obtained contributed considerably to elucidation of pathogenesis, diagnostics, and prevention of human and animal diseases and were exploited in breeding of genetically selected animals. Production of some biologically active compounds (vaccines, interferon, and monoclonal antibodies) is another field in which the in vitro culture has found wide applicability.

In concert with the increasing importance of this methodology, the number of workers using culture techniques multiplies at a fairly high rate along with the publications devoted to this topic. A great number of workers using in vitro culture methods are specialists in other areas of biological sciences, and culture represents for them only a source of material they work up further. To select a suitable in vitro system and to use it appropriately, however, these workers should be acquainted with at least the general principles conditioning successful culture as well as with the main laws ruling cell life in vitro.

One has to realize that in metazoan organisms, specialized, functionally interrelated populations of cells are living in mutual contact. A complexity of regulatory processes occurs among these populations that cannot be neglected under in vitro conditions. Ideally, cells in vitro should be provided with the same humoral factors and intercellular relations as those which influence them in the intact organism. Until now we have not completely succeeded in reestablishing such conditions in vitro. Currently, maintaining of whole organs outside the organism is far from being ideal. Good results have been obtained, however, with maintaining organ fragments or tissue pieces under in vitro conditions. However, the approach that is best worked out at the moment is a culture of dissociated cells freshly explanted from the organism or a culture of cell lines derived from them. These cell cultures represent relatively homogenous populations kept under controlled conditions, thus representing a suitable and simple model for studying intracellularly acting regulatory mechanisms as well as intercellularly functioning processes regulating growth and differentiation.

The following short survey deals predominantly with cell cultures and is aimed at supplying the very basic information on in vitro culture to those who in solving their particular problems will have to employ this methodology. The chapter certainly cannot be exhaustive; many of the progressive approaches that could offer cell culture are not mentioned here. Those seeking deeper information in this respect are referred to some of the numerous monographs dealing with this topic.[2-22]

II. BASIC CELL CULTURE METHODS

A. Preparation of Tissue and Cells[3,23]

For successful cultivation, the dissection of tissue should be performed under sterile conditions without mechanical damage. The tissue sample must be kept moist until the explantation into culture, which should follow as soon as possible. All solutions, media, vessels, and other equipment coming into contact with tissue have to be properly sterilized before use (filtration through membrane sterilizing filters, pore diameter 0.2 μm, autoclaving or dry heat). The dissected tissue is usually washed, cleaned, and minced into fragments (1 mm^3) which can be directly inoculated into culture vessels, or it is further worked up by mechanical, chemical, and enzymic methods. Mostly, a combination of methods is used for the complete dissociation of tissue into single cell suspension. The proper dispersion technique is selected according to the tissue used. Gentle pipetting of fragments or sieving through a nylon mesh is recommended for brain, spleen, thymus, and lymph nodes.[23,24] The most frequently used chelating agents are ethylendiaminetetraacetic acid (EDTA) or ethyleneglycol *bis*(β-aminoethylether) N,N′-tetraacetic acid (EGTA) (20 μg/mℓ). For enzymic treatment trypsin (125 to 250 μg/mℓ) or collagenase (0.1 to 0.3%) are convenient. If needed, combinations with other enzymes (pronase, hyaluronidase, and deoxyribonuclease) are also possible. The procedures used and applied to most tissues are modifications of the original method of Dulbecco and Vogt[25] and Rappaport.[26] For the disruption of some tissues, e.g., preparation of hepatocytes from liver, special techniques are available (collagenase perfusion technique[27-29]).

The resulting cell suspension is a mixed population of many cell types constituting the parent tissue. In order to obtain cultures consisting of a single cell type, several cell-separating methods such as density gradient centrifugation, electrophoresis, or affinity-column separation can be applied.[30,31] There is, however, also the possibility of enriching the cultured population in the required cell type by a proper selection of culture conditions, e.g., by pretreatment of the vessel surfaces or by the composition and pH of the medium.[32-34] Alternatively, cloning can be used for the same purpose.[35,36]

B. Culture Techniques

Cells originated from most tissues are anchorage dependent, i.e., they can be grown only when attached and spread on a suitable substrate. These cells are cultured as monolayers. Monolayer cell cultures[37] can be established either from tissue fragments or from single-cell suspensions.

After explantation, fragments submerged in a small quantity of medium attach to the vessel bottom. Cells migrating from them start to proliferate and form a layer over the surface area. Then, this primary culture has to be subcultured. The cells released from the bottom by agents similar to those used for the preparation of cell suspensions are dispersed in fresh medium and inoculated into the secondary culture (typically 5 to 20 thousands of cells per 1 cm^2). Remnants and fragments are discarded. If the primary culture started from a single-cell suspension instead of fragments, the cell monolayer is formed in a shorter culture period.

Cells in a monolayer culture reveal a characteristic growth pattern. After inoculation, there is a lag phase, the quiescent period with no cell division. Then cells pass through the log phase of growth in which the cell number increases exponentially. When the saturation density is reached, multiplication ceases, and the system reaches the stationary phase (Figure 1).

Unlike normal anchorage-dependent cells, a number of transformed cells can be cultured as free suspensions in slowly agitated spinner flasks (50 rpm), and modified media with low Ca^{2+} and Mg^{2+} content to avoid cell attachment and clumping[38] can be used. Recently, a microcarrier culture technique was developed also enabling anchorage-dependent cells to be cultured in suspensions. Microcarriers, the small beads made from biologically active components, provide a large surface area for cell anchorage. When suspended in a medium they enable growing of high amounts of normal cells in relatively small medium volume.[39] For biotechnology, the advanced large-scale cell culture systems using continuous chemostat, recycling, and perfusion culture were developed.[19,40-44] For special purposes there are various other important culture techniques which cannot even be enumerated here. For examples see References 45 through 61.

C. Evaluation of Cell Viability and Growth

Though very old, the most commonly used measurement of growth is the direct counting

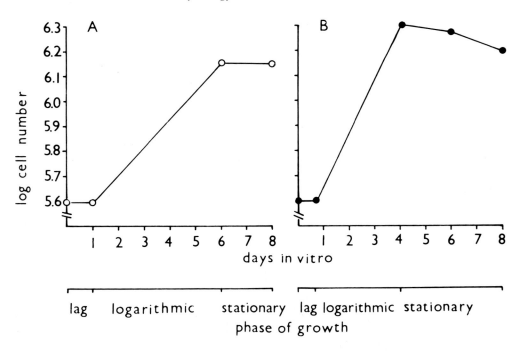

FIGURE 1.   Growth curve of diploid LEP 19 and heteroploid HeLa cell lines cultured under the same conditions. (A) LEP 19 — diploid human lung embryonal fibroblast; (B) HeLa — heteroploid human cervical carcinoma cells.

of cells suspended in a balanced salt solution in a hemocytometer. In combination with the dye exclusion test even the viability of cells can be evaluated by this procedure. The dye exclusion test is based on the assumption that viable cells with intact cellular membrane exclude certain dyes, e.g., eosin, Trypan blue, Alcian blue, and others.[62-64] The method of direct counting is convenient when only a few samples are to be evaluated. For routine enumeration of many cultures the electronic systems are preferred. Recently, the analysis of cell viability and growth was enriched by the advanced method of flow cytometry.[65]

Because of the correlation of protein and DNA biosynthesis with cellular growth both these parameters are frequently utilized to express the increase in cell mass during culture. These methods employ common analytical biochemical procedures adapted to cell culture.[64,66-69]

Alternatively, cell viability and growth can be evaluated by the ability of cells for clonal growth.[35] The highly diluted cell suspensions are plated on plastic dishes (100 cells per dish, 60 mm in diameter) or into microtest plate wells (one cell per 0.1- to 0.2-mℓ medium per one well). After the appropriate culture period the percentage of cells forming colonies (plating efficiency) is determined. The estimation of plating efficiency is obligatory in testing media and substrata quality. Cloning of mixed populations enables separation of different cell types and growing of pure cell populations. These populations are derived from a single parental cell and thus reveal minimal genetic variability which is advantageous if not necessary in specialized types of assay.[36]

D. Harvesting of Cells and Their Products

Harvesting in cell culture depends on the interest of the investigator as well as on the purpose which cells or their products should serve. If the aims are directed towards collecting the living cells, e.g., for storage in frozen state,[70] the cells are to be harvested[71] in the log phase of growth, suspended in medium (5 to 10 × 10^5 cells per milliliter) containing 10 to 20% fetal calf serum and 5 to 10% dimethylsulfoxide or glycerol, and slowly cooled in closed ampules. The frozen cells are recommended to be stored in liquid nitrogen which preserves the cell viability for prolonged time periods. Cryopreservation keeps cells for future reference and use without the changes in cellular characteristics frequently seen after long-term culture. Mostly, the cells grown in culture are utilized for biochemical studies in which gentle cell disrupting procedures are involved, allowing effective preparation of pure subcellular components with preserved integrity.[72,73] For selected purposes, e.g., for karyological examination, the cells should be harvested in a well-defined phase of the cell cycle. This requires specialized procedures which will be discussed later.

Culture medium can serve as a source of viruses, e.g., for preparation of vaccines, and as a source of cell products, e.g., interferons, interleukins, enzymes, growth factors, monoclonal antibodies, etc.[72,74-84] Purification of these products from the culture medium is of considerable advantage particularly in the case when serum-free culture systems are used. The methods applied are numerous and include biochemical and biophysical as well as other special approaches whose description is well beyond the scope of this introductory chapter.

III. SURVIVAL AND GROWTH REQUIREMENTS OF CELLS IN CULTURE

A. Environmental Factors

There are various environmental factors which can influence survival and growth of cultured cells. Among them the most important are temperature, pH range, dissolved gases concentration, buffering and osmolarity of the medium, humidification of the incubator, the nature of the surface on which the cells are grown, the sufficient supply of all essential nutrients and growth factors, and the absence of toxic and inhibitory factors.

Optimal temperature for most warm-blooded animal cells is near 37°C. The range of pH is rather narrow and occurs between 6.8 to 7.8 though it varies somewhat with cell type and medium used.[85-89] Generally, it is considered that atmospheric oxygen tension and 5% of CO_2 are suitable for cell proliferation. Most sodium bicarbonate-buffered media are designed for use with this CO_2 concentration.[89,90] These media when used in the open culture system should be additionally buffered since the loss of dissolved CO_2 during examination of dishes out of the incubator results in a rapid increase of pH. The most popular supplementary buffer is HEPES (N-2-hydroxyethylpiperazine-N'-2-ethanesulfonic acid) (10 to 30 mM).[85,90] The osmolarity of commercial media ranges from 250 to 330 mOsm/kg. The optimal values vary with cell type and animal species.[88-90] It should be mentioned that higher amounts of HEPES increase medium osmolarity. If needed, suitable values can be reached by lowering the sodium chloride concentration.[89] Osmolarity of the culture medium may alter because of evaporation. Therefore, relative humidity is to be maintained close to the saturation in the incubator. Routinely, cells are cultured in borosilicate or soda glass vessels or in specially treated polystyrene plastic flasks and dishes. However, substrata such as cellophane, Teflon®, silicone rubber, polycarbonate, and cellulose ester filters can be used as well.[37] Some culture systems require pretreatment of vessel surfaces with positively charged polymers such as poly-D-lysine[91,92] or biologically active components.[61,93] Nutritional requirements of cultured cells are quite complex and will be discussed separately. In this context the demands put on the quality of water, the major component of media, should be emphasized: only water of the highest purity, free of toxic organic as well as inorganic elements, can be used.[70,88] Treatment of tap water with a mixed-bed ion exchanger followed by two to three successive glass distillations will supply sufficiently ultrapure water for average cell culture operations.

B. Nutritional Requirements

A nutrient is defined as a substance that enters cells and is utilized as a substrate in biosynthetic pathways or in energy metabolism or serves as catalyst in these processes. Many nutrients also have a regulatory function. Nutrients in general are quite distinct from growth factors which are defined as nonnutritive substances that do not participate in biosynthesis, metabolism, or catalysis, but control proliferation in a regulative manner. These growth factors are similar in their mechanism of action to the classical peptide hormones.[94]

Nutritional requirements of cells in culture have been widely studied and reviewed.[10,13,95-97] Systematic studies gave rise to the commercially available low molecular mixtures of substances, e.g., medium 199,[98] MEM (Eagle's minimal essential medium),[99] DMEM (Dulbecco's modified MEM),[100] F 12 (medium mixture F 12 Ham),[101] or RPMI medium.[102] These media meet most of the qualitative demands of various cell types. They include energy sources, amino acids, vitamins, some other organic nutrients, and inorganic ions (Table 1). Traditionally, glucose (5 to 20 mM) serves as a source of energy. Alternatively, galactose or fructose,[90,103-105] glutamine,[106,107] pyruvate,[90,103] or other 2-oxocarboxylic acids[108,109] can be used for this purpose. There are 13 amino acids that are considered essential — arginine, cystine, glutamine, histidine, isoleucine, leucine, lysine, methionine, phenylalanine, threonine, tryptophan, tyrosine, and valine.[99] With increasing frequency the seven nonessential amino acids are being included — alanine, asparagine, aspartic acid, glutamic acid, glycine, proline, and serine.[91,110] In the presence of nonessential amino acids the requirements for essential amino acids are reduced. The B group vitamins[97,111-113] — biotin, folic acid, nicotinamide, pantothenic acid, pyridoxine, riboflavin, and thiamine — are required by most cells. Choline and mesoinositol, and vitamins C, B_{12}, A, and E are introduced for special studies and in serum-free culture systems.[97,113,114] Serum-containing media need not be supplemented by lipids since serum itself contains serum albumin-bound fatty acids and lipoproteins carrying phospholipids, triglycerides, and cholesterol.[97,113,115] Essential are unsaturated fatty acids (oleic, linoleic, linolenic, and aracidonic); cholesterol; cholesterol precursors; or phospholipids.[116] Further supplements improving media are nucleic acid precursors[113,114] and polyamines.[114] Major inorganic ions required are sodium, potassium, calcium, magnesium, phosphate, bicarbonate, and chloride. Their role is seen in maintaining the osmotic pressure and membrane potential, in establishing the buffering capacity, in promoting cell attachment, and in acting as co-factors for enzyme reactions.[86,113] Trace element requirements were defined during the development of serum-free media, and 15 trace elements have been established as essential or beneficial for animal cells — Co, Cu, I, Fe, Mn, Mo, Zn, Se, Cr, Ni, V, As, Si, F, and Sn. Naturally, serum-containing media need not be supplemented by these ions.[97,113,114]

It was widely accepted that the addition of serum (fetal calf, newborn calf, bovine, or horse serum; 0.5 to 20%) is essential for growth, since it provides culture with hormones, attachment and growth factors, and other necessary substances.[113,117] All the functions of this important fluid have not yet been clarified, obviously because of its complexity.[97,113] Unfortunately, there are also many disadvantages combined with the use of serum. Serum is not only an ill-defined part of medium, but also its quality varies substantially from one batch to another. In the presence of serum, differentiated cells are maintained in vitro with difficulties or not at all since they can be soon overgrown by quickly proliferating mesenchymal elements (fibroblasts). The ill-defined compounds of serum can interfere with experimental conditions in an uncontrollable manner. Thus, avoiding serum in the culture system is beneficial, if possible.[97,113,117]

Besides serum there are many other undefined media supplements (tissue extracts, lactalbumin hydrolysate, tryptose phosphate, peptone, etc.) used for improving culture conditions in some cases. However, their disadvantages are similar to those of serum.[97,118,119]

Standardization of culture conditions was improved by the use of serum-free medium containing purified serum growth-promoting proteins (GPP). They support proliferation of diploid and heteroploid cell lines including clonal growth as well as the establishment of primary cultures.[120,121] Commercially available GPPs (Institute of Sera and Vaccines, Prague, Czechoslovakia) are commonly used for serial culture of cell lines since they represent preparations of a standard composition and quality and are less expensive than serum.

Table 1
FORMULATIONS OF THREE WIDELY USED LOW MOLECULAR MEDIA

Components of the medium	MEM[a] (mg/ℓ)	DMEM[b] (mg/ℓ)	F 12[c] (mg/ℓ)
L-Amino acids			
Alanine	—	—	8.9
Arginine·HCl	105.0	84.0	211.0
Asparagine·H₂O	—	—	15.0
Aspartic acid	—	—	13.3
Cystein·HCl	—	—	31.5
Cystine	24.0	48.0	—
Glutamic acid	—	—	14.7
Glutamine	292.0	584.0	146.0
Glycine	—	30.0	7.5
Histidine·HCl	42.0	42.0	21.0
Isoleucine	52.0	104.8	3.9
Leucine	52.0	104.8	13.1
Lysine·HCl	73.0	146.2	36.5
Methionine	15.0	30.0	4.5
Phenylalanine	32.0	66.0	5.0
Proline	—	—	34.5
Serine	—	42.0	10.5
Threonine	48.0	95.2	11.9
Tryptophan	10.0	16.0	2.0
Tyrosine	36.0	72.0	5.4
Valine	46.0	93.6	11.7
Vitamins			
Biotin	—	—	0.007
Calcium D-pantothenate	1.0	4.0	0.48
Choline chloride	1.0	4.0	14.0
Folic acid	1.0	4.0	1.3
Inositol	2.0	7.0	18.0
Nicotinamide	1.0	4.0	0.04
Pyridoxal·HCl	1.0	4.0	0.06
Riboflavin	0.1	0.4	0.04
Thiamine·HCl	1.0	4.0	0.34
Vitamin B_{12}	—	—	1.36
Inorganic salts			
NaCl	6800.0	6400.0	7600.0
KCl	400.0	400.0	233.6
$NaH_2PO_4 \cdot 2H_2O$	150.0	125.0	—
$Na_2HPO_4 \cdot 7H_2O$	—	—	268.0
$MgCl_2 \cdot 6H_2O$	—	—	122.0
$MgSO_4 \cdot 7H_2O$	200.0	200.0	—
$CaCl_2$	200.0	200.0	33.3
$NaHCO_3$	2000.0	3700.0	1176.0
$Fe(NO_3)_3 \cdot 9H_2O$	—	0.1	—
$FeSO_4 \cdot 7H_2O$	—	—	0.83
$CuSO_4 \cdot 5H_2O$	—	—	0.002
$ZnSO_4 \cdot 7H_2O$	—	—	0.86
Other components			
n-Butyl-p-hydroxy benzoate	—	—	0.2
Glucose	1000.0	1000.0	1802.0
Hypoxanthine, sodium salt	—	—	4.1
Linoleic acid	—	—	0.08
Lipoic acid	—	—	0.21
Phenol red	10.0	15.0	1.2
Putrescine·2HCl	—	—	0.16

Table 1 (continued)
FORMULATIONS OF THREE WIDELY USED LOW
MOLECULAR MEDIA

Components of the medium	MEM[a] (mg/ℓ)	DMEM[b] (mg/ℓ)	F 12[c] (mg/ℓ)
Sodium pyruvate	—	110.0	110.0
Thymidine	—	—	0.73

<blockquote>
[a]   Eagle's minimum essential medium.[99]
[b]   Dulbecco's modification of Eagle's MEM.[100]
[c]   Ham's F 12 medium.[101]
</blockquote>

Lyophilized GPP can be stored at 4°C for at least 2 years without the loss of activity. The active component of GPP is a protein complex referred to as growth-promoting alpha-globulin (GPAG).[122,123] Its physiological effect on proliferation is very similar to that found in whole serum.[124-128] Serum-free medium supplemented with GPAG supports not only proliferation of all cells so far tested, it also supports the expression of differentiation of chick embryonal brain cells, myoblasts, and chondrocytes.[129-131] Because of its properties and nature, GPAG belongs to the family of growth factors, the hormone-like substances that regulate cell attachment, spreading, growth, and differentiation.[11]

Preparation of media from commercial low molecular mixtures is very simple. It consists of dissolving the powder in ultrapure water and equilibrating pH with sodium bicarbonate according to the producer's recommendation. This is followed by filtration through membrane sterilizing filters (pore diameter 0.2 μm). Before use, the appropriate quantity of serum is added. Diluted media can be stored at 4°C for a limited time period (about 14 days).

C. Serum-Free Cultivation

Since serum complicates further work, attempts have been made to develop serum-free media enabling advanced experimental approaches. In recent years, Sato,[132] Hayashi and Sato,[133] Bottenstein et al.,[134] and Barnes and Sato[135,136] have introduced several mixtures of hormones, supplementary nutrients, binding proteins, attachment factors, and extracellular matrix components applicable with distinct cell types. These mixtures are used as supplements for low molecular media, mostly DMEM/F 12 (1:1). They usually consist of insulin, transferrin, hydrocortisone, Se, and fibronectin. Many other components are frequently introduced because of special needs: albumin, steroid hormones, peptide hormones, thyroid hormones, prostaglandins, growth factors, and others.[6,10-12,137] Every individual cell type has been found to require only a small subset of hormone and growth factor supplements for growth in a defined medium. However, the requirements vary widely from cell to cell. Thus, very few predictions can be made about specific growth requirements for cell types that have not yet been studied. Suitable supplements supporting long-term culture of HeLa;[138] embryonal carcinoma;[139,140] pituitary,[141] melanoma,[142] and mammary carcinoma;[143] neuroblastoma;[144] and kidney epithelial[145] cell lines are listed in Table 2. Composition and preparation of serum-free media designed for cells of the endocrine system, for epithelial and fibroblastic cells, and for neuronal and lymphoid cells have also been published.[6]

It should be stressed that in the absence of serum the highest care possible must be taken in preparation of media, particularly with respect to the purity of chemical and water used; also, glass or polystyrene is not always the optimum substrate for all cultured cell types. Frequently, the surface of vessels should be pretreated using basic polymers (polylysine, polyornithine)[37,38,146,147] or different components of extracellular matrix.[93,146,148,149] In advanced culture systems, confluent monolayers of irradiated or killed stromal cells are used as feeder layers that enhance the maintenance of differentiated functions of epithelial cells.[61,146,150,151]

Table 2
GROWTH REQUIREMENTS OF SEVERAL CELL LINES IN SERUM-FREE CULTURE

	Human cervical carcinoma cell line HeLa	Embryonal carcinoma cell lines		Pituitary cell line GH$_3$	Melanoma cell line M$_2$R	Mammary carcinoma cell line MCF-7	Neuroblastoma cell line B 104	Kidney epithelial cell line MDCK
		PCC.4 aza-1	C17-S$_1$					
Medium applicable								
F 12	+							
F 12:DMEM (1:1)		+	+	+	+	+	+	+
Additional constituents								
Trace elements	+		Se only				Se only	
Insulin	5 µg/mℓ	10 µg/mℓ	5 µg/mℓ	5 µg/mℓ	5 µg/mℓ	250 ng/mℓ	5 µg/mℓ	5 µg/mℓ
Transferrin	5 µg/mℓ	5 µg/mℓ	10 µg/mℓ	5 µg/mℓ	5 µg/mℓ	25 µg/mℓ	100 µg/mℓ	5 µg/mℓ
Hydrocortisone	100 nM							
Testosterone								$5 \times 10^{-8}\ M$
Progesterone					10^{-8} or $10^{-9}\ M$		20 nM	
TSH-releasing hormone				1 ng/mℓ				
Parathyroid hormone				0.5 ng/mℓ				
Somatomedin C				1 ng/mℓ				
Luteinizing hormone releasing factor					1 ng/mℓ			
Follicle-stimulating hormone				0.5 µg/mℓ	0.5 µg/mℓ			
Triiodothyronine				$10^{-11}\ M$				
Prostaglandin T$_2$						100 ng/mℓ		$5 \times 10^{-12}\ M$
Prostaglandin E$_1$				1 ng/mℓ				
Fibroblast growth factor	10 ng/mℓ							
Epidermal growth factor	5 ng/mℓ					100 ng/mℓ		25 ng/mℓ
Fibronectin			5 µg/mℓ			7.5 µg/mℓ		
Fetuin		500 µg/mℓ						
Putrescin							100 µg/mℓ	
2-Mercaptoethanol		10 µg/mℓ						

D. Extracellular Matrix

The extracellular matrix is composed primarily of three categories of macromolecules: collagens, proteoglycans, and glycoproteins. Collagens constitute a group of genetically distinct but related proteins with specific tissue distribution.[152,153] Proteoglycans represent a highly polymorphic group of chemical entities that also have specialized tissue distribution.[154] Of the glycoproteins let us mention the two most important ones for tissue culture work: fibronectin and laminin. Fibronectin[155] has been identified as the major attachment glycoprotein of connective tissue, but it can be found in blood as well. Laminin[156] is present in basement membranes. These proteins mediate binding of cells to extracellular matrices, contributing thus to the formation of the supramolecular structures that surround most cells in the body. They are endowed with binding domains that exhibit a high specificity to a particular binding reaction (e.g., with individual collagen types, fibrin, cell surface, etc.). The extracellular matrix plays an important role in serum-free culture systems. When the bottoms of culture wells are covered with extracellular matrix components, cell attachment and spreading[61,93,148-150] as well as proliferation[61,136,158] are supported; moreover, the extracellular matrix affects differentiation of a number of cultured cell types.[61,93,159-161] The degree of specificity of different extracellular matrix components for different cell types has not yet been fully recognized. It appears, however, well established that multiple interactions occur on the cell surface side by side; in these interactions, for example, fibroblasts and epithelial cells are bound with a number of attachment proteins.[150]

The most widely used extracellular component in culture work is collagen. Originally, it was prepared from rat tail tendons by a method[162] which with a slight modification is still used. Tail tendons contain primarily type I collagen; though it is clear that this is not the type of collagen known to be in close association with, for example, epithelial cells in vivo, it can still be successfully used for culture of a variety of cell types.[61]

Preparation of collagen from rat tail tendons should proceed under sterile conditions. Extirpated and cleaned tendons are extracted with diluted acetic acid. From the extract, the undissolved remnants are separated by centrifugation. The supernatant represents a stock solution containing mainly collagen which can be stored at 4°C for several months. Before use, the appropriately diluted stock solution should be dialyzed against highly purified water to remove acetic acid. A small volume of resulting collagen solution is then spread over the vessel surface and simply allowed to dry.

Preparation of fibronectin, laminin, and other adhesive proteins (nectins) is slightly more complex. However, the importance of these adhesive proteins, especially in advanced serum-free culture systems, makes their preparation necessary. Basically, they are prepared by affinity chromatography methods.[163,164]

IV. SPECIAL ASSAY SYSTEMS AND TECHNIQUES

A. Cell Cycle Analysis

Cell cycle represents the set of events occurring between the two subsequent cell divisions. In most eucaryotic cells this period can be divided into four phases — prereplicative phase G_1, synthetic phase S (synthesis of DNA), postsynthetic phase G_2, and mitosis M. After mitosis the cell either enters the next cell cycle or stops its proliferation (G_0 state of quiescence)[165] (Figure 2). In exponentially growing populations, cells are distributed asynchronously throughout the cell cycle. Since metabolic processes are different in different phases, synchronized populations are often necessary.[166]

A traditional tool for cell cycle kinetic studies is autoradiography, which at present cannot be suppressed by any other method. It is based on detecting radioactivity through the formation of silver grains in a photographic emulsion spread over the labeled cells. The principle of this method is the same as photography except that the energy for conversion of silver bromide to metallic silver is derived from ionizing radiation rather than from photons

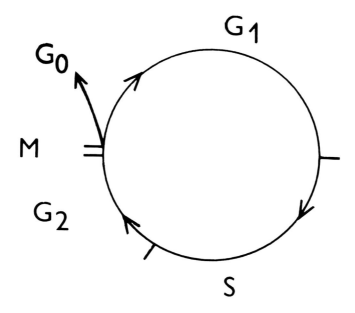

FIGURE 2.   Scheme of the cell cycle.

of light.[167,168] Usually, cells are labeled with ³H-thymidine which is incorporated into DNA during the S phase. From the data obtained by the analysis of autoradiographs, cell cycle parameters, e.g., mitotic and labeling index, growth fraction, and generation time, can be determined.[169]

During the last decade, cytokinetic studies were greatly facilitated by the development of flow cytometry.[65,170,171] This technique requires the preparation of a single cell suspension stained with a DNA-specific fluorescent dye. The cells are then processed one-by-one through a flow cytophotometer where the dye is stimulated to fluoresce. The resulting fluorescence is recorded for every cell as a measure of its DNA content. Thus, the percentage of cells in G_1, S, and G_2 + M phases can be determined. To distinguish G_2 and M cells the differential sensitivity of chromatin to denaturation can be employed.[172,173]

Application of flow cytometry provides a rapid and convenient means for answering questions related to cell cycle kinetics. However, the instrumental complexity and high hardware cost of this technology may limit its accessibility.

B. Evaluation of Cell Differentiation

The current model of differentiation is comprised of a population of stem cells out of which emerges a number of cells committed to differentiation. Between stem cells and terminally differentiated cells, a number of intermediate steps can be distinguished.[174-176] The development of a population to the differentiated state is accompanied by the appearance of characteristic signs in cell morphology and by the expression of specialized functions. Large divergence in these processes precludes the application of a universal method for their evaluation.

Light microscopy observation of living cells[177-181] (phase contrast, Nomarski interference contrast, and reflex constrast) offers preliminary information only. Widely used histological and histochemical methods are more informative; here the affinity of certain dyes to various cell components or products is employed.[178,180-185] Similarly, immunohistochemistry and fluorescence microscopy can be applied.[180,186-188] Ultramicroscopically specific cell constituents can be visualized.[177,180,186,189-192] Enzymes and specific cell products secreted into the culture medium can be measured biochemically.[181,185,193,194] Isotopic methods have also been employed.[177-179] In some cell types differentiation can be detected by their response to specific mediators or electric excitation.[195-197] Generally, the number of different cell types that can be distinguished reflects the number of methods used for the evaluation of the differentiated state. Several possibilities for estimating differentiation in cells derived from nerve tissue are shown in Table 3.

C. Transformation In Vitro

Transformation in cells is connected with the loss of susceptibility to the growth-controlling processes. Unlike normal diploid cells, transformed cells possess unlimited growth potential and anchorage-independent growth; express morphologic, karyotypic, and metabolic changes; and can grow tumors in susceptible animals.[212-215] It is well known that transformed cells arise if normal cells are exposed to chemical carcinogens or infection with tumor viruses. Chemical carcinogenesis is a multistage process as long as most carcinogens are not active per se but require metabolic activation. The chemical damage of DNA caused by an active carcinogen metabolite is to be fixed by subsequent cell division. Under appropriate circumstances the cell-bearing fixed lesion (initiated cell) can evolve to be neoplastic.[216] The transformation effect can be tested by the ability of cells to form colonies in soft agar.[217] Such methods can be used to evaluate the presence of carcinogens in environmental studies.

In some types of studies, pairs of normal cells and their transformed counterparts are required for comparative purposes. Pairs of such cell lines are available, for example, from the American Type Culture Collection, Rockville, Md. Under certain circumstances, however, it is necessary to prepare them in the laboratory by the transformation of selected cells with various strains of tumor viruses. RNA viruses transform almost all the cells in culture within a few days, while in general the DNA viruses transform only a small fraction of cells they infect.[212,213]

In vitro transformed cell lines and their normal counterparts are frequently used as a model system for studying oncogenic transformation. Their usefulness in understanding the biology of malignancy is, however, limited. Another model is represented by cell lines derived from human neoplasms. This model also exhibits several disadvantages; cell lines represent only a small population of the original tumor tissue since only few tumor cells are capable of vigorous growth in culture. Another problem is that the properties of the cells in culture may change as a function of passage. Despite this, the contribution of cell lines to present knowledge in this area is beyond question.[218]

D. Cell Fusion

The fact that two different cells can fuse and create a new entity has been known a long time. After treatment of cell membranes by fusion-promoting means, this phenomenon can be evoked intentionally. Classical treatment utilized the effect of enveloped viruses or lysolecithin and polyethyleneglycol.[78,219] In the last decade, the new fusion technique making use of the electric field was developed. It represents a universal, highly efficient method working much better under controlled conditions than previously used procedures.[220,221] The technical equipment is commercially available, but it is basically simple and can be easily substituted with a home-made device.[222]

Successful fusion results in the production of heterokaryons in which nuclei originated from both parental cells are included in common cytoplasma. From heterokaryons in which mitotic division takes place, hybrids are formed containing one nucleus with genetic material from both parental cells.[219] Hybrids are widely used in the analysis of gene expression, gene mapping, or virus detection.[219] Likewise, hybrids are important producers of biologically active compounds, e.g., monoclonal antibodies.[79,82,83,219] Monoclonal antibodies are secreted by hybridomas originated from the fusion of myeloma cells and spleen cells from appropriately immunized mice. The resulting cells reveal the multiplication ability of myeloma cells as well as the antibody-producing ability of spleen cells.[219] Recently, useful hybridomas

Table 3
EVALUATION OF DIFFERENTIATION IN CELL CULTURES DERIVED FROM NERVE TISSUE

Methodical approaches	Markers of differentiation	Ref.
Light microscopy	Nerve cells — bipolar or multipolar with long neurites arranged in bundles, a network of bundles interconnect cells over the whole culture surface	184, 187, 196—203
	Glial cells — flat, polygonal in shape, differentiate into cells with large or thin somas bearing many short branching processes (protoplasmic and fibrous astrocytes)	
Histology, histochemistry	Staining of Nissl substance, myelin, silver impregnation of neurites, presence of specific enzymes (acetylcholinesterase, cholinacetyltransferase)	182, 184, 196, 199, 201, 203
Immunochemistry; fluorescence microscopy	Binding of tetanus toxin, detection of intracellular or surface markers by monoclonal antibodies (presence of glial fibrillary acidic protein, S100 protein, presence of galactocerebroside)	187, 202, 204—208
Electron microscopy	Detection of intracellular markers by immunoperoxidase studies, formation of neurite bundles, myelin sheats, synaptogenesis	187, 196—198, 209, 210
Biochemistry, isotopic methods	Detection of specific proteins and enzymes, estimation of neurotransmitter (acetylcholine, catecholamines) synthesis and accumulation	193, 203, 206, 209, 211
Electrophysiology	Response to transmitters, membrane and action potential, synaptic transmission, etc.	195—197, 209, 211

were also prepared from human cells.[223-225] Monoclonal antibodies are highly specific and can be used for various diagnostic purposes as well as for the biotechnological preparation (interferon).[226] Advances in this area are closely related to progress in development of serum-free media and scaling up culture work.[19,227-232]

E. Chromosomal Examination[233,234]

In the mitotic phase of the cell cycle, nuclear DNA is organized into chromosomes with characteristic size and morphology. According to their appearance they can be classified and arranged into distinct groups to form a karyotype. The karyotype represents a characteristic feature of an animal species and is strictly controlled.

In primary culture the cells preserve the karyotype of the parent animal. With an increasing number of passages, however, an increasing number of cells with chromosomal variations can be observed. In established cell lines the chromosomal number varies from one cell to the other. It is either close to the normal diploid number (pseudodiploid line) or the range changes substantially forming a heteroploid line. Such variations not only include the number of chromosomes, but frequently their size, morphology, and structure are altered as well. Most established cell lines carry one or more characteristic chromosomes (marker chromosomes) which can be used as markers for cell identification.

The examination of chromosomes can be made in cells stimulated for growth and treated with mitotic inhibitors such as colcemid (N-deacetyl-N-methylcolchicine; 0.1 to 1.0 μg/mℓ) to increase the number of metaphase cells. These cells are processed to get specimens in which intact cells with preserved cellular membrane and well-spread chromosomes could be analyzed by light microscopy. Before examination chromosomes are stained by Giemsa stain. Solid staining is adequate for simple counting of chromosomes or for the identification of marker chromosomes. More detailed karyotypic examination requires banding methods be used. The banding techniques represent several procedures which have been developed for differential staining of certain regions of the chromosomes to give them a banded appearance. Photomicrographs of banded chromosomes serve for karyotyping. Individual chromosomes are cut out and arranged into the proper groups to which they belong. Karyotyping is an extremely valuable research tool in cell culture work, but it requires much experience. Unlike karyotyping, simple counting of chromosomes does not require special practice; this technique should be familiar to almost anybody involved in cell physiology as it is frequently needed.

V. EXAMPLES OF CELLULAR MODELS AND PRINCIPLES OF THEIR CHOICE

A. Primary Cell Cultures

Before beginning the work the investigator has to decide what kinds of cells should be used since some approaches require the use of freshly explanted, primary cell cultures while other experiments can be performed with cell lines whose growth requirements in culture are simpler and better understood.

A primary cell culture (Figure 3) is defined as a culture of cell population freshly dissociated from donor tissue and explanted in vitro. Such cells remember their in vivo origin and possess morphological and functional characteristics of their in vivo counterparts. Prolonged culture period, especially considerable number of passages, results in at least a partial loss of these characteristics. From this point of view, primary cultures mimic the in vivo situation much more closely than other systems. Progress in culture techniques in the last decade enabled the development of convenient culture systems in which various cells from almost all normal tissue and from some pathological specimens could be cultured. These cultured cells can be directly used not only for studying of physiological processes but also for acquiring new knowledge about intrinsic and extrinsic factors underlying pathogenesis of diseases. The cells are considered to be the lowest structural and functional entities integrating disease manifestation.[235] From the complexity of systems and questions studied, only a few can be mentioned here. Further details in this respect can be obtained from more specialized monographs and reviews.

One of the most intensively studied areas is the biology of vascular cells because the growth control of vascular smooth muscle cells (SMC) and endothelial cells (EC) is considered to be the key point in pathogenesis of atherosclerotic lesions[236,237] and hypertension.[238-240] Advanced in vitro systems for SMC and EC co-culture promise to contribute greatly to the elucidation of pathogenetic events resulting in disease manifestation.[45] The biology of vascular SMCs in culture was exhaustively reviewed.[21]

Another category of cells extensively studied in vitro is hepatic parenchymal cells, obviously because of the importance of liver and liver diseases. Adult hepatocytes represent a differentiated cell population which is known to respond in vivo to liver parenchyma injury by cell proliferation. These reversibly cycling cells have been found a suitable model in which induction of proliferation and/or differentiation triggered by hormones and hormone-like growth factors can be investigated.[241,242] In pharmacology and toxicology, hepatocytes can be used in following the metabolic degradation of drugs as well as in testing the carcinogenic effect of chemicals.[243] Coagulative necrosis and ischemic injury of cells are another field of pathophysiological research where hepatocytes became unavoidable.[235]

Since cancer belongs to the most serious human diseases, the interest of many cell scientists was turned to oncogenic transformation and tumor biology. As most human tumors are carcinomas, many workers deal with the tumors of epithelial origin. Contrary to the results obtained with normal human cells, the cells derived from tumor specimens grow in culture only poorly. Even now, culture techniques are not advanced enough to grow all the cells comprising a heterogenous tumor population. Similarly, only some explanted tumor specimens evolve a considerable degree of growth in vitro. Nevertheless, tumor cell lines could be derived which are used as tools for studies in tumor cell biology, biochemistry, and immunology. Their contribution to the present knowledge is far from negligible.[218,244] Current developments in cell model systems used in cancer research are reviewed in specialized monographs.[17,22]

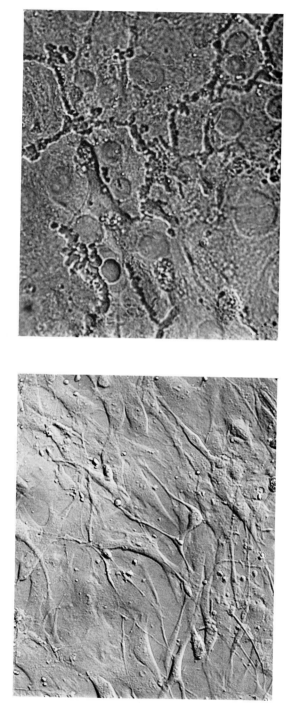

A

B

FIGURE 3. Primary cell cultures. Living cells, Nomarski interference contrast. (Magnification × 500.) (A) Human SMCs derived from aortal coarctation; (B) differentiating adult rat hepatocytes; and (C) heterogenous population of human laryngeal carcinoma.

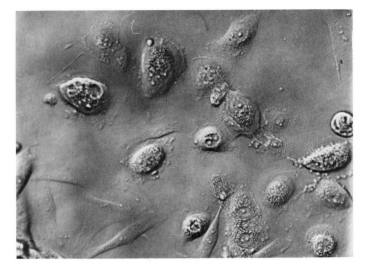

FIGURE 3C.

B. Cell Lines

A very popular model for cell investigation is either diploid or heteroploid cell lines (Figure 4). They have their origin in repeatedly passaged monolayers of primary cell cultures. Diploid cell lines possess normal diploid karyotype and limited lifespan in vitro. Heteroploid cell lines derived from them by spontaneous or induced transformation can be passaged permanently.[37,212,213,217] Cell lines provide a continuous supply of homogenous cellular material for biochemical studies. They are grown under controlled conditions. During the experiment, they can be visually controlled under light microscopy and directly manipulated if needed. Another advantage of established cell lines is that they can be stored frozen, and thus the same material is available for repeated examination.

Naturally, there are also some problems associated with the use of cell lines.[218] First, it is to be kept in mind that they represent only a small part of the parent tissue; the overall cell population present in such a tissue inevitably undergoes a selection in which cells capable of multiplication in vitro are accumulated. Moreover, the properties of cells in vitro may change remarkably as a function of passage. In spite of these facts, cell lines are extremely useful, have contributed greatly to the present knowledge in cell biology, and represent a model system to be used frequently in the future as well.

To obtain full benefit of the advantages cell lines can offer, the populations used for experiments should be standardized as fully as possible and subjected to extensive quality control procedures.[245-247] Many cell cultures contain undetected bacteria, yeast, viruses, or mycoplasmas. Some viruses are produced by them spontaneously and have an oncogenic potential. From this point of view, cell cultures represent a potential biohazard. Laboratory-acquired infection does exist, though it is not very frequent.[248] Among cell contaminants mycoplasmas are quite common and pertinacious. While mycoplasma infection is usually weak and relatively difficult to detect, infected cultures are already unacceptable for experimental use. Moreover, mycoplasma contamination expands rapidly attacking all cell lines maintained in the laboratory.[245,249]

In order to be protected from occasional infection, cell lines passaged in laboratories are continuously guarded by antibiotics.[250] Undoubtedly, in high risk situations (e.g., in primary cultures, especially in those derived from infected tissue or if a valuable cell line has to be saved) administration of antibiotics is quite necessary. In everyday laboratory work, their use should, however, be avoided whenever it is possible. By omitting antibiotics we actually

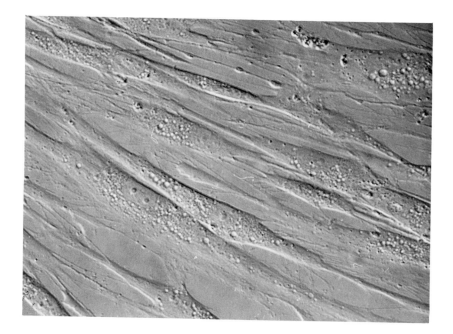

A

B

FIGURE 4.   Cell lines. Living cells, Nomarski interference contrast. (Magnification × 460.) (A) LEP 19 — diploid human lung embryonal fibroblasts; (B) HeLa — heteroploid human cervical carcinoma; and (C) E 7 — clone of mouse neuroblastoma C 1300.

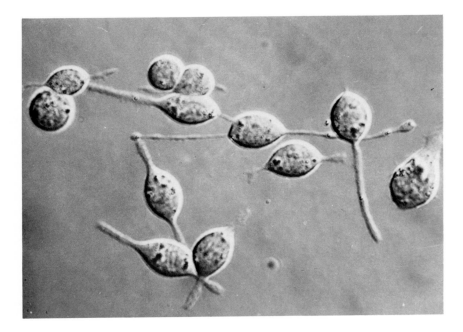

FIGURE 4C.

test the quality of our own work. In fact, in well-functioning laboratories contaminations are rare. Besides, permanent use of antibiotics can hide eventual infection which thus can propagate in culture undetected for a long period of time. There is also the apprehension that resistant bacterial contaminants can arise. In any case, antibiotics should be used with consideration.

It is well known that a significant number of cell lines are contaminated with other cells, especially with HeLa cells.[251-254] To avoid problems arising from infected or contaminated lines it is recommended to acquire cells only from sources where adequate control is guaranteed (American Type Culture Collection, Rockville, Md.; Human Genetic Mutant Cell Culture and Ageing Cell Culture Repositories, Camden, N.J.; Montreal Children's Hospital, Quebec, Canada; Center for Applied Microbiology and Research, Salisbury, England; Department of Medical Virology, Pasteur Institute, Paris, France). Necessary information can be obtained from an international data bank (CODATA Information Repository, National Institutes of Health, U.S.; Japan's Institute for Physical and Chemical Research — RIKEN; Faculte de Medicine de Nice, France). The catalogues available from these sources list the history of each cell line, a description of repository, and tests performed. The quality of cell line can be influenced by laboratory manipulation. For this reason, adequate tests must be performed repeatedly during long-term culture. If the source of the cell line is not quite reliable, the cells received should be kept in quarantine until sterility tests are completed.

Selection of the proper cell culture system is a necessary condition of successful cell in vitro research. The quality of cell line used can be just as critical. When respecting this fact as well as other fundamental principles of in vitro cell life, one can enjoy the work with cell cultures, finding it lovely, challenging, and easy.

REFERENCES

1. **Harrison, R. G.,** Observation on the living developing nerve fiber, *Soc. Exp. Biol. Med. Proc.,* 4, 140, 1907.
2. **Parker, R. C.,** *Methods of Tissue Culture,* 3rd ed., Hoeber, P. B., Ed., Academic Press, New York, 1961.
3. **Kruse, P. F. and Paterson, M. M., Eds.,** *Tissue Culture Methods and Applications,* Academic Press, New York, 1973.
4. **Paul, J.,** *Cell and Tissue Culture,* 5th ed., Churchill Livingstone, New York, 1975.
5. **Jakoby, W. B. and Pastan, I. H., Eds.,** *Cell Culture, Methods in Enzymology,* Vol. 58, Academic Press, New York, 1979.
6. **Barnes, D. W., Sirbasku, D. A., and Sato, G. H.,** *Cell Culture Methods for Molecular and Cell Biology,* Vols. 1—4, Alan R. Liss, New York, 1984.
7. **Harris, M.,** *Cell Culture and Somatic Variation,* Holt, Reinhart & Winston, New York, 1964.
8. **Willmer, E. N., Ed.,** *Cells and Tissues in Culture Methods, Biology and Physiology,* Vols. 1—3, Academic Press, New York, 1965—1966.
9. **Rothblat, G. H. and Cristofalo, V. J., Eds.,** *Growth, Nutrition and Metabolism of Cells in Culture,* Academic Press, New York, 1972.
10. **Waymouth, Ch., Ham, R. G., and Chapple, P. J., Eds.,** *The Growth Requirements of Vertebrate Cells in Vitro,* Cambridge University Press, Cambridge, 1981.
11. **Baserga, R., Ed.,** *Handbook of Experimental Pharmacology,* Vol. 57, Springer-Verlag, Berlin, 1981.
12. **Bing, D. H. and Rosenbaum, R. A., Eds.,** *Plasma and Cellular Modulatory Proteins,* Center for Blood Research, Boston, 1981.
13. **Sato, G. H., Pardee, A. B., and Sirbasku, D. A., Eds.,** *Cold Spring Harbor Conferences on Cell Proliferation,* Vol. 9, Cold Spring Harbor Laboratory, Cold Spring Harbor, N.Y., 1982.
14. **Murakami, H., Yamane, I., Barnes, D. W., Mather, J. P., Hayashi, I., and Sato, G. H., Eds.,** *Growth and Differentiation of Cells in Defined Environment,* Springer-Verlag, Berlin, 1985.
15. **Pollack, R., Ed.,** *Readings in Mammalian Cell Culture,* 2nd ed., Cold Spring Harbor Laboratory, Cold Spring Harbor, N.Y., 1981.
16. **Schweiger, H. G., Ed.,** *International Cell Biology 1980—1981, 2nd Int. Congr. on Cell Biology,* Springer-Verlag, Berlin, 1981.
17. **Dendy, P. P., Ed.,** *Human Tumours in Short Term Culture. Techniques and Clinical Applications,* Academic Press, London, 1976.
18. **Taub, M., Ed.,** *Tissue Culture of Epithelial Cells,* Plenum Press, New York, 1985.
19. **Mizrahi, A., Ed.,** *Advances in Biotechnological Processes,* Vol. 6, Alan R. Liss, New York, 1986.
20. **Schiff, L. J., Ed.,** *In Vitro Models of Respiratory Epithelium,* CRC Press, Boca Raton, Fla., 1986.
21. **Campbell, J. H. and Campbell, G. R., Eds.,** *Vascular Smooth Muscle in Culture,* Vols. 1—2, CRC Press, Boca Raton, Fla., in press.
22. **Webber, M. M. and Sekely, L. I., Eds.,** *In Vitro Models for Cancer Research,* Vols. 1—5, CRC Press, Boca Raton, Fla., in press.
23. **Bashor, M. M.,** Dispersion and disruption of tissues, *Methods Enzymol.,* 58, 119, 1979.
24. **Sensenbrenner, M., Booher, J., and Mandel, P.,** Cultivation and growth of dissociated neurons from chick embryo cerebral cortex in the presence of different substrates, *Z. Zellforsch.,* 117, 559, 1971.
25. **Dulbecco, R. and Vogt, M.,** Plaque and isolation of pure lines with poliomyelitis viruses, *J. Exp. Med.,* 99, 167, 1954.
26. **Rappaport, C.,** Monolayer cultures of trypsinized monkey kidney cells in synthetic medium. Application to poliovirus synthesis, *Soc. Exp. Biol. Med. Proc.,* 91, 464, 1956.
27. **Berry, M. M. and Friend, D. S.,** High-yield preparation of isolated rat liver parenchymal cells. A biochemical and fine structural study, *J. Cell Biol.,* 43, 506, 1969.
28. **Seglen, P. O.,** Preparation of isolated rat liver cells, in *Methods in Cell Biology,* Vol. 13, Prescott, D. M., Ed., Academic Press, New York, 1976, 29.
29. **Miyazaki, K., Takaki, R., Nakayama, F., Yamauchi, S., Koga, A., and Todo, S.,** Isolation and primary culture of adult human hepatocytes. Ultrastructural and functional studies, *Cell Tissue Res.,* 218, 13, 1981.
30. **Andrews, P. and Shortman, K.,** Zonal-unit-gravity elutriation. A new technique for separating large cells and multicellular complexes from cell suspensions, *Cell Biophys.,* 7, 251, 1985.
31. **Raber, J. M. and D'Ambrosio, S. M.,** Isolation of single cell suspensions from the rat mammary gland: separation, characterization, and primary culture of various cell populations, *In Vitro Cell. Dev. Biol.,* 22, 429, 1986.
32. **Barnes, D. and Sato, G. H.,** Methods for growth of cultured cell in serum-free medium, *Anal. Biochem.,* 102, 255, 1980.

33. **Baudyšová, M. and Michl, J.,** Serum-free medium is more beneficial than fetal calf serum-supplemented medium for morphological differentiation of primary chick embryo brain cells in culture, *Mol. Physiol.,* 2, 225, 1982.

34. **Tolar, M., Michl, J., Dlouhá, H., and Teisinger, J.,** Defined medium supplement with growth-promoting alpha-globulin supports the formation, differentiation and innervation of myotubes in cell cultures from chicken embryo, *Mol. Physiol.,* 3, 151, 1983.

35. **Reid, L. C. M.,** Cloning, *Methods Enzymol.,* 58, 152, 1979.

36. **Potten, C. S. and Hendry, J. H., Eds.,** Cell clones, in *Manual of Mammalian Cell Techniques,* Churchill Livingstone, Edinburgh, 1985.

37. **McAteer, J. A. and Douglas, W. H. J.,** Monolayer culture techniques, *Methods Enzymol.,* 58, 132, 1979.

38. **Ham, R. G. and McKeehan, W. L.,** Media and growth requirements, *Methods Enzymol.,* 58, 44, 1979.

39. **Thilly, W. G. and Lewine, D. W.,** Microcarrier culture: a homogenous environment for studies of cellular biochemistry, *Methods Enzymol.,* 58, 184, 1979.

40. **Spier, R. E. and Griffiths, J. B., Eds.,** *Animal Cell Biotechnology,* Vol. 1, Academic Press, London, 1985.

41. **Berg, G. J.,** An integrated system for large scale cell culture, in *Joint ESACT/IABS Meet. on the Production and Exploitation of Existing and New Animal Cell Substrates, Dev. Biol. Standard.,* Vol. 60, S. Karger, Basel, 1985, 297.

42. **van Wezel, A. L., van der Velden-de Groot, C. A. M., de Haan, H. H., van den Heuvel, N., and Schasfoort, R.,** Large scale animal cell cultivation for production of cellular biologicals, in *Joint ESACT/IABS Meet. on the Production and Exploitation of Existing and New Animal Cell Substrates, Dev. Biol. Stand.,* Vol. 60, S. Karger, Basel, 1985, 229.

43. **McLimans, W. F.,** Mass culture of mammalian cells, *Methods Enzymol.,* 58, 194, 1979.

44. **Acton, R. T., Barstad, P. A., and Zwerner, R. K.,** Propagation and scaling-up of suspension cultures, *Methods Enzymol.,* 58, 211, 1979.

45. **Jones, P. A.,** Construction of an artificial blood vessel wall from cultured endothelial and smooth muscle cells, *Proc. Natl. Acad. Sci. U.S.A.,* 76, 1882, 1979.

46. **Barrett, L. A., Mergner, W. J., and Trump, B. F.,** Long-term culture of human aortas. Development of atherosclerotic-like plaques in serum-supplemented medium, *In Vitro,* 15, 957, 1979.

47. **Lieberman, M., Adam, W. J., and Bullock, P. N.,** The cultured heart cells: problems and prospects, in *Methods in Cell Biology,* Academic Press, New York, 1980, 187.

48. **Lieberman, M., Horres, C. R., Shigeto, N., Ebihara, L., Aiton, J. F., and Johnson, E. A.,** Cardiac muscle with controlled geometry. Application to electrophysiological and ion transport studies, in *Excitable Cells in Tissue Culture,* Nelson, P. G., and Lieberman, M., Eds., Plenum Press, New York, 1981, 379.

49. **Honegger, P. and Mathieu, J.-M.,** Myelination of aggregating fetal rat brain cell cultures grown in a chemically defined medium, in *Neurological Mutations Affecting Myelination, INSERM Symp. No. 14,* Bauman, N., Ed., Elsevier/North-Holland Biomedical, Amsterdam, 1980, 481.

50. **Honegger, P. and Lenoir, D.,** Triiodothyronine enhancement of neuronal differentiation in aggregating fetal rat brain cells cultured in a chemically defined medium, *Brain Res.,* 199, 425, 1980.

51. **Sadler, T. W., Horton, W. E., and Hunter, E. S.,** Mammalian embryos in culture: a new approach to investigating normal and abnormal developmental mechanisms, in *Development Mechanisms: Normal and Abnormal,* Alan R. Liss, New York, 1985, 227.

52. **Rizzino, A. and Sherman, M. I.,** Development and differentiation of mouse blastocysts in serum-free medium, *Exp. Cell Res.,* 121, 221, 1979.

53. **Müller, H. J.,** Human in vitro fertilization and embryo transfer: expectations and concerns, *Experientia,* 41, 1515, 1985.

54. **Hodges, G. M. and Melcher, A. H.,** Chemically-defined medium for growth and differentiation of mixed epithelial and connective tissues in organ culture, *In Vitro,* 12, 450, 1976.

55. **Yang, J., Larson, L., and Nandi, S.,** Three-dimensional growth and morphogenesis of mouse submandibular epithelial cells in serum-free primary culture, *Exp. Cell Res.,* 137, 481, 1982.

56. **Weterings, P. J. J. M., Vermorken, A. J. M., and Bloemendal, H.,** Subcultivation of human hair follicle keratinocytes, *Exp. Cell Res.,* 139, 439, 1982.

57. **Gullino, P. M. and Knazek, R. A.,** Tissue culture on artificial capillaries, *Methods Enzymol.,* 58, 178, 1979.

58. **Leighton, J., Tchao, R., and Nichols, J.,** Radial gradient culture on the inner surface of collagen tubes: organoid growth of normal rat bladder and rat bladder cancer cell line NBT-II, *In Vitro Cell. Dev. Biol.,* 21, 713, 1985.

59. **Jensen, M. D., Wallach, D. F., and Lin, P.-S.,** Comparative growth characteristics of VERO cells on gas-permeable and conventional supports, *Exp. Cell Res.,* 82, 271, 1974.

60. **Litwin, J.**, The growth of human diploid fibroblasts as aggregates with cellulose fibres in suspension, in *Joint ESACT/IABS Meet. on the Production and Exploitation of Existing and New Animal Cell Substrates, Dev. Biol. Standard.*, Vol. 60, S. Karger, Basel, 1985, 237.

61. **Reid, L. M. and Rojkind, M.**, New techniques for culturing differentiated cells: reconstituted basement membrane rafts, *Methods Enzymol.*, 58, 263, 1979.

62. **Schrek, R.**, A method for counting the viable cells in normal and in malignant cell suspension, *Am. J. Cancer*, 28, 389, 1936.

63. **Jauregui, H. O., Hayner, N. T., Driscoll, J. L., Williams-Holland, R., Lipsky, M. H., and Galletti, P. M.**, Trypan blue dye uptake and lactate dehydrogenase in adult rat hepatocytes — freshly isolated cells, cell suspensions, and primary monolayer cultures, *In Vitro*, 17, 1100, 1981.

64. **Patterson, M. K. Jr.**, Measurement of growth and viability of cells in culture, *Methods Enzymol.*, 58, 141, 1979.

65. **Shapiro, H. M.**, *Practical Flow Cytometry*, Alan R. Liss, New York, 1985.

66. **Lowry, O. H., Rosenbrough, N. J., Farr, A. L., and Randall, R. S.**, Protein measurement with the folin phenol reagent, *J. Biol. Chem.*, 193, 265, 1951.

67. **Vytášek, R.**, A sensitive fluorometric assay for determination of DNA, *Anal. Biochem.*, 120, 243, 1982.

68. **Richards, W. L., Song, M.-K., Krutzsch, H., Evarts, R. P., Marsden, E., and Thorgeirsson, S. S.**, Measurement of cell proliferation in microculture using Hoechst 33342 for the rapid semiautomated microfluorometric determination of chromatin DNA, *Exp. Cell. Res.*, 159, 235, 1985.

69. **Mates, G., Daniel, M., and Walker, C.**, Factors affecting the reproducibility of a spectrofluorimetric assay for the enumeration of human venous endothelium in culture, *Cell Biol. Int. Rep.*, 10, 641, 1986.

70. **Douglas, H. J. and Dell'Orco, R. T.**, Physical aspects of tissue culture laboratory, *Methods Enzymol.*, 58, 3, 1979.

71. **Kirkpatrick, C. J., Melzner, I., and Göller, T.**, Comparative effects of trypsin, collagenase and mechanical harvesting on cell membrane lipids studies in monolayer-cultured endothelial cells and a green monkey kidney cell line, *Biochim. Biophys. Acta*, 846, 120, 1985.

72. **Zwerner, R. K., Wise, K. S., and Action, T. R.**, Harvesting the products of cell growth, *Methods Enzymol.*, 58, 221, 1979.

73. **Domsch, Ch. and Mersmann, G.**, Remarks on the differentiation of lysosomes from cultured human fibroblasts by silica gradient centrifugation, *Exp. Cell Res.*, 142, 482, 1982.

74. **Friedman, R. M.**, Induction and production of interferon, *Methods Enzymol.*, 58, 292, 1979.

75. **Bjare, U. and Räbb, I.**, Serum-free cultivation of lymphoid cells for virus and interferon production, in *Joint EXACT/IABS Meet. on the Production and Exploitation of Existing and New Animal Cell Substrates, Dev. Biol. Standard.*, Vol. 60, S. Karger, Basel, 1985, 349.

76. **Ihle, J. N., Keller, J., Henderson, L., Klein, F., and Palaszynski, E.**, Procedures for the purification of interleukin 3 to homogeneity, *J. Immunol.*, 129, 2431, 1982.

77. **Suda, T., Suda, J., Ogawa, M., and Ihle, J. N.**, Permissive role of interleukin 3 (IL-3) in proliferation and differentiation of multipotential hemopoietic progenitors in culture, *J. Cell. Physiol.*, 124, 182, 1985.

78. **Kennett, R. H.**, Cell fusion, *Methods Enzymol.*, 58, 345, 1979.

79. **Fazekas-de St. Groth, S. and Schneidegger, D.**, Production of monoclonal antibodies: strategy and tactics, *J. Immunol. Methods*, 35, 1, 1980.

80. **Spier, R. E. and Griffiths, J. B., Eds.**, *Animal Cell Biotechnology*, Vol. 2, Academic Press, London, 1985.

81. **Riesfeld, R. A. and Sell, S., Eds.**, Monoclonal antibodies and cancer therapy, in *UCLA Symposia on Molecular and Cellular Biology — New Series*, Vol. 27, Alan R. Liss, New York, 1985.

82. **Epstein, E. and Epstein, M.**, The hybridoma technology. I. Production of monoclonal antibodies, in *Advances in Biotechnological Processes*, Vol. 6, Mizrahi, A., Ed., Alan R. Liss, New York, 1986, 179.

83. **Epstein, E., Kobiler, D., and Epstein, M.**, The hybridoma technology. II. Applications of hybrid cell products — monoclonal antibodies and lymphokines, in *Advances in Biotechnological Processes*, Vol. 6, Mizrahi, A., Ed., Alan R. Liss, New York, 1986, 219.

84. **Rowlands, D. J.**, New advances in animal and human virus vaccines, in *Advances in Biotechnological Processes*, Vol. 6, Mizrahi, A., Ed., Alan R. Liss, New York, 1986, 253.

85. **Eagle, H.**, Some effects of environmental pH on cellular metabolism and function, in *Control of Proliferation in Animal Cells*, Clarkson, B. and Baserga, R., Eds., Cold Spring Harbor Laboratory, Cold Spring Harbor, N.Y., 1974, 1.

86. **Ham, R. G.**, Survival and growth requirements of nontransformed cells. C.III. Inorganic ions, physical chemistry and cell physiology, in *Handbook of Experimental Pharmacology*, Vol. 57, Baserga, R., Ed., Springer-Verlag, Berlin, 1981, 39.

87. **Taylor, W. G.**, Electrolytes, dissolved gases, buffers, and pH changes in tissue culture systems: an overview, in *The Growth Requirements of Vertebrate Cells In Vitro*, Waymouth, Ch., Ham, R. G., and Chapple, P. J., Eds., Cambridge University Press, Cambridge, 1981, 94.

88. **Waymouth, Ch.,** Major ions, buffer systems, pH, osmolarity, and water quality, in *The Growth Requirements of Vertebrate Cells In Vitro,* Waymouth, Ch., Ham, R. G., and Chapple, P. J., Eds., Cambridge University Press, Cambridge, 1981, 105.

89. **Ham, R. G. and McKeehan, W. L.,** Media and growth requirements. Physiological requirements, *Methods Enzymol.,* 58, 55, 1979.

90. **Waymouth, Ch.,** Studies on chemically defined media and the nutritional requirements of cultures of epithelial cells, in *Nutritional Requirements of Cultured Cells,* Katsuda, H., Ed., Japan Scientific Societies Press, Tokyo, 1978, 39.

91. **Ham, R. G. and McKeehan, W. L.,** Media and growth requirements. Modifications of the culture system that minimize the requirement for serum protein, *Methods Enzymol.,* 58, 49, 1979.

92. **McKeehan, W. L., McKeehan, K. A., and Ham, R. G.,** The use of low-temperature subculturing and culture surfaces coated with basic polymers to reduce the requirement for serum macromolecules, in *The Growth Requirements of Vertebrate Cells In Vitro,* Waymouth, Ch., Ham, R. G., and Chapple, P. J., Eds., Cambridge University Press, Cambridge, 1981, 118.

93. **Ruoslahti, E., Hayman, E. G., and Pierschbacher, M. D.,** Extracellular matrices and cell adhesion, *Atherosclerosis,* 5, 581, 1985.

94. **Ham, R. G.,** Survival and growth requirements of nontransformed cells. A.IV. Nutrient, growth requirement, growth factor, mitogen and hormone, in *Handbook of Experimental Pharmacology,* Vol. 57, Baserga, R., Ed., Springer-Verlag, Berlin, 1981, 15.

95. **Ham, R. G. and McKeehan, W. L.,** Media and growth requirements, *Methods Enzymol.,* 58, 44, 1979.

96. **Litwin, J.,** A survey of various media and growth factors used in cell cultivation, *Dev. Biol. Stand.,* 42, 37, 1979.

97. **Lambert, K. J. and Birch, J. R.,** Cell growth media, in *Animal Cell Biotechnology,* Vol. 1, Spier, R. E. and Griffiths, J. B., Eds., Academic Press, London, 1985, 85.

98. **Morgan, J. F., Morton, H. J., and Parker, R. C.,** Nutrition of animal cells in tissue culture. I. Initial studies on a synthetic medium, *Proc. Soc. Exp. Biol. Med.,* 73, 1, 1950.

99. **Eagle, H.,** Amino acid metabolism in mammalian cell culture, *Science,* 130, 432, 1959.

100. **Dulbecco, R. and Freeman, G.,** Plaque production by the polyoma virus, *Virology,* 8, 396, 1959.

101. **Ham, R. G.,** Clonal growth of mammalian cells in a chemically defined synthetic medium, *Proc. Natl. Acad. Sci. U.S.A.,* 53, 288, 1965.

102. **Moore, G. E., Gerner, R. E., and Franklin, H. A.,** Culture of normal human leukocytes, *JAMA,* 199, 519, 1967.

103. **Eagle, H., Barban, S., Levy, M., and Schulze, H. O.,** The utilization of carbohydrates by human cell culture, *J. Biol. Chem.,* 233, 551, 1958.

104. **Leibovitz, A.,** The growth and maintenance of tissue-cell culture in free gas exchange with the atmosphere, *Am. J. Hyg.,* 78, 173, 1963.

105. **Delhotal, B., Lemonnier, F., Couturier, M., Wolfrom, C., Gautier, M., and Lemonnier, A.,** Comparative use of fructose and glucose in human liver and fibroblastic cell cultures, *In Vitro,* 20, 699, 1984.

106. **Reitzer, L. J., Wice, M. B., and Kennell, D.,** Evidence that glutamine, not sugar, is the major energy source for cultured HeLa cells, *J. Biol. Chem.,* 256, 2669, 1979.

107. **Zielke, H. R., Ozand, P. T., Tildon, J. T., Sevdalian, D. A., and Cornblath, M.,** Growth of human diploid fibroblasts in the absence of glucose utilization, *Proc. Natl. Acad. Sci. U.S.A.,* 73, 4110, 1976.

108. **Neuman, R. E. and McCoy, T. A.,** Growth-promoting properties of pyruvate, oxalacetate and α-ketoglutarate for isolated Walker carcinosarcoma 256 cells, *Proc. Soc. Exp. Biol. Med.,* 98, 303, 1958.

109. **McKeehan, W. L. and McKeehan, K. A.,** Oxocarboxylic acids, pyridine nucleotide-linked oxidoreductases and serum factors in regulation of cell proliferation, *J. Cell. Physiol.,* 101, 9, 1979.

110. **Griffiths, J. B. and Pirt, S. J.,** The uptake of amino acids by mouse cells (strain LA) during growth in batch culture and chemostat culture: the influence of cell growth rate, *Proc. R. Soc. London, Ser. B,* 168, 421, 1967.

111. **Waymouth, Ch.,** Construction of tissue culture media, in *Growth, Nutrition and Metabolism of Cells in Culture,* Vol. 1, Rothblat, G. H. and Cristofalo, V. J., Eds., Academic Press, New York, 1972, 11.

112. **Morton, H. J.,** Known cellular growth requirements and the composition of currently available defined media, in *The Growth Requirements of Vertebrate Cells In Vitro,* Waymouth, Ch., Ham, R. G., and Chapple, P. J., Eds., Cambridge University Press, Cambridge, 1981, 16.

113. **Waymouth, Ch.,** Preparation and use of serum-free culture media, in *Cell Culture Methods for Molecular and Cell Biology,* Vol. 1, Barnes, D. W., Sirbasku, D. A., and Sato, G. H., Eds., Alan R. Liss, New York, 1984, 24.

114. **Ham, R. G.,** Survival and growth requirements of nontransformed cells. C.IV. Qualitative nutrient requirements, in *Handbook of Experimental Pharmacology,* Vol. 57, Baserga, R., Ed., Springer-Verlag, Berlin, 1981, 46.

115. **Yamane, I.,** Role of bovine albumin in a serum-free culture medium and its application, *Natl. Cancer Inst. Monogr.,* 48, 131, 1978.

116. **Ham, R. G.,** Survival and growth requirements of nontransformed cells. C.VI. Lipids and related substances, in *Handbook of Experimental Pharmacology,* Vol. 57, Baserga, R., Ed., Springer-Verlag, Berlin, 1981, 54.

117. **Ham, R. G.,** Survival and growth requirements of nontransformed cells. B.II. Inadequacy for nontransformed cells of classic assay systems, in *Handbook of Experimental Pharmacology,* Vol. 57, Baserga, R., Ed., Springer-Verlag, Berlin, 1981, 17.

118. **Rutzky, L. P.,** Peptone growth factors for serial cell proliferation in the absence of serum, in *The Growth Requirements of Vertebrate Cells In Vitro,* Waymouth, Ch., Ham, R. G., and Chapple, P. J., Eds., Cambridge University Press, Cambridge, 1981, 277.

119. **Taylor, W. G.,** Studies on the chemical nature of a growth-promoting agent, bacto-peptone dialysate, in *The Growth Requirements of Vertebrate Cells In Vitro,* Waymouth, Ch., Ham, R. G., and Chapple, P. J., Eds., Cambridge University Press, Cambridge, 1981, 258.

120. **Michl, J. and Řezáčová, D.,** Cultivation of mammalian cells in a medium with growth-promoting proteins from calf serum, *Acta Virol.,* 10, 254, 1966.

121. **Baudyšová, M., Spurná, V., Nebola, M., Vyklický, L. Jr., and Michl, J.,** Establishment of mouse neuroblastoma clone E 7 in serum-free medium, *Physiol. Bohemoslov.,* 33, 155, 1984.

122. **Michl, J.,** Metabolism of cells in tissue culture in vitro. I. The influence of serum protein fractions on the growth of normal and neoplastic cells, *Exp. Cell Res.,* 23, 324, 1961.

123. **Michl, J.,** Metabolism of cells in tissue culture in vitro. II. Long-term cultivation of cell strains and cells isolated directly from animals in a stationary culture, *Exp. Cell Res.,* 26, 129, 1962.

124. **Macek, M. and Michl, J.,** A contribution to the culturing of human diploid cells, *Acta Univ. Carol. Med.,* 10, 519, 1964.

125. **Michl, J.,** Proliferative capacity constant of metazoan cells in culture, *Cell. Biol. Int. Rep.,* 1, 427, 1977.

126. **Michl, J. and Spurná, V.,** Growth-promoting alpha-globulin. A pleiotypic activator, *Exp. Cell Res.,* 84, 56, 1974.

127. **Michl, J. and Spurná, V.,** Pinocytosis as an essential transport mechanism of metazoan cells in culture, *Exp. Cell Res.,* 93, 39, 1975.

128. **Michl, J. and Baudyšová, M.,** Pinocytosis of fluorescein-labelled growth-promoting alpha-globulin in human diploid cells, *Cell. Biol. Int. Rep.,* 5, 381, 1981.

129. **Baudyšová, M. and Michl, J.,** Serum-free medium is more beneficial than fetal calf serum-supplemented medium for morphological differentiation of primary chick embryo brain cells in culture, *Mol. Physiol.,* 2, 225, 1982.

130. **Tolar, M., Michl, J., Dlouhá, H., and Teisinger, J.,** Defined medium supplemented with growth-promoting alpha-globulin supports the formation, differentiation and innervation of myotubes in cell cultures from chicken embryo, *Mol. Physiol.,* 3, 151, 1983.

131. **Baudyšová, M. and Michl, J.,** Proliferation and differentiation of chondrocytes in cell culture, *Physiol. Bohemoslov.,* 32, 346, 1983.

132. **Sato, G. H.,** The role of serum in cell culture, in *Biochemical Actions of Hormones,* Vol. 3, Litwack, G., Ed., Academic Press, New York, 1975, 391.

133. **Hayashi, I. and Sato, G. H.,** Replacement of serum by hormones and permits growth of cells in a defined medium, *Nature,* 259, 132, 1976.

134. **Bottenstein, J., Hayashi, I., Hutchings, S., Masui, H., Mather, J., McClure, D. B., Ohasa, S., Rizzino, A., Sato, G. H., Serrero, G., Wolfe, R., and Wu, R.,** The growth of cells in serum-free hormone-supplemented media, *Methods Enzymol.,* 58, 94, 1979.

135. **Barnes, D. and Sato, G. H.,** Methods for growth of cultured cells in serum-free medium, *Anal. Biochem.,* 102, 255, 1980.

136. **Barnes, D. and Sato, G. H.,** Serum-free cell culture: a unifying approach, *Cell,* 22, 649, 1980.

137. **Ham, R. G.,** Survival and growth requirements of nontransformed cells. C.VII. Hormones, hormone-like growth factors, and carrier proteins, in *Handbook of Experimental Pharmacology,* Vol. 57, Baserga, R., Ed., Springer-Verlag, Berlin, 1981, 57.

138. **Hutchings, S. E. and Sato, G. H.,** Growth and maintenance of HeLa cells in serum-free medium supplemented with hormones, *Proc. Natl. Acad. Sci. U.S.A.,* 75, 901, 1978.

139. **Rizzino, A. and Sato, G. H.,** Growth of embryonal carcinoma cells in serum-free medium, *Proc. Natl. Acad. Sci. U.S.A.,* 75, 1844, 1978.

140. **Darmon, M., Bottenstein, J., and Sato, G. H.,** Neural differentiation following culture of embryonal carcinoma cells in a serum-free defined medium, *Dev. Biol.,* 85, 463, 1981.

141. **Hayashi, I., Larner, J., and Sato, G. H.,** Hormonal growth control of cells in culture, *In Vitro,* 14, 23, 1978.

142. **Mather, J. P. and Sato, G. H.,** The growth of mouse melanoma cells in hormone-supplemented, serum-free medium, *Exp. Cell Res.,* 120, 191, 1979.

143. **Barnes, D. and Sato, G. H.,** Growth of a human mammary tumour cell line in a serum-free medium, *Nature,* 5730, 388, 1979.

144. **Bottenstein, J. E.,** Serum-free culture of neuroblastoma cells, in *Advances in Neuroblastoma Research,* Evans, A. E., Ed., Raven Press, New York, 1980, 161.

145. **Taub, M., Chuman, L. Ü. B., Rindler, M. J., Saier, M. H., Jr., and Sato, G. H.,** Alterations requirements of kidney epithelial cells in defined medium with malignant transformation, *J. Supramol. Struct. Cell Biochem.,* 15, 63, 1981.

146. **Hawrot, E.,** Cultured sympathetic neurons: effects of cell-derived and synthetic substrata on survival and development, *Dev. Biol.,* 74, 136, 1980.

147. **McKeehan, W. L.,** Use of basic polymers as synthetic substrata for cell culture, in *Cell Culture Methods for Molecular and Cell Biology,* Vol. 1, Barnes, D. W., Sirbasku, D. A., and Sato, G. H., Eds., Alan R. Liss, New York, 1984, 209.

148. **Grinnell, F. and Feld, M. K.,** Initial adhesion of human fibroblasts in serum-free medium: possible role of secreted fibronectin, *Cell,* 17, 117, 1979.

149. **Kleinman, H. K., Cannon, F. B., Laurie, G. W., Hassell, J. R., Aumailley, M., Terranova, V. P., Martin, R. G., and DuBois-Dalcq, M.,** Biological activities of laminin, *J. Cell. Biochem.,* 27, 317, 1985.

150. **Yamada, K. M., Akiyama, S. K., Hasegawa, E., Humphries, M. J., Kennedy, D. W., Nagata, K., Urushihara, H., Olden, K., and Chen, W.-T.,** Recent advances in research on fibronectin and other cell attachment proteins, *J. Cell. Biochem.,* 28, 79, 1985.

151. **Malick, L. E., Tompa, A., Kuszynski, Ch., Pour, P., and Langenbach, R.,** Maintenance of adult hamster pancreas cells on fibroblastic cells, *In Vitro,* 17, 947, 1981.

152. **Harel, L.,** Diffusible factors in tissue cultures. C.I. The feeder effect, in *Handbook of Experimental Pharmacology,* Vol. 57, Baserga, R., Ed., Springer-Verlag, Berlin, 1981, 318.

153. **Bornstein, P. and Sage, H.,** Structurally distinct collagen types, *Annu. Rev. Biochem.,* 49, 957, 1980.

154. **Nimni, M. E.,** Collagen. Structure, function and metabolism in normal and fibrotic tissue, *Semin. Arth. Rheum.,* 13, 1, 1983.

155. **Hascall, V. C. and Hascall, G. T.,** Proteoglycans, in *Cell Biology of Extracellular Matrix,* Hay, E. D., Ed., Plenum Press, New York, 1981, 39.

156. **Hynes, R. O. and Yamada, K. M.,** Fibronectins: multifunctional modular glycoproteins, *J. Cell Biol.,* 95, 369, 1982.

157. **Timpl, R., Engel, J., and Martin, G. R.,** Laminin — a multifunctional protein of basement membranes, *Trends Biochem. Sci.,* 8, 207, 1983.

158. **Giguere, L., Cheng, J., and Gospodarowicz, D.,** Factors involved in the control of proliferation of bovine corneal endothelial cells maintained in serum-free medium, *J. Cell. Physiol.,* 110, 72, 1982.

159. **Parry, G., Lee, E. Y.-H., Farson, D., Koval, M., and Bissell, M. J.,** Collagenous substrata regulate the nature and distribution of glycosaminoglycans produced by differentiated cultures of mouse mammary epithelial cells, *Exp. Cell Res.,* 156, 487, 1985.

160. **Thivolet, H. C., Chatelain, P., Nicoloso, H., Durand, A., and Bertrand, J.,** Morphological and functional effects of extracellular matrix on pancreatic islet cell cultures, *Exp. Cell Res.,* 159, 313, 1985.

161. **Piper, H. M., Spahr, R., Probst, I., and Spieckerman, P. G.,** Substrates for the attachment of adult cardiac myocytes in culture, *Basic Res. Cardiol.,* 80 (Suppl. 2), 175, 1985.

162. **Bornstein, M. B.,** Reconstituted rat-tail collagen used as a substrate for tissue cultures on slips in Maximov slides and roller tubes, *Lab. Invest.,* 7, 134, 1958.

163. **Yamada, K. M. and Akiyama, S. K.,** Preparation of cellular fibronectin, in *Cell Culture Methods for Molecular and Cell Biology,* Vol. 1, Barnes, D. W., Sirbasku, D. A., and Sato, G. H., Eds., Alan R. Liss, New York, 1984, 215.

164. **Ledbetter, S. R., Kleinman, H. K., Hassell, J. R., and Martin, G. R.,** Isolation of laminin, in *Cell Culture Methods for Molecular and Cell Biology,* Vol. 1, Barnes, D. W., Sirbasku, D. A., and Sato, G. H., Eds., Alan R. Liss, New York, 1984, 231.

165. **Whitfield, J. F., Boynton, A. L., Rixon, R. H., and Youdale, T.,** The control of cell proliferation by calcium, Ca^{2+}-calmodulin, and cyclic AMP, in *Control of Animal Cell Proliferation,* Vol. 1, Boynton, A. L. and Leffert, H. L., Eds., Academic Press, Orlando, Fla., 1985, 331.

166. **Ashihara, T. and Baserga, R.,** Cell synchronization, *Methods Enzymol.,* 58, 248, 1979.

167. **Stein, G. H. and Yanishevsky, R.,** Autoradiography, *Methods Enzymol.,* 58, 279, 1979.

168. **Rogers, A. V.,** *Techniques of Autoradiography,* 3rd ed., Elsevier/North-Holland Biomedical, Amsterdam, 1979.

169. **Maurer-Schultze, B.,** Various autoradiographic methods as a tool in cell growth studies, in *Cell Growth,* Nicolini, C., Ed., Plenum Press, New York, 1982, 83.

170. **Gray, J. W. and Coffino, P.,** Cell cycle analysis by flow cytometry, *Methods Enzymol.,* 58, 233, 1979.

171. **Gray, J. W., Dolbeare, F., Pallavicini, M. G., Beisker, W., and Waldman, F.,** Cell cycle analysis using flow cytometry, *Int. J. Radiat. Biol.,* 49, 237, 1986.

172. **Darzinkiewicz, Z., Traganos, F., Sharpless, T., and Melamed, M. R.,** Different stainability of M vs G_2 and G_0 vs G_1 cells, in *Pulse Cytophotometry,* Lutz, D., Ed., European Press, Ghent, 1978, 269.

173. **Larsen, J. K., Munch-Petersen, B., Christiansen, J., and Jørgensen, K.,** Flow cytometric discrimination of mitotic cells: resolution of M, as well as G_1, S and G_2 phase nuclei with mithramycin, propidium iodide and ethidium bromide after fixation with formaldehyde, *Cytometry,* 7, 54, 1986.

174. **Potten, C. S.,** Cell replacement in epidermis (keratopoiesis) via descrete units of proliferation, *Int. Rev. Cytol.,* 69, 271, 1981.

175. **Sibatani, A.,** Concepts of "differentiation" in cell culture. How not to think of differentiation in relation to carcinogenesis, *Cancer Forum,* 6, 241, 1982.

176. **Lloyd, C. W. and Rees, D. A., Eds.,** *Cellular Controls in Differentiation,* Academic Press, New York, 1981.

177. **Miyazaki, K., Takaki, R., Nakayama, F., Yamauchi, S., Koga, A., and Todo, S.,** Isolation and primary culture of adult human hepatocytes, *Cell Tissue Res.,* 218, 13, 1981.

178. **Hunter, M. G., Magee-Brown, R., Dix, C. J., and Cooke, B. A.,** The functional activity of adult mouse Leydig cells in monolayer culture. Effect of lutropin and foetal calf serum, *Mol. Cell. Endocrinol.,* 25, 35, 1982.

179. **Fayet, G., Hovsépian, S., Dickson, J. H., and Lissitzky, S.,** Reorganization of porcine thyroid cells into functional follicles in a chemically defined serum- and thyrotropin-free medium, *J. Cell Biol.,* 93, 479, 1982.

180. **Nefussi, J.-R., Boy-Lefevre, M. L., Boulekbache, H., and Forest, N.,** Mineralization in vitro of matrix formed by osteoblasts isolated by collagenase digestion, *Differentiation,* 29, 160, 1985.

181. **Wu, R., Nolan, E., and Turner, C.,** Expression of tracheal differentiated functions in serum-free hormone-supplemented medium, *J. Cell. Physiol.,* 125, 167, 1985.

182. **Sensenbrenner, M., Booher, J., and Mandel, P.,** Histochemical study of dissociated nerve cells from embryonic chick cerebral hemispheres in flask cultures, *Experientia,* 29, 699, 1973.

183. **Van, R. L. R. and Roncari, D. A. K.,** Complete differentiation of adipocyte precursors. A culture system for studying the cellular nature of adipose tissue, *Cell Tissue Res.,* 195, 317, 1978.

184. **Panula, P., Rechardt, L., and Hervonnen, H.,** Observation on the morphology and histochemistry of the rat neostriatum in tissue culture, *Neuroscience,* 4, 235, 1979.

185. **Karasawa, K., Kimata, K., Ito, K., Kato, Y., and Suzuki, S.,** Morphological and biochemical differentiation of limb bud cells cultured in chemically defined medium, *Dev. Biol.,* 70, 287, 1979.

186. **Smith, G. H. and Vonderhaar, B. K.,** Functional differentiation in mouse mammary gland epithelium is attained through DNA synthesis, inconsequent of mitosis, *Dev. Biol.,* 88, 167, 1981.

187. **Roussel, G., Labourdette, G., and Nussbaum, J. L.,** Characterization of oligodendrocytes in primary cultures from brain hemispheres of newborn rats, *Dev. Biol.,* 81, 372, 1981.

188. **Hatfield, J. S., Skoff, R. P., Maisel, H., Eng, L., and Bigner, D. D.,** The lens epithelium contains glial fibrillary acidic protein (GFAP), *J. Neuroimmunol.,* 8, 347, 1985.

189. **Dardick, I., Poznanski, W. J., Waheed, I., and Setterfield, G.,** Ultrastructural observations on differentiating human preadipocytes cultured in vitro, *Tissue Cell,* 8, 561, 1976.

190. **Wilson, S. P. and Viveros, O. H.,** Primary culture of adrenal medullary chromaffin cells in a chemically defined medium, *Exp. Cell Res.,* 133, 159, 1981.

191. **Robine-Leon, S., Appay, M. D., Chevalier, G., and Zweibaum, A.,** Proliferation, differentiation and maturation of a mouse epidermal keratinocyte cell line, *Exp. Cell Res.,* 133, 273, 1981.

192. **Gebhardt, R., Jung, W., and Robenek, H.,** Primary cultures of rat hepatocytes as a model system of canalicular development, biliary secretion and intrahepatic cholestasis. I and II, *Eur. J. Cell Biol.,* 29, 68 and 77, 1982.

193. **Meyer, T., Burkart, W., and Jockusch, H.,** Choline acetyltransferase induction in cultured neurons: dissociated spinal cord cells are dependent on muscle cells, organotypic explants are not, *Neurosci. Lett.,* 11, 59, 1979.

194. **Simonneau, L., Herve, B., Jackquemin, E., and Courtois, Y.,** State of differentiation of bovine epithelial lens cells in vitro. Relationship between the variation of the cell shape and the synthesis of crystallins, *Cell. Differentiation,* 13, 185, 1983.

195. **Potter, D. D., Landis, S. C., and Furshpan, E. J.,** Dual function during development of rat sympathetic neurones in culture, *J. Exp. Biol.,* 89, 57, 1980.

196. **Nelson, P. G., Neale, E. A., and Macdonald, R. L.,** Electrophysiological and structural studies of neurons in dissociated cell cultures of the central nervous system, in *Excitable Cells in Tissue Culture,* Nelson, P. G. and Lieberman, M., Eds., Plenum Press, New York, 1981, 39.

197. **Romijn, H. J., Habets, A. M. M. C., Mud, M. T., and Wolters, P. S.,** Nerve outgrowth, synaptogenesis and biological activity in fetal rat cerebral cortex tissue in serum-free, chemically defined medium, *Dev. Brain Res.,* 2, 583, 1982.

198. **Bunge, M. B., Bunge, R. P., Peterson, E. R., and Murray, M. R.,** A light and electron microscope study of long-term organized cultures of rat dorsal root ganglia, *J. Cell Biol.,* 32, 439, 1967.

199. **Sensenbrenner, M., Jaros, G. G., Moonen, G., and Mandel, P.,** Effects of synthetic tripeptide on the differentiation of dissociated cerebral hemisphere nerve cells in culture, *Neurobiology,* 5, 207, 1975.

200. **Moonen, G.,** Variability of the effects of serum-free medium, dibutyryl-cyclic AMP or theophylline on morphology of cultured new-born rat astroblasts, *Cell Tissue Res.,* 163, 365, 1975.

201. **Sensenbrenner, M., Maderspach, K., Latzkovits, L., and Jaros, G. G.,** Neuronal cells from chick embryo cerebral hemispheres cultivated on polylysine-coated surfaces, *Dev. Neurosci.,* 1, 90, 1978.

202. **Petmann, B., Louis, J. C., and Sensenbrenner, M.,** Morphological and biochemical maturation of neurones cultured in the absence of glial cells, *Nature,* 251, 378, 1979.

203. **Ziller, C., Smith, J., Fauquet, M., and Le Douarin, N. M.,** Environmentally directed nerve cell differentiation: in vivo and in vitro studies, in *Development and Chemical Specificity of Neurons,* Vol. 51, Cuènod, M., Kreutzberg, G. W., and Bloom, F. E., Eds., Elsevier/North-Holland Biomedical, Amsterdam, 1979, 59.

204. **Alliot, F. and Pessac, B.,** A glial fibrillary acidic protein (GFA)-containing cell clone from mouse cerebella transformed "in vitro" by SV-40, *Brain Res.,* 216, 455, 1981.

205. **Hirn, M., Pierres, M., Deagostini-Bazin, H., Hirsch, M., and Goridis, C.,** Monoclonal antibody against cell surface glycoprotein of neurons, *Brain Res.,* 214, 433, 1981.

206. **Ghandour, M. S., Labourdette, G., Vincendon, G., and Gombos, G.,** A biochemical and immuno-histological study of S100 protein in developing rat cerebellum, *Dev. Neurosci.,* 4, 98, 1981.

207. **Raff, M. C., Brockes, J. P., Fields, K. L., and Mirsky, R.,** Neuronal cell markers: the end of the beginning, in *Development and Chemical Specificity of Neurons,* Vol. 51, Cuènod, M., Kreutzberg, G. W., and Bloom, F. E., Eds., Elsevier/North-Holland Biomedical, Amsterdam, 1979, 18.

208. **Schachner, M. and Willinger, M.,** Cell type-specific cell surface antigens in the cerebellum, in *Development and Chemical Specificity of Neurons,* Vol. 51, Cuènod, M., Kreutzberg, G. W., and Bloom, F. E., Eds., Elsevier/North-Holland Biomedical, Amsterdam, 1979, 23.

209. **Patterson, P. H., Potter, D. D., and Furshpan, E. J.,** The chemical differentiation of nerve cells, *Sci. Am.,* 239, 50, 1978.

210. **Privat, A., Marson, A. M., and Drian, M. J.,** In vitro models of neural growth and differentiation, in *Development and Chemical Specificity of Neurons,* Vol. 51, Cuènod, M., Kreutzberg, G. W., and Bloom, F. E., Eds., Elsevier/North-Holland Biomedical, Amsterdam, 1979, 335.

211. **Patterson, P. H.,** Environmental determination of neurotransmitter functions in developing sympathetic neurons, in *Development and Chemical Specificity of Neurons,* Vol. 51, Cuènod, M., Kreutzberg, G. W., and Bloom, F. E., Eds., Elsevier/North-Holland Biomedical, Amsterdam, 1979, 77.

212. **Pastan, I.,** Cell transformation, *Methods Enzymol.,* 58, 368, 1979.

213. **Perbal, B.,** Transformation parameters expressed by tumor-virus transformed cells, in *Advances in Viral Oncology,* Vol. 4, Klein, G., Ed., Raven Press, New York, 1984, 163.

214. **Fusenig, N. E., Breitkreutz, D., Dzarlieva, R. T., Boukamp, P., Herzmann, E., Bohnert, A., Pöhlmann, J., Rausch, Ch., Schütz, S., and Hornung, J.,** Epidermal cell differentiation and malignant transformation in culture, in *In Vitro Epithelial Cell Differentiation and Neoplasia,* Vol. 6, Smith, G. J. and Stewart, B. W., Eds., Cancer Forum, Australian Cancer Society, Sydney, 1982, 209.

215. **Williams, G. M.,** Neoplastic transformation in liver epithelial cell cultures, in *In Vitro Epithelial Cell Differentiation and Neoplasia,* Vol. 6, Smith, G. J. and Stewart, B. W., Eds., Cancer Forum, Australian Cancer Society, Sydney, 1982, 120.

216. **Nettesheim, P.,** Detection and analysis of cells involved in neoplastic development, in *In Vitro Epithelial Cell Differentiation and Neoplasia,* Vol. 6, Smith, G. J. and Stewart, B. W., Eds., Cancer Forum, Australian Cancer Society, Sydney, 1982, 150.

217. **Bouck, N. and Di Mayorca, G.,** Evaluation of chemical carcinogenicity by in vitro neoplastic transformation, *Methods Enzymol.,* 58, 296, 1979.

218. **Smith, H. S. and Dollbaum, Ch. M.,** Growth of human tumors in culture. A.II. Current state of technology for culturing tumor cells, in *Handbook of Experimental Pharmacology,* Vol. 57, Baserga, R., Ed., Springer-Verlag, Berlin, 1981, 454.

219. **Kennett, R. H.,** Cell fusion, *Methods Enzymol.,* 58, 345, 1979.

220. **Zimmermann, U., Vienken, J., Halfmann, J., and Emeis, C. C.,** Electrofusion: a novel hybridization technique, in *Advances in Biotechnological Processes,* Vol. 4, Mizrahi, A., Ed., Alan R. Liss, New York, 1985, 79.

221. **Teissie, J. and Rols, M. P.,** Fusion of mammalian cells in culture is obtained by creating the contact between cells after their electropermeabilization, *Biochem. Biophys. Res. Commun.,* 140, 258, 1986.

222. **Široký, J., Nebola, M., Přibyla, L., and Karpfel, Z.,** Hemokaryons for animal and plant cells generated by electrofusion, *Gen. Physiol. Biophys.,* in press.

223. **Croce, C., Linnenbach, A., Hall, W., Steplewski, Z., and Koprowsi, H.,** Production of human hybridomas secreting antibodies to measles virus, *Nature,* 288, 488, 1980.

224. **Olsson, L. and Kaplan, H. S.,** Human-human hybridomas producing monoclonal antibodies of predefined antigenic specificity, *Proc. Natl. Acad. Sci. U.S.A.,* 77, 5429, 1980.

225. **Bischoff, R., Eisert, R. M., Schedel, I., Vienken, J., and Zimmermann, U.,** Human hybridoma cells produced by electrofusion, *FEBS Lett.,* 147, 64, 1982.

226. **Secher, D. and Burke, D. C.**, A monoclonal antibody for large-scale purification of human leukocyte interferon, *Nature*, 285, 446, 1980.

227. **Murakami, H.**, Serum-free cultivation of plasmacytomas and hybridomas, in *Cell Culture Methods for Molecular and Cell Biology*, Vol. 4, Barnes, D. W., Sirbasku, D. A., and Sato, G. H., Eds., Alan R. Liss, New York, 1984, 197.

228. **Steimer, K. S.**, Serum-free growth of SP2/O-AG-14 hybridomas, in *Cell Culture Methods for Molecular and Cell Biology*, Vol. 4, Barnes, D. W., Sirbasku, D. A., and Sato, G. H., Eds., Alan R. Liss, New York, 1984, 237.

229. **Murakami, H., Shimomura, T., Ohashi, H., Hashizume, S., Tokashiki, M., Shinohara, K., Yasumoto, K., Nomoto, K., and Omura, H.**, Serum-free stired culture of human-human hybridoma lines, in *Growth and Differentiation of Cells in Defined Environment*, Murakami, H., Yamane, I., Barnes, D. W., Mather, J. P., Hayashi, I., and Sato, G. H., Eds., Kodansha, Tokyo, 1985, 111.

230. **Hagiwara, H., Ohtake, H., Yuasa, H., Nagao, J., Nonaka, S., Chigiri, E., and Aotsuka, Y.**, Proliferation and antibody production of human-human hybridoma in serum-free media, in *Growth and Differentiation of Cells in Defined Environment*, Murakami, H., Yamane, I., Barnes, D. W., Mather, J. P., Hayashi, I., and Sato, G. H., Eds., Kodansha, Tokyo, 1985, 117.

231. **Sato, S., Kawamura, K., Hanai, N., nad Fujiyoshi, N.**, Production of interferon and monoclonal antibody using a novel type of perfusion vessel, in *Growth and Differentiation of Cells in Defined Environment*, Murakami, Y., Yamane, I., Barnes, D. W., Mather, J. P., Hayashi, I., and Sato, G. H., Eds., Kodansha, Tokyo, 1985, 123.

232. **Yonezawa, Y.**, Recycle continuous system for mass propagation of animal cells, in *Growth and Differentiation of Cells in Defined Environment*, Murakami, H., Yamane, I., Barnes, D. W., Mather, J. P., Hayashi, I., and Sato, G. H., Eds., Kodansha, Tokyo, 1985, 131.

233. **Worton, R. G. and Duff, C.**, Karyotyping, *Methods Enzymol.*, 58, 322, 1979.

234. **Therman, E.**, *Human Chromosomes. Structure, Behavior, Effects*, 2nd ed., Springer-Verlag, New York, 1985.

235. **Farber, J. L.**, Biology of disease. Membrane injury and calcium homeostasis in the pathogenesis of coagulative necrosis, *Lab. Invest.*, 47, 114, 1982.

236. **Ross, R. and Glomset, J. A.**, The pathogenesis of atherosclerosis. I and II, *N. Engl. J. Med.*, 295, 369 and 420, 1976.

237. **Benditt, E. P. and Benditt, J. M.**, Evidence for a monoclonal origin of human atherosclerotic plaques, *Proc. Natl. Acad. Sci. U.S.A.*, 70, 1753, 1973.

238. **Bierman, E. L. and Albers, J. J.**, Lipoprotein uptake by cultured human arterial smooth muscle cells, *Biochim. Biophys. Acta*, 388, 198, 1975.

239. **Owens, G. K., Rabinovitch, P. S., and Schwartz, S. M.**, Smooth muscle cell hypertrophy versus hyperplasia in hypertension, *Proc. Natl. Acad. Sci. U.S.A.*, 78, 7759, 1981.

240. **Grünwald, J., Robenek, H., Mey, J., and Hauss, W. H.**, In vivo and in vitro cellular changes in experimental hypertension: electronmicroscopic and morphometric studies of aortic smooth muscle cells, *Exp. Mol. Pathol.*, 36, 164, 1982.

241. **Ichihara, A., Nakamura, T., and Tanaka, K.**, Use of hepatocytes in primary culture for biochemical studies on liver functions, *Mol. Cell. Biochem.*, 43, 145, 1982.

242. **Hasegawa, K., Watanabe, K., and Koga, M.**, Induction of mitosis in primary cultures of adult rat hepatocytes under serum-free conditions, *Biochem. Biophys. Res. Commun.*, 104, 259, 1982.

243. **Guillouzo, A., Guguen-Guillouzo, C., and Bourel, M.**, Leberzellen in Kultur: ihre differenzierten Funktionen und deren Verwendung für Stoffwechselstudien, *Triangle*, 20, 121, 1981.

244. **Leibovitz, A.**, The establishment of cell lines from human solid tumors, in *Advances in Cell Culture*, Vol. 4, Maramorosch, K., Ed., Academic Press, Orlando, Fla., 1985, 249.

245. **McGarrity, G.**, Detection of contamination, *Methods Enzymol.*, 58, 18, 1979.

246. **Wolf, K.**, Laboratory management of cell cultures, *Methods Enzymol.*, 58, 116, 1979.

247. **Chen, T. R.**, In situ detection of mycoplasma contamination in cell cultures by fluorescent Hoechst 33258 stain, *Exp. Cell Res.*, 104, 255, 1977.

248. **Barkley, W. E.**, Safety considerations in the cell culture laboratory, *Methods Enzymol.*, 58, 36, 1979.

249. **McGarrity, G. J., Vanaman, V., and Sarama, J.**, Cytogenetic effects of mycoplasmal infection of cell cultures: a review, *In Vitro*, 20, 1, 1984.

250. **Perlman, D.**, Use of antibiotics in cell culture media, *Methods Enzymol.*, 58, 110, 1979.

251. **Gartler, S. M.**, Apparent HeLa cell contamination of human heteroploid cell lines, *Nature*, 217, 750, 1968.

252. **Lavappa, K. S., Macy, M. L., and Shannon, J. E.**, Examination of ATCC stocks for HeLa marker chromosomes in human cell lines, *Nature*, 259, 211, 1976.

253. **Lavappa, K. S.**, Survey of ATCC stocks of human cell lines for HeLa contamination, *In Vitro*, 14, 469, 1978.

254. **Peterson, W. D., Jr., Simpson, W. F., and Hukku, B.**, Cell culture characterization: Monitoring for cell identification, *Methods Enzymol.*, 58, 164, 1979.

Chapter 6

ENVIRONMENTAL IMPACT IN ADULT AND DEVELOPING ANIMALS

Věra Nováková

TABLE OF CONTENTS

I. INTRODUCTION

The environment of living organisms is represented by a set of external ecological factors which the organism perceives and by which it is affected. If necessary, i.e., if the organism is motivated, it can respond to these factors. This necessity (motivation) is determined by the intrinsic state of the organism. The relationships between the environment and the organism are determined by the genetic program of the given species and by individually acquired experience. The proportion between these two components varies. When the animal is short lived, when it does not encounter its parents from which it could take experience, and when it lives in a small territory and in an only slightly changing environment, then it reacts mostly instinctively. On the other hand, when the animal has a long life span, when it develops slowly even after birth (hatching), it is subject to parental care and gradually acquires its own individual experience, and consequently, its reactions are a combination of the inborn and the learned. The organism thus becomes adapted to the environment. In biological literature the term adaptation is used to denote a change in the structure and function of the organism and its behavior. It is a result of an evolutionary process of the given species which is genetically encoded. It is species specific, irreversible, and encompasses the whole species. Every species has a typical way of obtaining food, its consumption, and excretion; equally species specific are such features as locomotion, protection against predators, relations with the members of the same species and other species, sexual and parental behavior, exploration, marking the territory, biological rhythms, etc. In another meaning the term adaptation denotes changes that are short term, arise at any age, are not species specific but individual specific, are reversible, and are not directly genetically encoded. They reflect the relationship of the individual to the momentary state of the environment. This process is a physiological adaptation known also as adjustment. A typical example is acclimation during which a change in one or more environmental stimulus elicits in the organism a change in homeostatic systems. During evolution, higher animals developed a relative independence of the organism from environment, which is ensured homeostatically. As a consequence, the organism acquired a higher variability of functions which allows it to maintain its stability in a wide range of environmental influences.[1]

Environmental stimuli of diverse modalities thus create, together with the organism, a feedback unit. The intensity of the stimuli in relation to the subject may be subthreshold, threshold, or suprathreshold and may even lead to overloading. The stimuli may be physical, chemical, and also those that arise through the interaction of individuals within the species or through the interaction of individuals belonging to different species. An agent which acts on a sensitive organism elicits both specific and unspecific responses. The quality and quantity of the response depend on the modality and intensity of the stimulus, on the actual state of the organism, and its previous experience with the stimulus. Discussions revolve around whether physical or social stimuli are more important.[2,3] The dispute cannot be resolved on a general level; it can only be stated that every stimulus, nociceptive or pleasant, external or internal, social or physical, participates equally in the unspecific activation of the brain. In the case of a specific activation each particular case has to be judged separately, taking into account the importance of the kind of species under study and its life mode. In social species the most important are undoubtedly social stimuli. In contact species such as rat or dog these stimuli must have the character of a physical contact, whereas in noncontact species their effect may be distant. In solitary species, (e.g., hamster, cat) these stimuli have a positive biological effect only in the periods of mating and rearing of the progeny. Indispensable among physical stimuli are those for which the given species has a sensory apparatus.

The origin and development of relationships between the environment and the organism as well as the disturbances occurring in these relations have become the subject of research in order to elucidate their intrinsic mechanisms and organization. From the experimental point of view the following questions are of prime importance.

1. What are the possibilities of experimental study and what animals are suitable as experimental objects?
2. What is the methodology of choice and what criteria should be used?
3. What are the mechanisms of development of a phenotype and what is the significance of experience acquired in the youth?

II. POSSIBILITIES OF EXPERIMENTAL STUDIES AND CHOICE OF EXPERIMENTAL ANIMALS

The first data in this particular field were obtained in the study of natural free-living species. They brought information about a species-specific behavior and description of elements of individual behavioral patterns and their sequence. The dependence of behavioral changes on environmental alterations was determined as well. Diurnal rhythms, the yearly cycle, overpopulation, intra- and interspecies relationships, etc. may serve as typical examples.[4] These results form a sound basis for formulation of further experimental problems and questions, for method selection, and particularly for breeding, housing, and feeding of laboratory animals. The observation of life in nature, however, provides only partial information that is usually restricted to behavioral study without the possibility to investigate underlying physiological, structural, and metabolic processes and the origin of pathological deviations; to analyze individual phenomena; and to explore functional regulations. Therefore, there is a need for designing experiments which could provide such possibilities.

As a rule, these experiments cannot be performed on animals captured in nature which are under stress, as the results would be distorted. The first, rather isolated, studies were, therefore, performed on domestic animals — dogs, cats, some birds, sheep, etc. These animals were found to be unsuitable for systematic study because their breeding was non-standard and very expensive. Therefore, the work with laboratory animals was begun because they can be procured in numbers sufficient for statistical evaluation, are easy to mark, and thus permit the conduction of prolonged longitudinal studies. Laboratory animals include mice, hamsters, guinea pigs, rabbits, gerbils, and laboratorized dogs, cats, and monkeys. However, most studies have been carried out on rats. Their breeding has the longest tradition and the widest empirical basis; their biology is also the best known. They reproduce well throughout the year, are adaptable and nonaggressive, and it is possible to select different strains that carry pathological features.

In behavioral studies it is necessary to remember that the resistance of experimental animals against stressors arising during the experiments depends on their handling in youth. A typical handling procedure consists of transferring young animals during the suckling and weaning periods from the litter into a cage containing no mother and no siblings. After 3 min the animals are returned to the litter. The transfer is carried out daily.[5-7] Laboratory experience has shown that every manipulation with the young animal is actually a handling in its impact. Manipulation with older animals is termed gentling.[8] The procedure, in both its short- and long-term forms, is very effective. The animal calms down and the subsequent experiment has no stressing effect. The use of both handling and gentling is common in behavioral studies and leads to the homogenization of the population. The manipulated animals are then suitable for chronic experiments.

A question that is often discussed is whether the laboratory rat has retained the species-specific behavior and properties of its wild ancestor and whether its relationship to the environment has not changed by the laboratory housing and breeding. Animals captured in nature were, therefore, experimentally studied in comparison with their laboratory counterparts. The factors under study included sensitivity to stress, growth, reproduction, learning ability, memory, nutritional requirements, etc. The results of these experiments were controversial and led to the formulation of different hypotheses about the effect of laboratory breeding on rats. The laboratory form is sometimes viewed as degenerate, and the results obtained with these animals then have a doubtful value when generalized. On the other hand, optimistic opinions were published concerning the genotype stability of this species.[9] The solution of this problem is essential for the study of organism/environment relationships and of the organism's adaptability, but it is difficult to get a simple answer.

Let us consider as an example the results of experiments designed to determine whether the laboratory rat has retained its ability to form communities such as those in which the wild form lives. Randomly selected adult males and females grown under conventional breeding conditions did not create a harmonious community. In a large dispensed housing space one female was always differentiated, which concentrated all young from all litters in her nest and prevented all other females from nursing their young. Most of the young perished. These experiments seem to indicate that laboratory breeding has erased the ability of the animals to form a community (Figure 1).

When mothers with older offspring (15- or 30-day-old) were transferred into a group of adult males and females, the young rats acquired an experience with community life. When adult, they were then capable of forming a species-specific community in which each mother reared her young. In addition, other members of the community also nursed the young. The community thus gave rise to a new healthy generation (Figure 2). Laboratory rats thus retained their ability to live in a community, but to be able to manifest this ability, they had to acquire an early experience with this life style.[10]

A different behavior was observed in Japanese spontaneously hypertensive rats. The females of this strain exhibited a disturbance of maternal behavior: they did not build a nest and did not retrieve the young. The defect persisted even when the progeny of hypertensive parents were reared by a normotensive female. They were incapable of creating the early experience with community life and acquiring normal maternal behavior. The directed selection for the pathological syndrome (hypertension) thus apparently involved genotype changes to such an extent that the genetic basis of social behavior was also changed.

Another complicating factor is the interstrain differences in the function and structure of some organs and in behavior,[11] not to mention sex-dependent differences that are demonstrated even outside the reproduction sphere.[12]

Laboratory rats thus do not form a homogenous population. For each type of experiment a suitable strain should be selected. Optimum results would be, of course, obtained by a comparative study of various strains and both sexes. The data would then be more valid. The same situation is found in laboratory mice; individual strains, usually maintained under inbreeding conditions, yield different results. Comparative study on mice has become quite common in experiments in all branches of biology.[13]

III. METHODOLOGY

Laboratory research is usually done on adult animals supplied by a breeding station without any further detailed information; in a better case, the investigation is also supplied with data about the strain, age, and nutritional regime. The breeding conditions differ in individual stations (housing, weaning time, nutrition, manipulation with the animals, light regime, health condition). This may be one of the reasons why different laboratories obtain different results even with the same methodological approach. As no unified breeding scheme exists for different countries and because of the above differences among individual breeding stations even within one country, research laboratories often establish their own breeding colonies that take into consideration the requirements of the investigators and enable obtaining of much more detailed information about the animals. A restriction on the possibilities of

FIGURE 1.   Females with no experience of community life formed an abnormal community. All young were in a single nest, and only one female took care of them, actively preventing other females from approaching the young.

FIGURE 2.   Females that acquired the experience with community life during the critical period had separate nests, and each of them took care of her own offsprings.

study of environmental influences on life processes is the fact that the adult animal becomes acclimated to new experimental conditions according to its previous experience with individual environmental stimuli.

A convenient approach was, therefore, found in studying the postnatal development of the phenotype. Experiments designed to modify the phenotype and to find biologically effective environmental stimuli affecting the given species and sex involve modeling of the life environment for the given animals according to the investigator's aim. This can be accomplished essentially in two ways: a stimulus or a number of stimuli are added to, or eliminated from, normal environment. The environment is then called impoverished or enriched.[4] As most experiments are performed in the laboratory the normal environment is considered to correspond to normal breeding conditions. This is a very vague definition because of the above differences in breeding conditions. However, these "normal" conditions have one thing in common: they do not respect the natural way of life of the wild forms in nature. Even so, studies under artificially standardized conditions have brought many important findings whose knowledge and application should aid in further research and in optimization of breeding conditions.

When selecting the stimuli which are to be manipulated in the experiment one should pose the question: "What should the subject be deprived of or enriched by?" Every species has its own requirements of the stimulus field. Since manipulation with these specific stimuli has a higher chance of success, a comparison of the wild and the laboratory forms is very useful.

IV. ENRICHED ENVIRONMENT

In laboratory experiments this condition is fulfilled, for instance, as follows:

1. Large groups of animals are left to live in a "free" environment, i.e., in a large indispensed space with tunnels, ramps, basins, and with various objects that serve as toys for the young;[14,15] the effect of social stimulation and the possibility of manipulation with inanimate objects, and thus of increased motoric activity, are studied
2. Life in a normal breeding cage connected with a running wheel, in which the animals can wander freely, gives information about the effect of increased somatosensory stimulation and motoric activity[15]
3. Life in large, mostly unisexual groups tests the significance of the social effect of individuals of the same sex[16]
4. Life in communities comprising individuals of either sex and of different age, analogous to that found in the wild forms; the community lives in a large dispensed space simulating the burrow system[10]
5. Sensory stimulation — visual, auditory, olfactory[2]
6. Handling[5-7]
7. Gentling[8]
8. Foot shocks by electric current[6]

Stimulation of a developing organism has a long-term or permanent effect[17,18] reflected in

1. Growth rate alterations
2. Lowering of emotional reactivity
3. Facilitation of learning and increased stability of memory traces
4. Optimization of brain maturation
5. Lowering of reactivity to stressors
6. Increased resistance against infectious diseases and other illnesses

Experiment data indicate that enrichment of environment with stimuli of different modalities represents an improvement of life conditions. It is here where the idea has arisen that laboratory conditions are overstimulative for the young, as most rodents live in burrows isolated from external influences.[19] Indeed, young wild rats live in a burrow composed of chambers and underground corridors which give admittance to relatively fewer physical stimuli as compared with the breeding colony or the special room with increased sensory stimulation. However, the crucially important feature in the development of young individuals in social species is the maternal care and, from the 15th day of age, also the care of other members of the community because the young rats begin to move outside the nest from this age.[20] The young laboratory rat is in this sense deprived since it lives only with its mother and siblings until weaning and is then randomly placed in a small monosexual group in which it spends its whole life. When the learning ability was compared in laboratory rats reared in a community and in those from a conventional breeding colony, the former group was found to be more advanced.[21] Thus the social stimulation affected positively brain development, and its attenuation could not be compensated even by a higher sensory input.[22]

V. DEPRIVED ENVIRONMENT

It is created by placing the animals in a stimulus-poor environment or eliminating some stimulus from the normal environment.[17,18,24]

Sensory deprivation causes in dogs, cats, and rats

1. Structural, metabolic, and bioelectric deviations in specific projection regions of the brain
2. Disturbances of perception of optical, olfactory, auditory, and nociceptive stimuli[23]
3. Disturbances of locomotion and partial motoric activity (in monkeys, cats, and dogs)
4. Disturbances in learning ability and memory

Visual deprivation caused changes similar to those evoked by surgical deafferentation. However, several hours of light stimulation after eye opening sufficed to reduce the effect of the subsequent visual deprivation. The effect of deprivation was also weakened in other types of stimuli when the young animal was allowed to acquire an early experience with these stimuli.

For young mammals and birds, isolation from mother has fatal consequences which are spontaneously either irreversible or repaired only with great difficulties. In nature such isolants would not survive. The following disturbances were found in monkeys subjected to this treatment:[25-27]

1. Defective reproduction
2. Disturbances of maternal behavior: ignoring their infants, chewing on infant's feet and fingers, crushing the infant's face on the floor
3. Abnormal behavior: exaggerated oral activities, self-clutching, apathy, and indifference to external stimulation

Contact of the isolant with relatively younger animals had a certain therapeutic effect in that the isolant gradually acquired their behavior with concomitant loss of the abnormal properties.

VI. MOTHER-YOUNG INTERACTIONS

Most experiments were performed on rodents. The term social influence is very broad; several typical features can be observed in rats and mice.

The mother and her young constitute a dyad in which both parts affect each other. Thus, maternal behavior cannot be conceived as altruistic. Its basis is inborn, its quality is formed postnatally by the early experience in the period when the young was the object of maternal care. In multiparous mothers this is supplemented by previous experience. The expressions of maternal behavior are species specific, and the variability of this behavioral pattern is given by whether the species is nidicolous (altritial) or nidifugous (precocial). *Rattus norvegicus* belongs to a nidicolous species in which the young after their birth depend for a long time on the mother's care as well as the care of the community and its older siblings (helpers at the nest). The mother-young relationship is crucial until weaning.

The mutual bond is manifold (for survey, see References 28 and 29):

1. Acoustic bond: vocalization is mostly in the ultrasound range. It was recorded in a stress situation and thus represents a distress call — as a reaction to cold, unknown stimuli tactile or olfactory, and probably also to hunger. It is audible only at a short distance (in rats up to 15 cm). This is an important adaptation phenomenon because a longer range audibility would mean a risk, both for the mother and for the whole litter, of being discovered by a predator perceiving these ultrasound signals. The frequency decreases with the age of the young, while the duration of the pulses increases with age. Although individual signals differ in their meaning they always contain a frequency of about 40 kHz. The vocalization of young rats persists until the 20th day of age. Young rats that were observed to fall prey to predators in nature were those that failed to vocalize, i.e., older or defective animals.[30,31]
2. Chemical bond: young rats are born with immature micturition and defecation reflexes. The female licks their anogenital area and evokes thus micturition and defecation. Urine is a source of sodium chloride for the lactating mother, which has a high demand for it in this period. This type of care ceases on the 19th day of the young rat's age, i.e., later than the spontaneous micturition and defecation. The mother's care about young males is more intensive and more frequent. A pheromone bond exists between the lactating female and the suckling infants. The pheromone is formed in the mother's cecum and its precursor is a deoxycholic acid. It is found in special feces produced by the mother and consumed by the young between day 14 and 27 after delivery (in the period of weaning). This represents a contact between the mother and the young in the period when the offspring move outside the nest and are no longer retrieved by the mother. In nature this phenomenon is important because it enables the young to differentiate lactating females from nonlactating ones and to orientate themselves by olfaction towards the nest (homing). The production of the pheromone is regulated by prolactin in the liver, in which the level of prolactin rises sharply before the 14th postnatal day.[32]
3. Thermotactile bond: it is vital for sucklings that lack a mature thermoregulatory system. The female builds a warm nest and regulates its temperature by intermittent stays in the nest.[33]
4. Nutritional bond: how much milk a young rat consumes and when a spontaneous weaning takes place in most young rats in nature and in laboratory breeding stations is not known. An indirect determination of these factors by observation is never precise, but they can be determined directly with the use of isotope methods. The lactating female is given ^{85}Sr which passes into the milk and is then taken in by the young. Whole body radioactivity found in the puppies is proportional to the amount of the consumed milk. The age of the last milk intake can be deduced from these data as well. Maximal milk intake in the rat was found to be around day 15, weaning taking place on the 28th day.[34] Thus weaning carried out on day 21 for commercial reasons is premature. Moreover, weaning is not an instantaneous event; it occurs gradually. On day 17 the young begin to eat solid food, and 2 days later they also drink water.

The intake of food gradually increases concomitantly with decreasing milk intake. In the last days the mother's milk has no nutritional importance, but suckling ensures the contact of the young with their mothers.[35] The time of weaning is relatively fixed, and its duration was found to be unaffected by an intermittent food deprivation and by the number of young in the litter. It is affected only by those factors that interfere with brain maturation.[36]

5.   Mechanical bond: this includes activities associated with the transport of the young, e.g., retrieving into the nest, transfer after disturbance or during threat by a predator, after a temperature change, etc. This is a ritualized activity. The mother seizes the young by the skin between shoulder and flank and evokes in it a state of rigidity, with extended forelegs and flexed hindlegs. This kind of transport was calculated to be accomplished at the lowest energy cost for the mother.[28]

6.   Learning bond: this is effective mostly in the period of weaning and in the juvenile age. In the presence of the mother and other members of the community the young animal acquires food behavior (food collection, type of food),[37] maternal care,[38] and other behavioral patterns, probably by playing and emulating. Although the maternal behavior is genetically based and is humorally and neurally controlled, the complete species-specific picture is acquired by the young animal at the time when it is the subject of maternal care (see Figure 3 and 4).

A disturbance in any of these bonds represents for the young rat an intervention that it will not survive or that will cause a defective subsequent development of the animal. The mother-offspring attachment is complex and none of the above bonds is replaceable. Premature weaning, even if survived by the young animals, deforms further development (cf. examples in Table 1; for survey see References 39 through 41).

On comparing the effects of an enriched and deprived environment, individual environmental factors can be defined as (1) irreplaceable, (2) important but replaceable, and (3) irrelevant. This knowledge can aid in the construction of a normal environment, i.e., an environment containing a set of irreplaceable and important stimuli. Apart from the theoretical knowledge, such experiments aim at creating a proposal for optimization of the life environment of species domesticated, laboratorized, and species living in captivity.

VII. SIGNIFICANCE OF EARLY EXPERIENCE FOR PHENOTYPE DEVELOPMENT

Some stimuli are particularly or exclusively effective at a certain age of the individual, i.e., at the time of the highest sensitivity — the critical period. Optimum development requires a synchronization between the age and the specific composition of the life environment in order to ensure a correct — species specific — expression of the genotype. A prototype of an early experience is the imprinting described for the first time by Thomas More in 1516 and 1518 and later systematically studied.[42-45] It is the formation of a specific social attachment of the young to their mothers with great significance for later social behavior. In precocial birds it takes place several hours after hatching. In some ungulates, e.g., in sheep, the formation of this attachment takes place within several days after birth; in dogs it takes place in the 3rd and 4th week of life during the socialization phase;[46] in chimpanzees, in the 5th and 6th week; in children, in the 5th and 6th month after birth (for review, see Reference 17).

Each function can have its critical period at a different age. A genuine critical period for visual experience exists shortly after birth in those species that are born with open eyes or after eye-opening in species that are born with closed eyes. A critical period has been described for painful stimuli to which no conditioned phobia is formed within the critical period.[4] In rats, handling is effective up to the 10th day after birth.[5] In this species perinatal

FIGURE 3.   Normally weaned female built a soft, warm nest for her
young (upper panel) and covered the young when leaving the nest (lower
panel).

influence is exerted by the organizational effect of hormones, especially gonadal ones, that
form the sex-dependent differences in a number of organs and tissues including brain and
the pituitary gland.[47]

The research of the organism-environment relationship should determine which developmental processes are stimulus independent and which are stimulus dependent. The aim
is to find a critical period for the formation of an early experience in the latter and to define
stimuli that may prevent its appearance. When the organism has no possibility to create the
whole repertoire of these early experiences it becomes defective and poorly adaptable to
changing life conditions. This is important not only for preventive medicine, but also for
the reintroduction into free nature of those species no longer living under natural conditions.

VIII. CONCLUSIONS

Despite the homeostatically ensured relative stability of their internal environment, higher
animals react sensitively to environmental changes and form, together with the environment,
a dynamic equilibrium. The quality of the response is given by the type of stimulus, its
intensity, and the intrinsic state of the organism. The expressions are species specific and

FIGURE 4. Prematurely weaned female did not build a soft, warm nest for her young and did not cover the young on leaving the nest. This type of nest building is typical for females that did not acquire an early experience in the weaning period. Daughters and granddaughters of prematurely weaned females built equally defective nests for their young, even if normally weaned themselves. The defect was transferred to further generations by tradition.

individually variable, sex- and age-dependent, and molded by early and previous experience. Important are stimuli of a physical and chemical nature; in species living in communities social stimuli are especially decisive.

Life environment of a certain species-typical composition has an organizational significance for the developing organism. It realizes the expression of the genotype and shapes the individually characteristic properties — structural, physiological, and behavioral; it affects reactivity to stressors and tendency to disease at later age. Early experience arises in the period of a maximum sensitivity of the organism (critical periods), is highly stable, and can be transferred to further generations. An irreplaceable factor in the youth of all mammals is the mother that acts as a food donor and a provider of substances that the young organism is incapable of producing. The mother also functions as a substitute for immature functions, a protector against predators, and a determinant of the future behavior of the young. The key to future life is an early age in which genetic information is expressed under the influence of the species-specific environment.

Table 1
LATE EFFECTS OF PREMATURE WEANING IN LABORATORY RATS[38-40]

Transient changes	Lasting changes
Aldosterone — production of aldosterone was diminished (at 23 to 30 days)	Life-span — shorter life span in females
Corticosterone — the level of corticosterone was lower (at 40 and 95 days)	Fertility — fertility in females was lower
Thyroid gland — the activity of the thyroid gland was increased (at 19 to 31 days); the weight of the thyroid gland was lower (at 90 days)	Androgens — the production of androgens was impaired
	Gastric ulcer and erosions — the resistance to gastric ulcer and erosions produced by stress was lower
Thyroxin — plasma thyroxin concentration was increased (at 22 to 31 days)	Learning — the learning ability was lower
	Memory trace — the long-lasting memory trace remained unstable
Enzymatic activity in liver — the metabolism of enzymatic activity in liver was altered (at 22 to 31 days)	Natality cycle — disturbances in yearly natality cycle in females
	Maternal behavior — altered features of the maternal behavior
	EEG — electroencephalographic characteristics were changed
	RNA — drop in the total brain level of RNA, lower total RNA content in brain cells

REFERENCES

1. **Jones, R. W.,** *Principles of Biological Regulation,* Academic Press, New York, 1973, 279.
2. **Rosenzweig, M. R., Bennet, E. L., Hebert, M., and Morimoto, H.,** Social grouping cannot account for cerebral effects of enriched environments, *Brain Res.,* 153, 563, 1978.
3. **Welch, B. L.,** Psychophysiological response to the mean level of environmental stimulation: a theory of environmental integration, in Symposium on Medical Aspects of Stress in the Military Climate, U.S. Government Printing Office, Bethesda, Md., 1965, 29.
4. **Dewsbury, D. A.,** *Comparative Animal Behavior,* McGraw-Hill, New York, 1978, 146.
5. **Denenberg, V. H.,** Readings in early experience: a consideration of the usefulness of the critical period hypothesis as applied to the stimulation of rodents in infancy, in *Early Experience and Behavior: the Psychobiology,* Newton, G. and Levine, S., Eds., Charles C Thomas, Springfield, Ill., 1968, 142.
6. **Levine, S.,** The infantile experience on adult behavior, in *Experimental Foundation of Clinical Psychology,* Bachroch, A. J., Ed., Basic Books, New York, 1962, 139.
7. **Levine, S. and Wetzel, A.,** Infantile experiences, strain differences and avoidance learning, *J. Comp. Physiol. Psychol.,* 56, 879, 1963.
8. **Weininger, O.,** The effects of early experience on behavior and growth characteristics, *J. Comp. Physiol. Psychol.,* 49, 1, 1956.
9. **Boice, R.,** Burrows of wild and albino rats: effects of domestication, outdoor raising, age, experience and maternal care, *J. Comp. Physiol. Psychol.,* 91, 649, 1977.
10. **Nováková, V. and Babický, A.,** Role of early experience in social behaviour of laboratory-bred female rats, *Behav. Process.,* 2, 243, 1977.
11. **Nováková, V., Šterc, J., and Vorlíček, J.,** Interstrain differences in the total RNA content in brain cells of the laboratory rat, *J. Hirnforsch.,* 26, 109, 1985.
12. **Daly, M. and Wilson, M.,** *Sex, Evolution and Behavior,* 2nd ed., Willard Grant Press, Boston, 1983, 92.
13. **Cohen-Salmon, C., Carlier, M., Roybertoux, P., Jouhaneaw, J., Semal, C., and Paillette, M.,** Differences in patterns of pup care in mice. V. Pup ultrasonic emissions and pup care behavior, *Physiol. Behav.,* 35, 167, 1985.
14. **Hebb, D. O.,** *The Organization of Behavior: a Neuropsychological Theory,* Academic Press, New York, 1949.

15. **Hymovitch, B.**, The effects of experimental variations of problem solving in the rat, *J. Comp. Physiol. Psychol.*, 45, 313, 1952.
16. **Rosenzweig, M. R. and Bennett, E. L.**, Effects of environmental enrichment or impoverishment on learning and on brain values in rodents, in *Genetics, Environment and Intelligence*, Oliverio, A., Ed., North-Holland Biomedical, Amsterdam, 1977, 163.
17. **McVickre-Hunt, J.**, Psychological development: early experience, *Annu. Rev. Psychol.*, 30, 103, 1979.
18. **Fox, M. W.**, Neurobehavioral development and the genotype-environment interaction, *Q. Rev. Biol.*, 45, 131, 1970.
19. **Daly, M.**, Early stimulation in rodents: a critical review of present interpretations, *Br. J. Psychol.*, 64, 435, 1973.
20. **Calhoun, J. B.**, The Ecology and Sociology of the Norway Rat, Publ. No. 1008, U.S. Department of Health Education and Welfare, Bethesda, Md., 1962, 22.
21. **Nováková, V.**, Community life and the development of avoidance reaction in the laboratory rat, *Act. Nerv. Super.*, 22, 241, 1980.
22. **Nováková, V.**, A social environment influences the total RNA content of the brain cells of young laboratory rats, *Physiol. Bohemoslov.*, 33, 362, 1984.
23. **Melzack, R.**, Early experience: a neuropsychological approach to heredity-environment interaction, in *Early Experience and Behavior: the Psychology and Development*, Newton, G. and Levine, S., Eds., Charles C Thomas, Springfield, Ill., 1968, 65.
24. **Fuller, J. L.**, Experimental deprivation and later behavior, *Science*, 158, 1645, 1967.
25. **Harlow, H. F., Harlow, M. K., and Hansen, E. W.**, The maternal affectional system of rhesus monkeys, in *Maternal Behavior in Mammals*, Rheingold, H. L., Ed., John Wiley & Sons, New York, 1963, 254.
26. **Harlow, H. F., Harlow, M. K., and Suomi, S. J.**, From thought to therapy: lessons from primate laboratory, *Am. Sci.*, 59, 538, 1971.
27. **Harlow, H. F.**, *Learning to Love*, Ballantine, New York, 1971.
28. **Rosenblatt, J. S. and Siegel, H. I.**, Physiological and behavioural changes during pregnancy and parturition underlying the onset of maternal behaviour in rodents, in *Parental Behaviour of Rodents*, Elwood, R. L., Ed., John Wiley & Sons, New York, 1983, chap. 2.
29. **Alberts, J. R. and Gubernik, D. J.**, Reciprocity and resource exchange, in *Symbiosis in Parent-Offspring Relations*, Rosenblum, L. A. and Moltz, H., Eds., Plenum Press, New York, 1983, 7.
30. **Hofer, M. and Shair, H.**, Ultrasonic vocalization during social interaction and isolation in 2-week-old rats, *Dev. Psychobiol.*, 11, 495, 1978.
31. **Noirot, E.**, Ultrasounds and maternal behavior in small rodents, *Dev. Psychobiol.*, 5, 371, 1972.
32. **Moltz, H.**, The ontogeny of maternal behavior in some selected mammalian species, in *The Ontogeny of Vertebrate Behavior*, Moltz, H., Ed., Academic Press, New York, 1971, 265.
33. **Adels, L. E. and Leon, M.**, Thermal control of mother-young contact in Norway rats: factors mediating the chronic elevation of maternal temperature, *Physiol. Behav.*, 36, 183, 1986.
34. **Babický, A., Ošťádalová, I., Pařízek, J., Kolář, J., and Bíbr, B.**, Use of radioisotope techniques for determining the weaning period in experimental animals, *Physiol. Bohemoslov.*, 19, 457, 1970.
35. **Babický, A., Ošťádalová, I., Kolář, J., and Bíbr, B.**, Onset and duration of the physiological weaning period for infant rats reared in nests of different sizes, *Physiol. Bohemoslov.*, 22, 449, 1973.
36. **Babický, A. and Nováková, V.**, Influence of thyroxine and propylthiouracil administration on the intake of maternal milk in sucklings of the laboratory rat, *Physiol. Bohemoslov.*, 34, 193, 1985.
37. **Galef, B. G. and Clark, M. M.**, Parent-offspring interaction determines the time and place of first ingestion of solid food by wild rat pups, *Psychon. Sci.*, 25, 15, 1971.
38. **Nováková, V.**, Significance of the weaning period for natality and maternal behaviour of laboratory rat, *Physiol. Bohemoslov.*, 26, 303, 1977.
39. **Křeček, J.**, Theory of critical developmental periods and postnatal development of endocrine function, in *Biopsychology of the Development*, Tobach, E., et al., Eds., Academic Press, New York, 1971, 233.
40. **Macho, L., Štrbák, V., and Strážovcová, A.**, Thyroid and adrenal function in prematurely weaned rats, *Physiol. Bohemoslov.*, 19, 77, 1970.
41. **Nováková, V.**, Time of weaning: its effect on the rat brain, *Academia*, Prague, 1976.
42. **Kevan, P. G.**, Sir Thomas More on imprinting: observations from the sixteenth century, *Anim. Behav.*, 24, 16, 1976.
43. **Spalding, D. A.**, Instinct, with original observations on young animals, *McMillan's Magazine*, 27, 283, 1983; reprinted from *Br. J. Anim. Behav.*, 2, 2, 1954.
44. **Lorenz, K.**, Der Kumpan in der Umwelt des Vogels, *J. Ornithol.*, 83, 137, 1935.
45. **Hess, E. H. and Hess, D. B.**, Innate factors in imprinting, *Psychon. Sci.*, 14, 129, 1969.
46. **Scott, J. P.**, The organization of comparative psychology, *Ann. N.Y. Acad. Sci.*, 223, 7, 1973.
47. **Harris, G. W.**, Hormonal differentiation of the developing central nervous system with respect to patterns of endocrine function, *Philos. Trans. R. Soc. Lond. Ser. B*, 259, 165, 1970.

Chapter 7

MATHEMATICAL MODELS OF PHYSIOLOGICAL PROCESSES

I. Dvořák

TABLE OF CONTENTS

I. LIST OF SYMBOLS

t   time

$\left.\begin{array}{l} x \\ y \\ z \end{array}\right\}$ the space coordinates

c_i input concentration of ethanol
c_o output concentration of ethanol
c_m internal concentration of ethanol in the liver
F   blood flow through the liver
P   the probability function
v_i inflow rate of ethanol
v_o outflow rate of ethanol
v_m rate of metabolism of ethanol in the liver
G_{v_m} the rate function
g_m the concentration function
V   volume of the liver
k_m rate constant of ethanol metabolism
V_{max} maximal rate of ethanol metabolism
$\tilde{c}_m$ steady value of the ethanol concentration c_m

II. MATHEMATICAL MODELING IN ANIMAL PHYSIOLOGY

A. Role of Mathematical Models in Biomedical Research

Mathematical modeling has a special position among the methods of animal physiology, as it does not serve for data acquisition. Its aim is — together with mathematical statistics — to scrutinize the data more efficiently, to draw from them the hidden essential information, and to incorporate them into the unified framework of broader knowledge. Alternatively, it may detect discrepancies in the data obtained that may not be visible on first inspection. Its final goal is to represent hypotheses on the nature of studied physiological process exactly and rigorously, i.e., in the form of mathematical relations between strictly defined variables. This representation is called a *mathematical model.*

Predictions for measurable quantities of interest can be calculated from the model. Comparison of these predictions with a particular set of data brings arguments for sustaining the original hypotheses if agreement is found or indicates the necessity of model reformulation in the opposite case. Since sophisticated mathematical methods can be used in the model analysis and powerful methods of mathematical statistics can be applied to model verification, the method of mathematical modeling improves efficiency of scientific reasoning.

Application of mathematical modeling in biosciences has a long history. The books of Lotka,[1] Volterra,[2] and Rashevski[3] are often mentioned as the first attempts to introduce this method. It was the mathematical model of neuronal membrane potential changes[4] that brought the Nobel prize to Hodgkin and Huxley in the early 1950s. Until the mid-1960s, however, mathematical modeling in biosciences was still in its infancy, being still more of a curiosity than a seriously developed and applied methodology. This situation was caused mostly by the overwhelming complexity of biological and physiological systems. Even the simplest mathematical model designed not to lose touch with reality leads almost surely to serious problems in solving model equations. In every physiological system there are very many quantities that can be measured, that exhibit complicated internal bonds between them. A realistic model inevitably yields a set of complex nonlinear equations. Sophisticated mathematical means for their investigation either had not been discovered by the mid-1960s or were not in the theoretical equipment of the biomedical researcher. In this precomputer era, moreover, even the numerical solution of realistic models was beyond the existing powers. Disregarding some rare exceptions we can say that all this limited mathematical modeling in biomedicine to production of metaphoric models that could serve as generators of questions and ideas (thanks for that!), but not as tools for serious analysis and interpretation of experimental data.

The situation subsequently changed during the late 1960s because of two technical reasons and one "philosophical" reason. The first technical reason were new methods of mathematical treatment of complex systems (e.g., the theory of qualitative solution of differential equations,[5,6] the theory of catastrophes,[7] the theory of stochastic processes,[8,9] or the theory of stochastic differential equations,[10]) that were developed and/or improved and introduced into routine application in theoretical biosciences. The second technical reason was the computer revolution that brought qualitatively new opportunities for numerical computing. New mainframe computers have made possible faithful simulation even of very complex systems and fast solution of large systems of highly nonlinear equations.

These explosive developments have had many positive, but also some negative, consequences. A large body of experience was accumulated during this period on constructing biomathematical models of varying complexity and detail. In some domains mathematical modeling has brought entirely new perspectives to the whole discipline; in addition to the neuronal potential generation model that was already mentioned, the theory of metabolism[11] or the modeling of biological populations[12] may be quoted. The Eigen's model of natural selection by hypercycles has introduced new horizons into the theory of evolution.

Mathematical modeling has bifurcated subsequently into several branches with methodologies modified by applied mathematical techniques (deterministic or stochastic approach, continuous or discrete interpretation, etc.). Such theoretical branches as mathematical ecology, mathematical genetics, or mathematical neurophysiology developed within various biomedical sciences.

On the other hand, it cannot be overlooked that this boom often followed paths of minor importance from the point of view of the mother biomedical science. Many hopes that application of cybernetics and mathematical modeling is going to solve the principal problems of organization of living matter were not fulfilled. Many models were constructed without clearly stated correspondence to known data, the models being their own *raison d'etre*. This resulted in widening the gap between theoretical modeling and experimental research in many areas.

The beginning of the 1980s was characterized by general sobering. The limits of mathematical modeling methods were realized, and some unjustified hopes were abandoned. The ambitions of constructing valid theoretical models of general biological phenomena were not abandoned entirely, but most of the models nowadays are smaller, with limited aims and clearly stated correspondence to experimental data.

Many published models concern animal or human physiology.[13] They range from very simple models to such monstrous systems as the Guyton model[14] of blood pressure regulation. Although mathematical models can be found in nearly all branches of physiology, the most frequently involved areas are general metabolism,[11,15] ventilation and blood circulation regulation, and neuron signal transmission[16] (see also Chapter 14).

B. Art of Mathematical Modeling

Literature devoted to application of mathematical methods in biology and physiology is abundant. Many textbooks are devoted to special mathematical techniques for formulation of models in physiology, e.g., calculus,[17,18] the theory of differential equations,[6,19] or the theory of probability.[9] Surprisingly few texts, however, tackle the general problem of how to construct the mathematical model of some given physiological phenomenon or process. The reason is obvious: no general way of constructing a mathematical model exists. Each mathematical model is unique. Until the availability of a unified mathematical theory of living systems, each model is only a partial description, reflecting both the nature of the studied processes or phenomena and the degree of their appreciation by the model creator.

In spite of these rather disappointing assertions in the preceding paragraphs, I shall try to summarize some basic rules for constructing the mathematical model of a physiological process. In order to be as illustrative as possible all the steps are accompanied by various versions of a very simple model of ethanol metabolism in the liver of a pig. It must be kept in mind, however, that all the rules presented are only heuristic. They have been proven useful in many cases, but they may upset us in a particular situation. In recognizing the limits of their applicability consists the essence of "the art of mathematical modeling".

III. CYCLIC CREATION OF THE MATHEMATICAL MODEL

The wide spectrum of mathematical models in animal physiology was already mentioned. In our considerations we limit ourselves only to models of physiological processes, i.e., to models describing the *dynamics* of some physiological system. Time is always the most important independent variable in these models, though possibly not the only one (see Section III.D).

Suppose the experimental arrangement for studying the process has been chosen and the data for model testing have been collected. The construction of the mathematical model of this process can be paraphrased by ten steps (see Figure 1):

1. Quantities important for characterization of the modeled process are determined. All these quantities are classified either as those that change during the studied process or as those that do not change. The former are called *variables*; thus, they are functions of time. The latter are called *parameters*. Although they do not change during one run of the experiment, they may change from one experimental arrangement to the other. They are introduced in order to also include into the model the influence of external (or internal) factors that remain (or are kept) constant during the experiment.

2. Calculus for formulating the mathematical model is chosen. This attributes a mathematical interpretation to the quantities chosen for representation of the studied system (we may consider them as deterministic or stochastic, continuous or discrete, etc.). Mathematical means in disposition determine this decision as well as the natural character of quantities in question. In this way the basic set of model variables and parameters is obtained.

3. The partial processes in the studied system, which influence the changes of the chosen variables, are determined.

4. Hypotheses on the nature of the studied processes and on their interactions are formulated.

5. The hypotheses are formalized in terms of model variables. The resulting set of mathematical relations represents (at a given stage) the desired mathematical model.

6. The model is identified, i.e., the values of its parameters are established, either by calculating using their relation to known constants or from experimental data (e.g., by regression analysis).

7. The model properties are analyzed, and predictions resulting from the model are calculated for those variables that allow comparison with experimental data.

8. The validity of the model for description of the studied process is evaluated by methods of mathematical statistics. More precisely, fits of two or more concurrent models to given experimental data are tested and compared.

9. If the model fit is poor, we should return to our hypotheses and modify them. This represents a completion of one revolution around the inner cycle on Figure 1. Sometimes the inappropriateness of the model is apparent immediately after calculating the model predictions. Then a predictions → new hypotheses short cut can be made.

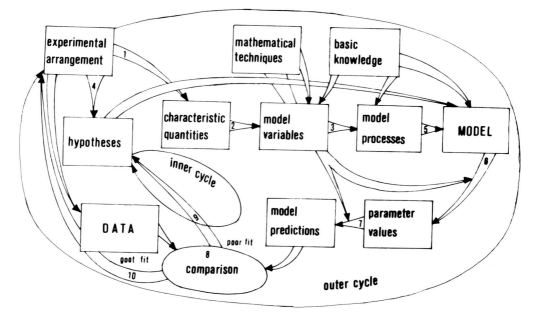

FIGURE 1. Cyclic model creation: inner cycle — hypotheses changing; outer cycle — rearrangement of the experiment.

10. If the fit is good, we may return to the original experimental arrangement and modify or extend it (this activity does not belong to mathematical modeling, of course). The outer cycle in Figure 1 is accomplished in this way. This may be repeated theoretically *ad infinitum*. Each revolution in our scheme (either around the inner or outer cycle) brings a deeper understanding of the modeled processes.

In the following paragraphs we explain steps 1 through 3 and step 5, and touch on step 7. Step 4 — hypotheses formation — is specific for each given problem, and its explanation does not belong to the methodology of mathematical modeling. Steps 6 and 8 belong to the realm of mathematical statistics. They are treated in the following chapter contributed by Havránek.[20] The last two steps — 9 and 10 — are self evident.

A. Example — Ethanol Metabolism by the Liver

Our theoretical reasoning will be illustrated through the whole paper by various versions of the mathematical model of ethanol metabolism in the liver of a pig. The model is described and studied in the paper by Keiding et al.[21,22] Detailed statistical analysis of its space-dependent version was published by Johansen.[23] For presentation in this chapter the simplest versions were chosen. Their aim is to illustrate various tricks of mathematical modeling techniques without bogging down into technically complicated calculations that may obscure the main ideas involved. If the reader is discontent because of disregard for many important biological details of the studied system, he should realize this didactic role of the model presented. He will be given the opportunity of observing in the following paragraphs that even such simple models bring nontrivial problems in their mathematical treatment. More sophisticated analysis can be found in the literature quoted above.

The basic scheme of the studied system and the arrangement of the experiment are schematically depicted in Figure 2. The liver is presented here as "bag" with blood inlet and outlet. Catheters are inserted into the blood stream at its entrance and exit. Ethanol is infused into the system at a steady rate, and the following three variables are measured

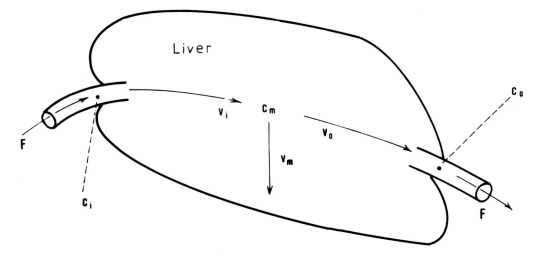

FIGURE 2.   Processes in the liver and arrangement of the experiment.

regularly: flow of blood through the liver F (in milliliters per minute $\doteq 1.66 \cdot 10^{-8}$ m³/sec) and the concentrations of ethanol at the inlet c_i and at the outlet c_o (both in moles per liter).

When the measurements on the pig liver are taken for a period of about 10 or 15 min (for the given rate of ethanol infusion), the steady state is reached. In this state the rate of ethanol inflow equals the sum of rates of ethanol metabolism in the liver and of ethanol outflow from the liver. Consequently, concentrations c_i and c_o are different, but they do not change in time. If infusion of ethanol is then increased, a new steady state is reached after approximately the same period. The experiment is performed usually for five to seven periods, each corresponding to a different level of ethanol infusion.

It is our aim to propose a mathematical model representing the described situation, which would allow an experimental verification by the measurements of F, c_i, and c_o in given time intervals.

B. Choice of Model Variables

In our simple case, the choice of model variables is to a large extent determined by the experimental arrangement itself. There is one natural independent variable — time, t. In the later stages of model development we may be interested also in space dependence of the model parameters and variables. We may, therefore, add space variables x, y, and z as optional independent variables. Quantities F and c_i are kept constant during each measurement. We shall regard them, therefore, as parameters. For the dependent variable we naturally accept the remaining quantity that changes during the experiment — c_o. We feel, however, that this quantity — external when we speak about the liver processes — may not be sufficient for creation of the mathematical model of ethanol metabolism. This model must — though to a limited extent — also consider internal processes of the liver. That is why we add a second model variable — internal liver ethanol concentration, c_m.

We must now attribute a mathematical interpretation to all these quantities (variables and parameters). We may classify them according to two different points of view:

1.   Whether they are continuous or discrete (concerns all the independent and dependent variables and parameters)
2.   Whether they are deterministic or stochastic (concerns only dependent variables and parameters)

When we interpret a variable or a parameter as a continuous quantity, we suppose that it can acquire as its value any real number. On the other hand, if we interpret them as discrete quantities, we suppose that they can acquire only whole multiples of a given quantum Δ. Although theoretically we may formulate the model also as a mixed model, i.e., some parameters and/or dependent and/or independent variables being discrete while others are continuous, it is impractical (and therefore seldomly used). We should, therefore, choose between purely continuous and purely discrete description. It is more habit than anything else to formulate the models of such physiological systems as the presented one as continuous models. The original reason is that they are often based on the laws of fluid mechanics or chemical kinetics, which are both formulated in terms of continuous variables. Discrete models are more often encountered in ecology, where the natural discreteness of time (one season or one generation period) and of population (by counting its members) comes at once to mind. We should remark, however, that discrete formulation may appear useful even for our model, when we want to apply numerical methods and computers for solution of our equations. In the following we shall interpret all parameters and independent and dependent variables as continuous.

The decision "deterministic or stochastic" is more profound, since it touches the very essence of our understanding of the natural processes. Repetition of a biological experiment gives, as a rule, similar, but never the same results. This naturally leads to formulation of the mathematical model of the process in question in probabilistic terms. Let us take the variable c_m as an example. Its probabilistic interpretation means that we define probability $P(c_m, t, F, c_i,$ other parameters) that at a given time instant t, for given values of parameters F, c_i (and other system parameters), the value of internal ethanol concentration will be within the interval $<c_m, c_m + dc_m>$, where dc_m means an infinitesimal increment of the concentration c_m. We try to describe how this probability depends on the values of the mentioned arguments (c_m, t, c_i, and other system parameters). This description, — if found, gives a relatively complete picture of the dynamics of the system. On the other hand, it is rather hard to complete and often very cumbersome to calculate with. That is why one is very often satisfied with a deterministic description, which usually reflects (though not always) the behavior of the most probable values of the probabilistic model. In this approach we directly search the functional dependencies of the type

$$c_m = g_m(t, c_i, F_i \text{ other parameters})$$

which reflect how the value of internal ethanol concentration depends on the time t and the values of model parameters. This dependence is given by the concrete form of the function g_m (for variable c_o the situation is analogous).

In the following we restrict our considerations only to deterministic models. A nice exposition of probabilistic mathematical models in biology may be found in the book by Iosifescu and Tautu.[8]

C. Choice of Model Processes

If the system variables and parameters are chosen, we must study their relations. In reality, each variable chosen for characterization of ethanol metabolism in the liver is influenced by many processes. We must choose only the most important of them, that means those that determine the qualitative features of the system dynamics. For our illustrative system, we see immediately that processes which cannot be neglected are inflow of ethanol into the liver, its metabolism within the liver, and its outflux from the liver. The rates of these processes influence the changes of our system variables c_m and c_o. We denote these rates v_i, v_m, and v_o, respectively. We shall keep in mind that these rates may again be functions of the space coordinates x, y, and z as well as functions of time. It is obvious that these rates are also functions of some system variables and parameters.

The reader may easily find himself the processes that were neglected in our reasoning. For instance, as blood pH is not considered a system variable in our model, no processes representing the pH influence on ethanol metabolism were included.

The chosen set of system variables, parameters, and processes is the cornerstone for development of our mathematical model.

D. Models with Lumped Parameters

We have already mentioned that time need not be the only independent variable of the model. Since every system studied occupies some portion of space, its properties depend also on the space coordinates x, y, and z. It is a matter of our decision whether we consider this dependence important for our model or not. In the former case, we create our (deterministic) model in such a way as to specify the time-space dependence of model variables (c_m taken again as an example

$$c_m = g_m(t, x, y, z, F, c_i, \text{other parameters})$$

The same we shall suppose for the rates of system processes. Taking v_m as an example, we write

$$v_m = G_{v_m}(t, x, y, z, c_o, c_i, F, \text{other parameters})$$

Such models we call models with distributed parameters. In the opposite case we neglect the dependence of model variables and parameters on the space coordinates. We consider the system studied as if its properties are lumped into one or several points in space and analyze only the time changes of system variables in these points. These models are called — as a counterpart to models with distributed parameters — models with lumped parameters.

The terms are relics from the epoch when the difference between parameters and variables was not so clearly recognized and both the terms were frequently used as synonyms. Although the parameters may also be space dependent, it is the spatial dependence of the *variables* that is important. The proper terms should therefore be ''models with distributed variables'' and ''models with lumped variables''.

The simplest model with lumped parameters is obtained if we suppose that the ethanol concentration within the liver c_m is homogenous. Then we may consider it only as a function of time and model parameters

$$c_m = g_m(t, F, c_i, \text{other parameters})$$

We shall suppose, moreover, that outlet concentration c_o is practically identical to c_m. By this trick there remains only one system variable — c_m. In an analogous way we shall suppose that the rates are only functions of time t and system variables and parameters. For instance, for v_m we shall write

$$v_m = G_{v_m}(t, c_o, c_i, \text{other parameters})$$

In the following paragraphs we shall study several models of this type.

1. Simple Linear Models

Formulating the basic equations of the model is, of course, the crucial point of the whole modeling process. In many physiological systems we can take advantage of their physical (or chemical) nature and apply the laws of physics (or chemistry) as the proper fundament. It is also the case of our illustrative system. Since the nature of the system is represented by a fluid (blood) flow through some reservoir (liver), where a chemical reaction (metabolism) takes place, we may apply a so-called balance equation, well known from thermodynamics or fluid mechanics (V is the volume of the liver)

$$V \frac{d\,c_m}{dt} = v_i - v_m - v_o \tag{1}$$

This equation may be verbally interpreted as follows: the change of the internal amount $(V \cdot c_m)$ of ethanol in the liver during some (infinitesimal) time interval equals the inflow of ethanol during this interval minus its outflow from the liver and minus its metabolism within the liver during this interval.

In the following steps we must specify the exact forms of dependence of the rates v_i, v_m, and v_o on the system variables and parameters. It is the point where the hypotheses on the nature of processes taking place in the studied system enter the play.

There is no hesitation as to how the formulas for v_i and v_o should look. Since the blood flow through liver F is steady, the influx of ethanol is simply

$$v_i = F\,c_i \tag{2}$$

Similar will be true for ethanol outflow. Since concentrations c_m and c_o were lumped together, the outflow rate is

$$v_o = F\,c_m \tag{3}$$

Thus, we reach a situation when various hypotheses considered may concern only the rate law for the metabolism of ethanol within the liver. We should therefore ask

$$v_m = ?$$

This law may be rather complicated in reality, even if we neglect the space distribution of the process. It is advisable, however, for passing the first revolution of the cyclic creation of the mathematical model as quickly as possible, to start with the simplest formulation possible, and to only successively include more complicating factors.

Let us contemplate how the formula for v_m may look. Applying basic knowledge, we first observe that the rate v_m does not depend on time directly — all its time dependence goes via its dependence on c_m (the same is true for v_i and v_o as we can observe from Equations 2 and 3). It is reasonable to suppose that dependence of v_m on c_m will have the form depicted in Figure 3 by the full line. For small c_m values, v_m will grow proportionally to the internal ethanol concentration c_m. When the value of c_m grows, v_m will saturate as the limiting effect for metabolic processes will be more and more pronounced. At the end, v_m will reach a value very close to the value V_{max}, which represents the maximal rate of metabolizing capacity of the liver. Whatever the increase of c_m may be, it cannot increase the value of v_m over V_{max}.

Extreme regimes of the system often represent the simplest cases. It is, therefore, advantageous to start investigation of the model by studying the extreme cases. There are two extreme cases in our system:

1. Proportional metabolism rate (k_m is the proportionality constant):

$$v_m = k_m\,c_m \tag{4}$$

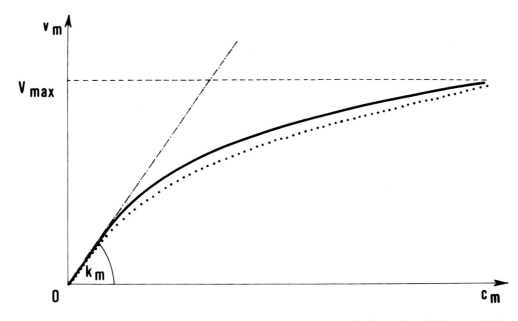

FIGURE 3.   Supposed dependence of the rate v_m on the concentration c_m. Full line = hypothetic curve, dashed line = constant metabolism rate approximation, dash-dotted line = proportional metabolism rate approximation, and dotted line = polynomial approximation.

This model (represented by the dash-dotted line in Figure 3) approximates well the supposed situation for small values of c_m.

2.   Constant metabolism rate:

$$v_m = V_{max} \tag{5}$$

This model (represented by the dashed line in Figure 3) approximates well the supposed situation for large values of c_m.

We shall deal with these two models one by one. Between them (for mediocre values of c_m) is the domain where neither the first nor the second model approximates faithfully the supposed situation. This problem we shall tackle in Section III.D.2.

a. Model with Proportional Metabolism Rate

Substituting Relations 2, 3, and 4 into the balance Equation 1 for v_i, v_o, and v_m, respectively, gives the linear differential equation

$$V \frac{d\, c_m}{dt} = Fc_i - k_m\, c_m - F\, c_m \tag{6}$$

for the unknown function c_m.

It is appropriate to insert a few words of explanation about the mathematical formalism underlying Equation 6. Readers familiar with calculus and elements of the theory of differential equations may skip the following text and continue reading from the paragraph containing Equation 7.

On the left-hand side of Equation 6 there is a time derivative of the variable c_m multiplied by a constant (liver volume). The time derivative can be understood as a linear prediction of the development of a given variable during the following (infinitesimally short) time interval. On the right-hand side of this equation there is a function of the same variable c_m. Equation 6 thus informs us how the very next development of the variable c_m depends on its instantaneous value. Equations of this type, that are relations between a function of some variable and (in general a function) of its derivative (or derivatives), are called *differential equations*.

It is worth pointing out that, contrary to algebraic equations where the solution is a number, solution of a differential equation is always a function. This function specifies how the model variable (c_m in our case) depends on the independent variable with respect to which the derivation was performed (time t in our case). Since the final formula specifying this function will also contain parameters of the equation (k_m, F, and c_i in our case), the solution of Equation 6 will have the form specified in general Formula 5 (k_m stands here for "other parameters").

In mathematical modeling of physiological systems we encounter many types of differential equations, and there are also many ways of classifying them. Possibly the most important classification lies in distinguishing ordinary differential equations (which contain only ordinary derivatives of the dependent variable) and partial differential equations (which contain partial derivatives). Equation 6 contains only one ordinary derivative of c_m, and it is therefore, an ordinary differential equation.

Partial differential equations result in mathematical formulation of models with distributed parameters. Since we do not study these models in detail here we shall not encounter partial differential equations in this chapter.

Another important classification is according to linearity of the given differential equation. Let us remind that the function f of the independent variable x is called linear, if it is of the form

$$f(x) = A x + B$$

where A and B are constants (parameters). An ordinary differential equation in its turn is called linear if its sides are a linear function of the model variable and its derivative (or derivatives). The left-hand side of Equation 6 is a linear function of dc_m/dt (multiplication by a constant), and also its right-hand side is a linear function of c_m (putting $A = k_m + F$, $B = F c_i$ we obtain the definition presented above). Hence, Equation 6 is a linear ordinary differential equation. Let us stress that linearity in this case concerns the variable, not parameters. By chance, both sides of Equation 6 are also linear functions of the model parameters, but this need not always be the case.

There is a large body of methods (see, for instance, Reference 6) for solving ordinary differential equations. For a while we put aside numerical methods that provide a table or a graph of values of the solution. Even the presentation of analytical methods that provide the explicit formula for calculating the values of the solution for given values of independent variable and parameters is beyond the scope of this contribution. Just to give the reader a general picture I mention the following: while there exists a unified theory for solving linear differential equations, nothing similar exists for nonlinear differential equations. Each nonlinear equation should be solved individually by proper application of methods fitted to the character of its nonlinearity. Most nonlinear differential equations are not solvable by analytic methods at all. Hence the importance of linear differential equations, which in general represent much easier mathematical problems. Unfortunately, mathematical models of real physiological processes lead to linear differential equations only exceptionally or in a very crude approximation.

As Equation 6 is a linear differential equation, its solution can be easily found. The reader may verify himself (by differentiating the solution and substituting it and its derivative into Equation 6) that the solution is

$$c_m(t) = \frac{F\,c_i}{F + k_m}\left(1 - e^{-\frac{F + k_m}{V}t}\right)$$ (7)

b. Model with Constant Metabolism Rate

Performing the same procedure as before but replacing Formula 5 instead of 4 for v_m in the balance equation, Equation 1, gives another linear differential equation for c_m:

$$V\frac{d\,c_m}{dt} = F\,c_i - F\,c_m - V_{max}$$ (8)

This equation is also easy to solve; the reader can again verify himself that the solution is

$$c_m(t) = \left(c_i - \frac{V_{max}}{F}\right)\left(1 - e^{-\frac{F}{V}t}\right)$$ (9)

c. Steady Values of Concentration c_m

It follows from the arrangement of our experiment (see Section III.D.1) that we are not interested at the moment in concentration changes during the transition period. It is the steady value of c_m that is of importance for comparison with experimental data. These values can be calculated for both models by taking the limits of c_m for time going to infinity (we shall denote steady value by a tilde). The model with proportional metabolism rate gives the formula

$$\tilde{c}_m = \lim_{t\to\infty} c_m = \frac{F\,c_i}{F + k_m}$$ (10)

while the model with constant metabolism rate gives

$$\tilde{c}_m = \lim_{t\to\infty} c_m = c_i - \frac{V_{max}}{F}$$ (11)

We can see that the two models give different predictions of dependence of the steady value $\tilde{c}_m$ on model parameters c_i and F. This dependence can be tested by fitting both models to experimental data (see Reference 20). As these two models represent only the simplest extreme cases, it is very probable that none of them will fit very well for the whole range of the parameter values.

2. Nonlinear Model

Let us suppose that really none of the models given by Solutions 6 and 7 (i.e., with predicted steady values of $\tilde{c}_m$ by Equation 10 or 11) fits well the experimental data. We must return again and change our hypotheses according to step 9 of our list. We already know that the approximations in Equation 4 or 5 are not satisfying; therefore, we must look for functions approximating the supposed dependence of v_m on c_m (full line in Figure 3) in the whole range of values of c_m.

a. Rational Polynom Nonlinearity

There are many candidates for approximation of the dependence $v_m = g(c_m)$, which is given by the full line in Figure 3. It is apparent that any function $g(c_m)$ approximating this curve must fulfill three conditions:

1. $g(0) = 0$ (the rate v_m is zero when c_m is zero)

2.

$$\frac{d\,g}{d\,c_m}(0) = k_m$$

(for small values of c_m g is close to the model with proportional metabolism rate)

3.

$$\lim_{c_m \to \infty} g(c_m) = V_{max}$$

(for large values of c_m g is close to model with constant metabolism rate)

When constructing the approximative function g it is only natural to profit as much as possible from the knowledge gathered during the preceding stages. That is the reason why we look for such a function g that would represent an extension of the two simpler extreme models. This means that the form of function g should be such that for some extreme values of its parameters it transforms into one of these two extreme models.

It is the function

$$v_m(c_m) = g(c_m) = \frac{V_{max}\,c_m}{c_m + \dfrac{V_{max}}{k_m}} \tag{12}$$

that fulfills all the requirements. The reader can verify himself that it fulfills conditions 1 through 3 listed above. Moreover, the following assertions are true:

$$\lim_{V_{max}/k_m \to 0} \frac{V_{max}\,c_m}{c_m + \dfrac{V_{max}}{k_m}} = V_{max}$$

$$\lim_{V_{max}/k_m \to \infty} \frac{V_{max}\,c_m}{c_m + \dfrac{V_{max}}{k_m}} = k_m\,c_m \tag{13}$$

This means that the important parameter reflecting the model complexity is the ratio V_{max}/k_m. For large values of this ratio the model Equation 12 is reduced to a model with proportional metabolism rate, while for its small values it is reduced to a model with constant metabolism rate, as we have desired.

Substituting Formula 12 for v_m into the balance equation, Equation 1 yields the differential equation

$$V\frac{d\,c_m}{d\,t} = F\,c_i - F\,c_m - \frac{V_{max}\,c_m}{c_m + \dfrac{V_{max}}{k_m}} \tag{14}$$

As the right-hand side of this equation is a nonlinear function of c_m, the whole Equation 14 is a nonlinear differential equation. Its solving is not so easy as in the preceding cases and cannot be, in fact, obtained in a closed form of an elegant mathematical formula. There are many methods (either analytical or numerical) for solving such an equation approximately. Since application of these methods would require rather space-consuming explanations, I shall omit it here. The interested reader may find detailed explanation in many textbooks (for example, References 5 and 6).

b. Steady Values of Concentration c_m

We can again take advantage of the fact that we are not interested in transient periods. The steady value of c_m may be determined from Equation 14 without really solving it. It is enough to realize that in the steady state c_m does not change with time; its time derivative is, therefore, zero. If we replace the time derivative on the left-hand side of Equation 14 by zero, we obtain an algebraic equation for calculating the steady value $\tilde{c}_m$ (the steady value is again denoted by the tilde):

$$0 = F c_i - F \tilde{c}_m - \frac{V_{max} \tilde{c}_m}{\tilde{c}_m + \dfrac{V_{max}}{k_m}} \tag{15}$$

This equation leads to a quadratic equation for $\tilde{c}_m$ which can be readily solved (the value of concentration must be a positive number; as the second root is always negative, it is omitted)

$$\tilde{c}_m = \frac{1}{2}\left(c_i - \frac{V_{max}}{k_m} - \frac{V_{max}}{F}\right) + \frac{1}{2}\sqrt{\left(c_i - \frac{V_{max}}{k_m} - \frac{V_{max}}{F}\right)^2 + 4c_i \frac{V_{max}}{k_m}} \tag{16}$$

It is immediately apparent that for $V_{max}/k_m = 0$ this formula is reduced to Formula 11 obtained from the model with constant metabolism rate. An analogical relation for the model with proportional metabolism rate does not hold, because the right-hand side of Formula 16 goes to infinity for $V_{max}/k_m \to \infty$. It is the consequence of multiplication of the whole of Equation 15 by the term $(c_m + V_{max}/k_m)$ during its solving. If we suppose V_{max}/k_m to grow over all limits, we cannot do this. In this situation we may obtain, however, a model with proportional metabolism rate directly from Equation 15.

Solution 16 gives better approximation for the dependence of the steady value $\tilde{c}_m$ on the model parameters c_i and F. Comparison of its fit to experimental data with that of Model 11 (such pairs of models are called *nested* because one is a special case of the other) will very likely show better fit for the nonlinear model.

One important note should be added here. Classification of the models as linear or nonlinear corresponds with the linearity or nonlinearity of the differential equations representing the model. This linearity or nonlinearity, however, does not imply necessarily the corresponding linearity or nonlinearity of the solution of the equation. Contrary, as we can observe from Formulas 7 and 9, even the solutions of linear differential equations are nonlinear functions (exponentials) of time and of the model parameters. This is even more true for solutions of nonlinear differential equations. As we can observe from Formula 16, even the steady solution of Equation 14 is a nonlinear function of the model parameters.

For the statistical verification of the model, however, it is not the model differential equation, but its solution which determines the linearity or nonlinearity of the problem involved. From this point of view, the model studied leads to *nonlinear regression* problems. A detailed explanation of the whole statistical scrutinization can be found in Chapter 8.

IV. ADVANCED MODELS AND OPEN PROBLEMS

After reading the preceding paragraphs the reader may have the impression that mathematical modeling is an easy procedure: a skilled researcher, versed in mathematical techniques, can propose the mathematical model of any studied system and develop it to his pleasure. This model, being the faithful mapping of the reality, will display transparently all the system dynamics. Unfortunately, nothing is further from reality than this opinion. In this section I am going to point out some problems of mathematical modeling and break down this comfortable illusion.

Let us recall that the mathematical model should be a tool that helps us represent knowledge extracted from the experimental data about the studied physiological system. It should also help us to understand the principles of organization of the system. If we want the model to represent vast and profound knowledge, we should incorporate into it as many factors as possible. Hence, with increasing detail, complexity also increases. On the other hand, if we want the model as a tool for understanding the principles underlying the system dynamics, we naturally expect it to be transparent, i.e., as simple as possible. How to match these contradictory requirements is the main contemporary problem of mathematical modeling.

In this contribution we have illustrated our reasoning by developing a simple mathematical model of a physiological process — metabolism of ethanol in the pig liver. This model is supposed to fit the experimental data at disposition. We must realize, however, that only very mild criteria have been applied to its testing. Rate of ethanol metabolism depends on many physical, chemical, and biological factors. Only the blood flow through the liver, input and output ethanol concentrations, and the rate law of biochemical ethanol removal were considered. No other factors were included into our model; its information "strength" is, therefore, very low. This is not a fault for a model used for didactic purposes. It is likely, however, that this would not satisfy us when we model real liver processes. We would then have to pass more revolutions around the outer cycle depicted in Figure 1 (proposing new experiments). The model would gain information strength with each revolution, but it would soon become very complex and cumbersome. It is the general feature of contemporary mathematical models that they in majority do not cope successfully with the contradiction between simplicity and representativity of the model.

The cause of this dilemma is embedded in the first step of our ten-item list. Variables and parameters that are chosen for the system description are those that are observable or measurable. Since we observe or measure the physical quantities (chemical quantities such as concentrations are numbered with physical ones in this case), the variables and parameters also have this nature. As there are many physical processes going on within even a very simple physiological system, there are many possible candidates for variables and/or parameters of the model. If we want a simple model, we must disregard many of them; that means we must neglect some elements of the studied system. The model is thus not representative. If we want a representative (i.e., detailed enough) model, we may easily arrive at tens or even hundreds of equations. Although there are numerical methods and computer programs for solving such systems, systems consisting of more than five equations are beyond any possibility of detailed theoretical analysis. Especially if the equations are nonlinear, which is often the case, instead of a simple and transparent model, we obtain a mess of interdependent relations. This representation can successfully simulate the behavior of the studied system, but it will not display its essence.

A natural question is whether a simple mathematical model of a system with such complex behavior as that exhibited by the physiological systems is still conceivable. Until very recently, it was generally believed that simple mathematical models have a solution corresponding only to a simple behavior of the modeled system. It was even taken for granted that it is not possible to construct a deterministic model describing the apparently aperiodic, chaotic behavior that we can often witness when we observe physiological systems in natural conditions. This belief was badly shaken by discovery of simple mathematical models with very complex solutions.[24] It was found that some systems of only three of four differential equations may have solutions that are indistinguishable (without sophisticated analysis) from the realization of a random process.[25] The necessary condition for this is the nonlinearity of the equations involved.

Physiological systems are not only the most complex systems encountered in nature, but also the systems exhibiting the most sophisticated internal organization. It is, therefore, conceivable to suppose that there may exist a few internal *essential variables* that order the dynamics of the whole system, the remaining variables being entirely governed by them. Models formulated in these variables should be simple, but, in spite of this, they should grasp the all important aspects of the dynamic structure of the system. We can hardly suppose, however, that these essential variables will be the measurable variables. It is more probable that these variables, important from the point of view of the control of system dynamics, will be abstract quantities. Their relation to the measurable quantities may be expressed by rather complex formulas. The essential variables should also bear a cumulative character, i.e., they should be only a few, but each should be a (complex) combination of contributions related to various measurable variables.

Models formulated in essential variables would represent an entirely new class of advanced mathematical models. Their construction, however, is the *terra incognita* of mathematical modeling. Since we agree that these variables cannot be measured, model structure cannot be defined from the very beginning according to our heuristic rules. The essential variables must be deduced by the model treatment by sophisticated mathematical methods. The first attempts to develop such methods were already published. Haken's[26] synergetic theory introduces an "order parameter" that has approximately the same meaning as our "essential variable". It was recognized that a control of the dynamics of physiological systems has hierarchical structure. Attention was, therefore, paid to developing models considering this fact.[27] Here also some essential variables emerge. The most advanced models from this point of view are models of metabolism.[11] Such variables as the pool variables or the slow and fast moieties that appear in these models are very close to our foreshadowed "essential variables", too. In spite of these partial successes most of the work in developing mathematical methods for elucidating the essential variables of modeled physiological systems still remains to be done.

V. CONCLUSIONS

Let us briefly summarize what we have learned about the mathematical modeling of physiological processes.

Creation of a mathematical model is a cyclic process which includes several steps. The simple example of ethanol metabolism in the liver served for their illustration. Starting from some basic scheme of the studied system (Figure 2), we have determined system variables (c_o, c_m) and system parameters (c_i, F). Practical reasons of easy calculation led us to interpret these variables and parameters as continuous and deterministic and to neglect their dependence on space coordinates. We applied the balance equation for fluid transport as the fundamental equation of our model. Our hypotheses on the nature of processes taking place within the liver were reflected by the form of a certain term (v_m) in this equation.

Our first concern was to study the simplest models possible. That is why we started with the extreme models. In the first one, we supposed that the rate of ethanol metabolism is constant; in the second one, we supposed that it is proportional to ethanol concentration. Both models have been represented by lineal differential equations that have been easy to solve. Their solution provides for each model the dependence of steady value of internal concentration of ethanol within the liver on the system parameters. Both models gave different results, and what is more important, they did not seem to fit well to experimental data. We have, therefore, returned to our hypotheses again in order to complete the inner cycle displayed in Figure 1.

Changing our hypotheses led to introduction of a nonlinear dependence of ethanol metabolism rate on its concentration. The nonlinear function was chosen in such a way as to contain both the preceding extreme models as special cases. The resulting nonlinear equation was not easy to solve. We were able, however, to determine the dependence of the *steady value* of ethanol internal concentration even for this case. Since this dependence can be expected to fit well with the experimental data, we have returned to the beginning, again completing the outer cycle depicted in Figure 1. In this way we have reached a workable mathematical model of the studied physiological system. This model is satisfactory for our purpose, but only very mild criteria were applied to its verification. I have pointed out that usually we cannot expect such simple models to be representative enough in modeling of real physiological systems. As a rule, we formulate the mathematical model of physiological system or process using the physical quantities as the model variables. In this situation the requirements of simplicity and representativity of the model contradict each other. The way out of the blind alley is possible formulation of the model in essential variables that may internally govern the dynamics of the system. These variables can be found only by sophisticated mathematical analysis of the problem. Development of mathematical methods for this analysis is still in the early stages, but first successes in this field indicate that we can expect further development of mathematical modeling methodology in the near future.

REFERENCES

1. **Lotka, A.,** *Elements of Mathematical Biology,* Dover Publications, New York, 1924.
2. **Volterra, V.,** *Leçons sur la Theorie Mathématique de la Lutte pour la Vie,* Gauthier-Villar, Paris, 1931.
3. **Rashevski, N.,** *Mathematical Biophysics. Physico-Mathematical Foundations of Biology,* Dover Publications, New York, 1938.
4. **Huxley, A. F. and Hodgkin, A. L.,** A quantitative description of membrane current and its application to conduction and excitation in nerve, *J. Physiol.,* 117, 500, 1952.
5. **Hale, J. K.,** *Ordinary Differential Equations,* McGraw-Hill, New York, 1969.
6. **Braun, M.,** *Applied Mathematical Sciences,* Vol. 15, Springer-Verlag, New York, 1983.
7. **Thom, R.,** *Structural Stability and Morphogenesis,* Addison-Wesley, Reading, Mass., 1975.
8. **Iosifescu, M. and Tautu, P.,** *Stochastic Processes and Applications in Biology and Medicine,* Editura Academeici, Bucharest, 1973.
9. **Gardiner, C. W.,** *Handbook of Stochastic Methods for Physics, Chemistry and the Natural Sciences,* Springer-Verlag, Berlin, 1983.
10. **Arnold, L.,** *Stochastic Differential Equations,* John Wiley & Sons, New York, 1974.
11. **Reich, J. G. and Sel'kov, E. E.,** *Energy Metabolism of the Cell. A Theoretical Treatise,* Academic Press, London, 1981.
12. **Okubo, A.,** *Diffusion and Ecological Problems: Mathematical Models,* Springer-Verlag, Berlin, 1980.
13. **Lucas, W. F.,** Life science models, in *Modules in Applied Mathematics,* Vol. 4, Marcus-Roberts, H. and Thompson, M., Eds., Springer-Verlag, New York, 1983.
14. **Guyton, A. C. and Coleman, T. G.,** Long-term regulation of the circulation. Interrelationships with body fluid volumes, in *Physical Bases of Circulatory Transport: Regulation and Exchange,* Reeves, E. B. and Guyton, A. C., Eds., W. B. Saunders, Philadelphia, 1967.
15. **Keleti, T. and Lakatos, S., Eds.,** *Mathematical Models of Metabolic Regulations,* Akademiai Kiadó, Budapest, 1976.
16. **Liberstein, H. M.,** *Mathematical Physiology. Blood Flow and Electrically Active Cells,* Elsevier, New York, 1973.
17. **Eason, G., Coles, C. W., and Gettinby, G.,** *Mathematics and Statistics for the Biosciences,* Ellis Hornwood, Chichester, 1980.
18. **Dudley, A. C. B.,** *Mathematical and Biological Interrelations,* John Wiley & Sons, Chichester, 1977.
19. **Batschelet, E.,** *Introduction to Mathematics for Life Scientists,* Springer-Verlag, Berlin, 1975.
20. **Havránek, T.,** Statistical evaluation of mathematical models of physiological processes, *Methods in Animal Physiology,* Deyl, Z. and Zicha, J., Eds., CRC Press, Boca Raton, 1988, chap. 8.
21. **Keiding, S., Johansen, S., Winckler, K., Tonnensen, K., and Tygstrup, N.,** Michaelis-Menten kinetics of galactose elimination by the isolated perfused pig liver, *Am. J. Physiol.,* 220, 1302, 1976.

22. **Keiding, S., Johansen, S., Midtboll, I., Rabol, A., and Christiansen, L.,** Ethanol elimination kinetics in human liver and pig liver in vivo, *Am. J. Physiol.,* 237, E316, 1979.
23. **Johansen, S.,** *Lecture Notes in Statistics,* Vol. 22, Springer-Verlag, New York, 1984.
24. **May, R.,** Simple mathematical models with very complicated dynamics, *Nature,* 261, 459, 1976.
25. **Schuster, H. G.,** *Deterministic Chaos. An Introduction,* Physic-Verlag, Weinheim, 1984.
26. **Haken, H.,** *Advanced Synergetics,* Springer-Verlag, Berlin, 1983.
27. **Nicolis, J. S.,** *Dynamics of Hierarchical Systems,* Springer-Verlag, Berlin, 1986.

Chapter 8

STATISTICAL EVALUATION OF MATHEMATICAL MODELS OF PHYSIOLOGICAL PROCESSES

T. Havránek

TABLE OF CONTENTS

I. INTRODUCTION

Statistical evaluation of mathematical models, though only one of the steps in the model building process,[1] is of vital importance as long as it is the step from speculation to critical science. Clearly, creative speculation and critical evaluation are two necessary parts not only of any model formulation process but of the development of scientific ideas in general.

Not every mathematical model can be evaluated statistically. A necessary condition is the possibility of transforming the model into an expression linking some phenomena that can be measured or observed in an experiment or in an observational study. We can say that for animal physiology experimental work is most typical. In the simplest case we are measuring one or more *design variables*, fixed by experimental conditions and usually considered to be nonrandom, and one *response variable*, the values of which consist of a systematic part influenced through the model by design variables and a random part expressing some disturbances and/or inherent biological variability. Even in experimental work, measured variables have to sometimes be considered that are not controlled by the design, but the influence of which is taken into account by the model. These variables can be called *noncontrolled explanatory variables*, and they are typical of observational studies that form a necessary tool in human physiology, epidemiology, or evaluation of growth models. Both kinds of nonresponse variables can be called *explanatory variables* since we expect that they can explain the obtained response after removing the random disturbances. Two facts are to be stressed: first, which variable is to be considered a response variable is a matter of choice depending, for example, on the possibility of controlling some variables or on the decision of which phenomenon represents the response in the time. Second, frequently the fair method is to consider not only one but more response variables. In statistics such cases are described as multiresponse models. For the sake of simplicity we restrict ourselves in this chapter to models with a single response variable, since the analysis of multiresponse models is from the formal and computational point of view much more complex. Despite this, applications of multiresponse models are in many real situations highly desirable, and they have become more common in recent years.

Let us mention that statistical terminology differs from the usual model building terminology,[1] where a response variable is called a system variable while other variables are called system parameters. In statistics, however, by parameters we mean unknown constants of the model, the values of which are to be estimated on the basis of a given data set.

We shall denote values of the response variable by y, and we shall consider a situation in which this response can be affected by a number of design or explanatory variables. If we have r such variables, their values will be denoted by $x_1, \ldots, x_r$. As a term covering all the above-mentioned cases, we shall use the neutral and, in statistics, common term *regressor variables*. This term seems to be better than the older though more frequently used term "independent variables", as the latter suggests some sort of interpretation that could be in many instances inappropriate.

Now we can write that our considered models will be of the form

$$y = \mu + f(x_1, \ldots, x_r; \ \beta_1, \ldots, \beta_p) + e \qquad (1)$$

where f is a function derived generally from some *a priori* considerations and usually fixed at the moment of the statistical evaluation of the model. By the Greek letters $\mu, \beta_1, \ldots, \beta_p$ we denote unknown parameters of the model. These parameters are unknown in the sense that we do not know *a priori* their numerical value. In the simplest case, regressor variables are considered to be nonrandom, fixed, for example, by experimental conditions. On the other hand, the error variable e is supposed to be a random variable having a zero mean, expectancy, $Ee = 0$. Moreover, its variance $\sigma^2 = VAR\ e$ is considered to be a finite but unknown number independent of the values of the regressors. Since e is a random error or disturbance, the response variable is a random variable having a "systematic" part depending on the regressors: $\mu + f(x_1, \ldots, x_r; \beta_1, \ldots, \beta_p)$ and a random part given by e. The systematic part, i.e., the expectancy of y for given values of regressors $x_1, \ldots, x_r$, consists of two members. The first member is the parameter μ called the intercept, and it is independent of $x_1, \ldots, x_r$. This parameter can be interpreted as a baseline value of the response. Clearly, in some cases it can be *a priori* equal to zero.

Instead of Equation 1, we can use a more concise notation and write shortly

$$y = \mu + f(\underline{x}; \ \underline{\beta}) + e$$

where $\underline{x}$ stands for $x_1, \ldots, x_r$ and $\underline{\beta}$ stands for $\beta_1, \ldots, \beta_p$.

Now it is necessary to stress that as a result of an experiment or observation, we obtain not only one value of the response variable y under a single set of regressor values $x_1, \ldots, x_r$, but a list of response values $y_1, \ldots, y_n$ under a corresponding sequence of regressor values $\underline{x}_1, \ldots, \underline{x}_n$, where $\underline{x}_i$ stands for $x_{i1}, \ldots, x_{ir}$, i.e., for a set (vector) of regressor values under which the ith response y_i was observed.

In the usual tabular form we can write here

$$
\begin{array}{c|ccc}
y_1 & x_{11}, & \ldots, & x_{1r} \\
\vdots & \vdots & & \vdots \\
y_i & x_{i1}, & \ldots, & x_{ir} \\
\vdots & \vdots & & \vdots \\
y_n & x_{n1}, & \ldots, & x_{nr}
\end{array}
\tag{2}
$$

where we have in one row the response value and regressor values under which the response was obtained.

A particular case is the situation in which the response is independent of the regressors. In such a case, the model is

$$y = \mu + e$$

where μ is the mean (or expectancy) Ey of the response. This model is sometimes called the *null model*. The mean μ is unknown and should be estimated from the observed values $y_1, \ldots, y_n$. In regular cases, this mean is estimated by the average or sample mean

$$\overline{y} = \frac{1}{n} (y_1 + y_2 + \ldots + y_n)$$

or in short $\overline{y} = \frac{1}{n} \sum_{i=1}^{n} y_i$ where we use the concise notation for the sum of values $y_1, \ldots,$ y_n, namely, $\sum_{i=1}^{n} y_i$ for $y_1 + \ldots + y_n$. The average $\overline{y}$ is an estimate for the unknown parameter μ; this fact is expressed in statistical notation by writing $\hat{\mu} = \overline{y}$. By $\hat{\mu}$ we denote an estimate of the parameter μ.

Now, we can estimate σ^2 by the sample variance

$$\hat{\sigma}^2 = \frac{1}{n-1} \sum_{i=1}^{n} (y_i - \overline{y})^2$$

or

$$\hat{\sigma}^2 = \frac{1}{n - 1} \sum_{i=1}^{n} (y_i - \hat{\mu})^2 \tag{3}$$

In the denominator we have the number of observations minus one, as one parameter, namely μ, was estimated. This variance can be interpreted as an estimate of the variance of the response not explained by the simple model, i.e., variance of the response around the mean value μ. In fact, $\overline{y}$ is just the value for μ minimizing Equation 3. If we now view this problem from another side, we can, for example, *a priori* suppose that the mean has a given value μ_0 and then compute a variance with respect to this value, i.e., with respect to this *restricted model*. This variance is $\hat{\sigma}_0^2 = \frac{1}{n} \sum_{i=1}^{n} (y_i - \mu_0)^2$. Now, somebody else can suggest another value for μ, say μ_1. The variance will be $\hat{\sigma}_1^2 = \frac{1}{n} \sum_{i=1}^{n} (y_i - \mu_2)^2$. Perhaps all of us will agree that if $\hat{\sigma}_1^2$ is considerably greater than $\hat{\sigma}_0^2$, we should prefer the model given by μ_0.

The reasoning in the general case is similar. As is usually the case of the null model, the values of parameters $\mu, \beta_1, \ldots, \beta_p$ are unknown and are to be estimated using some data of the form in Equation 2. Given these estimates, say $\underline{\hat{\beta}} = (\hat{\beta}_1, \ldots, \hat{\beta}_p)$ and $\hat{\mu}$, we can compute an estimate for σ^2, say

$$\hat{\sigma}^2 = \frac{1}{n - p - 1} \sum_{i=1}^{n} \{y_i - [\mu + f(\underline{x}_i; \underline{\hat{\beta}})]\}^2 \tag{4}$$

called *residual variance*, i.e., variance not explained by Equation 1 defined by the model function f. A very important question is whether, on the basis of the obtained value $\hat{\sigma}^2$, the model should be ruled out, i.e., whether the reduction of the variance of the response due the model is small. A similar question is if we have in the simplest case two models of the form in Equation 1 defined by two different functions f_1 and f_2 with estimated residual variances $\hat{\sigma}_1^2$ and $\hat{\sigma}_2^2$, which of these models is better, being closer to reality.

Note that in this general case we are estimating $p + 1$ parameters, hence in the denominator of the expression for residual variance, (Equation 4), we have $n - (p + 1) = n - p - 1$ instead of $n - 1$ in Equation 3.

A. Linear and Nonlinear Models

For technical reasons associated with the important task of estimating parameters, it is necessary to distinguish some particular classes of models, namely linear and nonlinear models. Note that in the statistical context these two terms have a different meaning from the meaning usual in model building[1] with differential equations.

Here we simply say that a model like Equation 1 is linear if it can be rewritten into the form

$$y = \mu + \beta_1 f_1(x_1, \ldots, x_r) + \ldots + \beta_p f_p(x_1, \ldots, x_r) + e \tag{5}$$

where functions $f_1, \ldots, f_p$ are considered to be known; unknown parameters are outside the functions. The simplest case is the simple multivariate linear regression model

$$y = \mu + \beta_1 x_1 + \ldots + \beta_r x_r + e$$

where the number of parameters is $r + 1$, i.e., it equals the number of regressors increased by the intercept term μ.

Some models built in this volume[1] for ethanol metabolism in liver are, in our present sense, linear. By statistical analysis we can evaluate, for example, two models linking the steady value of ethanol concentration in blood c as a response with the input concentration c_i and the blood flow F.

Under the first model (Equation 1^1), we can write

$$c = F c_i/(F + k_m) + e$$

with regressors c_i and F. Here k_m is the proportionality constant (parameter) in the model of proportional metabolic rate.[1] This model is linear if we are using the constant flow, i.e., if we put F = const. Then

$$c = [F/(F + k_m)] c_i + e$$

i.e., the response is y = c, parameter $\beta_1 = F/(F + k_m)$, and the regressor $x_1 = c_i$. Clearly, here is a restriction to the intercept μ that should equal 0 ($\mu = 0$).

The second model[1] (Equation 2), namely,

$$c = c_i - V_{max}/F + e$$

where V_{max} represents the maximal rate of metabolizing capacity of the liver, is linear even without the restriction concerning F. Here we have two regressors, c_i and $f_2(F) = 1/F$. Generally, the model has the form

$$c = \mu + \beta_1 c_i + \beta_2 f_2(F) + e$$

For physiological reasons values of parameters μ and β_1 are restricted to 0 and 1, respectively (and β_2 should be negative; these restrictions are due to the assumption of the constant metabolic rate). Under constant flow we obtain the model

$$c = \mu + \beta_1 c_i + e$$

where $\mu = V_{max}/F$ and β_1 is constrained to $\beta_1 = 1$.

The slightly generalized model with variable flow F, namely,

$$c = c_i - (V_{max}/k_m F)(F - k_m) + e$$

is again a linear model:

$$c = c_i - V_{max}/k_m - V_{max}/F + e$$

Here the constraint $\mu = 0$ is not considered; $\mu = -V_{max}/k_m$ could be nonzero (but negative).

A considerably greater effort is needed for statistical analysis of models that are not linear. A nonlinear model is simply every model that cannot be written in the form of Equation 5. Nonlinear models are not preferred by statisticians, but nevertheless they are, in real situations, quite frequent. Even in our simple example with ethanol metabolism[1] we have two such models that principally can be statistically evaluated.

For variable blood flow F, the first model (Equation 1), namely,

$$c = Fc_i/(F + k_m) + e$$

is nonlinear. It has the form

$$c = \mu + f(c_i, F; \beta_1) + e$$

with the constraint $\mu = 0$ and with $\beta_1 = k_m$ ($x_1 = c_i$, $x_2 = F$).

Similarly, the third model[1] (Equation 3):

$$c = \frac{1}{2}(c_i - V_{max}/K_m - V_{max}/F) - \frac{1}{2}\sqrt{[(c_iV_{max}/k_mV_{max}/F)^2 + 4V_{max}c_i/k_m]} + e$$

is not linear. It has the form

$$c = \mu + f_1(c_i, F;\ \beta_1) + f_2(c_i, F;\ \beta_1, \beta_2) + e$$

with $\beta_1 = v_{max}/2$, $\mu = -V_{max}/2k_m$, and $\beta_2 = V_{max}/k_m$. Here is clearly a constraint $\mu = (-1/2)\beta_2$.

As a compromise between linear and nonlinear models, frequently generalized linear models are used.[3] Such models are of the form in Equation 5 where instead of y on the left side we use a function of y, say g(y). This function is called a link function. The model has then the form

$$g(y) = \mu + \beta_1 f_1(\underline{x}) + \ldots + \beta_p f_p(\underline{x}) + e \tag{6}$$

B. Parameter Estimation

As we mentioned above, under the null model $y = \mu + e$, the estimate $\hat{\mu} = \bar{y}$ can be viewed as a result of minimalization of the sum of squared deviations of observed values $y_1, \ldots, y_n$ from the unknown value μ:

$$\sum_{i=1}^{n} (y_i - \mu)^2$$

This expression is minimized for given values $y_1, \ldots, y_n$ with respect to the unknown value μ, and in such a way that the estimate $\hat{\mu} = \bar{y}$ is obtained.

The same reasoning is used in the general case, where for given data of the form in Equation 2, the unknown parameters $\mu, \beta_1, \ldots, \beta_p$ are estimated by minimizing the sum of squares

$$\sum_{i=1}^{n} \{y_i - [\mu + f(\underline{x}_i;\ \underline{\beta})]\}^2 \tag{7}$$

In fact, in this way we try to minimize simultaneously the squared distance of the response values from their systematic part given by the model.

If the model is linear, then the estimation of $\mu_i, \beta_1, \ldots, \beta_p$ is usually tractable. The estimation task is then equivalent to solution of a system of linear equations for unknown $\mu, \beta_1, \ldots, \beta_p$.[4] There are some conditions for transformed values of regressors, i.e., values $f_1(x_1, \ldots, x_r), \ldots, f_p(x_1, \ldots, x_r)$, and for parameters $\beta_1, \ldots, \beta_p$ under which these equations for given data can be solved; in such a case parameters are estimable.

In the case of simple multivariate linear regression, it is necessary that $n > r + 1$, and a linear relation between regressors must not occur. It means that, for example, the case $x_1 = ax_2 + bx_3$ should be excluded. For numerical reasons even if x_{i1} equals approximately $ax_{i2} + bx_{i3}$, for each value $x_{11}, \ldots, x_{n1}$, the estimates of $\beta_1, \ldots, \beta_p$ can be highly unstable. They can dramatically change under a very small shift, e.g., in a value of only one of the response variables. This is the problem of co-linearity of regressors, and this problem has to be taken into account in evaluating models. What ''approximately'' means above is a matter of discussion — see the notion of tolerance used in appropriate computer programs.[5] Tolerance is $1 - R_j^2$, where R_j^2 is a squared multiple correlation coefficient of the jth regressor explained by a linear relation to other regressors, we see that the tolerance $1 - R_j^2$ is just the proportion of variance not explained by this relation. If the tolerance is less than, say, 0.01, the regressor is not used in computations because of its close linear relation to other regressors.

In a nonlinear case, an iterative procedure for estimating the parameters[4] must be used. In such a procedure, derivatives of the function f are generally used, submitted analytically by the user or computed numerically by a computer program. In any case, the procedure has to start with some given starting estimates of the parameters, and the success of it in a particular task depends on the quality, say fitness, of these starting points. A procedure can sometimes find completely false values of the parameters. In a simple case of a model

$$y = \mu + f_1(x_1; \ \beta_1, ..., \beta_p) + e \tag{8}$$

with one regressor, such case can be identified by a scatter plot (see Figure 1) containing the estimated values of the response, i.e., values

$$\hat{y}_i = \hat{\mu} + f_1(x_{i1}; \ \hat{\beta}_1, ..., \hat{\beta}_p) \tag{9}$$

In such a case the wrong solution obtained by a procedure can be easily identified; in a multivariate case the situation is much more complicated. Estimation of the parameters in a nonlinear multivariate case is a complicated task even for an experienced statistician using good software.

There are further complications in the task of parameter estimation which have, moreover, important effects for the testing process considered later: usually in real models there are some constraints to parameters, e.g., $\mu = 0$ or $\beta_1 = 1$ in our examples. Moreover, in many cases from a physical or other background concerning the models we can see that there are some additional constraints in the form of inequalities, e.g., $V_{max}/k_m > 0$ in Equation 3 considered above.

Then the task of minimizing the sum of squares

$$\sum_{i=1}^{n} \{y_i - [\mu + f(\underline{x}; \ \underline{\beta})]\}^2 \tag{10}$$

is not an unconstrainted task, but a task of constrained minimization which can be considerably more complicated. The effective number of parameters is then usually affected too.

Additionally, we have to mention that in some situations it is more appropriate to use another minimization criterion, e.g., to use absolute values instead of squares in Equation 7, i.e., to minimize

$$\sum_{i=1}^{n} |y_i - [\mu + f(\underline{x}_i; \ \underline{\beta})]| \tag{11}$$

The reason is that in this case the influence of values y_i "far" from $\mu + f(\underline{x}; \ \underline{\beta})$ is suppressed.

C. An Outline of Further Analysis

If the model parameters are estimated we are only at the beginning of the way to a critical evaluation of the model. We can say that the model is identified (in the words of Section VI from Chapter 7).[1] Now the model is to be evaluated; this step can be again placed in the context of the model building process (see Section VIII in Chapter 7).

First, the model is to be evaluated without questioning the model function f or the model

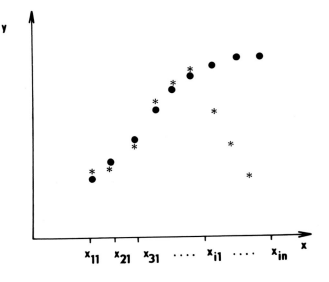

FIGURE 1.   •, with coordinates x_{i1}, y_i = estimated response, and *, with coordinates x_{i1}, y_i = observed response.

functions f_1, . . . , f_p in the case of a linear model. This problem reduces, in fact, to two parts, first being the *evaluation of parameters* and second the *evaluation of residuals*. Similar to the case of sample mean $\bar{y}$ for the null model where $\bar{y}$ is endowed by its standard error SEM = $\hat{\sigma}/;\sqrt{n}$, in the general case each estimate of a parameter is endowed by its estimated standard error. These standard errors depend clearly on the number of measured response values (on the number of observations) n. By standard error we can evaluate the quality of estimates, we can construct confidence intervals for them, or we can test if some parameters equal zero and thus can be excluded from the model. More generally, we can test whether a parameter equals to a given value. This is important in the case where there are, for some parameters, available *a priori* values obtained from some considerations during the model building process.[1]

For example, in the general form of Equation 2 above

$$c = \mu + \beta_1 c_i + \beta_2 f_2(F) + e$$

we can test whether really $\mu = 0$ and $\beta_1 = 1$ in accordance with our *a priori* belief concerning the model.

Estimated parameters are endowed not only by standard errors, but they are in general correlated. Again, this correlation structure is to be taken into account in evaluating the model, particularly the number of necessary parameters in the model.

These questions are described further in Section II.A.

The second part fo the problem is *evaluation of residuals*. If a model of the form y = $\mu + f(\underline{x}; \underline{\beta}) + e$ is evaluated, we compute the estimates $\hat{\mu}$, $\hat{\beta}_1$, . . . , $\hat{\beta}_p$. Now we can, for each response value, compute its contribution to the final sum of squares in Equation 7:

$$y_i - [\hat{\mu} + f(\underline{x}; \underline{\hat{\beta}})] \tag{12}$$

This quantity is called the ith residual. Inspection of residuals and other similar quantities can help us find some response values (or regressor values) that are outliers. Such values can be true values that are just influential, but on the other hand, they can be results of an error in measurement or of a shift in experimental conditions, etc. In any case, they can distort the estimates of parameters, their errors and consequently, the whole pattern of the model. The model can be then incorrectly evaluated (Section II.B is devoted to residuals and similar quantities).

The *quality of estimates* and hence of the whole evaluation of the model can be influenced heavily by the design of the experiment. Particularly here is a direct link between standard errors and the design of the experiment in the sense of planned values of regressors under which the response is measured. Repeated measurements of the response under each fixed regressor value are highly desirable. This topic is dealt in more detail in Section II.C.

Second, the model is to be evaluated with respect to the choice of the model function. As we mentioned above, the residual variance (Equation 4) plays a substantial role as a measure of goodness-of-fit of the model with observed data in this context. Questions regarding such global measures for evaluating the fit of a model with data sets will be discussed in detail in Section III.A. As we can note, the residual variance depends on the number of parameters involved; namely, its numerator,

$$RSS = \sum_{i=1}^{n} \{y_i - [\hat{\mu} + f(\underline{x}; \ \hat{\beta})]\}^2 \tag{13}$$

called the residual sum for squares (RSS), tends to be smaller for models with a greater number of estimated parameters (another question is those parameters, the values of which were obtained *a priori* due to some nonstatistical considerations in the model building process). This dependency should be taken into account if we are *comparing two or more possible models with respect to their performance for experimental data.*

In this stage of the statistical analysis it is reasonable to distinguish two cases, the first of them being considerably less complex than the second one. The first case is the case of *nested models* (Section III.B). Here one model can be considered as a specialization of the second one. A simple example is the linear regression, where $y = \mu + \beta x + e$, and the quadratic regression with $y = \mu + \beta_1 + \beta_2 x^2 + e$. If $\beta_2 = 0$, the latter specializes to the former. This case can be treated, for example, by testing just the null hypothesis that $\beta_2 = 0$.

A task that cannot be solved easily is *comparison of nonnested models* (Section III.C). Nonnested models are, for example, the quadratic regression model $y = \mu + \beta_1 x + \beta_2 x^2 + e$ and the trigonometric model $y = \mu + \beta_1 x + \beta_3 \sin x + e$. Clearly, to compare these models seems to be reasonable only in such a domain of the regressor values where the shapes of x^2 and $\sin x$ are similar. This particular case can be solved to a certain extent by introducing the third model, $y = \mu + \beta_1 x + \beta_2 x^2 + \beta_3 \sin x + e$, and testing two hypotheses in its frame. The first hypothesis, $\beta_3 = 0$, corresponds to the specialization to the first quadratic model; the second one, $\beta_2 = 0$, corresponds to the specialization to the second trigonometric model. For general nonlinear models, this task, even if frequent in practice, is usually solved inadequately since appropriate methods were developed only recently, and they are not covered by standard statistical software.[5]

Third, the model is not specified fully, but only a class of possible models is given at a step of the model building process. The task is, given a data set, to realize within this class a search for such models that give, in some sense, adequate explanation of the data (Section IV.A). Since the number of these models can be quite large, some further analysis is needed. In their context we can seek further for the simplest adequate models and/or models that are the best in some precisely defined sense. Additionally, some *a priori* knowledge concerning admissibility or nonadmissibility of some models from extramathematical or extrastatistical reasons should be included in the searching process. Sections IV.B and C are devoted to this third topic. Here, two basic techniques are presented, namely, stepwise searches and all-model searches.

If a mathematical model of a physiological process is even moderately complex, the task of its evaluation, including appropriate experiment design, becomes rather complicated to be treated by the physiologist alone. *Cooperation with an experienced statistician is unavoidable as well as the possibility of using flexible and carefully tested software.* In the present chapter some basic ideas are briefly explained, some of them relatively uncommon among nonstatisticians and some even among statisticians; a great deal of the references will be useful mainly for a consulting statistician, some of them being too complex to be used by the physiologist alone. At the end of this introduction we have to make important notice concerning statistical tests as the most commonly used statistical tools in physiology. In the present context these tests are used, but their role should not be overemphasized;[2] they are only auxiliary steps in the whole task of evaluating a mathematical model.

II. EVALUATION OF PARAMETERS

A. Standard Errors of Estimated Parameters

Like the mean, each estimate of a parameter is endowed by its standard error, say s_{β_i}. These standard errors are computed exactly in the case of a linear regression and approximately in the case of a nonlinear regression. If we assume that the disturbance e is normally distributed, we can use as a first approximation to the 95% confidence interval for the parameter β_i the interval

$$(\hat{\beta}_i - 1.96 s_{\beta_i}; \quad \hat{\beta}_i + 1.96 s_{\beta_i})$$

where 1.96 is the 0.975 quantile of the standardized normal distribution. If such an interval contains 0, we cannot reject the null hypothesis that $\beta_i = 0$. More exactly, under the normality assumption concerning the distribution of disturbances in Equation 1, quantiles of the Student's t distribution with $n - p - 1$ degrees of freedom are to be used.

In a good software,[5] covariances or correlations between estimates of the parameters β_1, . . . , β_p are computed. The role of these covariances or correlations is twofold. First, they are to be used in the task of establishing a confidence interval for a further response given some values of regressors;[6] second, high correlation between estimates can indicate an overparametrization of the model. In such a case, the model is too complex for the given data.

B. Residuals

For each value of the regressors $\underline{x}_i = (x_{i1}, \ldots, x_{ir})$ we can use estimated values of the parameters to estimate the value of the response using the following expression:

$$\hat{y}_i = \hat{\mu} + f(\underline{x}_i; \quad \hat{\beta}_1, \ldots, \hat{\beta}_p) \tag{14}$$

The difference between the observed value of the response y_i and the estimated value $\hat{y}_i$, namely,

$$y_i - \hat{y}_i \tag{15}$$

is called a *residual*. A summary statistic based on residuals is the RSS:

$$RSS = \sum_{i=1}^{n} (y_i - \hat{y}_i)^2 \tag{16}$$

which is equivalent to Equation 13. The residual variance is then

$$s_R^2 = RSS/(n - p - 1) \tag{17}$$

This value is an estimate of the variance of responses *not explained* by the model; it corresponds to the variance of the "noise" considered in Equation 1 (c.f. also Equation 4). Sometimes we speak about inherent variance of the phenomenon observed.

Residuals are important for model evaluation. If we plot residuals or squared residuals with respect to the response values, either observed or estimated, we can notice whether there is a trend in residuals, indicating, for example, that for large values of the response, the model works worse than for small ones (Figure 2A); sometimes the pattern can be more complicated (Figure 2B). If one of the regressor variables is a time variable, then usually it is very valuable to plot squared residuals against this variable (Figure 3). In this figure a "periodic" unexplained variance of the observed phenomenon can be identified.

The plots considered above form a rudimentary phase of the regression diagnostic.[7,8] Residuals can be used to detect individual response values that are outliers with respect to the model, i.e., such values for which the model is not fitting. To make such an analysis, residuals must be standardized to be mutually comparable. Standardized residuals are defined as

$$r_i = \frac{y_i - \hat{y}_i}{s_R \sqrt{(1 - v_i)}} \tag{18}$$

where $s_R = +\sqrt{s_R^2}$. We can see that residual standard deviation s_R is augmented in the numerator of Equation 18 by a factor $\sqrt{(1 - v_i)}$. This factor characterizes the position of the vector of regressor values of the ith case, say $\underline{x}_i = (x_{i1}, \ldots, x_{ir})$, between other vectors of regressor values $\underline{x}_1, \ldots, \underline{x}_{i-1}, \underline{x}_{i+1}, \ldots, \underline{x}_n$. The value v_i used in the above-mentioned factor lies between 0 and 1, and it is close to 1, for example, if $\underline{x}_i$ is well removed from the bulk of cases. Sometimes $1/v_i$ is called an effective number of cases defining y_i.

To be exact, let us say that v_i is a diagonal element of the matrix $X(X'X)^{-1}X'$ where X is the regressor part of Equation 2.

It is strongly recommended to use standardized residuals instead of ordinal residuals in plots since they are quite mutually comparable. If the value of such a residual is considerably greater than others, we can identify easily an outlier response. If normality of the "noise" in Equation 1 is assumed, then a test procedure for standardized residuals detecting outliers is provided[8] ($r_i^2/(n - p)$ following a β-distribution with mean $Er_i = 0$ and $VAR\ r_i = 1$).

Other forms of residuals called predictive residuals[8] are used to detect the ability to predict the ith response y_i on the basis of the information contained in the response values and the corresponding regressor values.

Also very important are additional measures, e.g., the Cook distance

$$D_i = r_i^2 v_i / [p(1 - v_i)] \tag{19}$$

measuring the influence of the position of $\underline{x}_i$ between regressor values $\underline{x}_i, \ldots, \underline{x}_n$ to the regression in question, i.e., to the model fit. Again, plots of values of such a measure against response values or time variables can give important information.

C. Design of Experiments for Reduction of Error of Estimated Parameters

For evaluation of models and/or parameters as well as for using models for prediction, it is very important to minimize variances (errors) of parameters. For example, if we consider the response under a simple univariate linear model, i.e., $y = \mu + \beta x + e$, and if we can

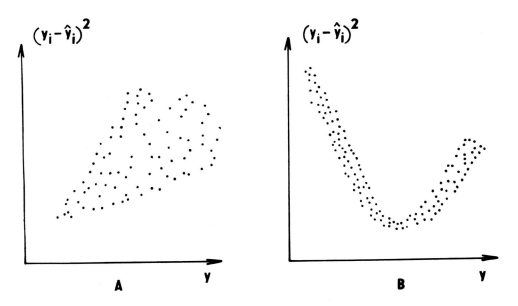

FIGURE 2.   Pattern of residuals. A = inhomogenous variance, B = lack of fit.

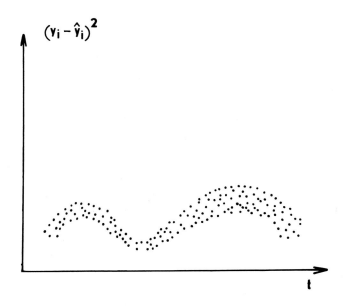

FIGURE 3.   Periodic unexplained variability.

make 2n measurements of the regressor in the interval $<x_{min}, x_{max}>$, then it is best to design the experiment in such a way that n measurements are done for the value x_{min} and n measurements are done for the value x_{max}. Such a design depends substantially on our ability to control values of the regressor and on our assumption that the model is linear, but it minimizes the standard error of estimates. Under such design, a "nonlinearity" of the form $y = \mu + \beta_1 x + \beta_2 x^2 + e$ cannot be detected.

We can see that there are two tasks: first, to design an experiment to reduce the variance of estimated parameters, and, second, to design an experiment for detecting differences between models; see the example above. For this particular task of comparing $\mu + \beta x$ vs. $\mu + \beta_1 x + \beta_2 x^2$, it mean using, roughly speaking, a three-point design with measurements of the response concentrated into three points of regressor values: x_{min}, $(x_{max} + x_{min})/2$, and x_{max}. The first task is important if we are convinced about a particular model and we are trying to estimate parameters (e.g., for prediction purposes) as well as possible. For simple linear models, such a design is tractable to suggest; for more complicated models, this task becomes very complicated and in many cases intractable. Usually a preliminary estimate of the residual or inherent variance is necessary, particularly for the second task. Nevertheless, in experimental practice one has to be aware of this task and its usefulness.

III. COMPARISON OF MODELS

A. Goodness-of-Fit Criteria

To compare models, we have to consider some goodness-of-fit criteria for evaluating the fit of a model with observed data. A general theory for construction of such criteria is the likelihood theory[3] or the information theory.[9] The basic notion is the *deviance* of a model from the given data set. Decreasing deviance means better goodness-of-fit. Under the likelihood assuming the normality of the disturbances in Equation 1, the deviance becomes just the RSS (Equation 16). Such a deviance depends only on the choice of the function f defining the model and the data observed.

Under the null model, RSS comes to be

$$RSS_o = \sum_{i=1}^{n} (y_i - \bar{y})^2 \tag{20}$$

i.e., the sum of squares of differences between response values $y_1, \ldots, y_n$ and the mean response $\bar{y}$. This sum of squares is related to the (unrestricted) variance of the response under the null model

$$s^2 = RSS_o/(n - 1) \tag{21}$$

For testing the hypothesis that the model does not contribute to explaining the response at all, i.e., $\beta_1 = \ldots = \beta_p = 0$, the test statistic

$$F = \frac{(RSS_o - RSS)/p}{s_R^2} \tag{22}$$

can be used under normality assumptions. If, for a given significance level, say $\alpha = 0.05$, F exceeds the $1 - \alpha$ quantile of the Fisher distribution with p and $n - p - 1$ degrees of freedom, we can conclude that the model significantly contributes to the explanation of the response. For linear models this test is an exact test; for a nonlinear case it must be considered an approximation.

RSS as well as residual variances are not standardized: they depend on the number of parameters involved in the considered model. Hence, they can hardly be used, without further assumptions, to compare various models and their fit, possibly on different data sets.

To avoid this inconvenience, some suggestions for another standardized measure of goodness-of-fit were presented in the statistical literature.[9] In these measures RSS and other quantities derived from them are augmented by the number of parameters involved. If RSS_m is the RSS of a model m and p_m the number of parameters of this model, then such a standardized goodness-of-fit measure can be expressed as

$$\frac{n}{2} \log \sigma^2 + \frac{RSS_m}{2\sigma^2} + \frac{d}{2} p_m + \text{const} \tag{23}$$

where d is a constant and σ^2 is the inherent variance of the studied response not explainable by any model.

If somebody estimates, for a given data set, this inherent variance for all models in question by a value $\hat{\sigma}^2$, then the general form becomes equivalent to the generalized Akaike information criterion

$$RSS_m + dp_m \hat{\sigma}^2 \tag{24}$$

Usually $d = 2$ is used, and then the criterion becomes $RSS_m + 2p_m \hat{\sigma}^2$, which is equivalent to

$$RSS_m/\hat{\sigma}^2 + 2p_m \tag{25}$$

In this last form we can see that goodness-of-fit relative to estimated inherent variance is penalized by twice the number of parameters. As adding, for example, one parameter surely improves the goodness-of-fit, this "automatic" improvement is penalized here by the factor two.

Using the idea that models are to be compared with respect to their predictive performance and that this performance increases with the number of cases (observations, response values), Mallows suggested another criterion that becomes[9]

$$RSS_m/\hat{\sigma}^2 + 2p_m - n \tag{26}$$

Here the goodness-of-fit is improved by the number of cases; this is useful if we compare models evaluated on different data sets. Clearly, for a fixed number of observations n, this criterion is equivalent to Equation 25. An open question is, for both criteria in Equation 24 and 26, the way of obtaining an estimate $\hat{\sigma}^2$ of the inherent variance. One way is to consider a general model in which all discussed models are nested (see later); such a general model is the most complex model under consideration for the phenomenon in question. Then σ^2 is estimated from the RSS_{m_g} of this general model m_g divided by $n - p_{m_g} - 1$. In some cases this general model can be easily found, or, in other words, there is agreement which model should be used as such a general model. If we are evaluating simple linear regression models with at most r regressors (see Equation 4), then the model

$$y = \mu + \beta_1 x_1 + \ldots + \beta_r x_r + e \tag{27}$$

is such a general model. Other particular models to be evaluated are obtained by excluding some of the regressors, i.e., some β-parameters are set equal to zero.

The usual assumption is that subsequent responses $y_1, \ldots, y_n$ are stochastically independent, that they are observations of a stochastically independent finite sequence of random variables. If now we suppose, as usual, the distributional homogeneity of disturbances (implying the zero mean and fixed variance), a time dependency can be involved only in the systematic part of the model $(\mu + f(\underline{x}; \underline{\beta})$. Time is then one of the regressors. In many situations, particularly in EEG analysis or other neurophysiological processes, this assumption is nonrealistic, and then another mathematical framework should be used, namely time series analysis. One of the simplest cases is an autocorrelation process where $y_t = \alpha_1 y_{t-1} + \ldots + \alpha_k y_{t-k} + e_t$. Here, a response in the time t depends on the k previous responses. Problems of model choice and evaluation of the goodness-of-fit for such models are closely related to the linear regression case discussed above, but the whole task is more complicated.[10]

B. Nested Models

If we have two models, say m_1 and m_2, such that m_1 can be obtained from m_2 by setting some parameters equal to zero, we say that m_1 is *nested* in m_2. For example, the model[1] m_1: $c = c_i - V_{max}/F + e$ is nested in m_2: $c = c_i - V_{max}/k_m - V_{max}/F + e$. Similarly, the simple linear model

$$m_1 : y = \mu + \beta_1 x_1 + e \tag{28}$$

is nested in the cubic model

$$m_2 : y = \mu + \beta_1 x_1 + \beta_2 x^2 + \beta_3 x^3 + e \tag{29}$$

All models just mentioned are multivariate linear models — linear in the parameters. There can be a linear model nested into a nonlinear one:[1]

m_1: $c = c_i - V_{max}/F + e$ is nested into m_3:

$$c = \frac{1}{2}(c_i - V_{max}/k_m - V_{max}/F)$$

$$- \frac{1}{2}\sqrt{\left[c_i - V_{max}/k_m - V_{max}/F)^2 + 4\frac{V_{max}}{k_m}c_i \right]}$$

We have to assume $V_{max}/k_m = 0$. Clearly a nonlinear model can be nested in another nonlinear model. Generally, we can view a nested model m_1 with respect to a model m_2 as a simpler model than m_2.

In the case of multivariate linear models with normal disturbances it is easy to test whether the nested model m_1 should be rejected in the frame of a more complex model m_2 as not satisfactorily explaining the response. In a more precise view, we are testing whether there is no difference between these two models in their goodness of fit. The null hypothesis tested is equivalent to the hypothesis that parameters contained in m_2 and not contained in m_1 are equal to zero. For Equations 28 and 29 above it means to test the hypothesis $\beta_2 = \beta_3 = 0$. Clearly, sometimes the exact formulation of this hypothesis could be more complicated, particularly in nonlinear models and models with constraints.

If $p_{m_2 - m_1}$ is the number of these parameters, RSS_{m_1} is the RSS under m_1, and RSS_{m_2} is the RSS for m_2, then this hypothesis should be tested by the F statistic

$$\frac{(RSS_{m_1} - RSS_{m_2})/p_{m_2 - m_1}}{RSS_{m_2}/(n - p_{m_2} - 1)} \tag{30}$$

which is to be compared with the $1 - \alpha$ quantile of the Fisher distribution with $p_{m_2 - m_1}$ and $n - p_{m_2} - 1$ degrees of freedom.

Note that in the numerator we have $RSS_{m_1} - RSS_{m_2}$, i.e., the difference between deviations under these models. The comparison of models using differences in deviations with the utilization of the additivity properties of deviations and additional techniques of the likelihood approach can be generalized to many other cases;[11] this approach seems to be promising.

For comparison of nested nonlinear models (or linear models into nonlinear, etc.) we can use the above test only approximately. In both cases one should be aware of the fact that due to some constraints an "effective" number of parameters (number of free parameters) can be less than the nominal one. This sometimes complicated question should be carefully considered in particular instances of models to be compared.

There can be some other criteria for comparing nested models; one can, for example, consider the case of the best prediction of future responses[12] and take into account both differences in the dimensionality of models as well as the estimated distance of the models. It is necessary to take into account that the increasing number of estimated parameters improves the fit to the given data set, but due to increasing inaccuracy of estimates it makes prediction worse.

C. Nonnested Models

Comparison of nested models, i.e., comparison of models m_1 and m_2, m_1 being nested into m_2, is relatively a simple task in comparison with the evaluation of relative merits of two models without any such relation between them. Models from our example of ethanol metabolism[1] m_1: $c = c_i - V_m/F + e$ and m_2: $c = Fc_i/(F + k_m) + e$ form a pair of such models; each of them has only one free parameter (V_{max} and k_m, respectively).

In the last 25 years some techniques for discriminating among nonnested models were developed,[13-20] but it seems that they are not frequently used in biological practice. These methods enable distinguishing among models that are, at first glance, hardly comparable. Consider[19] three models expressing dependency of the response y_t measured in some time moments t on two regressors measured either in the same time ($x_{t,1}$, $x_{t,2}$) or with the second regressor measured in the previous time moments ($x_{t,1}$, $x_{t-1,2}$). These models are

$$m_1: y_t = \mu^{(1)} + \beta_1^{(1)} x_{t,1} + \beta_2^{(1)} x_{t,2} + e_t^{(1)}$$

$$m_2: y_t = \mu^{(2)} + \beta_1^{(2)} x_{t,1} + \beta_2^{(2)} x_{t-1,2} + e_t^{(2)}$$

$$m_3: y_t = \mu^{(3)} x_{t,1} x_{t-1,2}$$

It is necessary to stress that values of parameters can be different in different models. The first model is linear; in the second model the response in time $t(y_t)$ depends on the value of the second regressor in the previous time moment ($x_{t-1,2}$). The third model is, moreover, nonlinear (but it can be considered as a generalized linear model in the sense of Equation 6).

Under this framework not only, for example, can discrimination between two different regressors[15] (m_1: $y = \mu^{(1)} + \beta^{(1)}x_1 + e$ and m_2: $y = \mu^{(2)} + \beta_2^{(2)}x_2 + e$) under normal error assumption be considered, but even such a question as whether data (or noises) are distributed along exponential or log-normal density can be decided. It corresponds to the question concerning the variability under the null model. In this case one can construct a test deciding the null hypothesis that data are distributed with density $\alpha^{-1}e^{-y/\alpha}$ against the alternative that they are distributed with density $(y2\pi\beta_2)^{-1/2} \exp[-(\log y - \beta_1)^2/2\beta_2]$ and another test for the opposite question.

The case of comparison of two regressors is relatively easy, and it is transformed in a way into a standard framework by nesting both models into a model m_3: $y = \mu^{(3)} + \beta_1^{(3)}x_1 + \beta_2^{(3)}x_3 + e$. For more general cases the developed tests are rather complicated. Some authors[20] consider particular cases, assuming that the values of parameters in concurrent models are known. Their results can be applied in practice only if data are randomly split, one part used for parameter estimation and the second for discriminating among models. A promising technique for discrimination of models seems to be the bootstrap technique[21] developed in the last years. This technique is unthinkable without the impact of the use of powerful computational technology.

IV. SEARCH FOR MODELS

A. A Class of Models

In many situations we have to consider a large family of models. One of the simplest cases of this nature is the case of linear multivariate regression (Equation 5) where individual models of the family are defined by sets of indexes of nonzero parameters. Let the maximal number of β-parameters be r, say r = 10. Then, for example, the model $y = \mu + \beta_2 x_1 + \beta_5 x_5 + e$ is defined by the set (1, 5); the model $y = \mu + \beta_1 x_1 + \beta_6 x_6 + e$ is defined by (1, 6); and $y = \mu + \beta_1 x_1 + \beta_6 x_6 + \beta_7 x_7$ is defined by (1, 6, 7). Models (1, 5) and (1, 6) are nonnested, and (1, 6) is nested in (1, 6, 7).

Generally in such a class of models one has a partial ordering, $\leqq$ corresponding to the nesting and/or simplicity relation. For example, (1, 6) $\leqq$ (1, 6, 7). This relation should be used for conclusions on models, because if the model (1, 6, 7) is tested and found not to explain the response statisfactorily and hence is rejected, then any simpler model should be rejected, too (e.g., the models [1]; [6]; [1, 6]; [6, 7]; etc.). Such a principle can be used if we are searching a class of models for models that cannot be ruled out on the basis of our observed data. Usually we are looking for models that are as simple as possible.

In any case, if the set of models considered is finite, an exhaustive search is in principle possible. In the above linear case it means to test each model defined by a subset of (1, . . . , 10). However, the number of such models can easily be large. Generally for ten possible parameters, in both the linear and nonlinear case we have $2^{10} = 1024$ models defined by setting some parameters equal to zero. However, if we consider 20 parameters, we obtain 1,048,576 models, and for 30 parameters we obtain 1,073,741,824 models. To evaluate all these models on a computer is hardly tractable. In practice, tasks with, say, 100 explanatory variables corresponding to 1.26×10^{29} models are not so rare as some may think.

Principally two kinds of techniques can be used to manage the task of finding "simple" models that could explain the observed data within a class of models. The first of them is the stepwise technique.

B. Stepwise Selection Procedures

Consider again the simple linear multivariate case in Equation 5. A stepwise selection procedure, e.g., the program BMDP2R[5] used as forward selection, starts with correlating all regressors with the response variable. Then a regressor is selected for which this correlation is the greatest. Let this regressor be, say, x_5. Then, a new, transformed response is defined, namely, $y - (\hat{\mu} + \hat{\beta}_5 x_5)$, and other regressors are correlated with this response. Again a regressor with the maximal correlation is selected and included in the model. Similarly, the procedure works further. At any step it is necessary to decide whether including the new regressor contributes to the explanation of the response. Techniques for nested models are usually used here. For example, if in the second step the regressor x_7 is to be included, we can test the hypothesis $\beta_7 = 0$ in the frame of the model $y = \mu + \beta_5 x_5 + \beta_7 x_7 + e$.

There are further refinements of this technique, e.g., backward or forward selection or stepwise selection combining features of both these approaches.[4,5,22] There is a considerable literature concerning, for example, the role of tests in such a stepwise procedure, including the question of the choice of appropriate significance levels.[23,24] It is effective to use these procedures for building models interactively[25] using a priori knowledge about the role of various regressors, etc. Some generalizations of stepwise techniques for nonlinear cases were developed.[26,27] Stepwise procedures are computationally tractable and can be used interactively, but they can miss a simple model which gives a reasonable fit or better fit than models selected since they do not consider the whole set of possible models.

C. All-Model Searches

We have to distinguish between two tasks. One task is to find a single best model or a prescribed number of best models. To define a "best" model we can consider maximalization or minimization of some of the criteria considered in Section III.A. Particularly for a linear case, techniques were developed for such a task considering but not evaluating all possible models;[28] such a technique is implemented in the BMDP9R program.[5] Comparative studies for these new techniques and the stepwise techniques were done[29] showing that on real data sets some small differences of results in favor of the new all-model searches appear. The published techniques need reasonable computing support, and they are substantially developed for the linear case only.

The second task is being given the data to find all models that fit reasonably well and/or explain to some extent the response, moreover, such that no simpler model explains the response as well. To perform this task we have to use substantially the simplicity ordering of models, even if it is used in the previous task, too, and the principle that if a model m_1 is rejected on the basis of the data, all simpler models $m_2 \leqq m_1$ are to be rejected as well without evaluating them in the data. Similarly if a model m_1 is not rejected then all more complex models $m_2(m_1 \leqq m_2)$ are not to be rejected without evaluating them on the data. The ideas regarding such an approach were gradually developed considering models for categorical data[30-33] or regression models.[34] This approach is rather general; it can be used in the linear as well as in the nonlinear case. For a reasonable computer relization a lattice structure of the partial ordering $\leqq$ of models is to be used;[35] hence, the inner technology of this approach can be rather complex, but the idea is simple and understandable.

REFERENCES

1. **Dvořák, I.**, Mathematical models of physiological processes, in *Methods in Animal Physiology*, Deyl, Z. and Zicha, J., Eds., CRC Press, Boca Raton, FL, 1988, chap. 7.
2. **Cox, D. R.**, The role of significance tests, *Scand. J. Stat.*, 4, 49, 1977.
3. **McCullagh, P. and Nelder, J. A.**, *Generalized Linear Models*, Chapman & Hall, London, 1983.
4. **Draper, N. R. and Smith, H.**, *Applied Regression Analysis*, 2nd ed., John Wiley & Sons, New York, 1981.
5. **Dixon, W. J., Ed.**, *BMDP — Biomedical Computer Programs*, University of California Press, Los Angeles, 1983.
6. **Dunn, O. J. and Clark, V. A.**, *Applied Statistics: Analysis of Variance and Regression*, John Wiley & Sons, New York, 1975, 263.
7. **Atkinson, A. C.**, Regression diagnostics, transformations and constructed variables, *J. R. Stat. Soc.*, B44, 1, 1982.
8. **Cook, R. D. and Weisberg, S.**, *Residuals and Influence in Regression*, Chapman & Hall, London, 1982.
9. **Atkinson, A. C.**, Likelihood ratios, posterior odds and information criteria, *J. Econometrics*, 16, 15, 1981.
10. **Anděl, J.**, Fitting regression models in time series analysis, *Statistics*, 13, 121, 1982.
11. **Whittaker, J.**, Model interpretation from additive elements of the likelihood function, *Appl. Stat.*, 33, 52, 1984.
12. **San Martin, A. and Spezzaferi, F.**, A predictive model selection criterion, *J. R. Stat. Soc.*, B46, 296, 1984.
13. **Cox, D. R.**, Further results on tests of separate families of hypotheses, *J. R. Stat. Soc.*, B24, 406, 1962.
14. **Jackson, O. A. Y.**, Some results on tests of separate families of hypotheses, *Biometrika*, 55, 355, 1968.
15. **Atkinson, A. C.**, A method for discriminating between models, *J. R. Stat. Soc.*, B32, 323, 1970.
16. **Sawyer, K. R.**, Testing separate families of hypotheses: an information criterion, *J. R. Stat. Soc.*, B46, 419, 1984.
17. **Sawyer, K. R.**, Certain generalizations of separate hypotheses tests, *Biometrika*, 72, 124, 1985.
18. **Sawyer, K. R.**, Multiple hypothesis testing, *J. R. Stat. Soc.*, B46, 419, 1984.
19. **Pesaran, M. H. and Deaton, A. S.**, Testing of non-nested nonlinear regression models, *Econometrics*, 46, 667, 1978.
20. **Borowiak, D. A.**, A multiple model discrimination procedure, *Commun. Stat. Theor. Methods*, 12, 2911, 1983.
21. **Efron, B.**, Comparing non-nested linear models, *J. Am. Stat. Assoc.*, 79, 719, 1984.
22. **Berk, K. N.**, Forward and backward stepping in variable selection, *J. Stat. Computer Simulation*, 10, 177, 1980.

23. **Sparks, R. S., Zucchini, W., and Coutsourides, K.,** On variable selection in multivariate regression, *Commun. Stat. Theor. Methods,* 14, 1569, 1985.
24. **Wilkinson, L. and Dallal, G. E.,** Tests of significance in forward selection regression with an F-to-enter stopping rule, *Technometrics,* 23, 377, 1981.
25. **Henderson, H. V. and Velleman, D. F.,** Building regression models interactively, *Biometrics,* 37, 391, 1981.
26. **Bunke, H. and Droge, B. A.,** A stepwise procedure for the selection of non-linear regression, *Statistics,* 16, 35, 1985.
27. **Green, J. R. and Al-Bayatti, M. F.,** Selection of regressor variables when EY is an unknown non-linear function, *Statistics,* 16, 15, 1985.
28. **Furnival, G. M. and Wilson, R. B.,** Regression by leaps and bounds, *Technometrics,* 16, 399, 1974.
29. **Berk, K. N.,** Comparing subset regression procedures, *Technometrics,* 20, 1, 1978.
30. **Hájek, P. and Havránek, T.,** *Mechanizing Hypothesis Formation,* Springer-Verlag, Heidelberg, 1978, chap. 8.
31. **Havránek, T.,** The GUHA method in the context of data analysis, *Int. J. Man-Machine Studies,* 15, 265, 1981.
32. **Havránek, T.,** A procedure for model search in multidimensional contingency tables, *Biometrics,* 40, 95, 1984.
33. **Edwards, D. and Havránek, T.,** A fast procedure for model search in multidimensional contingency tables, *Biometrika,* 72, 339, 1985.
34. **Cox, D. R. and Snell, E. J.,** The choice of variables in observational studies, *Appl. Stat.,* 23, 51, 1974.
35. **Edwards, D. and Havránek, T.,** A fast model selection procedure for large families of models, *J. Am. Stat. Assoc.,* 82, 205, 1987.

Techniques for Measuring and Monitoring Biological Functions

Chapter 9

TRACER METHODS

R. Poledne

TABLE OF CONTENTS

I. INTRODUCTION

Metabolite concentration data have furnished physiology with a large body of essential information. In spite of the fact that the concentration of any metabolite is the result of its entry and removal from a pool, the data obtained at one point display only a static character. To quantitate the production and clearance of any metabolite or hormone requires a kinetic study. Tracer methods are the only methods of choice to follow the dynamics of metabolic pathways in vivo.

Early tracer methods were introduced half a century ago utilizing the wide expansion of physics in the 1940s. Both tracer methods using radioactive and stable isotopes appeared almost at the same time.

The transport and transformations of metabolites from one place or state to another are usually nonlinear, and the kinetics depends on the amount of metabolites present.

The systems can be studied by nonisotope analytical or physical methods in terms of time-dependent temperature, flow, or chemical concentration changes; such changes are characteristic for a given system and accessible for determination. Most of the systems studied in nature usually do not follow strictly first-order kinetics. The principle of all tracer methods is simple: when a tracer substance is added to a tracee system, the kinetics of the tracer is always linear. For a tracer in a steady-state system, the rate constants are time independent.

What is a tracer? A tracer is a labeled form of a substance-tracee. Ideally, a radioactive or unusual stable isotope of an element is introduced to the molecule of the substance studied. Obviously, the behavior of such a labeled molecule is very close to that of the tracee. Radioactive (^{3}H, ^{14}C) or stable (D, ^{13}C) isotopes of carbon and hydrogen are most commonly built into the structure of the metabolite studied. This type of label is named homologous, and the differences in the behavior of the tracer and the tracee are minimal, depending on the isotope effect only, which for most purposes can be neglected. When a heterologous element is introduced into the molecule of the tracee for labeling (e.g., iodine), the behavioral identity of the tracer should be checked.

When a labeled atom is introduced into a molecule of a studied substance by a chemical reaction in vitro (organic synthesis, iodination, etc.), the resulting tracer is called exogenously labeled in different positions of its carbohydrate or polypeptide chain. Metabolites labeled in vivo after a precursor administration are named endogenously labeled. Lipoproteins labeled after i.v. application of ^{14}C-fatty acid or plasma proteins labeled after i.v. application of ^{3}H-amino acid can serve as typical examples.

Tracer methods allow study of a system as a black box. In this context the system is understood as a complex of metabolites and processes coupled mutually in a definite space surrounded by a relatively stable environment. This environment stability must be such as to allow neglect of the environmental influence on the system studied.

When we apply a tracer method and, consequently, postulate a formal mathematical expression of the system studied and this system is in a certain transient pseudolinear state, then after the tracer is applied, the tracer data obtained are linear and easily used for analysis. In this way all difficulties in the accessibility of data describing a biological system are overcome or minimized by tracer method application. Tracer kinetics is more simple to analyze compared to the analysis of behavior of a total biological system. However, the step that follows acquisition of experimental data, i.e., their interpretation, is more difficult. Tracer kinetics does not fully describe the kinetics of the tracee, and the rate constants obtained by tracer analysis must be interpreted in terms of tracee mass and its physiological significance.

II. DEFINITION OF TERMS

A compartment (I) is an anatomical, physiological, or hypothetical subdivision of a system throughout which the ratio of concentration of the tracer C* and the tracee C is uniform in any given time.

The rate of appearance (R_a) and the rate of disappearance (R_d) are the most common factors determined in tracer kinetic studies and represent the mass of substance (in moles) entering and leaving the compartment I per unit of time.

Net transport is the difference between two transports in opposite directions.

Residence time (t) is defined as the time that a particle of a substance is expected to spend in compartment I.

Fractional catabolic rate (FCR) is the reciprocal value of residence time:

$$FCR = \frac{1}{t} \tag{1}$$

Volume (V) of distribution represents the volume (in milliliter) in which the metabolite is distributed.

Total mass (M) is the quantity of a metabolite in a given volume of distribution V times metabolic concentration. It is also equal to the pool size (P) and can be calculated as the product of the rate of appearance and residence time:

$$M = P = R_a \cdot t \tag{2}$$

Turnover rate is also referred to as the flux rate, and in the steady state, it is equal to the rate of appearance R_a.

More detailed description of terms is presented in a monograph by Shipley and Clark[1] and a nomenclature article by Brownell et al.[2]

III. TRACER ADMINISTRATION

Principally, two techniques are used for in vivo tracer application; the bolus injection technique represents a flash injection of labeled material to a compartment where the tracer is to be distributed during a time interval that is negligible compared to the time of the metabolic study. Maximum specific radioactivity (when a radioactive isotope is used) or maximum enrichment (when a stable isotope is used) is observed at zero time (Figure 1A) and declines thereafter. A typical example is shown in Figure 2A where free fatty acids in plasma (considered as a compartment) are labeled after an i.v. flash injection of 1-^{14}C-palmitate in the rat. A small quantity of labeled palmitate (bound to rat albumin for injection) was administered to avoid the mass effect of injected fatty acids. As the release of free fatty acids from adipose tissue is in equilibrium with their uptake, the plasma compartment of free fatty acids is in the steady state. Disappearance of the tracer follows a multiexponential curve as labeled palmitate is extracted in the liver and other tissues. Next, triglycerides of the liver are labeled, and subsequently the triglycerides of very low density lipoproteins produced in the liver are released back to the plasma compartment, and their radioactivity can be used for very low density lipoprotein studies.[3,4] Bolus injection is preferred with radioactive isotopes, as the necessary stable isotope enrichment at zero time is difficult to obtain, and data interpretation is more complicated.[5]

Constant infusion technique is well suited for both stable and radioactive isotopes (Figure 1B). When the system of very low density lipoprotein triglycerides, described in the previous paragraph, is labeled by constant infusion of ^{14}C-glycerol or ^{3}H-glycerol, radioactivity of its plasma pool increases gradually (Figure 2B). After some time of infusion, a plateau is reached. Thereafter, liver triglycerides are labeled, and consequently the radioactivity of triglycerides of very low density lipoproteins increases.

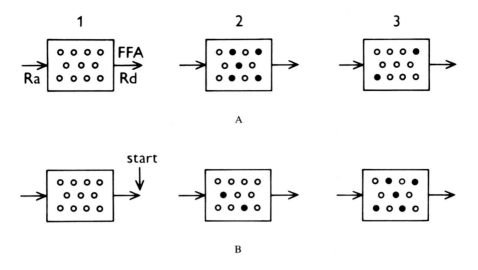

FIGURE 1.   Schematic representation of the labeling of a plasma-free fatty acid compartment in steady state. Unlabeled fatty acids — open circles, ^{14}C-palmitic acid — closed circles. (A) Bolus injection — specific radioactivity is highest after flush administration of the tracer and declines gradually. (B) Specific radioactivity increases gradually from the start of continual infusion.

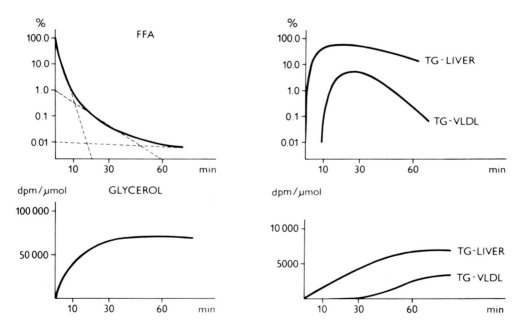

FIGURE 2.   Labeling of very low density lipoprotein triglycerides by ^{14}C-palmitic acid flush injection or by ^{3}H-glycerol continual infusion.

If the pool size of the metabolite under study is large compared to the amount of the labeled compound delivered by infusion, the plateau is reached only after a long period of time. In order to cut this period down, it is possible to use a combination of bolus injection and constant infusion; this procedure is referred to as the priming dose technique.

A tracer can be administered orally, i.p., or i.v. depending on the experiment design, size of the experimental animal, and the character of the tracee. Oral administration is used in nutritional studies or in experiments in which a labeled precursor is applied to obtain a biologically labeled tracer. I.v. administration is most widely used in in vivo physiological studies, but the chemical form of the tracer should be identical with a particular metabolite present in the intravascular space. I.p. administration is used in small laboratory animals in which i.v. administration is technically complicated. When a tracer of a low molecular mass is applied, the distribution of the labeled compound from the peritoneum to the total distribution volume of body fluids is quite fast, and the results obtained after i.v. and i.p. administration are almost identical.[6] Distribution of a tracer with a higher molecular mass in body fluids is more complicated and, consequently, the interpretation of data obtained may be more difficult.

IV. EXPERIMENTAL DATA INTERPRETATION

The easiest interpretation of tracer experimental data can be made by plotting sums of exponentials describing the time change of the specific radioactivity of the metabolite studied after a bolus injection of the corresponding tracer (Figure 2). If the simplifying assumptions, e.g., steady-state conditions and linearity of the system, are justified, the rates of the metabolite production R_a and its disappearance are equal

$$R_a = R_d \tag{3}$$

and the mass of the metabolite does not change with time

$$\frac{dM}{dt} = 0 \tag{4}$$

The experimental data of specific radioactivity of the metabolite studied can be plotted against time in a semilogarithmic scale; the slope of such a linear relation gives the rate constant. It applies also to experimental data described by multiexponential decrease of specific radioactivity. The calculation of the rate appearance R_a or fractional catabolic rate (FCR) is based on the area under the curve of specific radioactivity of tracer C^*, obtained after a bolus injection (see Figure 2). Function $C^*(t)$ is described by the sum of exponential functions

$$C(t) = a_1 + e^{b_1 t} + a_2 \cdot e^{b_2 t} + \ldots \tag{5}$$

and $\int_0^\infty C^*(t) \, dt$ is the area under this curve. If the total tracer input into the compartment studied is unity ($= 1$), the area under the curve is equal to the residence time t and the FCR is calculated using Equation 1.

When the turnover of the metabolite is studied after infusion at a constant rate R_a^* of the adequate tracer for a sufficient time to reach the plateau (Figure 2), then $R_a^* = R_d^*$ where R_d is the rate of labeled metabolite disappearance. The rate of metabolite production R_a is obtained from the known rate of tracer infusion and plasma concentration of tracer C^* and tracee C:

$$R_a = R_a^* \cdot \frac{C}{C^*} \tag{6}$$

Although this simple experimental data analysis possesses the advantage of unique and understandable interpretation, on the other hand, this analysis exploits only a limited amount of information contained in the tracer experimental data. Another advantage of using such mathematical relations is the independence of data obtained on a model structure.

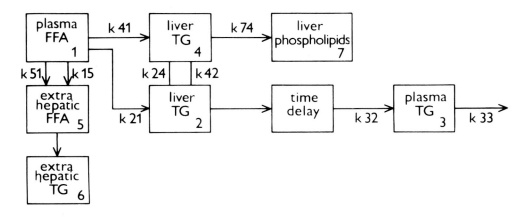

FIGURE 3.   Multicompartmental model of very low density lipoprotein triglycerides production (k32) and utilization (k33) in the rat after plasma-free fatty acid compartment labeling by flush injection of ^{14}C-palmitate.

More information can be obtained by compartmental analysis. It is based on a simulation of dynamics of the metabolite system by a compartmental model. Each compartment is assumed to contain a homogenous substance which can be changed or transported to another compartment. All models can be described by differential equations, but usually these equations cannot be solved in a simple closed form because of the system nonlinearity. However, it is possible to solve them with the aid of computers. This computer modeling appeared almost a quarter of a century ago.[7] Several specialized computer programs for computer modeling have been developed, of which SAAM[8] is most powerful.

To study a biological system, it is advisable to start with a simple two- or three-compartment model, to test a hypothesis, and to compare it with experimental data. Data fitting,[9] uniqueness of solution, and reliability range of the model are important steps of multicompartmental solution. The typical example of a multicompartmental model analysis arising from experimental data, described in the section on lipid metabolism, is presented in Figure 3. Fatty acids represented by ^{14}C-labeled palmitate and bound to albumin carrier are injected i.v. to rats. Their distribution volume and plasma volume are virtually identical. Fatty acid activity disappears from the plasma (compartment 1). Fatty acids are mainly transported into the unstable pool in the liver (compartment 2). There, triglycerides are synthesized and a lipoprotein complex is formed. After a delay of a few minutes, very low density lipoproteins containing labeled fatty acids begin to appear in the plasma (compartment 3). Part of the fatty acids introduced into the plasma are also taken up at other sites, mainly in the unstable fatty acid pool especially in the fat tissue (compartment 5). Part of these fatty acids is assumed to recirculate back into plasma by means of backward flow. A part of the activity of the liver soon relatively enters the stable hepatic lipid pool (compartment 4), producing a dynamic equilibrium between the stable (compartment 4) and unstable (compartment 2) pools. Part of the lipids from the stable hepatic pool becomes a structural part of the cell walls, forming a separate compartment 7. The remaining part of activity enters the stable pool at other fatty acid uptake sites (compartment 6).

Readers are recommended to see reviews for details.[10,11] The body of computer modeling of tracer experimental data has been considerably enlarged. This approach to experimental data interpretation has greatly contributed to a better understanding of metabolic regulations in spite of considerable criticism of the opponents of this technique; the pioneering works of Berman[7] and Berman and Weiss[8] on the growing influence of multicompartmental analysis need to be mentioned.

V. CARBOHYDRATE METABOLISM

A. Tracers

Glucose labeled with deuterium, ^3H, ^{13}C, and ^{14}C is used in studies of carbohydrate metabolism. Carbon-labeled glucose contains ^{14}C in positions 1 (1-^{14}C-glucose) or 6 (6-^{14}C-glucose) or in all carbon chain positions (uniformly labeled U-^{14}C-glucose). Each of these tracers is reversible which means that labeled lactate, alanine, glycerol, and other compounds arising during glucose breakdown recycle to the original labeled pool of plasma glucose. In addition, all these metabolites are water soluble and easily transportable across membranes. Total radioactivity of the plasma glucose pool, used for glucose kinetic studies after labeled glucose administration, is influenced by the radioactivity of the above-mentioned metabolites. Consequently, the results of glucose turnover are always underestimated and a correction for recycling must be introduced.[12] Carbon-labeled glucose is used for turnover studies as well as for glucose oxidation measurements.

Tritium can be introduced into any of the six positions of the carbon chain of glucose: 1-^3H-glucose, 3-^3H-glucose, 4-^3H-glucose, and 6-^3H-glucose are in use. It is necessary to stress that turnover data obtained by using the tracer labeled in position 2 are overestimated by futile cycling through hexose-6-phosphate isomers. The application of 3-^3H-glucose has the advantage of minimizing the effect of label recycling. In this case the main part of radioactivity of intermediary metabolites in plasma samples is present in the form of ^3H_2O (through conversion of dihydroxyacetone-3-phosphate to glyceraldehyde-3-phosphate).

Uniformly labeled glucose is preferentially used for the study of interconversion of carbohydrate substrate to tissue lipids[13] and tissue specificity of this process.[14]

Also stable isotope-labeled glucose is used to follow carbohydrate metabolism in man. In this case, deuterium or ^{13}C uniformly labeled glucose is used.

B. Applications

The glucose-insulin regulation system has been studied through the oral glucose tolerance test for years. More recently, two nonisotope methods, the clamp technique[15] and the minimal model of i.v. glucose tolerance,[16] allowed estimation of glucose clearance along with a description of glucose action. In both these methods, however, it is difficult to separate the utilization and production processes.

The application of the tracer method overcomes the disadvantage of the nonisotope approach. Labeled glucose is administered orally or i.v., and the plasma compartment can be labeled by a single injection or constant infusion. Also primed constant infusion is used to achieve a plateau-specific radioactivity or isotope enrichment in a shorter time.

When a single injection of a tracer is applied, monitoring of the disappearance processes is possible. In order to determine the glucose metabolic clearance and turnover rate, a simple noncompartmental analysis (Figures 3 and 4) of plasma glucose-specific radioactivity can be used. Both these parameters are model independent as the result does not depend on a particular structure of the model designed to study glucose metabolism. On the other hand, the noncompartmental approach to glucose kinetics seems to be inadequate as this simplification does not include all metabolic pathways and exchanges of the labeled substrate.

The first simplified model for glucose utilization was developed in the 1950s to study glucose metabolism[17] and was renewed by Segre et al.[18] in the 1970s. An example of a compartmental model of glucose kinetics is shown in Figure 4B where compartment 1 corresponds to the plasma glucose pool in a noncompartmental model (Figure 4A), compartment 2 is rapidly equilibrated (insulin independent), and compartment 3 is slowly equilibrated (insulin dependent). The fractional rate constants k_{02} and k_{03} describe the irreversible removal of glucose from compartments 2 and 3. It is necessary to stress that calculation of the mean residence time, distribution volume, and fraction rate constants are the data which depend on this model structure.

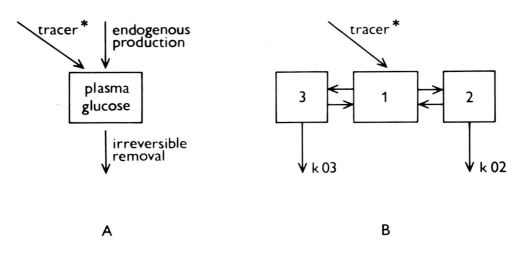

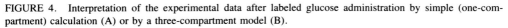

FIGURE 4.   Interpretation of the experimental data after labeled glucose administration by simple (one-compartment) calculation (A) or by a three-compartment model (B).

Recently, a revised "minimal model" of glucose disappearance has been described in which the tracer is injected along with unlabeled glucose. The data obtained from this model are proposed to be more precise, even though the block diagram is less complicated.[19] To determine the rate of glucose utilization by different tissues, i.v. administration of a bolus of 2-deoxy-1-³H-glucose can be used, and radioactivity accumulation of the unmetabolizable 2-deoxy-1-³H-glucose-6-phosphate measured.[20] This method can also be employed in combination with euglycemic clamp for the determination of glucose utilization in normoinsulinemic and hyperinsulinemic states.[21]

VI. LIPID METABOLISM

A. Tracers

Fatty acids labeled with ¹⁴C were used in the 1960s to obtain basic in vivo information concerning fatty acid metabolism and transport. All these data, however, are devaluated as lipid metabolism must be regarded as the metabolism of the lipoprotein complex. Cholesterol and fatty acid synthesis has been studied by the application of several labeled precursors (acetate, glucose, pyruvate, and mevalonate-labeled ¹⁴C), but the interpretation of experimental data is complicated due to an unknown dilution ratio of the labeled precursor in the intracellular space. Only the method exploiting tritiated water as tracer[22] is reliable in the determination of cholesterol and fatty acid synthesis in absolute value.

Iodinated lipoproteins are used as exogenously in vitro labeled tracers in studies of lipoprotein metabolism.[23] Also ¹⁴C-sucrose and ⁹⁹ᵐTc have been used for exogenous lipoprotein labeling. The disadvantage of exogenously labeled lipoproteins involves the possible modification of the lipoprotein structure by isolation (by ultracentrifugation in appropriate density) and in vitro labeling. This disadvantage is overcome by using endogenous labeling of apoprotein B with ³H-leucine or ⁷⁵Se-methionine.[24,25]

B. Applications

Although a considerable amount of information is available on the regulation of cholesterol synthesis,[26] the only correct data can be obtained by the method described by Turley et al.[22] While this method using i.v. application of a high dose of tritiated water and determination of cholesterol radioactivity after a short time interval is very accurate and easy to perform, high contamination of the laboratory by expired water, in spite of extremely careful manipulation, always occurs.

Two tracer methods are used to estimate absorption of alimentary cholesterol. The fecal isotope-ratio method employs simultaneous oral administration of ^{14}C-cholesterol and ^{3}H-sitosterol assuming that the plant sterol (sitosterol) is not absorbed. Cholesterol absorption is determined from the difference in the ^{14}C/^{3}H ratio in the dose administered and in feces.[27] As collection and analysis of feces make this approach rather complicated, an alternative method (called the plasma ratio method) was introduced by Zilversmit.[28] Two tracers are administered simultaneously: total body cholesterol is labeled by i.v. injection of ^{14}C-cholesterol, and alimentary cholesterol is administered in the form of the ^{3}H-labeled compound. The percentage of absorption can be calculated from the plasma ratio of ^{3}H/[14]C.

Total body cholesterol kinetic data can be obtained after i.v. or oral administration of labeled cholesterol. Simple two-compartment models of total body cholesterol turnover using analysis of serum cholesterol radioactivity decay after tracer application were described[29] and later developed to a more complicated stage.[30] The most sophisticated mathematical analysis of compartmental cholesterol kinetics was presented by Dell and Ramakrishnan.[31] It should be stressed that all these models have a single solution only when various arbitrary assumptions (concerning rates of alimentary and nonalimentary cholesterol entry) are made; obviously not all these assumptions are fully justified.

When ^{14}C-palmitate is administered i.v., specific radioactivity follows a multiexponential curve (Figure 2). Labeling of very low density lipoprotein triglycerides follows labeling of liver triglycerides as such. Metabolism of very low density lipoproteins can be revealed by multicompartmental analysis of such experimental data. When a minimal three-compartmental model is used,[3] many discrepancies between the model and experimental data are found. Therefore, the multicompartmental model has to be applied. Another example of such a multicompartmental model is presented in Figure 3 and described earlier. The data obtained from the multicompartmental analysis fit well with the experimental data. The crucial fact to be respected here is the time delay of the production of very low density lipoproteins after liver triglycerides labeling.

Alternatively, labeling of very low density lipoproteins by ^{3}H-glycerol was described.[32] The kinetics of very low density lipoproteins labeled by glycerol is supposed to be less complicated compared to palmitate labeling because of lower recycling of the former label.[32]

Although different ways of lipoprotein labeling can be used, currently all experimental data regarding triglyceride metabolism are analyzed with the model constructed by Zech et al.[33] For a critical review of plasma lipoprotein kinetics, see Phair.[34]

The main apoprotein studied in connection with atherosclerosis is apoprotein B because of the very close relationship of its plasma concentration and clinical complications of atherosclerosis: kinetics of its metabolism is currently being paid considerable attention. Although labeling of the protein part of this lipoprotein was described in the 1950s, later refinement of a reliable iodinating technique[35] by Bilheimer et al.[36] has accelerated extensive research in apoprotein B turnover. Only limited attention is paid to other apoproteins (A and E).

VII. PROTEIN METABOLISM

A. Tracers

The metabolism of proteins and individual amino acids can be studied with a broad range of labeled substrates and compounds. Principally, all elements present in an amino acid molecule can be substituted by radioactive (^{3}H, ^{14}C) or stable (D, ^{13}C, ^{15}N, ^{18}O) isotopes for a tracer study. Already the early studies using iodinated proteins (^{125}I, ^{131}I) and determination of excreted radioactivity in urine[37,42] have established this technique as a valuable tracer method for measurement of protein metabolism. Sulfur-containing amino acids can be labeled by using ^{35}S or [75]Se.[24,25]

B. Applications

The flux of individual amino acids can be determined by i.v. infusion of labeled amino acids, e.g., ^{14}C-alanine, ^{15}N-glycine, ^{13}C-leucine, etc.[38] When amino acid oxidation is measured, a high specific radioactivity of tracer isotope or high enrichment of a stable isotope must be reached because of the dilution of labeled CO_2 in the unlabeled bicarbonate pool;[5] therefore, the method of priming dose infusion is advisable. It should be remembered that due to the differences in recycling and metabolic fate of carbon and nitrogen, the flux data obtained with amino acids labeled by these two elements are not identical.[39] Also interconversion of amino acids can be studied by individually labeled amino acids application.[40]

Although the first whole body protein turnover study is almost 40 years old,[41] the problem of obtaining meaningful data remains open. Some investigators still feel that the sum of many processes of protein synthesis and catabolism cannot be described by a single, simplified parameter of whole body protein turnover and believe that the old nonisotopic method of nitrogen balance is more reliable. On the other hand, the nitrogen balance study also displays systematic errors, and it is difficult to accept it as an absolute reference method for a synthesis-catabolism balance study.

The majority of studies on protein turnover performed in the 1960s used the method described by Berson and Yalow[37] and an iodinated protein (predominantly albumin) as tracer. Complicated interpretation of the experimental data and the physiologically low level of information considerably devaluates such results. All tracer methods used for whole body protein turnover measurement were summarized by Waterlaw.[39]

When bolus injection of ^{14}C- or ^{13}C-labeled amino acid is used for a turnover study, compartmental analysis of the multiexponential decay curve is used. The main point of controversy is the definition of the physiological meaning of all mathematically derived pools. This interpretation problem has been partly solved in stochastic analysis of the data obtained after the constant infusion of ^{15}N-labeled amino acids. Stochastic analysis used in human studies simplifies the model only to a single metabolic pool indirectly labeled with ^{15}N-glycine which is in equilibrium with whole body protein.[43] Total protein turnover is calculated from the rate of ^{15}N-urea excretion after the system has reached a plateau. Although several assumptions must be incorporated in the formulation of such a simplified model,[43] its physiological validity was proved in a very precise experiment in the rat.[44]

VIII. CONCLUSION

Tracer methods using both stable and radioactive isotopes have contributed substantially during the past 2 decades to our understanding of metabolic regulations under physiological and pathological conditions. As a matter of fact, tracer methods applied in whole animals or in clinical studies in man represent an irreplaceable tool to verify in vivo the physiological significance of the tremendous body of information obtained in vitro. The application of tracer methods in vivo is intimately combined with the computer compartmental analysis of data obtained. The broad range of commercially available compounds together with access to computer technology resulted in a wide application of tracer methods to follow the metabolism of carbohydrates, lipids, and proteins.

REFERENCES

1. **Shipley, R. A. and Clark, R. E.,** *Tracer Methods for In Vivo Kinetics: Theory and Application,* Academic Press, New York, 1972.
2. **Brownell, G. L., Berman, M., and Robertson, J. S.,** Nomenclature for tracer kinetics, *Int. J. Appl. Radiat. Isot.,* 19, 249, 1968.

3. **Baker, N.**, The use of computers to study rates of lipid metabolism, *J. Lipid Res.*, 10, 1, 1969.
4. **Potůček, J., Hjek, M., and Brodan, V.**, Reliability range of model parameters and its application to biological models, *Appl. Math. Modelling*, 3, 199, 1979.
5. **Wolfe, R. R.**, Determination of substrate kinetics: bolus injection technique, in *Tracers in Metabolic Research*, Wolfe, R. R., Ed., Alan R. Liss, New York, 1984, 27.
6. **Poledne, R., Petrásek, R., and Vavrečka, M.**, Lipogenesis in vivo from acetate-1-^{14}C in the mouse, *Physiol. Bohemoslov.*, 17, 21, 1968.
7. **Berman, M.**, A postulate to aid model building, *J. Theor. Biol.*, 229, 1963.
8. **Berman, M. and Weiss, M. F.**, *SAAM Manual*, Department of Health Education and Welfare Publ. No. 78-180, National Institutes of Health, Bethesda, Md., 1978, 196.
9. **Berman, M.**, The application of multicompartmental analysis to problems of clinical medicine, *Ann. Intern. Med.*, 68, 423, 1968.
10. **Boston, R., Greif, P. C., and Berman, M.**, Conversational SAAM: an interactive program for kinetic analysis of biological systems, *Comp. Progr. Biomed.*, 13, 111, 1981.
11. **Groth, T.**, Biomedical modeling, in *Medinfo 77, IFIP*, Shires and Wolf, Eds., North-Holland, Amsterdam, 1977, 775.
12. **Cowan, J. S. and Hetenyi, G., Jr.**, Glucoregulatory responses in normal and diabetic dogs recorded by a new tracer method, *Metabolism*, 20, 360, 1971.
13. **Baker, N. and Huebotter, R. J.**, Compartmental and semicompartmental approaches for measuring glucose carbon flux to fatty acids and other products in vivo, *J. Lipid Res.*, 13, 716, 1972.
14. **Baker, N., Mead, J., Jr., and Kannan, R.**, Hepatic contribution to newly made fatty acids in adipose tissue in rats and inhibition of hepatic and extrahepatic lipogenesis from glucose by dietary corn oil, *Lipids*, 16, 568, 1981.
15. **De Fronzo, R. A., Tobin, J. D., and Anders, R.**, Glucose clamp technique: a method for quantifying insulin secretion and resistance, *Am. J. Physiol.*, 237, E214, 1979.
16. **Bergman, R., Ider, V. Y., Bowden, C. R., and Cobelli, C.**, Quantitative estimation of insulin sensitivity, *Am. J. Physiol.*, 236, E667, 1979.
17. **Baker, N., Shreeve, W. W., Shipley, R. A., Inceffy, G. E., and Miller, M.**, C^{14} study in carbohydrate metabolism. I. The oxidation of glucose in normal human subjects, *J. Biol. Chem.*, 211, 575, 1954.
18. **Segre, G., Turco, G. L., and Vercolle, G.**, Modeling blood glucose and insulin kinetic in normal, diabetic and other subjects, *Diabetes*, 22, 94, 1973.
19. **Cobelli, C., Pacini, G., Toffolo, G., and Sacca, L.**, Estimation of insulin sensitivity and glucose clearance from minimal model: new insights from labeled IUGTT, *Am. J. Physiol.*, 250, E591, 1986.
20. **Ferré, P., Leturque, A., Burnol, A. F., and Girard, J.**, A method to quantify glucose utilization in vivo in skeletal muscle and white adipose tissue of anesthetized rat, *Biochem. J.*, 228, 103, 1985.
21. **Leturque, A., Ferré, P., Burnol, A. F., Kande, J., Maulard, P., and Girard, J.**, Glucose utilization rates and insulin sensitivity in vivo in tissues of virgin and pregnant rats, *Diabetes*, 35, 172, 1986.
22. **Turley, S. D., Andersen, J. M., and Dietschy, J. M.**, Rates of cholesterol synthesis and uptake in the major organs of the rat in vivo, *J. Lipid Res.*, 22, 551, 1981.
23. **Bilheimer, D. W., Eisenberg, S., and Levy, R.**, The metabolism of very low density lipoprotein proteins, *Biochem. Biophys. Acta*, 260, 212, 1972.
24. **Eaton, R. P., Allen, R. C., and Schade, P.**, Beta-apolipoprotein secretion in man: investigation by analysis of ^{75}Se-labeled amino acids incorporation into apoprotein, in *Lipoprotein Kinetics and Modeling*, Berman, M., Grundy, S. M., and Howard, B., Eds., Academic Press, New York, 1982, 77.
25. **Fisher, W. R., Zech, L. A., Bardalaye, P., Warnke, G., and Berman, M.**, The metabolism of apolipoprotein B in subjects with hypertriglyceridemia and polydisperse LDL, *J. Lipid Res.*, 21, 760, 1980.
26. **Dempsey, M. E.**, Regulation of steroid biosynthesis, *Annu. Rev. Biochem.*, 43, 967, 1974.
27. **Borgström, B.**, Quantification of cholesterol absorption in man by fecal analysis after the feeding of a single isotope-labeled meal, *J. Lipid Res.*, 10, 331, 1969.
28. **Zilversmit, D. B.**, A single blood sample dual isotope method for the measurement of cholesterol absorption, *Proc. Soc. Exp. Biol. Med.*, 140, 863, 1972.
29. **Samuel, P. and Perl, W.**, Long term decay of serum cholesterol radioactivity: body cholesterol metabolism in normals and in patients with hypercholesterolemia and atherosclerosis, *J. Clin. Invest.*, 49, 346, 1970.
30. **Goodman, D., Noble, R. P., and Dell, R. B.**, Three-pool model of the long-term turnover of plasma cholesterol in man, *J. Lipid Res.*, 14, 178, 1973.
31. **Dell, R. B. and Ramakrishnan, R.**, A mathematical model for cholesterol kinetics, in *Lipoprotein Kinetics and Modeling*, Berman, M., Grundy, S. M., and Howard, B., Eds., Academic Press, New York, 1982, 313.
32. **Farquhar, J. W., Gross, R. C., Wagner, R. M., and Reaven, G. M.**, Validation of an incompletely coupled two-compartment nonrecycling catenary model for turnover of liver and plasma triglyceride in man, *J. Lipid Res.*, 6, 119, 1965.

33. **Zech, L. A., Grundy, S. M., Steinberg, D., and Berman, M.,** Kinetic model for production and metabolism of very low density lipoprotein triglycerides, *J. Clin. Invest.,* 63, 1262, 1979.
34. **Phair, R. D.,** Models of plasma lipoprotein triglycerides kinetics: a critical review, in *Lipoprotein Kinetics and Modeling,* Berman, M., Grundy, S. M., and Howard, B. V., Eds., Academic Press, New York, 1982, 221.
35. **McFarlane, A. S.,** Efficient trace-labeling of protein with iodine, *Nature,* 182, 53, 1958.
36. **Bilheimer, D. W., Eisenberg, S., and Levy, R. I.,** The metabolism of very low density lipoprotein proteins. Preliminary in vitro and in vivo observations, *Biochim. Biophys. Acta,* 260, 212, 1972.
37. **Berson, S. A. and Yalow, R. S.,** Distribution and metabolism of ^{131}I labeled proteins in man, *Fed. Proc., Fed. Am. Soc. Exp. Biol.,* 16, 13S, 1957.
38. **Nissen, S. and Haymond, M. W.,** Effects of fasting on flux and interconversion of leucine and alpha ketoisocaproate in vivo, *Am. J. Physiol.,* 241, E72, 1981.
39. **Waterlaw, J. C., Garlick, R. J., and Millward, D. J.,** *Protein Turnover in Mammalian Tissues and in the Whole Body,* North-Holland, Amsterdam, 1978.
40. **Haymond, M. W. and Miles, J. M.,** Branched chain amino acids as a major source of alanine nitrogen in man, *Diabetes,* 31, 86, 1982.
41. **Sprinson, D. B. and Rittenberg, D.,** Rate of interaction of amino acids of diet with tissue proteins, *Biol. Chem.,* 180, 715, 1949.
42. **Matthews, C. M. E.,** The theory of tracer experiments with ^{131}I-labeled plasma proteins, *Phys. Med. Biol.,* 2, 36, 1957.
43. **Picou, D. and Taylor-Roberts, T.,** The measurement of total protein synthesis and catabolism and nitrogen turnover in infants in different nutritional states and receiving different amounts of dietary protein, *Clin. Sci.,* 36, 283, 1969.
44. **Stein, T. P., Leskiw, M. J., Wallace, H. W., and Blakemore, W. S.,** Comparison of methods for the measurement of human protein synthesis, *Biochem. Med.,* 16, 211, 1976.

Chapter 10

MEMBRANE TRANSPORT IN VITRO

K. Janáček and R. Rybová

TABLE OF CONTENTS

I. INTRODUCTION

Transport across membranes is such a ubiquitous phenomenon in physiology that it is difficult to imagine an experiment with living cells which would not involve, to some extent, transport phenomena. From this it is seen that most experimental treatments of cells, tissues, and organs may bring, in principle, some information on membrane transport.

Still, there are several general approaches especially useful in transport studies. Chemical analyses of cells and tissues may be used to establish the amounts of substances which entered or left the cells. The use of tracers, most conveniently of radioactive isotopes, enables us to study transport phenomena even in steady states, when there is no net transport of substances across membranes, but their unidirectional fluxes, of course, continue. Finally, bioelectrical or electrophysiological, methods are of utmost importance in studies of transport of ions. Measurements of the so-called membrane potential (actually, the voltage across the membrane) make it possible to appreciate an important driving force in the transport of ions, and the flux of an individual ionic species can be sometimes identified with a measurable electrical current. Whole volumes together with practical schooling in laboratories would be necessary to properly explain all intricacies of the above methods. The aim of the present chapter is to give only basic information about the principles of the individual approaches. May it encourage the reader to look for more complete information in up-to-date monographs and review articles!

II. CHEMICAL ANALYSES IN TRANSPORT STUDIES

Most of the chemical elements can be determined by some of the three rather generally applicable analytical techniques, i.e., flame photometry, spectrography, and atomic absorption spectroscopy. Instruments of ever-increasing sophistication are available on the market, and literature delivered or recommended with them is the best guide to using them.

There are some notable exceptions for which the above universal approaches are not suitable, i.e., halogens, sulfur, nitrogen, and oxygen. The content of chloride anions may be conveniently determined by potentiometric (micro) titration; that of sulfate anions, by various spectrophotometric methods; nitrogen, for example, by the Conway microdiffusion analysis[1] after being transformed into ammonium ions by the procedure by Kjeldahl; and oxygen, again spectrophotometrically (especially continually in media) or by an oxygen electrode. Sensitive electrodes[2] are now commercially available for monitoring concentrations of various ions and compounds. Details of all the above methods may be found in biochemical and bioelectrochemical monographs.

Minute amounts of a number of elements may be determined by an extremely sensitive method known as neutron activation analysis, since these elements may be transformed by a flow of slow neutrons into radionuclides that can be assessed with a great precision. It is suitable to perform such activation on a chromatographic paper, since carbon, oxygen, and hydrogen of the cellulose cannot be activated by neutrons.

One of the most modern analytical methods, both convenient and rapid, is the new plasma spectrophotometry, based on the measurement of emission in the state of plasma, by which dozens of elements may be determined in a single sample in a few minutes.

Finally, there is a large choice of chromatographic and spectrophotometric techniques by which diverse organic compounds may be determined.

It should be added that due attention should always be paid to suitable extraction or mineralization procedures for experimental cells and tissues, which should prevent volatile components from escaping and unstable compounds from decomposing. Thus, for example, when the content of chloride anion is to be determined in a sample, a several-days-long extraction with a very dilute sulfuric acid is to be preferred over mineralization by boiling with concentrated nitric acid, during which some HCl may escape, etc.

Table 1
RADIOACTIVE ISOTOPES MOST
COMMONLY USED IN TRANSPORT
STUDIES

Element	Mass number of isotope	Half-life	Mode of decay
H	3 (tritium T)	12.3 years	β^-
C	14	5730 years	β^-
Na	22	2.6 years	β^+, γ
Na	24	15.0 hr	β^-, γ
P	32	14.3 days	β^-
S	35	87 days	β^-
Cl	36	300,000 years	β^-
K	42	12.4 hr	β^-, γ
Ca	45	165 days	β^-
Fe	59	45 days	β^-, γ
Rb	86	18.8 days	β^-
I	125	60 days	γ
I	131	8.1 days	β^-, γ

III. RADIOACTIVE ISOTOPES

Isotopes are used in transport studies to (1) determine transmembrane fluxes and to characterize the uptake kinetics, (2) to measure exchange of substances in steady states where the concentrations do not change, (3) to perform compartmental analysis, which is described in more detail later in this chapter, (3) to assess the effect of inhibitors, (4) to learn the fate of transported ions or organic substrates (in combination with the usual biochemical methods of separation and identification or, for example, histoautoradiography), etc.

Unidirectional flux is defined as the amount of substance transferred per unit time across a unit area from one side of the barrier to the other. It is expressed in moles m^{-2} sec^{-1} or sometimes in a more telling way in moles cm^{-2} sec^{-1}. The flux of the label (the so-called tracer flux) divided by the specific activity is equal to the unidirectional flux.

There are a number of radioactive isotopes which can be used in transport studies, and some of the most common are given in Table 1 together with their half-lives and modes of decay. The assay of radioactive isotopes is usually carried out on gas-flow counters, liquid scintillators, or crystal scintillators.[3]

Samples for gas-flow counters are dried on metal planchets and then inserted, one after another, into a chamber continuously flushed with methane or another gas ensuring low quenching of ionizing radiation. Emission of an ionizing particle starts a chain reaction, the gas becomes conductive, and a current pulse through a high-voltage tungsten wire in the chamber is registered. With samples of unequal weight and soft emitters, e.g., ^{3}H, ^{14}C, and ^{35}S, empirical corrections for self-absorption should be carried out.

The gas-filled devices are not suitable for determination of γ-emitters as photons mostly do not interact with the molecules of the diluted gas. In some crystals of high-density, light equivalents (which can be photomultiplied for detection) are emitted after interaction with photons. Very hard and penetrant emitters, e.g., ^{24}Na or ^{42}K, are therefore, most suitably assayed in a well-shaped crystal of sodium iodide activated with thallium. Self-absorption here is negligible; corrections for the decay of short-lived isotopes are necessary.

When the liquid scintillation method is used, the samples are introduced into a mixture of organic solvents; toluene ensures a high efficiency of the measurement; methanol, ethanol, or ethyleneglycol improve the miscibility of the mixture with water, in which the sample is dissolved. For good miscibility with water, dioxane is also often used as the organic solvent; its lower efficiency in energy transfer from the label to the scintillation molecules can be increased by naphtalene. Usually two scintillation substances are present in the mixture, 2,5-diphenyloxazole as the primary scintillator and 1,4-di-2-(5-phenyloxazolyl) benzene as the secondary one. Flashes of light are emitted upon interaction with β-particles, amplified by photomultipliers and counted. It is often of convenience to determine the radioactivity of samples on chromatographic papers or microbial filters directly by the liquid scintillation method. The orientation of paper pieces in scintillation vials has only a minute influence (except samples with ^3H) on the efficiency of determination. Corrections for quenching by various compounds present, color of the samples, or solid inclusions (paper strips) are to be made individually.

IV. INTERPRETATION OF EXPERIMENTALLY MEASURED PARAMETERS

A. Initial Rates and Saturation Kinetics

Initial rates of, for example, entry of a substance into cells or tissues, as opposed to its net inflow, depend on the concentration of the substance at one side of the membrane only (at the other side it is negligible) and are useful in determining the character of the transport process. Thus, if cells or tissues are suspended at time zero in a medium with a concentration c_o of a substance permeating the membrane by simple diffusion, the initial rate of entry (the initial influx) will be

$$J_{init} = Pc_o \tag{1}$$

where P is the permeability constant.

If, on the other hand, the influx proceeds by a saturable mechanism (by mediated diffusion or active transport), the initial influx will be given by the Michaelis-Menten formula,

$$J_{init} = J_{max} \frac{c_o}{K_{0.5} + c_o} \tag{2}$$

where J_{max} is the maximum unidirectional flow, corresponding to a full saturation of the transport mechanism, and $K_{0.5}$ is the half-saturation constant (the concentration at which J_{init} = $J_{max}/2$).

How to determine the initial influx experimentally? Certainly a minute but measurable amount of a substance which entered the cells during the shortest possible time interval gives a good approximation (expressed, say, in moles per kilogram of fresh weight per minute — knowing the surface to volume ratio, the figure can be transformed into the usual units of moles per square centimeter per second, since the fresh weight and volume of cells approximately coincide). However, there are better ways of estimating the slope of the time dependence of the substance entry at time zero, using two or more initial, evenly spaced points. Thus, for example, if the total inflow at the time t is y_1 and that at the time 2t is y_2, the initial influx may be calculated as

$$J_{init} = \frac{4y_1 - y_2}{2t} \tag{3}$$

as shown in the following example.

Time (min)	Inflow of the substance (mol kg^{-1})
0	0
2	8
4	14

$$J_{init} = \frac{4 \cdot 8 - 14}{2 \cdot 2} = \frac{18}{4} = 4.5 \text{ mol kg}^{-1} \text{ min}^{-1}$$

Now, plotting J_{init} against different concentrations c_o, the straight line for simple diffusion can be easily recognized and evaluated; with saturable transport various linearization procedures are recommended for estimation of J_{max} and $K_{0.5}$. In the procedure of Lineweaver and Burk reciprocal values of initial fluxes are plotted against reciprocal values of the corresponding concentrations, i.e., Equation 2

$$J_{init} = J_{max} \frac{c_o}{c_o + K_{0.5}}$$

is plotted in the reciprocal form

$$\frac{1}{J_{init}} = \frac{K_{0.5}}{J_{max}} \frac{1}{c_o} + \frac{1}{J_{max}} \tag{4}$$

as shown in Figure 1.

Woolf and Hofstee suggested a different procedure which changes less the statistical weight of individual points and is based on the following transformation:

$$J_{init} = J_{max} \frac{c_o}{c_o + K_{0.5}} = J_{max}\left(1 - \frac{K_{0.5}}{c_o + K_{0.5}}\right)$$

$$= J_{max}\left(1 - \frac{K_{0.5}}{c_o} \frac{c_o}{c_o + K_{0.5}}\right) = J_{max} - K_{0.5} \frac{J_{init}}{c_o} \tag{5}$$

The evaluation of parameters is obvious from Figure 2.

B. Kinetics of Tracer Exchange

In a steady state, concentrations of chemical substances in individual physiological compartments (homogenous regions of space separated from the rest by membranes) do not change. Still, the exchange of these substances between different compartments (e.g., cells and their surroundings) can be studied and their unidirectional fluxes determined using radioisotopes, added to some of the compartments in minute but measurable amounts. The driving force of tracer exchange is its "specific activity" (concentration of the isotope c^* divided by the total concentration of the substance c), and unless it is everywhere the same, an observable net transfer of tracer occurs. The rate of change of the isotope concentration in compartment i communicating with compartment o is given by

$$\frac{dc_i^*}{dt} = \frac{J_{oi}}{V_i} A \frac{c_o^*}{c_o} - \frac{J_{io}}{V_i} A \frac{c_i^*}{c_i} \tag{6}$$

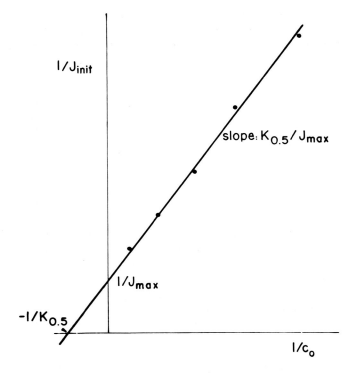

FIGURE 1.   Plot of data according to Lineweaver and Burk.

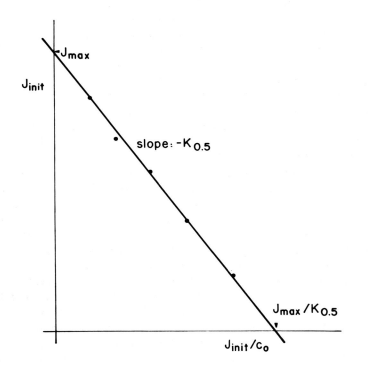

FIGURE 2.   Plot of data according to Woolf and Hofstee.

where V_i is the volume of the compartment, A its surface, J_{oi} is unidirectional influx, and J_{io} unidirectional efflux. Simple experimental conditions are preferred, during which, for example, c_o^* is negligibly low: the compartment is prelabeled with the isotope, and then the isotope is washed out into a sequence of fresh media. The solution of the above equation is then

$$c_i^* = c_{i,t=0}^* \exp \left(- \frac{AJ_{io}}{V_i c_i} t \right) \tag{7}$$

and can be linearized simply by taking natural logarithms

$$\ln c_i^* = \ln c_{i,t=0}^* - \frac{AJ_{io}}{V_i c_i} t \tag{8}$$

Our experimental system frequently represents more than one compartment in parallel or in a series. The wash-out curve of isotope then corresponds to a sum of exponential terms and cannot be simply linearized. However, if the rates of individual processes are sufficiently different, $\ln c^*$ becomes a linear function of time as soon as the rest of the exponential terms become negligibly small. The slowest exponential can thus be evaluated, subtracted from the experimental data, and the whole process repeated. Intricacies of this analysis can be found, for example, in the book by Kotyk and Janáček[4] or other, still more specialized monographs.[5,6]

V. BIOELECTRICAL MEASUREMENTS

A. Split Chambers and the Short-Circuit Current Technique

Measurement of differences of electrical potential across epithelial layers in vitro is relatively easy and free of theoretical objections. This is so since many of these layers have sufficiently large areas and can be mounted into split chambers, usually made of transparent material such as perspex, and there they separate two media of well-defined composition. Epithelial layers of poikilotherms, e.g., frog skin[7] or toad bladder, are especially easy to work with, since they survive for many hours at room temperature in simple aerated salt solutions. Transepithelial difference of the electrical potential is then simply the voltage between two solutions of identical or analogous composition and can be measured by a millivoltmeter and a pair of electrodes connected to the solution by bridges, calomel electrodes and KCl bridges being the usual choice.

The split-chamber approach reached its highest perfection in the ingenious short-circuit current technique by Ussing and Zerahn[8] and Rehm.[9] The principles of this technique are obvious from the drawing in Figure 3. To the circuit for measurement of the electrical potential difference (with ends of bridges very close to the layer surfaces) another circuit is added, in which voltage derived from a battery by a potentiometer drives current of a required direction and intensity across the layer. Two current electrodes (usually silver-silver chloride) are used for the purpose, with ends of the bridges rather far from the surfaces, so that the density of the electric current across the epithelial layer is approximately the same everywhere. Bridges connecting the silver-silver chloride electrodes with media in the chamber may be prepared from polyethylene tubing and filled with 1 to 3% agar gel in 3 M KCl or Ringer solution; agar gel liquefied by heating can be aspirated into the tubing with a syringe or water pump. Constrictions at the end of the bridges prevent agar gel from escaping. If required, silver-silver chloride electrodes may be placed directly in the media, since the solubility product of the silver chloride is low ($1.73 \cdot 10^{-10}$ at 25°C) and the concentration of silver cations in media containing chlorides still lower. On the other hand, sometimes it

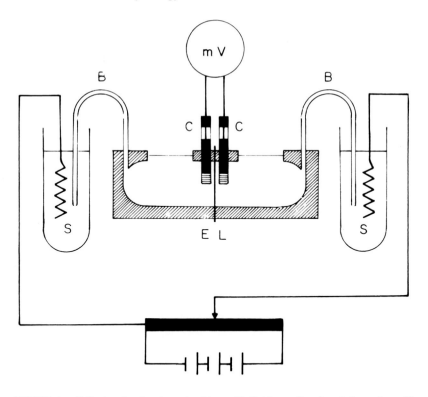

FIGURE 3.   Split chamber for short-circuiting epithelial layer. C: calomel electrodes, mV: millivoltmeter, S: silver-silver chloride electrodes, and B: bridges.

may be convenient to also connect the calomel electrodes with media by agar bridges. The ends of these bridges should again be close to the layer surfaces, to prevent measuring of a potential drop due to passage of current across an appreciable thickness of the media.

The current circuit makes it possible to change rather arbitrarily the electrical potential difference between the two media; the slope of the voltage-current characteristics is a measure of the epithelial layer resistance. When the current just compensates for the transfer of charge by active transport across the layer, the electrical potential difference between the two media drops to zero. When the media at the two surfaces of the layer are of identical composition it is obvious that all current under such total short-circuiting corresponds to an active transport of ion(s) across the layer. Isotopes are used to identify the ions actively transported. For instance, Ussing and Zerahn used a double-labeling technique (^{24}N and ^{22}Na) to identify active transport of sodium ions as the source of short-circuit current across frog skin.[8]

It is rather obvious that the interpretation of short-circuit current measurement is different when we are interested in the function of the whole complex wall of an organ performing, as a black box, transport between two solutions bathing its surfaces, and when we seek to find the active transport across the layer of epithelial cells only or even only across the cell membrane at which the pump is located. The details of the latter approach, related to parallel leaks and resistances, were discussed by Rehm[9] and Tai and Tai.[10] It should also be always kept in mind that coupled electrically silent transport, not accessible to electrical measurements, may be present.

B. Microelectrode Technique

The unique monograph *Microelectrode Methods for Intracellular Recording and Iono-phoresis* by Purves[11] is heartily recommended to everybody seriously interested in these matters. See also Heinz[12] and Garcia-Diaz et al.[13] for further reading. The following is an attempt to explain basic principles of the microelectrode technique.

The device commonly called a microelectrode by electrophysiologists is, in fact, a microbridge drawn from a borosilicate capillary into a fine tip with a diameter of less than 1 μm and filled usually with 3 M potassium chloride. The tip has to be as fine as that; in order not to damage a cell seriously its diameter should be less than 0.5% of the diameter of the cell. Concentrated potassium chloride solution has two advantages. First, it is rather conductive. Second, potassium cations and chloride anions have practically identical mobilities in water, and this property often reduces the liquid-junction potential at the tip of the electrode, called the tip potential, to a value of only a few millivolts. Still, the tip potential should be measured in a solution approximating the composition of the cell interior, and the data should be corrected for it. The electrical circuit is completed by two reference electrodes. For precise measurements calomel electrodes are recommended as the most stable; one of them is connected by a thin agar bridge with the interior of the microelectrode, another is in contact with the experimental medium. Sometimes silver-silver chloride electrodes are used, i.e., two pieces of silver wire covered with chloride, one of them dipped directly into the microelectrode. The measurement itself can be carried conveniently with a millivoltmeter with a high input resistance and recorded; for rapid changes an oscilloscope is necessary.

For preparation of microelectrodes (or micropipettes, as empty microelectrodes are sometimes called) commercially available microelectrode pullers are used. Borosilicate capillaries are locally softened by an electrically heated platinum loop, and two pieces of microelectrodes are torn apart by an electromagnetic pull. Careful adjustments of the heating and pull by trial and error allow us to modify the shape and tip diameter of the electrodes.

Filling of microelectrodes may be achieved in several ways. Microelectrodes kept in a suitable holder with their tips downwards may be immersed in a flask with methanol boiling under reduced pressure effected by a water pump. To remove air bubbles the suction is repeatedly interrupted; the bubbles shrink and leave the capillaries. Methanol in the capillaries is then replaced by distilled water and water by 3 M KCl, the different densities of these liquids making the exchange complete within a few hours. A long storage of microelectrodes in solutions is not recommended as the tips become hydrated and lose their solidity. More modern methods of filling use capillary forces. Glass fiber, preferentially of the same borosilicate, is introduced into the capillary before pulling the microelectrodes (commercially available capillaries have such fiber fused to the wall). A suitable diameter of the fiber is about 100 μm. After pulling, the fiber remains fused with the wall in the tip, and it is sufficient to inject the filling solution into the microelectrode by a syringe — it advances to the tip by capillarity and fills it within a few seconds.

Finally, a suitable micromanipulator has to be chosen to introduce the microelectrode into the cell. Instruments working on mechanical, pneumatic, or hydraulic principles are available commercially.

C. Resistance of Cell Membranes Measured with Microelectrodes

Ohm's law is used to estimate the resistance of cell membranes. If current of known density (in A·cm^{-2}) passes through the membrane at the point impaled by a microelectrode, a change in the membrane potential ΔE mV is recorded. The ratio of this change to the current density is then the membrane resistance (in Ω·cm^2). The transfer of electrical charge across the membrane should be kept as low as possible to prevent ill-defined polarization effects. For this reason, short, square-wave impulses of low intensity are usually injected into a cell; from the distorted shape of the response the capacity of the membrane may be determined. From the steady-state value of the response ohmic resistance may be calculated.

Evaluation of current density at the point of electrical potential monitoring is rarely a trivial problem (unless current is injected, for example, into the center of a spherical cell).

Decrease of the current density with the distance from the point of current injection by a current microelectrode is described by appropriate solution of the so-called cable equation. In elongated structures such as nerve or muscle fibers the situation is relatively simple, since the decay of the current density with the distance is exponential.

For details[14,15] and further extensions[16] see the appropriate literature.

D. Stimulation, Voltage Clamp, Ionophoresis

Injection of current through microelectrodes may not only serve to measure the resistance and capacitance of membranes, but also the stimulation of excitable membranes. Excitability is highly developed in membranes of nerve and muscle fibers. The basis of excitability is the opening of further ion-selective channels in the membrane; many of these channels open when the electrical field in the membrane is changed, and this may be achieved by an appropriate injection of current. Also, the current necessary to maintain the membrane potential at a certain level may be monitored by means of microelectrodes. This current is mediated by a transport of ions across the membrane; by changing the ionic composition of media it is possible to find out which ionic species are the main carriers of electrical charge across the membrane.

This electrophysiological approach is known as the voltage-clamp technique,[17] and its basis is a circuit with a negative feedback which brings to zero any deviation from the required value of the membrane potential. The short-circuit current technique discussed above is a special case of the voltage clamp. By adjusting the value of the membrane potential and concentrations of ions in the medium, the difference of the electrochemical potential for a given ionic species can be more or less arbitrarily changed and the character of the transport studied. For instance, if the difference of the electrochemical potential across the membrane for a given ionic species is zero, these ions can carry measurable current (which can be identified with isotopes) only as a result of an active transport (transport coupled directly or indirectly to metabolism).

Finally, the injection of current through a microelectrode may serve to ionophoresis. Current flowing through a microelectrode inserted into a cell may carry into the cytoplasm minute amounts of active substances, dyes, etc.[11]

E. Ion-Selective Microelectrodes

Whereas microelectrodes for measuring membrane potentials are mere salt microbridges, i.e., conductive pathways to the cell interior, ion-selective microelectrodes are real electrodes that can be used to measure intracellular activities of cations or anions.[2,18-21] Their preparation is similar to that of the ordinary glass microelectrodes. Sometimes they are made from selective glasses (as, for example, the electrode for intracellular pH measurement); their tips are then sealed, and the surfaces of the rest of the microelectrodes are isolated by suitable coats. Most often some 200 to 300 μm of the microelectrode is filled with a liquid containing a specific ion carrier. Specific "cocktails" (i.e., ready mixtures for filling the microelectrodes) for estimation of activities of H^+, Na^+, K^+, Mg^{2+}, and Ca^{2+} are now commercially available.

When inserted into a cell, the ion-selective microelectrodes measure not only the activity of a given ion, but also the membrane potential (the difference of the electrical potential across the membrane). Hence, the membrane potential has to be measured separately with an ordinary microelectrode, and a correction must be made.

F. Alternative Methods for Measuring Membrane Potentials

There are approaches enabling us to measure membrane potentials without microelectrodes which can be applied to elongated structures such as nerve or muscle fibers. They may be considered as methodological improvements of the original attempts to measure membrane potentials with extracellular electrodes, one of them closed to the intact membrane, the other in the proximity of a membrane injured by dissection or depolarized by concentrated potassium chloride. These so-called injury potentials represent only a minute fraction of the real membrane potential due to a short circuit in the external medium. Somewhat better results were obtained when an insulation by air or mineral oil was applied between the two points; however, underneath such insulation there was always the moist and conductive surface of the membrane. A radical improvement was achieved by the technique of the sucrose gap,[22] in which the membrane is washed by a solution of pure sucrose in deionized water.

Whereas the sucrose gap is suitable for elongated structures, membrane potentials of large spherical cells, such as snail neurons, may be measured by Kostyuk's and Krishtal's method.[23] Small apertures in plastic tubings are placed in contact with the cell surfaces, cell membranes are ruptured by suction, and conductive pathways to the cell interior are established. Apart from the membrane potential measurement, also perfusion of the cell interior and thus control of its ionic composition are possible.

Finally, some cells or intracellular organelles are so small that the microelectrode technique can hardly be applied to them, let alone Kostyuk's and Krishtal's technique. The membrane potential can then be calculated from the stationary Donnan distribution of a radioisotope-labeled lipophilic (and thus permeable) ion, e.g., triphenylmethylphosphonium, tetraphenylphosphonium, or tetraphenylarsonium.[24,25]

G. Study of Individual Channels in Membranes

Channels, or ion-selective pores, are formed by integral proteins penetrating the whole thickness of a membrane. Transport of ions across individual channels, its voltage, or agonist dependence may be studied by modern electrophysiological methods. There are two different approaches to this problem; one of them is the noise analysis,[26,27] the other, the patch clamp method.[28-30]

By the statistical analysis of noise, fluctuations in the number of channels open at a given instant of time are studied. Both the amplitude of an individual channel current and the average time for which the channel remains open are in principle accessible by this indirect approach.

Real progress was achieved by the patch-clamp method which allows the monitoring of current which passes through even a single channel. The method is based on principles analogous to those mentioned above in connection with the voltage-clamp technique and Kostyuk's and Krishtal's method of membrane potential measurement. A micropipette is brought into contact with a membrane, and by application of a slight mouth suction a small area of the membrane (a patch) is separated and adheres to the capillary tip. The area of the patch is only 10 μm^2 and may encompass only a few channels (or only a single one). Both inside-out and outside-out patches may be obtained depending on the mode of interaction of the micropipette (diameter of the tip 1 to 2 μm) with the membrane. The perfect seal between the patch and the micropipette is apparent from a very high resistance in the circuit (of the order of 10^9 Ω).

VI. STUDIES WITH MEMBRANE VESICLES

It is possible to investigate transport processes in isolation from metabolism in well-defined media on both sides of the barrier and to learn the effects of external as well as of internal factors under conditions preset at will with membrane vesicles.

A classical example is erythrocyte ghosts; later animal physiologists aimed at preparing tight vesicles mostly from epithelial tissue to get the opportunity of separately studying transport properties of luminal and basolateral membranes. As the physiological role of these two membranes is different, they differ also to a certain degree in their chemical composition, e.g., protein-lipid ratio, surface charge density, etc., which enables one to separate the membrane material after homogenization of the tissue.[31,32] Sealed vesicles are formed spontaneously under conditions used for isolation of membrane fractions.

The most successful separation for luminal membrane fractions was achieved by differential precipitation with calcium or magnesium ions. The surface-charged groups of basolateral membranes and of cellular organelles form cross-linked bonds with the ions and precipitate. In the procedure introduced first by Schmitz et al.[33] $CaCl_2$ was added to a final concentration of 10 mM to intestinal tissue homogenized in 0.05 M buffered mannitol. The clear supernatant obtained after centrifugation of the Ca^{2+}-treated homogenate contained fragments of the luminal membrane which sedimented at 20,000 × g.

Basolateral membrane vesicles were successfully prepared by the relatively quick and simple separation on the self-generating Percoll (modified colloidal silica) density gradient.[34,35] This method substituted the previously used and rather lengthy procedures of differential centrifugation followed by sucrose density gradient centrifugation or separation by free-flow electrophoresis.

However, the preparation procedures of tight vesicles from epithelial membranes mostly have not yet yielded homogenous material, this being reflected among other factors in the differing size of the vesicles. The heterogeneity also concerns the orientation of the vesicles (right-side-out orientation of the membrane was obtained with luminal vesicles, whereas in basolateral vesicles the membrane is rather randomly oriented) and differences in membrane composition which correspond to the histological differences *in situ* (e.g., to the topographical variability of enzyme protein concentrations in the brush-border membrane of the small intestine). The cross-contamination by membraneous material from other cell fractions is not rare. Therefore, one has to be cautious when interpreting the results; especially the measurements of transport rate (e.g., V_{max}) are influenced by vesicle heterogeneity. Three-dimensional cell fractionation based on sedimentation coefficient, density, and partitioning in a dextrane-polyethylene glycol two-phase system was suggested[36] to decrease the cross-contamination of the membrane fractions.

Moreover, changes in permeability to ions such as H^+ or K^+ as compared with intact cell membrane permeabilities were clearly observed, pointing to the fact that the transport properties of the membrane may be changed during isolation procedures. Nevertheless, studies on vesicles are helpful, for example, in distinguishing transport characteristics on the cis- and trans- sides of the membrane or systems residing in the luminal and basolateral membranes; in characterizing the stoichiometry of cotransport mechanisms; or in assessing the contribution of transmembrane potential difference to a transport process. Useful information on Na^+/H^+ and Na^+/Ca^{2+} antiporters and Na^+-cotransport systems (e.g., Na^+-glucose, Na^+-lactate, or Na^+-inorganic anion, amino acid transport, etc.) has been obtained with isolated membrane vesicles, the methods of their preparation being currently improved.

REFERENCES

1. **Conway, E. J.,** *Microdiffusion Analysis and Volumetric Error,* Crosby Lockwood & Son, London, 1957.
2. **Koryta, J.,** *Ion-Selective Electrodes,* Cambridge University Press, Cambridge, 1975.
3. **Hendee, W. R.,** *Radioactive Isotopes in Biological Research,* John Wiley & Sons, New York, 1973.
4. **Kotyk, A. and Janáček, K.,** *Cell Membrane Transport,* 2nd ed., Plenum Press, New York, 1975.
5. **Atkins, G. L.,** *Multicompartment Models for Biological Systems,* Methuen, London, 1969.
6. **Jacquez, J. A.,** Tracers in the study of membrane processes, in *Membrane Physiology,* Andreoli, T. E., Hoffman, J. F., and Fanestil, D. D., Eds., Plenum Press, New York, 1980, 147.

7. **Clarkson, T. W. and Lindemann, B.**, Experiments on Na transport of frog skin epithelium, in *Laboratory Techniques in Membrane Biophysics,* Passow, H. and Stämpfli, R., Eds., Springer Verlag, Heidelberg, 1969, 85.

8. **Ussing, H. H. and Zerahn, K.**, Active transport of sodium as the source of electric current in the short-circuited isolated frog skin, *Acta Physiol. Scand.,* 23, 110, 1951.

9. **Rehm, S. W.**, Ion transport and short-circuit techniques, in *Current Topics in Membranes and Transport,* Vol. 7, Bronner, F. and Kleinzeller, A., Eds., Academic Press, New York, 1975, 217.

10. **Tai, Y.-H. and Tai, C.-Y.**, The conventional short-circuiting technique under-short-circuits most epithelia, *J. Membr. Biol.,* 59, 173, 1981.

11. **Purves, R. D.**, *Microelectrode Methods for Intracellular Recording and Ionophoresis,* Academic Press, London, 1981.

12. **Heinz, E.**, *Electrical Potentials in Biological Membrane Transport,* Springer-Verlag, Heidelberg, 1981.

13. **Garcia-Diaz, J. F., Stump, S., and Armstrong, W. McD.**, Electronic device for microelectrode recordings in epithelial cells, *Am. J. Physiol.,* 246, C339, 1984.

14. **Rybová, R.**, Nerve, in *Cell Membrane Transport,* 2nd ed., Kotyk, A. and Janáček, K., Eds., Plenum Press, New York, 1975, 428.

15. **Schanne, O. F. and Ruiz P.-Ceretti, E.**, *Impedance Measurements in Biological Cells,* John Wiley & Sons, Chichester, 1973.

16. **Frömter, R.**, The route of passive ion movement through the epithelium of *Necturus* gall-bladder, *J. Membr. Biol.,* 8, 259, 1972.

17. **Smith, T. G., Jr.**, Various methods of voltage clamping membrane potential with microelectrodes, in *Current Methods in Cellular Neurobiology,* Vol. 3, Barker, J. L. and McKeloy, J. F., Eds., John Wiley & Sons, Chichester, 1983, 39.

18. **Kessler, M., Clark, L. C., Lubbers, D. W., Silver, I. A., and Simon, W.**, *Ion and Enzyme Electrodes in Biology and Medicine,* Urban and Schwarzenberg, Munich, 1976.

19. **Thomas, R. C.**, *Ion-Sensitive Intracellular Microelectrodes,* Academic Press, London, 1978.

20. **Ammann, D.**, *Ion-Selective Microelectrodes,* Springer-Verlag, Berlin, 1986.

21. **Baumgarten, C. M.**, Methods for monitoring myocardial potassium and chloride with ion-selective microelectrodes, in *Methods in Studying Cardiac Membranes,* Vol. 2, Dhalla, N. S., Ed., CRC Press, Boca Raton, Fla., 1984, 213.

22. **Boev, K. and Golenhofen, K.**, Sucrose-gap technique with pressed-rubber membranes, *Pflügers Arch.,* 349, 277, 1974.

23. **Kostyuk, P. G. and Krishtal, O. A.**, Separation of sodium and calcium currents in the somatic membrane of mollusc neurones, *J. Physiol.,* 270, 545, 1977.

24. **Salzberg, B. M.**, Optical recording of electrical activity in neurones using molecular probes, in *Current Methods in Cellular Neurobiology,* Vol. 3, Barker, J. L. and McKeloy, J. F., Eds., John Wiley & Sons, Chichester, 1983, 139.

25. **Demura, M., Kamo, N., and Kobatake, Y.**, Determination of membrane potential with lipophilic cations. Comparison of estimated values with various phosphonium ions, *Biochim. Biophys. Acta,* 812, 377, 1985.

26. **De Felice, L. J.**, *Introduction to Membrane Noise,* Plenum Press, New York, 1981.

27. **Mathers, D. A.**, Electrical noise in biological membranes, in *Current Methods in Cellular Neurobiology,* Vol. 3, Barker, E. L. and McKeloy, J. F., Eds., John Wiley & Sons, Chichester, 1983, 101.

28. **Sakmann B. and Neher E.**, *Single-Channel Recording,* Plenum Press, New York, 1983.

29. **Auerbach, A. and Sachs, F.**, Patch-clamp studies of single ion channels, *Annu. Rev. Biophys. Bioeng.,* 13, 269, 1984.

30. **Rae, J. L. and Levis, R. A.**, Patch voltage clamp of lens epithelial cells: theory and practice, *Mol. Physiol.,* 6, 115, 1984.

31. **Murer, H. and Kinne, R.**, The use of isolated membrane vesicles to study epithelial transport processes, *J. Membr. Biol.,* 55, 81, 1980.

32. **Murer, H., Biber, J., Gmaj, P., and Stieger, B.**, Cellular mechanisms in epithelial transport: advantages and disadvantages of studium with vesicles, *Mol. Physiol.,* 6, 55, 1984.

33. **Schmitz, J., Preiser, H., Maestracci, D., Ghosh, B. K., Cerda, J. J., and Crane, R. K.**, Purification of the human brush border membrane, *Biochim. Biophys. Acta,* 323, 98, 1973.

34. **Scalera, V., Storelli, C., Storelli-Joss, C., Haase, W., and Murer, H.**, A simple and fast method for the isolation of basolateral plasma membranes from rat small-intestinal epithelial cells, *Biochem. J.,* 186, 177, 1980.

35. **Sacktor, B., Rosenbloom, I. L., Liang, C. T., and Cheng, L.**, Sodium gradient- and sodium plus potassium gradient-dependent L-glutamate uptake in renal basolateral membrane vesicles, *J. Membr. Biol.,* 60, 63, 1981.

36. **Mircheff, A. K.**, Empirical strategy for analytical fractionation of epithelial cells, *Am. J. Physiol.,* 244, G347, 1983.

Chapter 11

METHODS IN ENERGY BALANCE, FOOD INTAKE, AND NUTRITION

Z. Deyl and M. Adam

TABLE OF CONTENTS

I. INTRODUCTION

Energy balance of organisms can be reached through regulatory adjustments of metabolic expenditures and randomly determined food intake. On the other hand, regulation may result from an adjustment of the feeding controlling mechanisms to changing energetic demands. Whatever the effectors involved, the balance between energy input and output is very accurate as demonstrated by the constancy of body weight throughout maturity.

Because the physiological balance between sources of energy (nutriments) and energy output involves practically all organs directly or indirectly, the literature on this topic belongs to different branches of physiology and biochemistry, and the overall number of references is enormous.[1] The methodological background of energy balance and food intake covers (1) energic expenditures (this can be done by either direct or indirect calorimetry) and (2) information about energy influx which is obtained from measurements about the amount and quality of food accepted. As methods of food analysis are well beyond the scope of the present volume, the reader is, in this respect, deferred to more specialized monographs.[2,3]

II. DIRECT CALORIMETRY

Direct calorimetry[4-6] purely serves the purpose of measuring heat production; it has to be kept in mind that about one quarter of body heat is dissipated by moisture vaporization which can be measured by absorption (using e.g., sulfuric acid, perchlorate, etc.) while the remaining three quarters is emitted by radiation, conduction, and convection and can be measured by adsorption in water.

A calorimeter is a well-insulated box, just large enough to hold the subject. The chamber interior is kept at constant temperature by water circulating through pipes attached to the ceiling. The heat absorbed by the water is computed from the amount of water flowing per unit time and from the temperature difference betwen ingoing and outgoing water. In compensation calorimetry, one chamber holds the animal while another, similar chamber has electric resistance wires to produce exactly the same amount of heat as that emitted by the subject in the other chamber. Finally, it is also possible to apply differential calorimetry in which the rate of total heat supply is maintained constant — first, with the calorimeter empty, and second, with the animal inside.

III. INDIRECT CALORIMETRY

Indirect calorimetry is based on the fact that normally oxygen consumption and carbon dioxide production are closely related to heat production. Since the energetic equivalent of oxygen consumed and carbon dioxide produced varies with the substance oxidized, it is theoretically necessary to know the composition of the food oxidized. The amount of protein oxidized is computed from urinary nitrogen excretion. Assuming that protein contains 16% nitrogen and that all urinary nitrogen is derived from protein oxidation, the protein catabolized is estimated by multiplying the urinary nitrogen by 6.25. These assumptions are fully acceptable though not literally true. The relative amounts of fat and carbohydrate oxidized are determined from the nonprotein respiratory quotient (RQ). The RQ is the ratio of moles of CO_2 produced to moles of oxygen consumed. For the oxidation of carbohydrates RQ equals unity; for mixed fat, is 0.71. Short-chain fats have an RQ nearer 0.8; long-chain fats, nearer 0.7. The RQ for mixed protein is around 0.81 (e.g., being 0.83 for the amino acid alanine).

Table 1 indicates the percentages of fats and carbohydrates oxidized and the energetic equivalents of oxygen and carbon dioxide for different nonprotein RQ values. Table 2 illustrates the method of computing heat production employing the RQ values and urinary

Table 1
THERMAL EQUIVALENTS OF O_2 AND CO_2 AND THE CORRESPONDING PERCENTAGES OF FAT AND CARBOHYDRATES OXIDIZED FOR DIFFERENT RQ

RQ	O_2 J/ℓ	CO_2 J/ℓ	CO_2 J/g	% O_2 consumed by Carbohydrates	% O_2 consumed by Fat	% Heat produced by oxidation of Carbohydrates	% Heat produced by oxidation of Fat
0.70	19.606	28.008	14.259	0	100	0	100
0.71	19.623	27.640	14.071	1.0	99.0	1.1	98.9
0.72	19.673	27.326	13.912	4.4	95.6	4.8	95.2
0.73	19.723	27.020	13.757	7.85	92.2	8.4	91.6
0.74	19.778	26.727	13.606	11.3	88.7	12.0	88.0
0.75	19.786	26.439	13.460	14.7	85.3	15.6	84.4
0.76	19.882	26.163	13.318	18.1	81.9	19.2	80.8
0.77	19.936	25.886	13.180	21.5	78.5	22.8	77.2
0.78	19.983	25.619	13.042	24.9	75.1	26.3	73.7
0.79	20.037	25.363	12.912	28.3	71.7	29.9	70.1
0.80	20.087	25.108	12.782	31.7	68.3	33.4	66.6
0.81	20.138	24.861	12.657	35.2	64.8	36.9	63.1
0.82	20.189	24.619	12.535	38.6	61.4	40.3	59.7
0.83	20.242	24.389	12.414	42.0	58.0	43.8	56.2
0.84	20.292	24.158	12.297	45.4	54.6	47.2	52.8
0.85	20.347	23.937	12.184	48.8	51.2	50.7	49.3
0.86	20.397	23.719	12.075	52.2	47.8	54.1	45.9
0.87	20.447	23.502	11.966	55.6	44.4	57.5	42.5
0.88	20.502	23.297	11.862	59.0	41.0	60.8	39.2
0.89	20.552	23.091	11.757	62.5	37.5	64.2	35.8
0.90	20.602	22.891	11.652	65.9	34.1	67.5	32.5
0.91	20.652	22.694	11.552	69.3	30.7	70.8	29.2
0.92	20.702	22.502	11.456	72.7	27.3	74.1	25.9
0.93	20.753	22.313	11.360	76.1	23.9	77.4	22.6
0.94	20.807	22.133	11.268	79.5	20.5	80.7	19.3
0.95	20.857	21.953	11.175	82.0	17.1	84.0	16.0
0.96	20.907	21.778	11.088	86.3	13.7	87.2	12.8
0.97	20.962	21.610	11.000	89.8	10.2	90.4	9.6
0.98	21.012	21.439	10.916	93.2	6.8	93.6	6.4
0.99	21.062	21.276	10.832	96.6	3.4	96.8	3.2
1.00	21.117	21.117	10.749	100	0	100	0

nitrogen production in a dog. In most cases it is not advisable to apply the whole calculation as outlined above, as the RQ values have not always the rigorous significance given in Table 1. Thus, for example, cattle and other ruminants produce large quantities of CO_2 in the digestive tract by anaerobic bacterial fermentation and by liberation of CO_2 from bicarbonates. Naturally this extrametabolic CO_2 cannot be distinguished from respiratory metabolism CO_2. Under such conditions oxygen consumption is the best measure of heat production. Excessive CO_2 may also be liberated under conditions of acidosis and overventilation in general. A low RQ may also result from incomplete oxidation formation of sugars from proteins, fats, and other substances. A survey of heat production in albino rats is shown in Table 3. While the range of caloric equivalents of CO_2 is relatively wide, the range of the energetic equivalent of O_2 is relatively narrow (19.66 to 20.92 J), an extreme range of 7% or a deviation of about 3.5% from the mean value considering RQ 0.82 which is within the limits of experimental error in metabolism measurements. Furthermore, since the average RQ of a protein is 0.82 which corresponds to the average caloric value of O_2 of 20.18 J, no correction need be made for protein metabolism.

The simplest and probably the most versatile method for measuring energy metabolism

Table 2
COMPUTING ENERGY METABOLISM AND NONPROTEIN RQ FROM THE URINARY N AND RESPIRATORY EXCHANGE IN A DOG (AN EXAMPLE)

1.	g urinary N excreted/hr	0.136
2.	g "protein" oxidized ([1] × 6.25)	0.850
3.	g CO_2 associated with protein oxidation ([1] × 9.35)	1.272
4.	g O_2 associated with protein oxidation ([1] × 8.49)	1.15
5.	ℓ CO_2 associated with protein oxidation ([3] × 0.5087)	0.647
6.	ℓ O_2 associated with protein oxidation ([4] × 0.6998)	0.805
7a.	Total g CO_2 exhaled/hr	6.75
7b.	Total ℓ CO_2 exhaled/hr ([7a] × 0.509)	3.44
8a.	Total g O_2 consumed/hr	6.17
8b.	Total ℓ O_2 consumed/hr ([8a] × 0.6998)	4.32
9.	ℓ nonprotein CO_2 ([7b] − [5])	2.79
10.	ℓ nonprotein O_2 ([8b] − [6])	3.52
11.	Nonprotein RQ ([9]/[10])	0.79
12.	Overall RQ ([7b]/[8b]) (indicating that overall and nonprotein RQ are not likely to differ much)	0.80
13.	J/ℓ O_2 at given RQ (0.79)	20.037
14.	Nonprotein energy (J) ([10] × [13])	70.53
15.	Protein energy (J) ([1] × 110.8)	15.07
16.	Total energy by direct calorimetry (J) ([14] + [15])	85.6
17.	J by direct calorimetry (J)	87.70

Note: 1 g urinary N derived from protein is associated with the consumption of 5.91 ℓ or 8.49 g O_2, the production of 4.76 ℓ or 9.35 g CO_2, and production of 110.8 J 1 g O_2 = 0.6998 ℓ, 1 g CO_2 = 0.509 ℓ. The oxidation of 1 g fat (tripalmitin) is associated with consumption of 2.01 ℓ O_2 and production of 1.41 ℓ CO_2; the oxidation of 1 g starch is associated with the consumption of 0.83 ℓ O_2 and production of 0.83 ℓ CO_2. The nonprotein RQ is estimated by deducting the ℓ and protein CO_2 produced (g urinary N × 4.76) from the total ℓ CO_2 produced, and the ℓ of protein O_2 consumed (g urinary N × 5.9) from the ℓ of total O_2 consumed.

is measuring the rate of oxygen consumption as fed from a calibrated oxygen container and computing the heat production by the energetic value of oxygen. This method was adopted for measuring the energy metabolism in farm animals (cattle, horses, swine, sheep, and dogs) and was used for measuring heat production even in elephants.

The indirect methods for measuring heat production in animals are divided into the:

1. Closed-circuit type involving rebreathing the same air after removing its CO_2 by circulating through alkali and replacing the consumed O_2 by fresh oxygen[7]
2. Open-circuit type[8] involving the circulation of outside air through the system.

A. Closed-Circuit Chamber System for Small Animals
The apparatus consists of (1) a constant-temperature cabinet, (2) a burette system (8-ℓ

Table 3
GROWTH AND METABOLISM OF ALBINO RATS

Litters (average per rat)

Age (days)	Body weight (g)	J^a/day	J/m^b/day
1	5.64	6.32	1908
2	6.21	8.75	2494
3	7.07	9.67	2569
4	8.37	9.75	2310
5	9.74	11.30	2431
6	10.7	13.01	2657
7	12.2	14.56	2787
8	13.3	12.47	2222
9	14.2	10.67	1828
10	15.5	13.47	2176
11	16.3	12.43	1950
12	17.4	14.27	2142
13	18.9	16.99	2418
14	19.8	16.95	2339
15	20.8	17.74	2360
16	21.7	19.83	2577
17	22.9	18.91	2381
18	23.6	19.79	2452
19	24.4	20.21	2444
20	25.4	22.13	2611
21	28.9	27.32	2983
22	32.1	29.37	2987

Females (average per rat)

Age (days)	Body weight (g)	J^a/day	J/m^b/day
24	34.0	33.05	3259
26	39.6	43.10	3854
28	47.0	51.88	4226
30	52.2	61.92	4674
32	60.8	69.04	4728
+32	50.5	56.90	4372
34	68.8	81.17	5033
+34	57.2	51.88	3682
36	76.6	82.80	4992
+36	75.9	70.30	4176
40	93.3	37.10	5063
+40	85.6	77.80	4276
45	112	111.7	5192
+46	104	95.8	4682
50	119	109.6	4899
+51	118	94.1	4247
55	138	105.0	4330
+56	127	97.1	4176
60	153	114.6	4535
65	162	115.5	4272
+66	149	97.9	3816
74	171	113.0	4130
+76	162	93.7	3448
85	180	107.1	3699
+87	178	96.2	3347
100	192	113.0	3837
+101	174	96.2	3469
120	197	113.4	3682
+121	185	99.6	3372

Males (average per rat)

Age (days)	Body weight (g)	J^a/day	J/m^b/day
24	35.5	33.5	3226
26	39.1	47.7	4343
28	44.7	52.7	4406
30	51.7	63.2	4661
32	60.4	68.6	4728
+32	56.7	51.5	3674
34	68.4	73.6	4535
+34	58.7	62.8	4385
36	78.1	79.9	4711
+36	74.2	69.0	4163
40	101.1	102.3	5038
+40	89.5	84.1	4519
45	128	117.2	5012
+46	118	101.7	4577
50	147	122.6	4807
+51	137	97.9	4025
55	167	118.8	4305
+56	161	112.1	4146
60	201	128.9	4146
65	221	138.5	4192
+66	209	115.5	3628
74	249	140.6	3958
+76	242	126.4	3619
85	269	137.7	3690
+87	264	132.2	3586
100	293	141.8	3607
+101	276	133.5	3527
120	310	141.0	3456
+121	296	131.4	3335

Table 3 (continued)

GROWTH AND METABOLISM OF ALBINO RATS

a   The heat production was calculated on the assumption that 1 ℓ of oxygen had a heat equivalent of 19.7 J for the litters and 20.5 J for the older rats; a value of 19.7 J was also assumed for the fasted rats.

b   Surface area was computed from the following equation: surface area in m^2 = 0.0011 (weight in g).

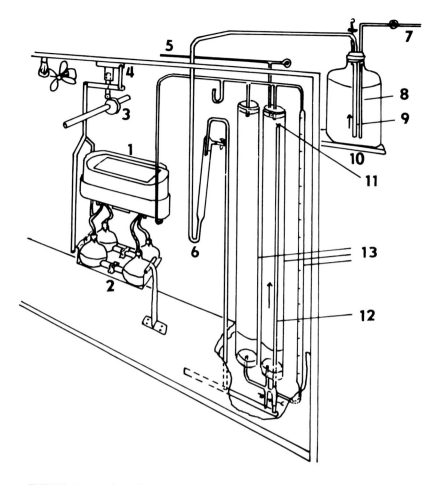

FIGURE 1. A schematic representation of the Regnault-Reiset apparatus for closed circuit measurements. 1 — animal chamber; 2 — CO_2 absorbers; 3 — rotating shaft; 4 — rocking mechanism; 5 — inlet tubing from the oxygen tank; 6 — equilibrator for water-dissolved oxygen; 7 — inlet from the oxygen spirometer; 8 — oxygen inlet tube; 9 — siphon tube; 10 — manostat; 11 — outlet of the siphon tube; 12 — siphon tube; and 13 — burette system.

capacity), (3) a manostat, and (4) CO_2 absorbers. The burette system is calibrated to show what volume of oxygen corresponds to a 1-mℓ volume of the measuring burette. There are also some accessories needed: they include a pressure gauge, an equilibrator (which adjusts temperature of water in the manostat to that of the chamber and oxygen concentration of the water to that prevailing at chamber temperature), and an oxygen spirometer which keeps air out of the top of the manostat by connection with pure oxygen (Figure 1).

Because of the danger of the temperature-dependent volume changes of oxygen measured, the apparatus is kept in a constant temperature cabinet. The variation in this room temperature should not exceed 0.1°C.

The rate of oxygen consumption is measured by the rise of the water in the burette. As the oxygen from the burette system is consumed, it is replaced by water which flows from the manostat whenever the pressure at the siphon outlet falls below that at the inlet. In the manostat the bottom of the oxygen inlet is at such a level that the water overpressure plus the pressure of air above the water level is equal to the pressure at the siphon tube outlet. Carbon dioxide is consumed in a set of $Ba(OH)_2$-containing bottles.

The rate of carbon dioxide production is determined by assaying the amount absorbed in the $Ba(OH)_2$-containing bottles.

The oxygen consumption measurements should not start until after the animal has been in the chamber for 30 min in order to accustom the animal to the new environment, bring the system to the standard temperature (usually 30°C), and establish an equilibrium between the absorbing rate in the CO_2 absorbing bottles and the CO_2 production rate. The air in the chamber contains about 1% CO_2 which remains practically constant throughout the whole trial.

B. Closed-Circuit Spirometric Mask for Large Animals

The overall assembly of the apparatus is shown in Figure 2. In principle the apparatus is a spirometer in which the rate of oxygen consumption is measured by the rate of decline of the oxygen bell. The air is circulated freely through the porous soda lime in one direction by the valves. The oxygen bell floats freely on a water seal and is counterbalanced by a weight. The decline of the bell is automatically recorded. The oxygen bell is of a size to produce an oxygen consumption of reasonable slope (for human and most domestic animals, 1 mm of height should correspond to about 20 mℓ of oxygen). The rate of oxygen consumption is computed from the slope of the graphic record. If solid CO_2 absorbers are used, then the RQ value cannot be estimated. Another complicating factor here is that in the case of ruminants that exhale some methane, this gas can accumulate in the oxygen bell and may distort the results; in most cases, however, the error due to the methane accumulation usually does not exceed 3%.

C. Open-Circuit System for Small Animals

This is a very old method that can be generally used for small laboratory animals, e.g., mice, rabbits, fowl, or laboratory rats. The system itself consists of a respiration chamber and a set of water and carbon dioxide absorbers. Sodium or barium hydroxide are used for absorbing CO_2, while concentrated sulfuric acid, magnesium perchlorate, or any other water-absorbing substance can be used for removing humidity.

The air is freed from water and carbon dioxide before it enters the respiratory chamber and is drawn through the chamber and absorbers by a pump. The air leaving the chamber deposits the present moisture and carbon dioxide in another set of absorbers. This part of the incoming oxygen is retained in the system after it has been converted to carbon dioxide. While the animal loses weight during the experiment (water and carbon dioxide loss), the respiration chamber together with the absorbers of the outcoming air gain in weight. The gain represents oxygen consumed, and the weight increase of the absorbers refers to moisture vaporized and carbon dioxide produced.

D. Open-Circuit Chamber and Mask Methods for Large Animals

Naturally for large animals, and today even for the small ones, balancing a considerable part of the system is not too handy. The air coming into the chamber is assumed to contain 0.031% CO_2 and 0.939% O_2; the outgoing air is analyzed for the content of the respective gases. Moreover, the rate of the passing air is measured by a gas meter. The percentages of oxygen decrement and carbon dioxide increment in the outgoing air together with the information about the amount of air that has passed through the chamber inform us about the amount of oxygen consumed and carbon dioxide produced.

The disadvantages of this method are seen in the time-consuming gas analysis of the chamber air, which is outdoor air slightly contaminated with expired air. Any slight error in this analysis is greatly increased in the computed heat production. The difficulty is increased if the expired air contains combustible gases, as it does in ruminant-exhaled air.

Instead of whole chamber, air analyses is also possible and, perhaps, more advantageous

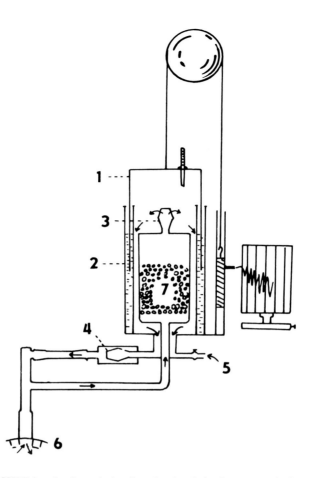

FIGURE 2. A schematic drawing of a closed circuit apparatus for large animals. 1 — spirometer bell; 2 — water seal; 3, 4 — valves; 5 — oxygen inlet; 6 — mouth piece; and 7 — soda lime.

using an open-circuit mask system involving gas analysis. This method is based on collecting the expired air into a bag or into a spirometer over a short period. The analysis of directly expired air containing several percent carbon dioxide increment and oxygen decrement is very much simpler than that of chamber air containing a fraction of a percent of CO_2. The analysis of the chamber air has to be at least ten times more precise (to 0.002%) than directly expired air (0.02 to 0.05%).

IV. BODY MASS AND SURFACE AREA MEASUREMENT

Since simple body mass (W) is not a good reference base for metabolic balance studies and since metabolism tends to vary more nearly in proportion to surface area or to $W^{2/3}$, one should be able to get data about body surface. In most cases the above assumption, i.e., that surface area varies with two thirds power of the total body mass, suffices. However, this is strictly true for geometrically similar bodies of constant specific mass. Evidently, small and large, young and old, and fat or thin animals, especially of different species, are not geometrically similar and not of constant gravity. To get more precise data one can use W^b as the reference base, where the coefficient b has to be established experimentally. For this purpose one needs to measure the surface area of an animal. In practice this is done by

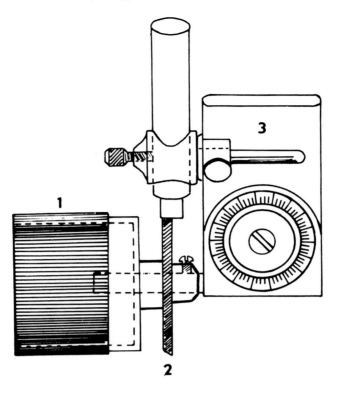

FIGURE 3.   A schematic representation of a simple surface integrator.
1 — roller; 2 — marking crayon; and 3 — revolution counter.

a surface integrator. Such a device can be quite simple as illustrated in Figure 3. A roller is joined with an ordinary revolution cou, ter, and a crayon used to mark parallel lines over the animal's body is mounted to it.

Body surface area can be also measured by some of the coating techniques. The main requirements are to affix the tape, flat and smooth, over the entire surface of the skin and to remove this mold without altering the area. This operation requires tremendous time and patience, not speaking about practice and skill. Accordingly, it became customary to use (in humans) height and weight formulas or nomograms based on direct measurements.[9]

A number of body volume and surface area measurement techniques were developed in connection with epidemiologic nutritional studies in underdeveloped countries. These techniques are hardly applicable directly to animal physiology; nevertheless, they are listed here briefly for completeness; readers interested in these problems should consult anthropometric literature.

Stereophotogrammetry[10,11] is based on photomapping techniques with two cameras. The contour lines are plotted from the stereo pictures with the aid of a drawing machine. This method is accurate, but it requires at least two overlapping photographs of the subject. It is also noticeable that the cost of equipment for this category of measurements is quite high.

Another method for measuring the surface area and volume of the human body is based on the multiplex projector system which is used for plotting area maps. A life-size image is reconstituted; in practice, the method requires a specialist to plot the contour lines and is limited to the face; measurement of a whole body is impractical.[12]

Photometry, using a photodermoplanimeter, determines the radiation surface area of the human body.[9] The radiation surface is ascertained by the reverse method of measuring the subject's absorbing surface when that surface has been made 95% absorbent to visible light by means of a pigment. This method is simple, but the results do not represent the total surface area of the body, and moreover, the method does not measure body volume.

In the cyrtographometric determination[13] a grid of shadows is projected with parallel beams of light, placing a small light source at the focal point of a spherical mirror. In this method the quantitation is difficult because the contours are too wide, and the shaded areas too large.

The monophotogrammetric method[14] makes use of illuminating the subject from either side by lights that shine through colored transparent strips of equal width. The color transparency resulting from photography is projected onto a drafting cloth, and an isopleth map is traced. The volume and surface area are calculated from the demarcation lines that are comparable to contour lines, but are frequently obscured by shadows because of the lateral source of illumination.

Yet another system is based on the Moiré interferences of a grid with its shadow cast onto the surface of a subject to visualize contour lines *in situ*.[15]

V. FOOD INTAKE AND NUTRITION

It is quite obvious that the nutrients consumed by the animal represent one part of the energy balance. The composition of diets naturally differs from one animal species to the other and moreover, it is frequently purposefully changed because of the experiment.

The measurement of *ad libitum* food intake in intact animals under controlled conditions provides the basis for all additional types of study. An adequate reference to normal food intake is necessary along with data about the cumulative intake (over 6, 12, and 24 hr); the size of the meals; and meal-to-meal intervals. Laboratory techniques that permit the automatic recording of meal patterns and the comparison of responses throughout the diurnal cycle are available.[16] It is necessary to stress that all species studied so far exhibited a typical feeding pattern.[17,18] Laboratory rats do not eat continuously or at random. During a meal, intake occurs at a constant rate interrupted only by short pauses. Individual meals are separated by nonfeeding periods. When a dark-light periodicity is introduced, rats take larger and more frequent meals at night and smaller and more distant meals during the day. In average, rats ingest five to eight meals of 1 to 6 g each at night. Maximum food intake per one meal usually reaches 2 g.[19]

Two methods are in use to identify the peripheral stimulus to eat: the first approach consists of changing various aspects of the nutritional status and energy metabolism through manipulation or administration of agents capable of eliciting or inhibiting the feeding response. The other approach is based on temporal correlations among spontaneous feeding (or the strength of the response) and variations of some parameters of the nutritional steady state. This, in particular, refers to hypo- or hyperglycemia or fat store repletion and depletion.

A. Effect of Exogenous Insulin

An insulin injection elicits in rats a dose-dependent augmentation of the subsequent cumulative food intake over the following 4- to 6-hr period.[20,21] The feeding response is stimulated after a 30- to 40-min delay period. During this latency period feeding is inhibited. Instead of a single injection, continuous infusion can also be made use of.[22,23] The elevation of food intake is achieved through more frequent meals, though no augmentation of meal size occurs. Under continuous i.v. infusion (0.1 to 0.6 IU/hr), a dose-related elevation of the 12-hr cumulative intake is observed reaching twice as much as the basal intake and more. These data, however, refer to day feeding; if the same experiments are carried out during the night, only a 25% increase in food intake is to be expected.[22] A gastric or i.v. load of glucose prior to insulin administration and the resulting hyperglycemia prevent insulin-induced hypoglycemia and overeating. The short-term overeating induced by insulin is later compensated for by reduced food intake, and the 24-hr cumulative intake is maintained.[24]

Chronic insulin administration (24 hr i.v. infusion 0.2 IU/hr/kg) induces both decreased intake and weight loss after some latency period. Stimulation of sustained hyperinsulinemia by chronic i.v. glucose administration representing 25 to 50% of control intake also induces a delayed and sustained reduction in oral food intake.[23] This effect is due to chronic hyperinsulinemia. The mechanisms of insulin action are anything but easy to interpret. On the contrary, two daily injections of long-acting insulin produce uncompensated hyperphagia which results in obesity.[25] An explanation of the mechanisms involved was reviewed by LeMagnen.[1] It is possible to conclude that unlike the acute stimulation of food intake by insulin, the feeding suppression and consequent weight loss obtained with chronic hyperinsulinemia are probably due to the cumulative action of insulin upon the CNS.

B. Effect of Exogenous Glucagon

A glycogen injection into normoglycemic animals produces hyperglycemia through elevation of glucose production in the liver followed by augmentation of the rate of glucose uptake.[26,27] In slightly deprived animals such an injection delays the onset of the first meal during the day. However, when presentation of food is delayed by about 3 hr, intake is augmented for some time. Intraportal glucagon administration is more efficient in delaying food intake than intrajugular administration.

C. Effect of Glucose Analogs

I.p. or i.v. administration of 2-deoxy-D-glucose, 3-methyl-D-glucose, or 5-thio-D-glucose (300 to 500 mg i.p., rats) results in stimulation of feeding.[28-34] Such a treatment shortens the latency to feeding and increases the amount eaten during the subsequent 5 hr after antimetabolite administration in fed or slightly deprived rats. On the contrary, during the night 2-deoxy-D-glucose administration reduces food intake for the subsequent 6 hr in fed and previously deprived rats.

Glucose infusion that exaggregates the induced hyperglycemia blocks the stimulatory effect of 2-deoxy-D-glucose on feeding (similarly it blocks the effect of insulin).[35] Hyperglycemia induced by 2-deoxy-D-glucose is unsuppressed after adrenalectomy; the stimulatory effect upon feeding in adrenalectomized animals is enhanced.[34] The antagonism between the hyperglycemic and feeding effects of 2-deoxy-D-glucose is further stressed by the fact that the blood threshold needed for the 2-deoxy-D-glucose concentration to elicit a feeding response it positively correlated to the blood glucose level. However, a dissociation between sites responsible for the hyperglycemic and feeding effect is supported by the finding that the thresholds needed for eliciting these two effects are different.[36]

As has been mentioned above, adrenalectomy abolishes the 2-deoxy-D-glucose-induced hyperglycemia; however, it does not affect the increase of plasma-free fatty acids induced by the antimetabolite.

D. Effect of Diabetes

Though a separate chapter is devoted to diabetes within this volume, let us be reminded that in this context, rats made diabetic as a result of alloxan or streptozotocin treatment are in a chronic state of hypoinsulinemia and hyperglycemia and become hyperphagic after some latency.[37,38] Currently, it is accepted that in this situation, because glucose is not replaced by other energy metabolites at the level of potential chemosensors involved in feeding, failure in glucose uptake and utilization by cells are the causes of diabetic overeating. The development of obesity is, however, compensated in part by the urinary loss of carbohydrates and by the impairment of lipid synthesis from carbohydrates due to insulinopenia. Insulin administration suppresses overeating in animals kept on a carbohydrate-rich diet.[39]

E. Caloric Compensation

It appears well demonstrated that single or continuous intragastric, intraduodenal, or i.v.

loads of nutriments are combined with a decrease in oral food intake.[40] After a single intragastric administration of nutriments the compensatory reduction of oral food intake occurs over the subsequent 6 hr.[41] In situations where maximum food compensation is required by intragastric administration it is recommended to apply a sequential food administration pattern that reproduces the normal meal pattern of the animal. Continuous i.v. feeding is also compensated for by reduced oral intake, but only solutions of sugars that can be metabolized are effective.[42] Compensation is only partial and can be improved by adding insulin to the glucose infusion. In order to be able to follow food intake either orally or after, for example, intragastric of i.v. infusion, it is necessary to follow concurrent blood parameters. These techniques permit the experimenter to draw blood either at regular intervals or continuously in a free-moving animal. Such possibilities are obtained by making use of chronically implanted intraportal or intrajugular catheters. This methodological approach was applied for the first time by Steffens[21] and Strubbe and Bouman.[43]

F. Liporegulatory Mechanisms

It is generally recognized that there exists a liporegulatory mechanism that influences the utilization and storage of body fats and governs their repletion and depletion.[44] Animals made overweight by forced feeding become hypophagic at the end of the forced feeding period and lose weight until they return to their initial body weight. Generally, in rats the residual food intake during weight loss is nocturnal and coincides more or less with the day feeding pattern. Conversely, after a forced weight loss caused by fasting or prolonged food restriction, rats become hyperphagic as soon as food is available *ad libitum*. Hyperphagia, in this case, refers to increased food intake during the daytime.

It has to be remembered that regulation of lipogenesis and lipolysis follows a circadian, centrally programed pattern. Also the rate of glucose utilization and insulin release exhibits a circadian rhythm. Regarding glucose and insulin, opposite trends during the day and night occur in human and laboratory rats: glucose-stimulated insulin release and tissue sensitivity to this hormone are higher at night in rats.[45]

G. Influencing of Food Intake through the CNS

Experiments done with neocortex and extrahypothalamic structures clearly indicate the involvement of hypothalamic and limbic structures in nutritional homeostasis.[46] Thus, for example, intraventricular administration of pentobarbital sodium elicits hyperinsulinemia followed by augmented food intake.[47] This is interpreted as the effect of the drug on ventromedial nuclei. Indeed, bilateral electrolytic lesion of ventromedial nuclei or ventromedial hypothalamus results in hyperphagia and obesity in most mammals and birds.[46,48]

Glucose or glucose plus insulin administration into rat lateral ventricle causes reduced food intake.[49] Fructose or insulin alone is without effect. On the contrary, chronic intraventricular insulin administration induces hypophagia and weight loss.[50] Generally, the effect is analogous to that of chronic i.v. insulin treatment.

Intraventricular administration of phlorizin (a glucose utilization blocker) produces sustained hyperphagia.[51] Administration of glucose antimetabolites into the CNS elicits similar effects to those observed that systemic administration (2-deoxy-D-glucose and 5-thio-D-glucose).[52,53] By administering low doses of 2-deoxy-D-glucose which are ineffective in changing either feeding or plasma insulin and glucose levels, it is possible to obtain an elevation of free fatty acids in plasma.

A glucose microinjection induces a delayed long-term reduction in food intake.[54] A local implant of insulin strongly reduces the first food intake during the night in rats, deprived before of food for 12 hr.[55] Chronic microinjection of insulin into ventromedial nuclei in rats leads to persistent hypophagia and body weight loss.[56] The effect of 2-deoxy-D-glucose implants is a matter of controversy.

Local anesthesia of ventromedial nuclei introduced by procaine hydrochloride induces hyperphagia and shortens the meal-to-meal intervals.[57] Administration of barbiturates results in a short-term (10-min) stimulation of feeding.[58] There are a number of other agents that are capable of introducing hyperphagia in animals through their neurotoxic effect. Gold thioglucose or colchicine may serve as typical examples.[1] Local administration of 6-hydroxydopamine induces damage to the dorsal noradrenergic bundle leading to hyperphagia and obesity.

There are also some surgical methods that can be exploited in studying the amount of food consumed.[59] A parasagittal cut anterior and lateral to ventromedial nuclei produces a weak and transient hyperphagia. The destruction of ventromedial nuclei increases the effect; posterior coronal cut has the same effect as the lesion of ventromedial nuclei. A cut between the ventromedial hypothalamus and lateral hypothalamus does not prevent the induction of hyperphagia induced, for example, by electrolytic damage to ventromedial nuclei. A lesion to the lateral hypothalamus leads to aphagia; an intact lateral hypothalamus is not necessary to achieve the effects of ventromedial nuclei destruction; it is only necessary to make the rat able to eat. In rats recovered from lateral hypothalamic aphagia, hyperphagia is observed after a medial lesion or reappears if the ventromedial nuclei lesion has been done before the lateral lesion. The metabolic symptoms of ventromedial damage persist when aphagia is induced by a following lesion of the lateral hypothalamus.

H. Miscellaneous Experimental Approaches

Overnutrition in mature laboratory animals is achieved most easily by forced feeding through a cannula inserted into the stomach or by other approaches just described. For fetal overgrowth the model has to be more sophisticated: fetal overgrowth occurs in the infant born of a diabetic mother.[60] Excess adipose tissue is found coincidentally with excessive lean body mass, and visceral enlargement has been recorded as well. Various theories concerning this excessive growth have ranged from a possible role of fetal growth hormone and the excessive activity of the adrenal gland to hyperplasia, hypertrophy, and excess activity of the islet cells within the pancreas.[61] The latter appears most acceptable. The reason for the excess release of fetal pancreatic insulin can be ascribed to fluctuations in maternal and fetal glucose, together with increased responsiveness to amino acid levels within the fetal circulation. The model has been described with pregnant primates (*Macaca mulata*): a single dose of streptozocin is administered i.v. in midgestation resulting in ablation of pancreatic β-cells.[61]

Animal undernutrition (cutting down the cumulative daily intake to one half) is frequent in aging studies as it offers extension of animal life.[62] As a matter of fact, this is the only model that offers increased mean survival (applicable to the laboratory rat, some invertebrates, and particularly worms and *Drosophila*). Undernutrition also can be achieved by increasing the number of animals in one litter (see Chapter 6) or by implanting the blastocyst at an unfortunate site in the uterine horn, where the blood flow is poorer than elsewhere.[63] This can be done with pigs, where in large litters this occurs spontaneously and one of the piglets is born weighing only one third to one half as much as its littermates.

REFERENCES

1. **LeMagnen, J.,** Body energy balance and food intake. A neuroendocrine regulatory mechanism, *Physiol. Rev.,* 63, 314, 1983.
2. **Anon.,** Official Methods of Analysis of the Association of Official Analytical Chemists, et 11th Association of Official Analytical Chemists, Washington, D.C., 1970.

3. **Schormüller, J.**, *Handbuch der Lebensmittelchemie, Analytik der Lebensmittel,* Springer-Verlag, Berlin, 1965.
4. **Brody, S.**, *Bioenergetics and Growth,* Reinhold Publishing, New York, 1945, 307.
5. **Lavoisier, A. L.**, Expériences sur la respiration les animaux et sur les changements qui arrivent à l'air en passant par leur poumons, *Mem. Acad. Sci. Inst. France,* 1777.
6. **Ganong, W. F.**, *Review of Medical Physiology,* Lange Medical, Los Altos, Calif., 1973, 230.
7. **Regnault, V. and Reiset, J.**, Récherches chimiques sur la respiration des animaux des diverses classes, *Ann. Chim. Phys.,* 26, 299, 1849.
8. **Pettenkofer, M. and Voit, C.**, *Ann. Chem. Pharm.,* 1862; as quoted in **Brody, S.**, *Bioenergetics and Growth,* Reinhold Publishing, New York, 1945, 335.
9. **van Graan, C. H.**, The determination of body surface area, *S. Afr. Med. J.,* 43, 952, 1969.
10. **Maruyasu, T., Oshima, T., Yanagisawa, S., Hasebe, Y., and Matsuyama, Y.**, Measurement of human body by stereophotogrammetry, *Ningen Kogaku,* 4, 258, 1967.
11. **Hertzberg, H. T. E., Dupertuis, C. W., and Emanuel, I.**, Stereophotogrammetry as an anthropometric tool, *Photogram. Eug.,* 23, 942, 1957.
12. **Burke, P. H.**, Stereophotogrammetric measurement of normal facial asymmetry in children, *Hum. Biol.,* 43, 536, 1971.
13. **Roche, A. F. and Wignall, J. W. G.**, The cyrtographometer: a new instrument for recording contours, *Am. J. Phys. Anthropol.,* 20, 521, 1962.
14. **Pierson, W. R.**, Monophotogrammetric determination of body volume, *Ergonomics,* 4, 213, 1961.
15. **Terada, H.**, A new apparatus for stereometry: moiré contourograph, in *Nutrition and Malnutrition,* Roche, A. F. and Falkner, F., Eds., Plenum Press, New York, 1974.
16. **Pokrovsky, V. and LeMagnen, J.**, Réalisation d'un dispositif d'energistrement graphique continu et automatique de la consommation alimentaire du rat blanc, *J. Physiol. (Paris),* 55, 318, 1963.
17. **Armstrong, S.**, A chronometric approach to the study of feeding behavior, *Neurosci. Behav. Rev.,* 4, 27, 1980.
18. **LeMagnen, J.**, Interactions of glucostatic and lipostatic mechanisms in the regulatory control of feeding, in *Hunger: Basic Mechanisms and Clinical Implications,* Novin, D., Wyrwicka, W., and Bray, G. A., Eds., Raven Press, New York, 1976, 89.
19. **LeMagnen, J. and Devos, M.**, Parameters of the meal pattern in rats: their assessment and physiological significance, *Neurosci. Behav. Rev.,* 4 (Suppl. 1), 1, 1980.
20. **Booth, D. A. and Brookover, T.**, Hunger elicited in the rat by a single injection of bovine crystalline insulin, *Physiol. Behav.,* 3, 439, 1968.
21. **Steffens, A. B.**, Plasma insulin content in relation to blood glucose level and meal pattern in the normal and hypothalamic hyperphagic rats, *Physiol. Behav.,* 5, 147, 1970.
22. **Larue-Achagiotis, C. and LeMagnen, J.**, The different effects of continuous night and daytime insulin infusion on the meal pattern of normal rats. Comparison with the meal pattern of hyperphagic hypothalamic rats, *Physiol. Behav.,* 22, 435, 1979.
23. **Vanderweele, D. A., Pi-Sunyer, F. X., Novin, D., and Bush, M. J.**, Chronic insulin infusion suppresses food ingestion and body weight gains in rats, *Brain Res. Bull.,* 5 (Suppl. 4), 7, 1980.
24. **Booth, D. A. and Pitt, M. E.**, The role of glucose in insulin-induced feeding and drinking, *Physiol. Behav.,* 3, 447, 1968.
25. **Panksepp, J., Pollack, A., Krost, K. P., Meeker, R., and Ritter, M.**, Feeding in response to repeated protamine zinc insulin injections, *Physiol. Behav.,* 14, 487, 1975.
26. **Holloway, S. A. and Stevensson, J. A. F.**, Effect of glucagon on food intake and weight gain in the young rat, *Can. J. Physiol. Pharmacol.,* 42, 867, 1964.
27. **Vanderweele, D. A., Haraczkievicz, E., and DiConti, M. A.**, Pancreatic glucagon administration, feeding, glycemia and liver glycogen in rats, *Brain Res. Bull.,* 5 (Suppl. 4), 17, 1980.
28. **Brown, J.**, Effects of 2-deoxyglucose on carbohydrate metabolism: review of the literature and studies in the rat, *Metabolism,* 11, 1112, 1962.
29. **Frohman, L. A., Muller, E. E., and Cocchi, D.**, Central nervous system mediated inhibition of insulin secretion due to 2-deoxyglucose, *Horm. Metab. Res.,* 5, 21, 1973.
30. **Frohman, L. A. and Nagai, K.**, Central nervous system-mediated stimulation of glucagon secretion in the dog following 2-deoxyglucose, *Metabolism,* 25 (Suppl. 1), 1449, 1976.
31. **Booth, D. A.**, Modulation of the feeding response to peripheral insulin, 2-deoxyglucose or 3-O-methyl-glucose injection, *Physiol. Behav.,* 8, 1069, 1972.
32. **Likuski, H. J., Debons, A. F., and Cloutier, R. J.**, Inhibition of gold thioglucose-induced hypothalamic obesity by glucose analogues, *Am. J. Physiol.,* 212, 669, 1967.
33. **Ritter, R. C. and Slusser, P.**, 5-Thio-D-Glucose causes increased feeding and hyperglycemia in the rat, *Am. J. Physiol.,* 238 (*Endocrinol. Metab.,* 1), E141, 1980.
34. **Thompson, D. A. and Campbell, R. G.**, Experimental hunger in man: behavioral and metabolic correlates of intracellular glucopenia, in *Central Mechanisms of Anorectic Drugs,* Garattini, S. and Samanin, R., Eds., Raven Press, New York, 1978, 437.

35. **Stricker, E. M. and Rowland, N.,** Hepatic versus cerebral origin of stimulus for feeding induced by 2-deoxy-D-glucose in rats, *J. Comp. Physiol. Psychol.*, 92, 126, 1978.
36. **Smith, G., Gibbs, P. J., Strohmayer, A. J., and Stokes, P. E.,** Threshold doses of 2-deoxy-D-glucose for hyperglycemia and feeding in rats and monkeys, *Am. J. Physiol.*, 222, 77, 1972.
37. **Booth, D. A.,** Some characteristics of feeding during streptozocin-induced diabetes in the rat, *J. Comp. Physiol. Psychol.*, 80, 238, 1972.
38. **DeCastro, J. and Balagura, S.,** Meal patterning in the streptozocin-diabetic rat, *Physiol. Behav.*, 15, 259, 1975.
39. **Friedman, M. I.,** Hyperphagia in rats with experimental diabetes mellitus: a response to a decreased supply of utilizable fuels, *J. Comp. Physiol. Psychol.*, 92, 109, 1978.
40. **Thomas, D. W. and Mayer, J.,** Meal size as a determinant of food intake in normal and hypothalamic obese rats, *Physiol. Behav.*, 21, 113, 1978.
41. **Booth, D. A.,** Satiety and behavioral caloric compensation following intragastric glucose loads in the rat, *J. Comp. Physiol. Psychol.*, 78, 412, 1972.
42. **Quatermain, D., Kissileff, H., Shapiro, R., and Miller, N. E.,** Suppression of food intake with intragastric loading: relation to natural feeding cycle, *Science*, 173, 941, 1971.
43. **Strubbe, J. H. and Bouman, P. R.,** Plasma insulin levels in the unanesthetised rat during intracardial infusion and spontaneous ingestion of graded loads of glucose, *Metabolism*, 27, 341, 1978.
44. **LeMagnen, J., Devos, M., Gaudilliere, J. P., Louis-Sylvestre, J., and Tallon, S.,** Role of a lipostatic mechanism in regulation by feeding of energy balance in rats, *J. Comp. Physiol. Psychol.*, 84, 1, 1973.
45. **LeMagnen, J., Devos, M., and Larue-Achagiotis, C.,** Food deprivation induced parallel changes in blood glucose, plasma free fatty acids and feeding during the two parts of the diurnal cycle in rats, *Neurosci. Biobehav. Rev.*, 4 (Suppl. 1), 17, 1980.
46. **Braun, J. J.,** Neocortex and feeding behavior in the rat, *J. Comp. Physiol. Psychol.*, 89, 507, 1975.
47. **Woods, S. C. and Porte, D., Jr.,** Insulin and the set point regulation of body weight, in *Hunger: Basic Mechanisms and Clinical Implications*, Novin, D., Wyrwicka, W., and Bray, G. A., Eds., Raven Press, New York, 1976, 273.
48. **Grossman, S. P.,** Neurophysiologic aspects: extrahypothalamic factors in the regulation of food intake, *Adv. Psychosom. Med.*, 7, 49, 1972.
49. **Woods, S. C. and Porte, D., Jr.,** Effect of intracisternal insulin on plasma glucose and insulin in dog, *Diabetes*, 24, 905, 1975.
50. **Woods, S. C., Lotter, E. C., McKay, L. D., and Porte, D., Jr.,** Chronic intracerebroventricular infusion of insulin reduces food intake and body weight of baboons, *Nature (London)*, 282, 503, 1979.
51. **Glick, Z. and Mayer, J.,** Hyperphagia caused by cerebral ventricular infusion of phlorizin, *Nature (London)*, 219, 1374, 1968.
52. **Muller, E. E., Panerai, A., Cocchi, D., Frohman, L. A., and Mantegazza, P.,** Central glucoprivation: some physiological effects induced by intraventricular administration of 2-deoxy-D-glucose, *Experientia*, 29, 874, 1973.
53. **Slusser, P. G. and Ritter, R. C.,** Increased feeding and hyperglycemia elicited by intracerebrovascular 5-thioglucose, *Brain Res.*, 202, 474, 1980.
54. **Epstein, A. N.,** Reciprocal changes in feeding behavior produced by intrahypothalamic chemical injections, *Am. J. Physiol.*, 199, 969, 1960.
55. **Hatfield, J. S., Millard, W. J., and Smith, C. J. V.,** Short-term influence on intraventromedial hypothalamic administration of insulin on feeding in normal and diabetic rats, *Pharmacol. Biochem. Behav.*, 2, 223, 1974.
56. **Nicolaidis, S.,** Mécanismes nerveaux de l'équilibre énergétique, *J. Annu. Diabetol. Hotel*, 153, 1978.
57. **Larkin, P. R.,** Effect of ventromedial hypothalamic procaine injections on feeding lever pressing and other behavior in rats, *J. Comp. Physiol. Psychol.*, 89, 1100, 1975.
58. **Maes, H.,** Time course of feeding introduced by pentobarbital injections into the rats VMH, *Physiol. Behav.*, 24, 1107, 1980.
59. **Ahlskog, J. E., Randall, P. K., and Hoebel, B. G.,** Hypothalamic hyperphagia: dissociation from hyperphagia following destruction of noradrenergic neurons, *Science*, 399, 1975.
60. **Cheek, D. B., Maddison, T. G., Malinek, M., and Coldbeck, J. H.,** Further observations on the corrected bromide space of the neonate and investigation of water and electrolyte status in infants born in diabetic mothers, *Pediatrics*, 28, 861, 1961.
61. **Cheek, D. B., Brayton, J. B., and Scott, R. E.,** Overnutrition, overgrowth and hormones (with special reference to the infant born of the diabetic mother), in *Nutrition and Malnutrition*, Roche, A. F. and Falkner, F., Eds., Plenum Press, New York, 1974, 47.
62. **Ross, M. H.,** Length of life and caloric intake, *Am. J. Clin. Nutr.*, 25, 834, 1972.
63. **Widdowson, E. M.,** Changes in pigs due to undernutrition before birth, and for one two and three years afterwards, and the effects of rehabilitation, in *Nutrition and Malnutrition*, Roche, F. and Falkner, F., Eds., Plenum Press, New York, 1974, 165.

Chapter 12

DATA ACQUISITION AND COMPUTER ANALYSIS IN ACUTE AND CHRONIC EXPERIMENTS (CIRCULATION)

T. L. Smith and A. A. Khraibi

TABLE OF CONTENTS

I. INTRODUCTION

The cardiovascular system is one of the most widely studied organ systems being investigated today. Too often, however, we become mired in the exhaustive investigation of one component of that system (i.e., myocardial mechanics, erythrocyte ion fluxes, blood pressure) and ignore all of the other components which comprise the whole. In fact, the cardiovascular system is so heterogeneous that one laboratory can only evaluate a small part of the whole. This chapter will deal with acute and chronic measurements of the cardiovascular system as it is broken down into two large components; the delivery system (evaluated by measurements of cardiac output and arterial pressure) and the business end of the circulation (microvascular measurements).

The discussion within this chapter will be limited for the most part to rats. The reader may find that the measurement techniques of the macrocirculation (heart and great vessels) are somewhat less sophisticated than those used in larger species of experimental animals. This is true, especially for continuous monitoring techniques, and is due in great part to the size constraints imposed by the rat.

The use of the rat as an experimental animal has many advantages over larger animals, however. Rats are less expensive and easier to maintain, and many different strains (Sprague-Dawley, Wistar, Fischer, etc.) and models (Okamoto spontaneously hypertensive rats, Wistar-Kyoto) are available for study. Most importantly, these inbred strains and substrains are genetically so similar that between-animal variation is greatly reduced and statistical inferences are, therefore, increased. The trade-off between decreased sophistication of instrumentation and increased statistical inference is probably fairly even.

II. MODELS AND METHODS

A. Macrocirculatory Techniques

Techniques for the evaluation of the macrocirculation are fairly well established, particularly in larger animals but also in rats. The macrocirculation is evaluated by monitoring the cardiac output and arterial (and venous) pressure(s). These variables allow one to assess the pump (heart) and conduits (major vessels) involved in the delivery functions of the cardiovascular system. Techniques for monitoring macrocirculatory function may be acutely or chronically applied.

Cardiac output may be measured by a variety of methods. The direct Fick technique is certainly applicable[1] as are indicator dilution techniques.[2] The latter would include radio-labeled microspheres which must be used with caution, however, since the suspension media may adversely affect the measurement of cardiac output.[3] Although all of these techniques can be perfected to yield accurate results, they still are limited in that they require chronic, in-dwelling catheters (not an insignificant point) and are restricted to occasional measurements which span some sampling period. These techniques are, therefore, not optimal for measuring transient phenomena.

Cardiac output may also be measured by techniques which yield beat-by-beat information. These would include electromagnetic and ultrasonic flowmetry. Electromagnetic flowmetry will be discussed here, as ultrasonic flowmetry will be addressed by others elsewhere within this book. Electromagnetic flowmetry in rats has been used fairly widely.[4-7] Ledingham and Pelling[4] were the first to apply such techniques to the study of cardiac output in rats.[4] Their design was very elegant, but required that the investigator produce his own flow probe, a task quite beyond most of us. Pfeffer and Frohlich[5] and Pfeffer et al.[6] were another prominent research group to use electromagnetic flowmetry in rats. They used commercially available probes in a number of elegant studies. All of these investigations were performed acutely in anesthetized animals. The next phase in the use of electromagnetic flowmetry in rats was to chronically implant suitable probes and perform measurements in unanesthetized animals. This last technique will be the focus of this section.

Electromagnetic flow probes suitable for chronic implantation are commercially available (Carolina Medical Electronics, King, N.C., Skalar, The Netherlands) in sizes which allow instrumentation of animals as small as 50 g. These probes can be implanted through an intercostal thoracotomy in a procedure which requires about 1.5 hr. The techniques for performing the implantation have been described in detail elsewhere.[7] A short description of the procedure is included here, however, for clarity's sake.

Anesthesia appropriate for surgical intervention must be used for all chronic implantation procedures. Although we have used a variety of anesthetic agents in the past (chloral hydrate, pentobarbital Na, ether, methoxyflurane), we have finally found one with which we are comfortable. A 1:1 mixture of ketamine HCl (100 mg/mℓ) and xylazine (200 mg/mℓ) injected i.m. at a dose rate of 0.1 mℓ/100 g body weight produces good anesthesia for periods of 30 to 40 min. This combination has a wide margin of safety so that supplemental dosages may be administered safely if the procedure takes longer than expected. This anesthesia is characterized by minimal respiratory depression and by good maintenance of body core temperature. For procedures that one knows will take 1 to 1.5 hr, one can give an initial injection of ketamine/xylazine together with an i.p. injection of pentobarbital Na (15 mg/ kg).

Animals must be artificially respired during the flow probe implantation procedure. This requires that the animal be intubated, a simple procedure to learn. The animal is anesthetized and placed on its back. The upper incisors are held down with an elastic, and a strong light is placed over the throat. The tongue is pulled out and away from the field of view. The larynx is then readily visualized, and an endotracheal tube made of polyethylene tubing (PE100) can be passed into the trachea. With a little practice the entire procedure can be performed within seconds. The animal can then be ventilated using a rodent respirator.

Flow probe implantation for cardiac output determination is accomplished via an intercostal thoracotomy at the level of the third intercostal space. The ascending aorta is dissected free of surrounding tissue over a length of 4 mm at a distance some 3 mm from the heart. The electromagnetic flow probe is placed on the ascending aorta and secured there with a silicone rubber key. The probe head is held in place by tightening the ribs around it. Overlying muscle layers are reapproximated, and the flow probe cable is routed s.c. to the midscapular area. The flow probe connector is then sutured to the back of the neck with sutures anchored in dacron mesh placed s.c. beneath the connector. A negative pressure system operating at -20 cm of water (vacuum) is used via a silicone rubber chest tube to evacuate the chest cavity and restore a negative intrapleural pressure.

Animals usually recover from this operative procedure very rapidly and regain their preoperative weights within 2 to 3 days. Measurements of cardiac output can then be performed readily by connecting the animal to an electromagnetic flowmeter. Recovery times vary from individual to individual. To be certain that the animal is fully recovered and that valid results are obtained, we routinely wait 5 to 7 days postoperatively before collecting cardiac output data.

The age at which the flow probe is implanted will, in part, determine the longevity of the preparation. The normal cause of failure in animals instrumented in this way is aortic rupture. If animals are instrumented when very small, the probe becomes excessively constrictive as the animal (and its aorta) increase in size. For example, an animal instrumented at 75 g would require a 5.0-mm circumference probe. By the time that animal weighed 200 g, it would require a 6.5-mm circumference probe, and the 5.0-mm probe would cause premature rupture. If, however, an animal is instrumented after it has reached maturity, then the preparation may be expected to survive at least 6 weeks. Some techniques used in larger species to increase probe longevity also work well in rats. An excellent example is the use of cushioning materials to reduce erosion of the aorta on the proximal (myocardial) edge of the probe surface. Silicone rubber can be easily applied to the margins of the probe lumen and aid in prolonging life expectancy.

A desirable aspect of this technique is that both acute and chronic observations may be made upon conscious, instrumented animals. One may make both short-term measurements on selected days or, with a little computer hardware, continuous measurements for extended periods of days or weeks. Power requirements for these devices are such that it is not presently possible to use telemetry techniques. The animals must therefore be connected to the flowmeter via lightweight cables. For acute measurements this means that some handling artifact could be introduced. In acute studies it is advisable to let the animal rest quietly in its cage for at least a couple of hours prior to the collection of hemodynamic data. This allows the animal to calm down, and the hemodynamic values will have returned toward baseline. Continuous (24-hr) monitoring techniques require a tether or shielded connector cable between the rat and flowmeter. This is also a novel experience for the rat, and it should be allowed to accommodate to this situation for 2 to 3 days prior to the collection period.

In addition to measurements of cardiac output it is also crucial to be able to measure arterial pressure. This has not been easy to do on a chronic basis. Successful long-term catheters in the past have relied upon special catheter designs and implantation techniques which are not often easily mastered.[1,8] We have worked upon catheter designs a great deal, experimenting with ''ideal'' materials and placements. In short, we have found that material is apparently less important than sterility, a point well made by Popp and Brennan.[9] Catheter potency may be maintained for long (6-week) periods of time via continuous infusion of sterile, heparinized Ringers solution. We have used this technique successfully on catheters of differing materials and placements. In our laboratory we use both carotid and tail artery placement of catheters. Carotid artery catheters are somewhat easier to implant, but result in the loss of the left carotid sinus. In addition, the catheters can sometime result in the formation of thrombi (particularly if they protrude too far into the aorta) which then can break off and move to the kidneys causing renal infarcts.

Tail artery catheters have an advantage over carotid artery catheters in that they do not affect the number of functional baroreceptors nor do they generally produce renal infarcts. These catheters can, however, generate thrombi which can lodge in the femoral branches producing necrosis in the legs. In short, there is no ''ideal'' catheter available for use in rats currently available. At best, the present catheters are a compromise, capable of reliable results for only a couple of months in the hands of most investigators.

In addition to arterial and venous catheters, one can also utilize atrial and left ventricular catheters. Although these catheters can be put in place via an intravascular route, they are generally fairly large with respect to the lumen of the vessel. This greatly impedes flow and enhances thrombus formation. These catheters are best placed via direct access to the atria or ventricles through a thoracotomy. The catheters can be made from a variety of materials, but polyethylene or silicone rubber probably work the best. A small disc of silicone rubber glue is put on the catheter a short distance from the tip and is used to hold anchoring sutures to the heart and helps seal the entry site of the catheter. These catheters are placed into the chambers via stab wounds produced by hypodermic needles. Anchoring sutures of 7-0 silk are placed through the discs and attached to the heart muscle. The catheters exit the chest through the back wall of the thorax by using a trochar made from a segment of hypodermic tubing. These catheters allow direct measurements of ventricular or atrial pressures for several weeks.

B. Microcirculatory Techniques

Long term microcirculatory techniques, at least in the rat, are somewhat more limited than are hemodynamic techniques. At present there are only a couple of techniques available

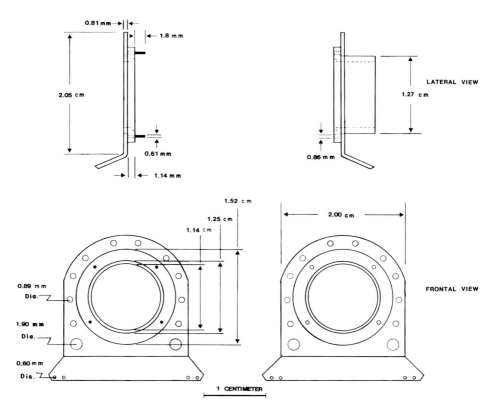

FIGURE 1. Chronically implantable microcirculatory chamber. Materials are polycarbonate with stainless steel pins.

for chronic microvascular observations in the rat. These are all basically similar and utilize a chronically implanted chamber on the animal's back.[10,11] Microvascular observations of arterioles, venules, and capillaries may then be made for periods of 6 to 8 weeks in conscious animals. The chamber used in this laboratory[11] is implanted on the back and allows visualization of preexisting microvasculature in the skeletal muscle of the cutaneous maximus. The chamber is shown in Figures 1 and 2. It is made primarily of polycarbonate and is very lightweight. The plastics do not conduct heat well, so the rat is not subjected to a constant heat loss via its chamber during the implantation lifespan.

Microvascular observation techniques in conscious animals present a problem; in order to make quantitative measurements one has to use either photo- (or video-) micrography or microphotography. Each technique has its disadvantages. Microphotography utilizes close-up or macro lenses used in conjunction with extensions to increase their magnification. This produces a rather weighty and unwieldy camera which must then be used to observe the microvascular chamber apparatus while the animal is unanesthetized (i.e., looking around its cage and not cooperating with the photographer). With this technique one can record on a daily (or more frequent) basis the microvascular changes occurring during an experiment. One is limited by the magnification capabilities of conventional lenses to the study of larger (60 μ or greater) vessels. Furthermore, this macrophotography is inherently limited to static measurements and yields no information about the vasomotor capabilities of the microvascular bed being investigated. It "sees" only a tiny temporal window no longer than the shutter speed of the camera.

Photo/video micrography differs from microphotography in that it can use much higher magnifications (which allow visualization of capillaries) and can employ videorecording

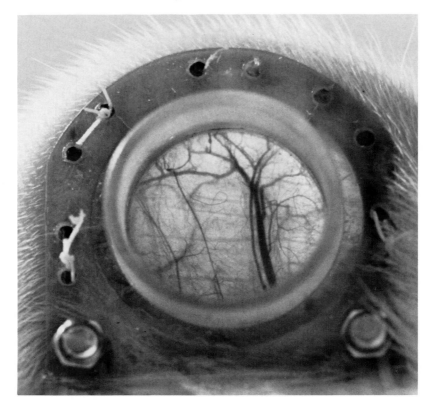

FIGURE 2. Microcirculatory chamber, 2 weeks postimplantation. Note the preexisting microvascular network sandwiched within.

techniques for the characterization of transient fluctuations which cannot be evaluated with a single sample. The main disadvantage that one encounters with this technique is that it requires that the rat be restrained in order to perform observations upon the compound microscope. The chamber and the rat are immobilized so that movement artifacts (even from respiration) do not make viewing impossible. This restraint is not necessarily a disaster; the rat can easily be conditioned to restraint. The investigator is left with a sense of uneasiness, however, since the animal is being subjected to an artificial environment during the observation period.

The implantation of these chambers is not difficult technically, but does require strict adherence to sterile technique. The animal is anesthetized, and the area of implantation is prepared by shaving, depilation, and application of surgical scrubs and antiseptic solutions. The skin on the back is then lifted and supported by a C-shaped frame via silk ligatures. The skin on opposing sides is marked and carefully removed from underlying skeletal muscle (cutaneous maximus). The skeletal muscle is then sandwiched between the chamber halves. The chamber is held together with synthetic suture materials and small nylon bolts. The animals recover from this operative procedure quite rapidly, and, since the chamber allows one to observe preexisting microvasculature, a long recovery period (several weeks) to allow for the ingrowth of new vessels is not necessary.

After the animal has been instrumented with a microvascular chamber it is best to wait for 4 to 5 days so that the effects of surgery are reduced. Once the animal has healed, measurements of the microvasculature may be made using either acute or chronic experimental interventions. In this manner acute experiments may be performed in conscious, chronically instrumented animals. If one is careful in his selection of interventions, it is possible to use the same animal for more than one study. A requirement, however, is that the sequence of interventions be randomized to avoid possible bias introduced by the order in which the interventions are performed.

The chronic microvascular chambers are best suited at present for the examination of responses due to systemic administration of pharmacologic agents or stimulation of whole body physiologic mechanisms. Work is underway that will allow the examination of local microvascular alterations in response to local interventions. The systemic hemodynamic observations allow one to correlate microvascular events with changes in central hemodynamics. The present system does allow one to make repeated observations of the same microvascular network or vessels for several weeks. The preparation demonstrates a very exciting range of microvascular activities characterized by a high incidence of vasomotion of arterioles as well as a great deal of plasticity, an ability to structurally change in response to appropriate stimuli. For example, one can detect microvascular changes in the proportions of open vs. closed arterioles and veins within a 24-hr period after the administration of vasoactive agents. The overall caliber of vessels can also change within a short period (24 hr). The total number of microvessels available to serve the tissue can be significantly altered within just 3 days. Therefore, the overall ratio of arterioles to venules, and potentially the fluid dynamics at the microtissue level, can be altered within a short time period.

III. DATA ACQUISITION AND COMPUTER ANALYSIS

This section deals with the acquisition of data from chronically instrumented animals using traditional as well as computer techniques. The animal/transducer interface does not differ to a great degree if one is using traditional methods such as strip chart recorders or computer acquisition of data. This section can be divided into techniques for systemic hemodynamics and those for microvascular measurements.

A. Systemic Hemodynamics

A system has been developed in our laboratory whereby cardiovascular variables can be monitored for 2 or more months continuously in the conscious, freely moving rat. This monitoring system minimizes the interaction between the experimenter and the animal, besides making it possible to monitor the rat in its home cage. Observations may be made of any possible trends in cardiovascular parameters during light and dark cycles. Too often, these trends are obscured by experimenter manipulations of the animal during the lights-on (quiescent) time period.

There are two basic systems currently in use in our laboratory. The first utilizes only pressure information via chronically implanted catheters. This system allows us to screen a large number of animals with minimal instrumentation and set-up time. The second system utilizes animals implanted with ascending aortic electromagnetic flow probes and arterial catheters. The latter system allows us to perform fairly comprehensive investigations of systemic hemodynamics in chronically instrumented, conscious animals.

The first system allows the continuous monitoring of systolic arterial pressure (SAP), diastolic arterial pressure (DAP), pulse pressure (PP), mean arterial pressure (MAP), heart rate (HR), PP $\times$ HR (an index of cardiac output, assuming constancy of aortic compliance), and baroreceptor reflex index (BARO). One can see that this allows one to evaluate changes in heart rate elicited by changes in arterial pressure. These variables are updated and displayed on a video monitor each minute. The averages are computed and printed out each hour along with the coefficient of variation for each of the above variables during that hour.

Information is gathered from the instrumented animal in the following manner. The arterial catheter exits the animal and passes through a protective spring to a hydraulic swivel located outside the cage of the rodent. This spring protects the catheter and provides torque for rotation of the hydraulic swivel. The swivel is connected to a flow-through pressure transducer. The other port of the pressure transducer is connected to a syringe pump which infuses 2 mℓ of heparinized (60 IU/mℓ) Ringer's solution per day through the pressure catheter. This infusion is sufficient to prevent clotting or platelet adhesion on the tip of the catheter, but does not dampen the pressure pulse wave.

The pressure tranducer is connected to an instrumentation amplifier which sends a high-level (0 to 5 V) signal to an analog-to-digital (A/D) converter. This A/D converter is located within a microcomputer located in an adjacent room. This arrangement prevents disturbance to the animal from the sound of the printer and the movement of persons within the laboratory. The microcomputer is connected to a video monitor and printer. The microcomputer used in this system is either a Sanyo MBC 555 or a P.C.'s Unlimited, I.B.M. compatible unit. Programing is done in Turbo Pascal for maximal speed. This system allows one to monitor the above variables in 16 rats simultaneously. The sequence of calculations is shown in Figure 3.

Acquisition of systemic hemodynamic data may be performed using a common technique regardless of whether acute or continuous monitoring is desired. The scheme for this technique is similar to that above.

Essentially, the system consists of three parts: the instrumented animal, the transduction device(s), and the microcomputer. The instrumented animal has already been addressed. The transduction devices are electromagnetic flowmeters and instrumentation amplifiers for Wheatstone-bridge applications. The microcomputers can be any of a host of models equipped with A/D converters and programed in a suitably fast language (e.g., turbo Pascal). An example of the results obtainable may be seen in Figure 4. Note the marked diurnal variation present in some parameters.

3. Microvascular Measurements

Measurements made of the microvasculature can be categorized as either static or dynamic. Static measurements would include photomicrography and microphotography. Obviously, these static techniques could be used to document dynamic processes. One simply obtains information about the status of the microvasculature at one point in time. Not only could this be in the form of a photograph or slide, it could also be a documentation by direct visual quantitation performed by an observer carefully making measurements through the microscope and recording them in notebooks. Although this would be the most basic form of data collection, it is perhaps also the most subjective and potentially the most vulnerable to procedural errors. The use of photographic techniques does allow the investigator to retain a lasting representation of the microvascular network being examined. Even though the static techniques describe the microvasculature for only a brief instant, the sequential use of these techniques can capture and detail dynamic phenomena such as those involving angiogenesis.

Photographic analysis is presently the method of choice for static analysis applications. The use of fine grain films with large format cameras allows the investigator to produce enlargements with sufficient resolution for analysis. Although morphometric techniques can be used in the analysis of microvascular changes, they do not discriminate between arteries and veins nor can they discriminate between changes in vessel length as opposed to vessel number. The latter drawback is particularly inconvenient if one wishes to assess the microvascular resistance. Obviously, changes in number would alter the number of parallel pathways whereas changes in length would alter resistance in the opposite direction. A preferable alternative is a structural analysis of the microvasculature.

Structural analysis of the microcirculation is the study of vascular networks based upon their branching patterns, function (artery, vein, or capillary), and physical properties, e.g., length and diameters of individual vessels.[12] This form of analysis allows one to determine the mechanisms whereby microvascular phenomena occur. For example, a microvascular response to overperfusion might well be a decrease in the number of small arterioles with an increase in the number of venules. Morphometric analysis might see no difference in the

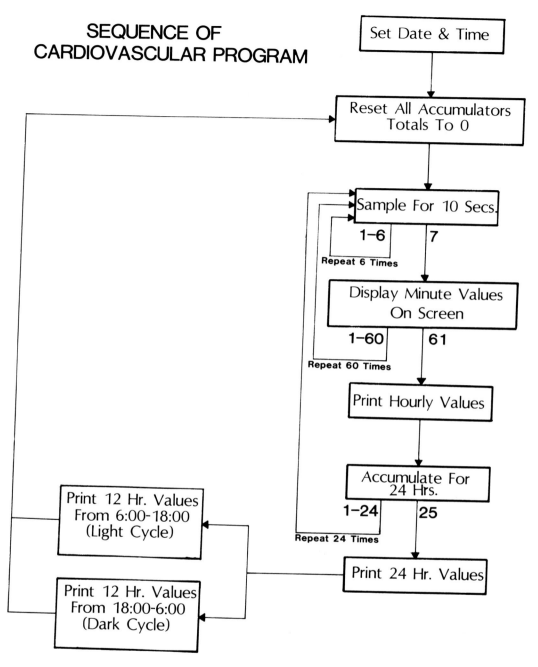

FIGURE 3. Sequence of events in a cardiovascular data collection program.

overall vascular density. In fact, the alteration in the ratio of small arterioles to small venules could drastically impact the dynamics of exchange and reabsorption at the tissue level. In addition, one could observe a physiologic response characterized by an increase in overall length of vessels. Using morphometric techniques one could interpret this as an increase in vascular number and therefore cross-sectional area (which would reduce microvascular resistance) when in fact, the cross-sectional area was unchanged and the microvascular resistance was actually increased.

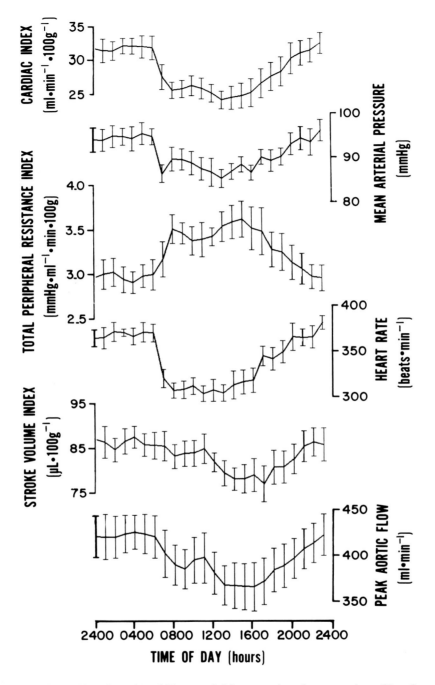

FIGURE 4.   Hemodynamic variables recorded from conscious, instrumented rats (N = 6) over a 24-hr period. Each point represents 60 measurements (one per minute) averaged for all animals.

Although there are a number of techniques for structural analyses, the one with which we have the greatest familiarity is the use of digitizing tablets in conjunction with microcomputers. These tablets work upon an electromagnetic principle and are capable of distinguishing 1 part in 11,000. This allows one to analyze a photomicrograph using a cursor to define points corresponding to length and diameter. These points are then digitized and entered as X/Y coordinates. The microcomputer can then analyze these points and compile them in files corresponding to sizes (diameters and lengths), function, and branching order. The vascular networks can then be reconstructed and analyzed systematically.

Dynamic analysis of the microvasculature requires the use of either cinematographic or video techniques. These media allow one to analyze microcirculatory events either "on-line" with video techniques or "off-line" via video recordings or movies. Documentation of microcirculatory events such as vasomotion, microvascular reactivity to neuronal or pharmacologic stimuli, and microrheological phenomena, among others, is possible. In addition, one can perform microvascular measurements of velocity and diameter using video techniques. Systematic applications of these techniques yields valuable network information which gives insight as to the behavior of the entire microvascular unit under investigation.

IV. CONCLUSIONS

The study of the cardiovascular system of rats is important in that rats are widely used as models of many cardiovascular diseases, both primary and secondary in nature. These studies may now be performed in conscious animals, thus avoiding compromising effects of anesthetics and acute surgical procedures. This allows the investigator to arrive at statistical conclusions of greater validity with respect to normal physiological function. Furthermore, measurements of systemic hemodynamics may be performed 24 hr/day using microcomputers. Such on-line procedures make possible the analysis of diurnal patterns as well as the variance in those patterns.

Some microcirculatory measurements may also now be performed in conscious rats for relatively long periods of time. This is important because anesthetics can alter those measurements making interpretation exceedingly difficult. The experimental design can also be such that an animal serves as its own control, thus making statistical comparisons of drug or temporal effects much easier.

REFERENCES

1. **Popovic, V. P. and Kent, K. M.,** 120 day study of cardiac output in unanesthetized rats, *Am. J. Physiol.,* 207, 767, 1964.
2. **Stanek, K. A., Smith, T. L., Murphy, W. R., and Coleman, T. G.,** Hemodynamic disturbances as a function of the number of microspheres injected, *Am. J. Physiol.,* 245, H920, 1983.
3. **Flaim, S. F., Morris, Z. Q., and Kennedy, T. J.,** Dextran as a radioactive microsphere suspending agent: severe hypotensive effect in the rat, *Am. J. Physiol.,* 235, H587, 1978.
4. **Ledingham, J. M. and Pelling, D.,** Cardiac output peripheral resistance in experimental renal hypertension, *Circ. Res.,* 20/21 (Suppl. 2), II187, 1967.
5. **Pfeffer, M. A. and Frohlich, E. D.,** Electromagnetic flowmetry in anesthetized rats, *Am. J. Appl. Physiol.,* 33, 137, 1972.
6. **Pfeffer, M. A., Frohlich, E. D., Pfeffer, J. M., and Weiss, A. K.,** Pathophysiological implications of the increased cardiac output of young spontaneously hypertensive rats, *Circ. Res.,* 24/25, (Suppl. 1), I235, 1974.
7. **Smith, T. L. and Hutchins, P. M.,** Central hemodynamics in the developmental stage of spontaneous hypertension in the unanesthetized rat, *Hypertension,* 1, 508, 1979.
8. **Weeks, J. R. and Jones, J. A.,** Routine direct measurement of arterial pressure in unanesthetized rats, *Proc. Soc. Exp. Biol. Med.,* 104, 646, 1960.
9. **Popp, M. B. and Brennan, M. F.,** Long term vascular access in the rat: importance of asepsis, *Am. J. Physiol.,* 241, H606, 1981.
10. **Pappenfuss, H., Gross, J., Intaglietta, M., and Treese, F.,** A transparent access chamber for the rat dorsal skin fold, *Microvasc. Res.,* 18, 311, 1979.

11. **Smith, T. L., Osborne, S. W., and Hutchins, P. M.,** Long term micro and macrocirculatory measurements in conscious rats, *Microvasc. Res.,* 29, 360, 1985.
12. **Schmid-Schoenbein, G., Zweifach, B. W., and Kovalcheck, S.,** The application of steriological principles to morphometry of the microcirculation in different tissues, *Microvasc. Res.,* 14, 303, 1977.

Selected Physiological Methodologies

Chapter 13

GROWTH AND DEVELOPMENT REGULATION

V. Štrbák and J. Knopp

TABLE OF CONTENTS

I. REQUIREMENTS FOR A GOOD MODEL

In spite of the rather extensive discussion about models and modeling in Chapter 1 let us stress here that in practice a good experimental model can be considered such in which the experimental group differs in a single parameter from the controls. If necessary, several control groups should be used (absolute control, controls injected with the vehiculum, etc.). It is also necessary to consider the adequacy of the model. It is, e.g., not possible to use early weaning of rats as the model of simple undernutrition since other parameters besides feeding are changed in the nest after the mother's withdrawal. Similarly, early weaning cannot be used to study the stress response during maturation because of concomitantly occurring other changes that are of varying importance like nutritional conditions, psychological alterations, thermoregulatory changes, etc. To tell the truth, it is quite difficult to create a satisfactory experimental model, particularly in ontogenetic studies, with only a single changing parameter. The comparison of different age groups of rats often reflects not only age effects, but also dietary changes (milk, pellets), the effect of a new social environment (after the mother's withdrawal from the nest, transferring of animals into new cages, and their separation according to sex), change in the nest temperature, etc.

II. ONTOGENIC STUDIES AS AN EXPERIMENTAL TOOL

An ontogenic study represents a specific approach to the analysis of physiological phenomena. At first, information gathered should describe and clarify developmental processes as such. On the other hand, developmental models may help us considerably in our understanding of a number of other fundamental physiological mechanisms. There are problems that are better accessible to study during early stages of individual development when some of the regulatory or interfering factors are still not operative. This offers the oportunity to study some functions in living organisms and/or in tissues and cells in relatively clear models. The presence of extremely high levels of thyroliberin (TRH) in the pancreas of newborn rats can serve as an illustrative example.[1] This neuropeptide, regulating pituiary thyrotropin secretion is present in the pancreas of adult animals in very low concentrations due to a high TRH degrading activity, making its study difficult.[2,3] TRH was therefore characterized and its localization in B cells demonstrated in the pancreas of newborn animals. Moreover, negative correlation of TRH levels with pancreatic TRH degrading activity during maturation called attention to the crucial role of degradation in the regulation of TRH content in the pancreas.[1] At some stages of development many tissues can be considered naive - they have not yet experienced the influence of some regulatory factors. The first response to a stimulus differs frequently from that after repeated exposure. Some specific studies can be done only because of the absence of certain regulatory influences.

III. MODEL TYPES AND METHODS CHOICE

A. Normal Growth and Development (Maturation)

Physiology of a young organism is much more complex than that of mature animals. Somatic and functional changes that occur during the early developmental phases of life are greater than those at any later stage. These changes are called maturation and represent a genetically determined (autonomous) functional and morphological differentiation, which is basically irreversible. Maturation includes physical growth (mainly an increase in cell number and size) and functional differentiation (the initiation of a new function or functional units). Initiation of many functions is vitally important for some stages of development (e.g. respiration after birth). The genetic preposition of readiness, however, often precedes the appearance of the function itself by a certain period of time. This lag period between the readiness of the genetic material to release specific information and the beginning of the special function is the first and most important part of critical periods or sensitive phases in development.[4] Along the ontogenic time axis, many limited sensitive phases occur. It has been speculated that the sensitive phase mirrors a transitory activation of genes.[4]

1. Longitudinal Studies

Studies of normal growth and development supply the basic information essential for further studies in developing organisms. In a longitudinal study, values obtained from the same individuals are compared at different ages thus giving reliable information on the age effect and minimizing the effects of interindividual variation. The number of individuals available for repeated measurements are, however, almost always limited. At the beginning of this type of study it is necessary to clarify how representative the group studied is and what part of the population it represents.

The choice of experimental animals may substantially affect the validity of the results. For example, if we are going to study the postnatal growth in eight Wistar rat pups born to one mother and fed by her milk we may get a homogenous group with low variability of parameters. All the pups are subject to the same pre- and postnatal medium. A comparison with pups from other litters might, however, reveal substantial differences. If 32 pups delivered by 4 mothers are equally distributed into 4 litters, each litter represents a population of 4 pairs with different prenatal histories. The validity of data in this case is obviously more general. A comparison of pairs from different mothers inside one litter gives information on the role and extent of prenatal differences. On the other hand, a comparison of siblings born to one mother and reared with different mothers will inform us on the role and extent of postnatal influence on the parameter studied. Moreover, results may be different when 4, 8, or 14 pups are reared with one mother because of different milk intake. There are also other conditions which should be defined for the population measured: commercial breeding stations perform weaning of rats usually at 21 days of age. It has been shown, however, that at this age rat pups drink significant amounts of maternal milk and that physiological weaning culminates at 28 days in the rat.[5] Early weaning results in many endocrine changes, some of them permanent.[6] For any measurements during normal development the ''normal'' situation should be carefully defined.

There are, basically, two approaches used in longitudinal studies: (1) the growth parameters are averaged at definite preselected time intervals from the same set of individuals and (2) physiological data are evaluated separately for every individual studied. The former approach is used for e.g., drawing the average growth curve of the population and for preparing the standards or for studying the relationship between growth and the studied variable(s).[7] Such data can be, also naturally, obtained by means of a cross-sectional study. This approach is, however, entirely misleading if the problem is to trace the course of changes in an individual subject over time, to identify as many events as possible which may be relevant to the individual in determining his future development, and to construct velocity standards. In fact, the averaging process not only conceals irregularities of growth rate in developing individuals but it can also considerably distort common trends when these occur at different ages in different subjects.[7] The sudden change at puberty and the different ages of onset can only be brought about by securing information from a cohort on successive occasions and by studying the main features of the growth curve of every individual. In such a case one has to use the other approach, i.e., to analyze the growth curves of individual animals separately, without averaging.

Following the two different approaches are two categories of statistical methods that we can use to analyze data collected in longitudinal research. The first one concerns methods capable of testing and comparing a set of parameters by fitting models linear in their constants to the average growth of the cohorts over time. The other concerns methods capable of analyzing individual growth curves.[7]

2. Cross-Sectional Studies

This type of research compares different individuals of different ages. Results greatly depend on the extent of the interindividual variation of parameters measured and their possible mutual compensation resulting from a high number of observations. Nevertheless, this approach also offers some advantages: parameters to be compared can be measured in the same moment in different age groups thus avoiding an effect of some uncontrolled factors when the experiments are carried out over an extremely long period of time.

In conclusion, cross-sectional studies are usually designed to obtain an inventory of the surveyed population. Longitudinal designs are specially used to study individual growth and development models. They are, however, more expensive, time consuming, and only partly up to date: the initial data are per se out of date.[8]

B. Factors Affecting Growth and Development

1. Genetic Factors

Normal growth and development is genetically determined (autonomous). The role of genes and regulatory mechanisms can be studied at different levels of organization starting from the molecular level up to the level of the integrated organism.

a. Studies of Genetic Factors at the Level of the Integrated Organism

Different strains of experimental animals have their own patterns of growth and development. Genetically affected strains are particularly difficult to compare with other animals: multiple errors may affect growth and development. However, what can be done is to determine patterns of inheritance for special features (recessive, dominant, x-linked, etc.) which may help to localize the error. Moreover, with the knowledge of the error the physiologist can study and prevent possible development of secondary consequences. A comparison of these animals with other strains or even heterozygotes can be misleading because of the occurrence of several, often independent differences (errors). Thus, e.g., Snell dwarf mice are characterized by a genetically determined congenital lack of pituitary growth hormone, thyrotropin and prolactin.[9] Okamoto and Aoki[10] isolated from Wistar rats a colony of "spontaneously hypertensive rats" with elevated plasma TSH level.[11] We could show in the progeny of spontaneously hypertensive and normotensive rats that the genetic control of the blood pressure in these rats is not dependent on (can be dissociated from) the genetic control of the thyroid function.[12]

2. Endogenous Factors Involved in Gene Expression

During the development of a new organism new cells and tissues appear as the product of differentiation of the fetus. Some of these cells and tissues with the material they produce are themselves important factors for the regulation of growth and development of the whole organism. Again, the role of these factors can be studied at different levels.

a. Methods at Cellular and Molecular Levels

Methods used for studying growth and development regulation at the cellular and molecular levels exploit the wide variety of biochemical techniques including such specialized approaches as gene manipulation or advanced separation and isolation techniques. The description of such methodologies is clearly beyond the scope of the present volume and the reader seeking more detailed information along these lines is directed to specialized monographs.

For reasons specified above we shall limit ourselves to the description of various possibilities that can be fructified in this type of research rather than describing the methodical details.

b. Methods at Tissue and Integrated Organism Levels

Growth and development regulation at the molecular and cellular levels requires the action of a number of hormones, both of peptidic and steroid nature. Growth hormone, which is the first to be mentioned, is a pituitary polypeptide of MW 21,500. It is necessary for normal growth and development, while its abnormal production is associated with numerous disease states. Another category of growth regulating substances are the somatomedins or insulin-like growth factors (IGFs) that comprise a family of peptides circulating in plasma and stimulating DNA synthesis in a variety of cultured cells. Two human IGFs have been characterized: IGF-X, a 70-amino acid basic protein, playing a fundamental role in postnatal mammalian growth as the major mediator through which growth hormone (GH) exerts its biological effects. The function of IGF-II, a 67-amino acid neutral peptide, is less clear. In order to investigate gene expression of these hormones, different experimental model systems were used. Any similar system is of value if it can be used to determine the mechanism of gene expression and to analyze the control of this process. However, there is no simple way to determine whether a component, which is limiting in vitro, would also be limiting in vivo.

Isolation of rat growth hormone complementary DNA (cDNA) and subsequent characterization of the chromosomal gene in rat and humans have prompted studies on the mechanisms of regulation of these genes. The sequence of the GH gene has been completely determined[13] and the GH gene serves as an excellent model for studying complex control mechanisms since it is expressed and regulated in the clonal rat pituitary tumor cell lines G/C and GH_3.[14]

Glucocorticoids and related steroids represent an important class of hormones regulating cellular processes during both differentation and development. Further, GH production in cells is also stimulated by thyroid hormones. This reguation seems to operate at the level of receptors in the nucleus which bind the hormones and are associated with nuclear chromatin complexes. Specific triiodothyronine (T_3) receptors have been also identified in the nucleus of cultured pituitary cells. In vitro translation studies showed that both steroid and thyroid hormones regulate GH production by promoting an increase in its mRNA levels.[15,16] The mechanisms regulating GH synthesis by thyroid and steroid hormones are quite complex and contradictory results may be obtained. A marked elevation of GH levels in several GH-producing cell lines by thyroid hormone was found when serum from a hypothyroid calf was used instead of normal fetal calf serum.[17] On the other hand, glucocorticoids are only effective in producing a stimulation of GH synthesis[18] when combined with T_3. Moreover, the multihormonal response has been found to be greater than the stimulation by either hormone alone. Owing to this a synergistic effect on GH synthesis by glucocorticoid and thyroid hormone is combination has been proposed.

In order to investigate the intracellular events controlling GH gene expression, cDNA clones and genomic clones were isolated and used to study the organization of the gene and to examine its regulation.[19] A successful regulation of expression was obtained when the GH gene was transferred into a mouse using a novel retroviral vector system.[20] In another study it was shown that both glucocorticoids and thyroid hormone independently regulate transcription of the GH gene, which in turn appears to result in an increase in the levels of GH nuclear RNA precursors.[18] Hormonal induction of nuclear precursors was followed by an increased accumulation of newly synthetized GH mRNA. RNA blot analysis suggested that the synthetic pathway of rat mRNA is complex, involving up to six potential nuclear RNA precursors. In this process a modulation of mRNA half life cannot be excluded.[18] The physiological importance of this regulation was established by demonstrating that the transcription of the rat GH gene can also be induced in vivo. A four fold induction 15 min after dexamethasone treatment showed that this induction is physiologically relevant and not restricted to pituitary cell lines.

The use of cDNA probes to investigate regulation of GH synthesis at the RNA level yielded additional results that are difficult in their interpretation. Glucocorticoids alone but not the other hormones[21,22] have been observed to be effective here.[18,21] Recently, effects of glucocorticoids and thyroid hormone on GH synthesis were re-examined by a new technique in a medium capable of maintaining GH-producing cells without the use of animal serum. It was shown that a synthetic glucocorticoid, dexamethasone, significantly affects GH transcription in GH_3 or GC cells in the presence of T_3, while T_3 itself is effective alone.[17] Based on the latter study a further conclusion was made implying a dual level of control of GH synthesis by dexamethasone and triiodothyronine. A rapid fivefold stimulation of transcription is followed by a tenfold increase in GH mRNA stability, which accounts for a severalfold increase in the concentration of cytoplasmic mRNA. It was suggested that dexamethasone only enhances the pre- and posttranscriptional induction of the GH gene.

Very little is known about IGF-I biosynthesis, due to its low content in tissues and also because no cultured cell lines produce significant quantities of this peptide in contrast to IGF-II. It was estimated, by molecular sieve chromatography, that the protein responsible for IGF-I activity extracted from rat liver has a higher molecular weight (approximately 30,000) than that extracted from plasma (8000), the larger molecular weight material was suggested to represent an IGF-I precursor. The amino acid sequence derived from a human IGF-I cDNA clone supports the assumption of a larger precursor.[23] Two different IGF-I cDNAs isolated from human liver library were characterized. The nucleotide sequences of these cDNAs suggest the existence of two different IGF-I protein precursors with only one IGF-I gene in the human genome. These observations suggest a second level of regulation, consisting in different processing of the gene product. It is also possible that the amino and carboxyl peptides have a specific biological function.[23]

The role of defined substances and organs (tissues) can be studied by classical methods: by ablation of the organ studied and by following the reversibility of observed effects after the animals were treated with products of the respective organ. Another approach is to study the role of a substance after its administration to tissue culture or into the living organism. The latter approach is particularly suited for studying the effects of exogenous substances. On the other hand, neutralization of an endogenous substance with, e.g., an antibody (Figures 1 and 2) as well as inhibition or stimulation of endogenous secretion of biologically active substances (hormones) by pharmacological interventions (neurotransmitter agonists, antagonists, releasing hormones), represent more physiological methods in studying the role of endogenous substances.

A good model for the study of the development of hormonal growth dependence in the rat was developed in Nicoll's laboratory.[26]

Xiphoid cartilages were removed from 18 to 20-day-old fetuses and from 2-, 9-, and 30-day-old rats, transplanted under the kidney capsule of an adult syngeneic female hosts and grown for 11 days. The xiphoids were weighed before becoming implanted and at the end of incubation. The growth capacity of transplanted xiphoids declined progressively with increasing donor age. The tissue also showed an age-related increase in dependence on thyroid hormones for growth. The xiphoid transplants from fetal rats grew equally well in the euthyroid and hypothyroid hosts, but those from 30-day-old showed a near-total dependence on thyroid hormones for growth.[26]

Another example of the study of hormonal dependence in growth and development is the use of fetal transplants in developmental neuroendocrinology.[27] These studies revealed a considerable degree of neuronal differentiation to occur after transplatation of fetal hypothalamus into adult host brain. Transplantation in general is a powerful tool for investigating perinatal influence of hormones on neural development as long as the endocrine environment of the adult host can be easily manipulated. The possibility that such transplants improve recovery processes in host brains, however, exists.

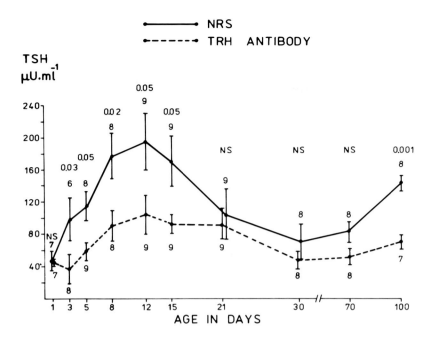

FIGURE 1. The effect of TRH antibody on plasma TSH during ontogenesis. Undiluted rabbit antiserum or normal rabbit serum (NRS) was injected 2 hr prior to decapitation. Excess binding of TRH (thyroliberin) was shown in the plasma of all age groups. Immunoneutralization of endogenous TRH resulted in a decrease of plasma thyrotropin. In the age groups without significant effect of the antibody low hypothalamic TRH secretion (1 day-old) and/ or more important role of other regulating factors are supposed. (From Štrbák, V., Angyal, R., Jurčovičová, J., and Randušková, A., *Biol. Neonate*, 50, 91, 1986. With permission.)

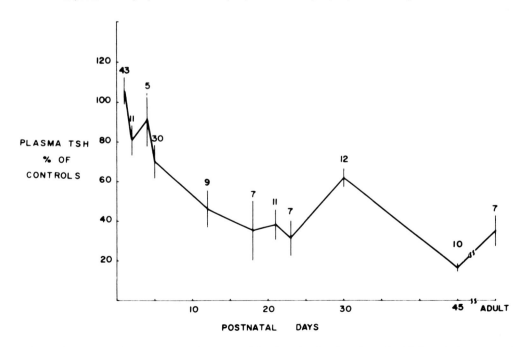

FIGURE 2. The fall in plasma TSH from control levels 4 hr after hypothalamic ablation, during postnatal ontogenesis in the rat. Note similar effects of the surgical technique and specific antibody treatment shown in the previous figure. (From Štrbák, V. and Greer, M. A., *Endocrinology*, 15, 488, 1979. With permission.)

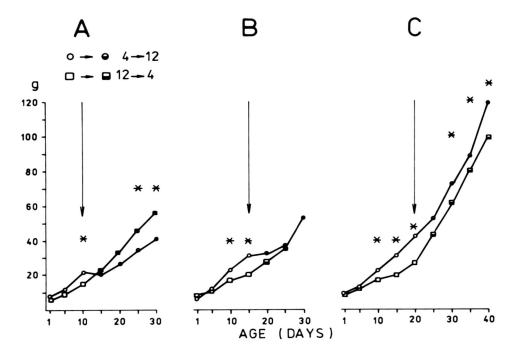

FIGURE 3.   Body weight of growing rats from litters with 4 or 12 animals per lactating female and after a change in the numbers of sucklings on postnatal day 10 (A), 15 (B), and 20 (C). Arrows indicate the day of the exchange, asterisks indicate significant difference. (P < 0.05 or less).

3. Ecological Factors Affecting Growth and Development

Among the ecological factors which may affect growth and development, nutritional effects are studied most frequently. A good animal model for the study of the effects of prenatal nutrition on the growth and development was introduced in Dobbing's laboratory.[28,29] Undernourished mother rats were fed about half the amount of the diet taken by control mothers during pregnancy and lactation. Whole litters were fostered between mothers 5 days after birth on the following way: (1) between control mothers (control groups), (2) from the control to an underfed mother (infantile growth retardation), and (3) from the underfed to the control mother (small for dates rats). After weaning, all young had free access to a diet. Another model based on experimental placental insufficiency resulting in fetal growth retardation probably also includes an effect of several nonnutritional factors.[30,31] In the Wigglesworth model,[32] the uterine vessels are ligated at the cervical end of the uterine horn. If performed unilaterally, the vessel ligation results in intrauterine death or intrauterine growth retardation, and the nonligated side can be used as control.[33]

There are several ways on how to change the nutritional level in infant laboratory animals. By modifying the size of litters reared with a lactating rat it is possible to induce under and overnutrition resulting in altered growth and development of pups.[34,35] Usually, pups are distributed perinatally into litters with 4 (overfed), 8 (control), 12 to 14 (underfed), or even 18 (severely undernourished) animals suckled by one lactating mother. Interesting experiments could be done with this model to explore the time of the critical period for growth velocity. In Macho's laboratory the number of animals was adjusted to 4 or 12 suckling pups per one mother on the 2nd day of life.[6] At the age of 10, 15, or 20 days the animals were exchanged (Figure 3).

It was observed that after the exchange on the 10th postnatal day the neonatally underfed animals could accelerate their growth in the small litter. On the other hand, those overfed in the first postnatal days that were transferred later in large litters retarded in body weight gain. If the exchange occurred on the 15th postnatal day the retarded animals were able to catch up on growth at the age of 25 days. When the exchange was carried out on the 20th postnatal day, persistent differences in body weight were found. This is of particular interest, because at the age of 30 days all the animals were weaned and distributed into cages with the same number of animals and pelleted diet and tap water were available ad libitum.

An interesting model in which by manipulating the solid food and water intake the weaning period can be altered was introduced by Babický et al.[36] Water and solid food were placed on top of the neighboring cage where only the rat mother could reach them. Under these circumstances the onset of the natural weaning period for infant rats was artificially postponed.[36]

Another model on how to induce overeating and subsequent obesity in young rats is based on bilateral destruction of the ventromedial hypothalamic nuclei by use of a stereotactic instrument and electrocoagulation.[37] The authors were able to perform the procedure with 10-day-old animals. This procedure requires, however, rather expensive equipment and surgical skill. Moreover, experimental animals show neuroendocrine deficits and behavioral disturbances (hyperemotionality-aggression, grooming neglect) which makes them difficult to handle and complicates the interpretation of results.[37]

To study the effect of defined diets on growth and development a model with artificial rearing of pups was proposed. Intragastric cannulae were permanently implanted at the age of 4 days, the pups were placed into plastic containers which floated freely in a water bath mantained at 42°C. The cannulae were connected to syringes containing rat milk substitute and the syringes were mounted on an infusion pump.[38,39] Stressful handling, limited motor activity, and total absence of maternal care make application of this model limited for special studies. A similar model with the rat pups fed by tubes and kept with a foster mother has also been used.[40] The pups are freely moving except for the feedings by the tube every 2 hr. The foster nonlactating mothers serve several functions: (1) they stimulate stomach emptying by physical manipulation of the pup and by allowing the pups to suckle, and (2) the foster mothers stimulate evacuation of urine and feces, prevent dehydration, and maintain body temperature of the pups by providing an immediate environment of high humidity and relatively constant temperature.

Another way to produce undernutrition in infant rats is the removal of the lactating mother from the litter for several hours daily.[41,42] In this way undernourishment can be produced with the experimental animals weighing about half that of controls at the age of 10 days and about 25% only by day 60.[42] Their development is also retarded. Again, the situation is not simple, the model is not clear. Intermittent mother removal results not only in decreased milk intake but stress from the absence of the mother, changes in the body temperature, urine retention, and possibly other factors also affecting the pups.

In this context a few words should be said about premature or early weaning. Infant rats survive permanent absence of the lactating mother after the 15th postnatal day. However, weaning at this age results in a significant decrease of body weight in young rats.[43] If weaned by postnatal day 18, the decrease of body weight is only transient, short lasting.[44] At this age an abrupt change in the quality of the diet is the major mechanism of induced changes. At a younger age, the animals are also undernourished during the first days after weaning. In both cases, however, the removal of the mother is associated with stress (increased adrenal activity for 1 to 2 days). Evidently the model is not clear because it includes to many changing parameters: changed caloric intake, change in the diet, stress, change in the nest temperature, and new social relations.

IV. SOURCES OF EXPERIMENTAL MISTAKES AND HOW TO OPTIMALIZE EXPERIMENTAL CONDITIONS

The model itself can be a source of errors if it includes too many uncontrolled or changing conditions which makes interpretation of results impossible or too complicated. Another category of errors arises from the use of a model in an inappropriate situation.

The way in which the experiment is carried out (even when using the optimal model) can result in a failure because of a number of reasons. Typical examples are the transfer of animals into another cage or room immediately before an experiment designed to study catecholamines or corticoids in plasma, low temperature during such an experiment (if the experiment has not been designed to study the effect of environmental temperature), noise during the experiment, etc. In ontogenic studies, experimental and control animals should come from the same litter if it is allowed by the experimental model. Nevertheless, different treatment inside a litter with the same mother may still occur. Perinatal random distribution of newborn animals into litters is of much help. However, differences in postnatal maternal care may also influence the results. It has been proved that the number of suckling pups in a litter affects their adrenal and thyroid activity as well as metabolism.[8] It is an advantage to adjust the number of newborn pups in a litter to eight animals (except for models studying the effect of the number of pups in a litter). A comparison of animals from small-size nests with those from large ones may give misleading results, which may be erroneously ascribed to differing experimental conditions. Early weaning also affects hormonal activity in the thyroid and adrenals.[6,44] The animals should be weaned regularly at the end of the natural weaning period, optimally by day 28 to 30. When animals from different litters are to be compared, the animals should descend from the same father. In contrast, if inbred animals are used in experiments, their experimental behavior may be too specific and the interpretation limited.

Although largely ignored the interpretation of results is probably the most frequent source of errors; also usually a relatively small volume of information is used. The more simple the experimental situation the smaller chance for misleading interpretation. Complicated models lead to divergent explanations of the same results. We usually accept that the same results are differently interpreted by a physiologist, psychologist, biochemist, nutritional physiologist, endocrinologist, etc., in belief that the others are wrong since they lack information the other has. Sometimes, we admit that there may be a piece of the truth in each of these interpretations. Under similar circumstances we would better admit that the experimental model used is too complicated and will hardly give any positive answer to our specific question.

REFERENCES

1. **Artan-Spire, S., Wolf, B., and Czernichow, P.,** Developmental pattern of TRH-degrading activity and TRH content in rat pancreas, *Acta Endocrinol.,* 106, 102, 1984.
2. **Wolf, B., Aratan-Spire, S., and Czernichow, P,** Hypothyroidism increases pancreatic thyrotropin-releasing hormone concentrations in adult rats, *Endocrinology,* 114, 1334, 1984.
3. **Aratan-Spire, S., Wolf, B., and Czernichow, P.,** Effects of hypo- and hyperthyroidism on pancreatic TRH-degrading activity and TRH concentrations in developing rat pancreas, *Acta Endocrinol.,* 106, 209, 1984.
4. **Stave, U.,** Maturation, adaptation and tolerance, in *Perinatal Physiology,* Stave U., Ed., Plenum, New York, 1978, chap. 3.
5. **Babický, A., Ošťadalová, I., Pařízek, J., Kolář, J., and Bíbr, B.,** Use of radioisotope techniques for determining the weaning period in experimental animals, *Physiol. Bohemoslov.,* 19, 457, 1970.

6. **Macho, L.,** The effect of early weaning on thyroid and adrenal gland function, in *Development of Thyroid and Adrenal Function During Ontogenesis,* Macho, L., Ed., VEDA, Bratislava, 1979, 68.

7. **Marubini, E.,** Review of models suitable for the analysis of longitudinal data, in *Human Growth and Development,* Borms, J., Hauspie, R., Sand, A., Susanne, C., and Hebbelinck, M., Eds., Plenum, New York, 1984, 753.

8. **Van't Hof, M. A.,** Up-to-date cross-sectional information in longitudinal designs, in *Human Growth and Development,* Borms, J., Hauspie, R., Sand, A., Susanne, C., and Hebbelinck, M., Eds., Plenum, New York, 1984, 767.

9. **Webb, S. M., Lewinski, A. K., Steger, R. W., Reiter, R. J., and Battke, A.,** Deficiency of immunoreactive somatostatin in the median eminence of Snell Dwarf mice, *Life Sci.,* 36, 1239, 1985.

10. **Okamoto, K. and Aoki, K.,** Development of a strain of spontaneously hypertensive rats, *Jpn. Circ. J.,* 27, 282, 1963.

11. **Kojima, A., Takahashi, Y., Ohno, S., Sato, A., Yamada, T., Kubota, T., Yamori, Y., and Okamoto, K.,** An evaluation of plasma TSH concentration in spontaneously hypertensive rats (SHR), *Proc. Soc. Exp. Biol. Med.,* 149, 661, 1975.

12. **Albrecht, I. and Štrbák, V.,** The role of the thyroid gland function in the genesis of hereditary hypertension, in *Hormones and Development,* Macho, L., and Štrbák, V. Eds., VEDA, Bratislava, 1979, 445.

13. **Barta, A., Richards, R. I., Baxter, J. D., and Shine, J.,** Primary structure and evolution of rat growth hormone gene, *Proc. Natl. Acad. Sci. U.S.A.,* 78, 4867, 1982.

14. **Tashjian, A. H., Yasumura, Y., Levine, L., Sato, G. H., and Parker, M. L.,** Establishment of clonal strains of rat pituitary tumor cells that secretes growth hormone, *Endocrinology,* 82, 342, 1968.

15. **Evants, G. A. and Rosenfeld, M. G.,** Cell-free synthesis of a prolactin precursor directed by mRNA from cultured rat pituitary cells, *J. Biol. Chem.,* 251, 2842, 1976.

16. **Samuels, H. H., Horkowitz, Z. D., Stanley, F., Casanova, J., and Shapiro, L. E.,** Thyroid hormone controls glucocorticoid action in cultured GH_2 cells, *Nature,* 268, 254, 1977.

17. **Diamond, D. J. and Goodman, H. M.,** Regulation of growth hormone messenger RNA synthesis by dexamethasone and triiodothyronine. Transcriptional rate and mRNA stability changes in pituitary tumor cells, *J. Mol. Biol.,* 181, 41, 1985.

18. **Evans, R. M., Birnberg, N. C., and Rosenfeld, M. G.,** Glucocorticoid and thyroid hormones transcriptionally regulate growth hormone gene expression, *Proc. Natl. Acad. Sci. U.S.A.,* 79, 7659, 1982.

19. **Page, G., Smith, S., and Goodman, H.,** DNA sequence of the rat growth hormone gene: location of the 5' terminus of the growth hormone mRNA and identification of an internal transposon-like element, *Nucleic Acid Res.,* 9, 2087, 1981.

20. **Doehmer, J., Barinaga, M., Vale, W., Rosenfeld, M. G., Verma, I. M., and Evans, R. M.,** Introduction of rat growth hormone gene into mouse fibroblasts via a retroviral DNA vector: expression and regulation, *Proc. Natl. Acad. Sci. U.S.A.,* 79, 2268, 1982.

21. **Spindler, S. R., Mellon, S. H., and Baxter, J. D.,** Growth hormone gene transcription is regulated by thyroid and glucocorticoid hormones in cultured rat pituitary tumor cells, *J. Biol. Chem.,* 257, 11627, 1982.

22. **Yaffe, B. and Samuels, H. H.,** Hormonal regulation of the growth hormone gene. Relationship of the rate of transcription to the level of nuclear thyroid hormone-receptor complexes, *J. Biol. Chem.,* 259, 6284, 1984.

23. **Rotwein, P.,** Two insulin like growth factor messenger RNAs are expressed in human liver, *Proc. Natl. Acad. Sci. U.S.A.,* 83, 77, 1986.

24. **Štrbák, V., Angyal, R., Jurčovičovà, J., and Randušková, A.,** Role of thyrotropin releasing hormone in thyroid stimulating hormone and growth hormone regulation during postnatal maturation in female Wistar rats, *Biol. Neonate,* 50, 91, 1986.

25. **Štrbák, V. and Greer, M. A.,** Acute effects of hypothalamic ablation on plasma thyrotropin and prolactin concentrations in the suckling rat: evidence that early postnatal pituitary-thyroid regulation is independent of hypothalamic control, *Endocrinology,* 15, 488, 1979.

26. **Cooks, P. S., Yonemura, G. U., and Nicoll, C. S.,** Development of thyroid hormone dependence for growth in the rat: a study involving transplanted fetal, neonatal and juvenile tissues, *Endocrinology,* 115, 2059, 1984.

27. **Paden, C. M., Silverman, A. J., Stenevi, U., and McEwen, B. S.,** The use of fetal hypothalamic transplants in developmental neuroendocrinology, in *Neural Transplants,* Sladek, J. R., Jr., and Gash, D. M., Eds., Plenum, New York, 1984, 283.

28. **Adlard, B. P. F., Dobbing, J., and Smart, J. L.,** An alternative animal model for the full-term small-for-dates human baby, *Biol. Neonate,* 23, 95, 1973.

29. **Smart, J. L., Adlard, B. P. F., and Dobbing, J.,** Further studies of body growth and brain development in "small-for-dates" rats, *Biol. Neonate,* 25, 135, 1974.

30. **Gruenwald, P.,** Chronic fetal distress and placental insufficiency, *Biol. Neonate,* 5, 215, 1963.

31. **Hill, D. E., Meyers, R. E., Holt, A. B., Scott, R. E., and Cheek, D. B.,** Fetal growth retardation produced by experimental insufficiency in the Rhesus monkey. II. Chemical composition of the brain, liver, muscle and carcass, *Biol. Neonate,* 19, 68, 1971.
32. **Wigglesworth, J. S.,** Experimental growth retardation in the foetal rat, *J. Pathol. Bacteriol.,* 88, 1, 1964.
33. **De Prins, F. A. and Van Assche, F. A.,** Intrauterine growth retardation and development of endocrine pancreas in the experimental rat, *Biol. Neonate,* 41, 16, 1982.
34. **Kennedy, G. C.,** The effect of age on the somatic and visceral response to overnutrition in the rat, *J. Endocrinol.,* 15, 19, 1957.
35. **Widdowson, E. M. and McCance, R. A.,** Some effect of accelerating growth. I. General somatic development, *Proc. Res. Soc. Lond. (Biol),* 152, 188, 1960.
36. **Babický, A., Pavlík, L., Pařízek, J., Bíbr, B., Kolář, J., and Oštádalová, I.,** Maternal milk intake by infant rats temporarily denied access to other food sources, *Physiol. Bohemoslov.,* 24, 67, 1975.
37. **Fisher, R. S., Almli, R. C., and Parsons, S.,** Infant rats: VMH damage and the ontogeny of obesity and neuroendocrine dysfunction, *Physiol. Behav.,* 21, 369, 1978.
38. **Sonnenberg, N., Bergstrom, J. D., Ha, Y. H., and Edmont, J.,** Metabolism in the artificially reared rat pup: effect of a atypical rat milk substitute, *J. Nutr.,* 112, 1506, 1982.
39. **Smart, J. L., Stephens, D. N., and Katz, H. B.,** Growth and development of nutrition, *Br. J. Nutr.,* 49, 497, 1983.
40. **Czajka, D. M., Miller, S. A., and Browning, A. M.,** Studies of the protein requirement of the neonate rat, *J. Nutr.,* 103, 1608, 1973.
41. **Bass, H. N., Netsky, M. G., and Young, E.,** Effect of neonatal malnutrition on developing cerebrum, *Arch. Neurol.,* 23, 289, 1970.
42. **Krigman, M. R. and Hogan, E. L.,** Undernutrition in the developing rat: effect upon myelination, *Brain Res.,* 107, 239, 1976.
43. **Kraus, M., Křeček, J., and Popp, M.,** The development of corticosterone production by the adrenal gland in normally and prematurely weaned rats, *Physiol. Bohemoslov.,* 16, 120, 1967.
44. **Macho, L., Štrbák, V., and Strákzovcová, A.,** The effect of premature weaning on thyroid and adrenal gland functions in the rat, in *Hormones in Development,* Hamburgh, M. and Barrington, E. J. W., Eds., Appleton-Century-Crofts, New York, 1971, 801.

Chapter 14

MATHEMATICAL MODELS OF GROWTH, DEVELOPMENT AND AGING*

S. Ďoubal and P. Klemera

TABLE OF CONTENTS

---

* *Symbols:* **Latin—** A = quadratic regulation area; b_i = constant; k = ultimate size; k_G = Gompertz constant; m = mass of system; n_T = state of timing system; n_w, m_w = constants of Weiboull distribution; n,p = parameters of growth; number of elements or states; p_i = probability of microstate i; Q = thermal energy; R = reliability; S = entropy; t = time; T = temperature; x = size of organism; y = output; and y_0 = desired output. **Greek—** α_i, β_i = constants; β, β' = growth constant; δ = failure rate of elements; Λ, λ, λ_i = failure (mortality) rate; λ_0 = initial failure rate; and T = constant of integration.

I. INTRODUCTION

The development of sciences to exactness inevitably claims the use of mathematical methods and models. This fact has been apparent in physics since the 17th century and became evident in chemistry some 100 years ago. Nevertheless, in biology, certain impediments of the extensive use of mathematical models still exist. The sources of obstacles are of an objective as well as of a subjective nature. Extreme complexity and intricacy of biological systems, and the stochastic nature of biological processes are the objective problems. The subjective obstacle is in the traditional way of thinking and education of biologists. We believe that the key to overcoming these problems is an exact solution of problems of complexity. The above-discussed considerations are fully relevant to the problems of modeling of growth, development, and aging.

A general tool for the design of mathematically based complex system models is provided by *system theory* and the methodically related *theory of cybernetical modeling*. The principles of system theory are used in the models which are designed on the basis of the theory of reliability and in those made on the basis of control theory. Similarly, the description of the dynamics of growth and mortality could be taken for a specific class of models of dynamic systems. The aim of this type of model is usually the description of dynamics of systems; they do not analyze their structure.

The philosophy of state-space models has risen from the physical sciences. These models describe and predict dynamics and development of the state of the system on the basis of measuring the actual values of the so-called state variables. As is similar to some of the growth and mortality models, the state-space models remain on the descriptive or behavioral level of modeling.

An important class of models of growth, development, and aging is derived from the principles of thermodynamics. The main advantage of this class of model is the logical consistency, and the ability to explain the causes of macroscopic phenomena by means of processes on a molecular level. More extensive application of the thermodynamical principles in biology had been limited in the past by the structural complexity of biological systems and by the nonequilibrious nature of the majority of biological processes. The origin of the thermodynamics of dissipative structures and especially the development of network thermodynamics enables us to overcome both obstacles and combine thermodynamic principles with the system and cybernetic approach.

The field of mathematical modeling is highly heterogenous, and the methods applied are often remarkably different in individual cases. Above it, the number of models is permanently growing and, simultaneously, new methods are applied. Therefore, it is particulary difficult to put forward a representative survey of the main classes of mathematical models of growth, development, and aging in a single chapter. In addition, the extent of this chapter excludes the detailed discussion of principles and theoretical backgrounds of existing models. We chose, hence, those classes of mathematical models which we considered to be perspective and/or particularly important.

The discussion as well as the explanation of these models is also of necessity brief. The reader can find additional information in the references included at the end of the chapter.

II. SYSTEM THEORY AND MODELING

A. Systems

System theory offers an exactly defined and general enough frame for the construction of mathematical models of development, growth, and aging. The advantage of the "system approach" lies in eliminating both the vagueness of nonsystem models and the indefinitness of their interpretation.[1,2]

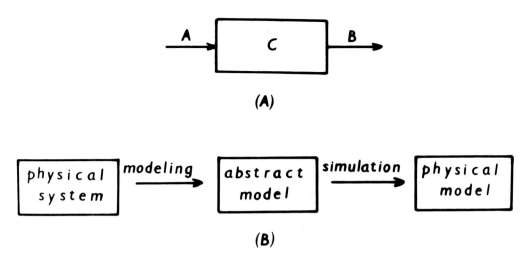

FIGURE 1. (A) Graphic representation of (unstructured) system; A inputs, B outputs, C internal states. (B) Modeling of physical system under study yields to an abstract model of it; behavior of the system can be simulated by suitable physical model.

In the elementary case, the *system* (dynamic system) is specified by the set A of input states, the set B of output states, and the set C of internal states (Figure 1A). A description of *behavior* of the system is complete if the internal state at the time t_1 and the course of input states from t_1 to t_2 make it possible to determine the course of output states from t_1 to t_2 and to predict the internal state at time t_2. If this prediction is unique, the system is called a *deterministic* one; provided only probabilistic prediction is possible, we speak of a *stochastic* system.[3]

The output of one system can act upon the inputs of other systems. By this *coupling* of unstructured systems (i.e., of *elements*), it is possible to build systems of arbitrary complexity. The way of mutual coupling of elements building the complex system is called the *structure* of the system. The structure of any system and the behavior of all its elements fully determine the behavior of the complex system, though only implicitly.[4]

Evidently, it is possible to define various systems on any physical object, even when the same phenomenon is studied. Variability of these systems can depend on the level of object organization on which the problem is analyzed. For example, studying the growth of an organism, we can be interested in the effect of proliferation dynamics upon the mass of the organism in one case, or in the dependence of the mass upon anabolic and catabolic processes on the subcellular level in another case. The elements of these systems will be different, and so will the structure and inputs of them.

The possibility of dividing a complex system into subsystems, sub-subsystems, etc. and of solving the overall problems hierarchically is very important. This way it is possible to divide the problems of the complex system into the solving of simpler problems on partial levels, and then only pass to the level of an all complex system.

B. Modeling and Simulation

By *modeling*, the setting up of such an abstract system is understood that describes adequately enough the properties of the physical object under study, i.e., a system that is an *abstract model* of the physical system (Figure 1B). The term *structural model* is used if the structure of the abstract model and the behavior* of its elements correspond to the

* In the cybernetic sense, see Section II. A.

physical system. Such a type of model has an explanatory value. On the other hand, an unstructured abstract model is called a *model of behavior*, and its value is only descriptive and possibly predictive; the same holds for a structured abstract model corresponding to the physical system in its external behavior, but not in the structure and in the behavior of elements.

The theoretical study of properties of an abstract model is often difficult and/or insufficient. Then *simulation* of the system is approached (Figure 1B), i.e., experiments are made with a suitable physical realization (*physical model*) of the abstract system — today, usually in the form of computer program.[4]

When judging and comparing models, it is important to distinguish their mutual isomorphy and homomorphy. For an *isomorphic* model M of system S, there is a one-to-one correspondence between both systems (their states, behavior, structure, etc.). For a *homomorphic* model, there exists mapping of system S onto model M, but this mapping may be simplifying in the sense that some (in S distinguished) entities fuse in M.

Modeling is one of the most basic knowledge processes. Purposeful mathematical modeling of biological systems steps forward, as a rule, in the sequence formation of a model of behavior → formation of an homomorphic structural model → formation of an isomorphic structural model → formation of a model on more detailed level of resolution, and so on. The decision at which link to stop depends on the question answered by the modeling and, certainly, on the complexity of the systems under study.[5]

III. THEORIES OF GROWTH AND MORTALITY

A. Theories of Growth

Quantitative descriptions of growth by mathematical formulas as have appeared in the work of Bertalanffy,[6] Richards,[7] Nelder,[8] Turner,[9] and others can be considered to be, from the system theory point of view, models of behavior. In other words, as we saw earlier in Section II of this chapter, they do not model structures of systems, but perform only formal mathematical descriptions of the dynamics of the systems. Nevertheless, the design of these models could contribute to subsequent modeling of their structures.

The aim of models of growth is to find universal growth functions which could, via the choice of appropriate values of their parameters, approximate dynamics of growth of organisms or populations.

An attempt to synthesize existing growth theories was made by Turner.[9] This model presumes that the rate of growth of an organism or population is jointly proportional to a monotonically increasing function of the distance between the original and the actual size of the system, and to a monotonically increasing function of the distance between the actual size and the ultimate size of the system. These two functions are supposed to be the power ones. The fundamental rate equation has been derived on the basis of these presuppositions.

$$\dot{x} = \frac{\beta}{k^n} x^{1-np}(k^n - x^n)^{1+p} \tag{1}$$

where x is the size of the organism or population; k is the ultimate size; $\dot{x} = dx/dt$ is rate of change of size; β is the so-called intrisic growth constant; n,p are parameters of growth function; for n holds $n > 0$, for p holds $-1 < p < t$.

Solving the previous equation we obtain the generic growth function

$$x = \frac{k}{\{1 + [1 + \beta np(t - \tau)]^{-1/p}\}^{1/n}} \tag{2}$$

where τ is the constant of integration.

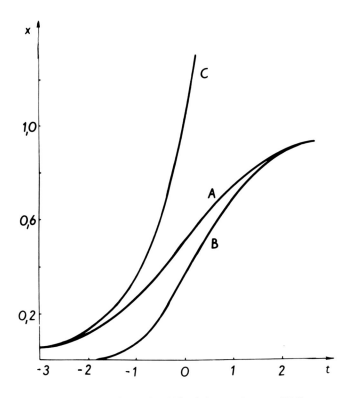

FIGURE 2.   Curves of growth. (A) Logistic growth curve; (B) Gompertz growth curve; and (C) Malthus growth curve for $\tau = 0$, $k = 1$, and $\beta = 1$.

A remarkable number of well known curves could be considered a special or limiting case of the generic function mentioned above. The limiting case for $p \to 0$ is the so-called generalized logistic function:[6,9]

$$x = \frac{k}{\{1 + \exp - [\beta n(t - \tau)]\}^{1/n}} \qquad (3)$$

Similarly, for $n = 1$ we obtain the so-called hyperlogistic function:

$$x = \frac{k}{1 + [1 + \beta p(t - \tau)]^{-1/p}} \qquad (4)$$

The widely used logistic function (line A in Figure 2) could be considered a limiting case of the generic Equation 2 for $p \to 0$ and $n = 1$:

$$x = \frac{k}{1 + \exp[-\beta(t - \tau)]} \qquad (5)$$

The so called hyper-Gompertz function[9]

$$x = k \exp\{-[\beta'_p(t - \tau)]\}^{-1/p} \qquad (6)$$

and Gompertz growth function (line B in Figure 2)

$$x = k \exp\{-\exp[-\beta'(t - \tau)]\} \qquad (7)$$

are also limiting cases of the generic function.

In the case of k → ∞, np → 0, we obtain Malthuss' law of geometric growth (line C in Figure 2)

$$x = \exp[(t - \tau)] \qquad (8)$$

The identification and parameter estimation of the above-mentioned models require the knowledge of data on changes in size or mass of the organisms or population.

These data are usually easy to obtain. The correct determination of the type of growth function as well as the accuracy of parameter estimation depends on the number of individuals in the sample (experimental group, population) and on the homogenity of it.

Data evaluation could be performed by standard statistical procedures[10] (regression methods, tests of statistical significancy, rank-sign tests, etc.). The models discussed in this section are used for the quantification of dynamics of growth and enable one to predict further development of the system. The assesment of parameters of growth function could provide some information on the structure of system and on the nature and causes of growth. Nevertheless, as we mentioned previously, these models are the models of behavior. Consequently, their contribution to understanding the growth process is limited.

B. Theories of Mortality

The most striking feature of aging is the fact that the risk of death increases with age. The consequence of this fact on the population level is an increasing mortality rate. The changes of mortality are far more salient than changes of other phenomena associated with the aging process (decrease in functional capacity, accuracy of synthesis, etc.). Relatively simple laws hold for the changes of the mortality rate in the human population as well as in populations of a remarkable number of species living in laboratory or domestic environments. Apparently, these rates reflect an activity of the mechanism of aging. This fact is the particular interest of gerontologists,[11] and several theories of mortality were developed in the past. Similar to theories of growth, the theories of mortality are largely models of behavior.

The concept of mortality is not used uniformly. We believe that the most universal and exact definition of mortality can be expressed analogously to the definition of the failure rate used in the theory of reliability.[12,19]

Let R(t) be the probability that the organism does not die until the time t. Then the mortality rate λ (further only mortality) may be defined as follows:

$$\lambda = -\frac{d\,R(t)}{dt}\frac{1}{R(t)} \qquad (9)$$

The course of the mortality rate v. time is called the mortality curve. The above-mentioned definition is universal enough for the biological as well as demographical applications. The formula 9 enables the transformation of survival curves to mortality curves, provided the survival curvs are obtained from sufficiently large groups to be considered the courses of probability R(t) v. time.

A characteristic mortality curve for a human population is shown in Figure 3. The curves of mortality of domestic and laboratory animals are of a similar character. In all probability, the mortality curves for all mammals (living in domestic conditions) have the shape shown in Figure 3A. Similar course of mortality is observed in some other species. The mortality curves of *Drosophila imagoes,* for example, differ from the course shown in Figure 3 only by the absence of part II, perhaps because the larval stadium has not been taken into account. From the gerontological point of view, parts III and IV in Figure 3 are of particular interest.

The exponential increase of mortality rate in the adult period of life (part III, Figure 3)

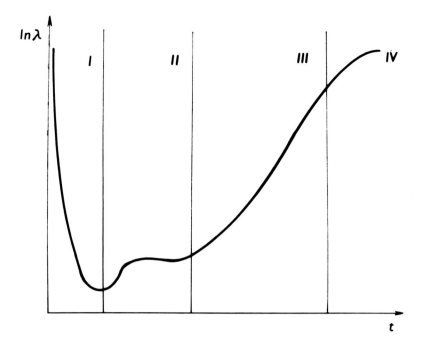

FIGURE 3. Characteristic course of mortality during the life of an organism. Period I corresponds to early childhood (childhood diseases), period II is the era of low mortality, period III is the era of validity of Gompertz law, and period IV is the period of high age.

was noted one and a half centuries ago by Gompertz.[13] The so-called Gompertz law may be written as

$$\lambda = \lambda_o \exp(k_G t) \tag{10}$$

Makeham supplemented the previous formula by an additive constant, which reflects the existence of catastrophic, non-age-dependent, deaths. For the contemporary European populations, the Makeham constant is negligible, as well as for animals living under good conditions.

Besides the Gompertz law, the so-called Weiboull distribution of probability of survival is occasionally used for the approximation of mortality curves. The mortality function has in this case the following form:

$$\lambda = \lambda_o + n_W t^{mw} \tag{11}$$

$$\lambda = n_W t^{mw} \tag{12}$$

where λ_0, n, and m are constants.

The influence of environmental conditions on mortality curves deserves special attention. The improvement of living conditions results in the shift of mortality curves according to Figure 4.[11,15] This character of relation between the mortality curves is believed to be caused by environmental influences, which do not interfere with the aging process. This opinion is also supported by the experiments on germfree and conventional groups of *Drosophila* which were performed in our laboratory and which resulted in a relation between mortality curves as shown in Figure 4.

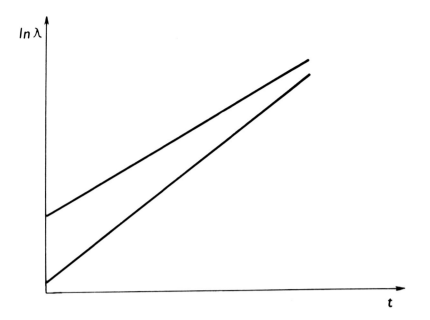

FIGURE 4.   Mortality rate v. age in a semilogarithmic plot (Gompertz period); characteristic relation between mortality curves observed in organisms living in different environment.

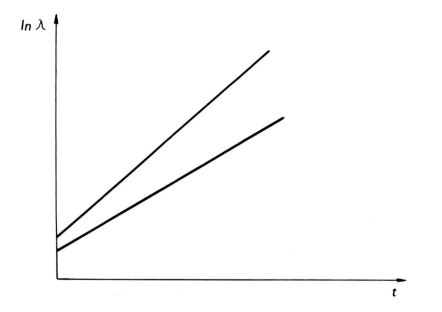

FIGURE 5.   Mortality rate v. age in a semilogarithmic plot (Gompertz period); relation between mortality curves when the aging rate is modified.

Nevertheless, some other experiments result in the relation given in Figure 5. This relation was found in mice after exposure to busulfanum 1,4-bis-(methylsulfonyloxy)-*n*-butanum (a drug with a mutagenic effect); in Drosophila exposed to a relatively great dose (400 Gy) of X-rays; and in *Drosophila* bred in different temperatures. Great doses of ascorbic acid (antioxidant substance) administered permanently resulted in lowering of mortality, and the relation between the mortality curves was also of the character shown in Figure 5. We believe that these findings support the idea that the divergence in the mortality curves (Figure 5) indicates the changes in the rate of aging.

The data necessary for the estimation of parameters of the mortality models can be obtained from either longitudinal or cross-sectional studies (see also Section IV). The statistical evaluation can be performed by means of standard statistical procedures.[10] Nevertheless, these procedures are relevant provided the fluctuations are of a stationary and ergodic character. In the case of longitudinal experiments, the satisfaction of stationary hypotheses is the crucial problem. In other words, the statistical parameters of environmental fluctuations have to be independent on time shift. Hence, the expeimental conditions have to be strictly the same in the course of the experiment.

The cross-sectional studies are often the only available method of obtaining the mortality curves of a human population. In this case, the crucial problem is the satisfaction of the ergodic presupposition. The consequence of ergodic presumption is the claim of genetical identity of all age groups and, above it, that environmental conditions should be the same in the course of their lives.

There are apparent practical difficulties in verification of ergodic conditions. It is therefore advisable that interpretations of the results of cross-sectional experiments be carried out with caution.

The advantage of theories of mortality is the easy availability and unambiguity of data. Their main limitation is the fact that they are models of behavior and, therefore, they do not explain the causes of changes in mortality during life. This shortcoming is overcome in structurally based models of mortality;[15,16] an example of this type of model is presented in Section IV of this chapter. Another approach is based on the state-space principle. These models propose the explanation of changes of mortality by changes of the physiological state of the organism (see Section VI). A further limitation of the theories of mortality is the time interval of their validity. The Gompertz law can be applied only to the period of adulthood (period III in Figure 3). Similarly, the Weiboull approximation is convenient for this period. Whereas the non-Gompertz character of mortality curves in childhood is considered to be obvious, the decrease of rate of growth of mortality in high age[16,17] is somewhat surprising. Though the demographic data on the mortality in high age in the human population are unreliable and rare, they exhibit convincingly enough that the mortality in high age increases slower than is predicted by the Gompertz law. There are even some indications that the mortality in the human population at the age of about 100 years does not increase at all; in some cases a decrease was observed. Considering the fact that mortality reflects the activity of the basic mechanisms of aging, the explanation of the above-discussed phenomenon is of outstanding interest. Note that this tendency was found also in such simple organisms as *Drosophila* (Figure 6). In this context, the work of Pakin[18] is of special interest. The validity of the Gompertz law is discussed in that paper and found to be controversial. Nevertheless, the findings could be an artifact caused by the nonergodic character of analyzed data.

IV. RELIABILITY MODELS OF AGING

The theory of reliability was developed for the needs of engineering. Its methodology, however, is essentially universal. Consequently, its potential can be explored in the analysis of the reliability of biological systems, especially in the modeling of the aging process.[19,20] The methodology of the theory enables one to assess the reliability of the system on the basis of its structure and reliability of its elements. The theory of reliability also proposes methods of evaluation of reliability and methods of prediction of future reliability of systems. Further, the theory serves as a tool to the design of systems with prespecified reliability.

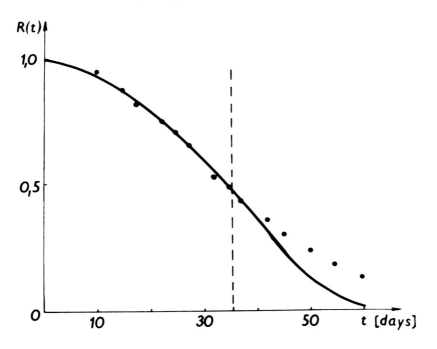

FIGURE 6.   Survival curves of Drosophila. The full line is a theoretical curve calculated on the basis of presumption of validity of Gompertz law. For the regression, the data on survival from time interval 0 to 35 days were used. The deviation from Gompertz law in "high age" is apparent.

All the basic concepts and criteria of the theory of reliability are based upon the concept of failure. A quantification of reliability is also based on the concept of failure. The most frequently used criteria of reliability are the probability of nonfailure and failure rates. The probability of nonfailure is the probability that failure does not occur until a time t. The failure rate is defined on the basis of probability of nonfailure by formula 9, where probability R(t) is in this case equal to the probability of nonfailure.

For the purpose of a reliability analysis, it is important to know the relation between the reliability of elements and the reliability of the whole system. This relation depends upon the reliability structure of the system. In many cases the structure of system is built up from three types of basic structures, namely from the serial (Figure 7), modular (Figure 8), and parallel (Figure 9) arrangements. In the case of serial structure, the elements of a system operate in such a way that the correct function of the system requires a correct function of all elements. The modular and parallel structures are, on the contrary, redundant structures.

Some ideas and principles of the theory of reliability can be immediately applied to gerontology. On the other hand, the specificity of biological systems has to be taken into account. The main problems are the complexity of biological systems, the wide range of introducing redundancy and repairs, and the structural changes of biological systems in the course of life of the organism.

A possible way for overcoming of these obstacles was analyzed in our previous work.[15] The so-called general modular system was introduced to the basic structures. The relation between reliability of elements and reliability of the whole system is, in the general modular system, supposed to be stochastic. This approach to reliability analysis is more adequate to the nature of biological systems.

The data required for design of reliability models depends on the definition of failure. If the death of an organism is considered to be the failure, then the discussion presented in

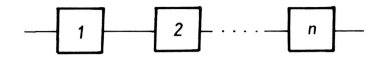

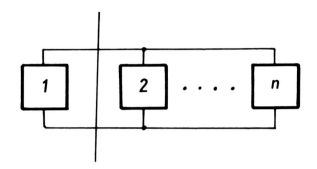

FIGURE 7. Structure of a serial system.

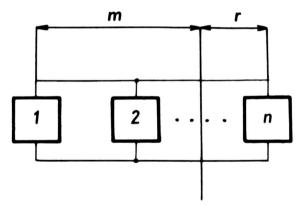

FIGURE 8. Structure of a parallel system. Elements 2,3 . . . n are the reserve.

FIGURE 9. Structure of a modular system. The elements in a modular system are identical. The total number of elements is m + r, where r is the reserve. For correct operation of a modular system, it is necessary that at least m elements do not fail.

Section III is valid. If we take the occurrence of some dysfunction or disease for the failure, then there will be problems of availability and reliability of data because of the unreliability of statistical information.

Most frequently, the data on mortality or failure rate are required for design of reliability models. The construction of mortality (or failure) rate curves claims longitudinal experiments and numerous groups (several hundreds of individuals) of experimental animals. Thus, the experiments are extraordinarily demanding. The choice of suitable model organisms is important in this context. The demand of a great number of individuals in an experiment can be reduced by the possibility of obtaining the mortality curves from the survival curves. The following transformation formula is obtained from Equations 9 and 10:

$$\ln[-R(t)] = -\frac{\lambda c}{k} [\exp(k_G t) - 1] \qquad (13)$$

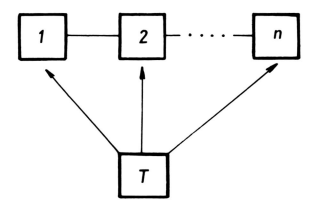

FIGURE 10.   Model of a reliability structure of organism.

This formula enables one to obtain parameters λ_0 and k of the Gompertz law from data (R(t)) on the survival of the organism. The standard regression methods can be used for this purpose. The transformation allows a reduction in the necessary number of individuals in an experiment from several hundreds to several tens.

The theory of reliability is a tool for modeling of the aging process. The models designed on the basis of the theory of reliability are capable of explaining the changes in mortality by the inner structure of the organism. For illustration, we now introduce an example of such a model.

Let the model satisfy Equation 10 and let its reliability change according Figure 3, parts III and IV. Further, let the mortality curves of the model satisfy the rules according Figures 4 and 5. Our aim is to find such a structure, which corresponds to the above-mentioned behavior as well as to the real structure of the organism.

The proposed structure of the model is shown in Figure 10. The subsystems in Figure 10 correspond to the subsystems in organisms which are of importance to life. The structure is serial (Figure 7). Considerations about tendencies in mortality curves result in the assumption that the model has to include the influence of a single timing system T (Figure 10), which is not in serial relation to life-important subsystems. Suppose that the state (n_T) of the timing system (that perhaps can be described by its performance or by the number of its elements) controls the failure rates λ_i of subsystems

$$\lambda_i = f_i(n_T) \tag{14}$$

As to Equation 14, it is reasonable to expect an acceptable approximation in the form

$$\lambda_i = \frac{\alpha_i}{n^{\beta_1} + b_i} \tag{15}$$

where α_i, β_i, and b_i are constants. The values of coefficients α_i and β_i reflect the influence of the environment on mortality.

Consider now the most simple case of the deterioration of the timing system. If the failure rate γ of elements of timing system is constant, then the decrease in number n of elements of timing system is as follows:

$$n = n_o \exp(-\gamma t) \tag{16}$$

For mortality λ of the system holds

$$\Lambda = \sum_{i=1}^{n} \lambda_i \tag{17}$$

For $n^{\beta_i} >> b_i$ gives combination of Equations 15, 16, and 17 in following equation

$$\Lambda = \sum_{i=1}^{n} \alpha_i n_o^{-\beta_i} \exp(\gamma \beta t) \tag{18}$$

If $\beta_1 = \beta_2 = \ldots = \beta$, we obtain the well-known Gompetz law

$$\Lambda = \Lambda_o \exp(k_G t)$$

where

$$\Lambda_o = \sum_{i=1}^{n} \alpha_i n_o^{-\beta}$$

$$k_G = \gamma \beta \tag{19}$$

Changes in coeficient γ represent the changes in the rate of aging. These changes modify the mortality curves according Figure 5. Changes in coefficients α_i reflect changes of environment and result in the relation between mortality curves presented in Figure 4.

The model also exhibits cessation of the growth of mortality in the high age. When $n^{\beta_i} << b_i$ (the timing system is exhausted), the mortality of system is further constant.

Consequently, this model is at least a good model of behavior. If there is also a structural analogy between the model and organism, then the model could contribute to understanding the nature of the process of aging.

V. MODELING OF AGING OF BIOLOGICAL CONTROL SYSTEMS

Feedback control systems are observed at all levels of biological organization, from the molecular to the supracellular, or even to the social one. The control systems are responsible for the stability of an organism, and the homeostasis of an organism is based on the activity of numerous control systems. The application of the principles of control theory is relatively frequent[23-27] in physiology. The use of the methods of control theory in gerontology is perspective as well. Models of changes of performance of the control processes during aging, models of changes in behavior of aging systems in general, and methods for identification of age-dependent damages in feedback systems could be very useful.

In accordance with the principles of control theory,[28-30] the basic structure of the control system is shown in Figure 11. The feedback loop is responsible for keeping the output of a system on a desired value. The typical feedback regulator consists of feedback receptors, controller, and actuator (effectors).

Control theory makes use of the methodology of system theory and, accordingly, provides a theoretical basis of synthesis (seeking for structure, whereas the behavior is predeterminated), analysis (derivation of behavior from the structure of the system and from the behavior of elements), identification (derivation of a general description of behavior on the basis of measuring input-output relations), and of modeling. The control theory also introduces methods of evaluation of performance of the control.

As pointed out earlier, the identification, the evaluation of performance of the control,

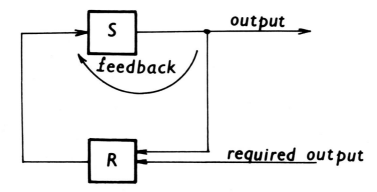

FIGURE 11. Fundamental structure of a control system.

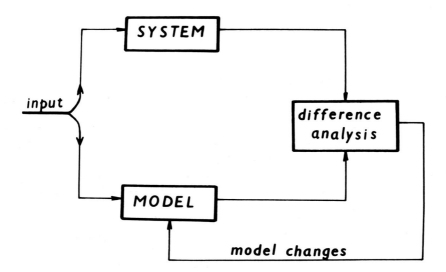

FIGURE 12. Identification of systems. The principle of the method of the adaptive model.

and the modeling in general are of particular interest of gerontology. The identification is, in all cases, based on analysis of output-input relations. In the case of linear systems (systems, the behavior of which can be described via linear differential equations), we can obtain the equations we search for by means of analysis of output response on deterministic input signals.[30,31] The accuracy of the methods is limited and depends crucially on disturbances.

The identification can be further performed on the basis of anlysis of statistical parameters of input and output signals. These methods are essentially universal and can be applied also to identification of nonlinear systems.[30]

The method of the adaptive model (Figure 12) is based on different principles. In this method, the input signals are the same in the real system and in the model. The difference between the responses of both systems is eliminated via changes of parameters of the model or through modification of its structure.

Several methods of evaluation of performance of the control process exist. The most frequently used are maximum control deviation, control time, and quadratic regulation area. The last one can be determined by the formula

$$A = \int_o^t (y - y_o)^2 \, dt \qquad (20)$$

where y is output, and y_0 is desired output.

The above-mentioned criteria are based on an analysis of output response to standard disturbance.

The advantage of the above-discussed methodology is the existence of elaborate theory which can contribute to a more exact analysis of changes in aging organisms. On the other hand, some obstacles still exist in this field. For instance, the biological systems are essentially nonlinear and the character of fluctuations is often nonstationary and nonergodic. Consequently, there are serious problems in the identification of systems. Further, the relevant data for modeling are often difficult to obtain. Above it, the structure of biological systems changes frequently in the course of life. We can then conclude that application of the control theory for the modeling of aging is desirable, but theoretically as well as experimentally complicated.

VI. THERMODYNAMIC MODELS

The exceptionally important class of models of growth, development, and aging is based on thermodynamic principles. Modern thermodynamics introduced methods of explanation of causes and dynamics of evolution of biological systems. The consistency of the theory is accompanied with a complexity of theoretical methods.

Thermodynamics has been developing from a classical phenomenological approach to thermodynamic systems near equilibrium, through statistical interpretation of basic terms,[31,32] and further to linear nonequilibrium thermodynamics and to the thermodynamics of dissipative structures.[34,35] The network of thermodynamics[36-38] results from the combination of the principles of thermodynamics with the principles of cybernetics. Thus, the main disadvantage of thermodynamics (difficult modeling of structurally complex systems) was overcome and the base of the thermodynamic modeling of complex systems was introduced.

The key concepts of thermodynamics are the *thermodynamic system* and *entropy*. The thermodynamic system is a macroscopic system (the parts of space with a great number of elements), the elements of which interact mechanically or electromagnetically.[33,34]

The entropy S is a quantity through which the evolution of thermodynamic system can be determined. For reversible processes holds

$$dS = \frac{dQ}{T} \qquad (21)$$

where S is entropy, Q is thermal energy added to the system, and T is absolute temperature.

Statistically, the entropy is defined as follows:

$$S = -k \sum_{i=1}^{n} p_i \ln p_i \qquad (22)$$

where p_i is the probability of existence of microstate i.

For nonequilibrial systems and irreversible processes holds

$$dS = d_{ir} S + d_{eq} S \qquad (23)$$

where $d_{ir}S$ is the entropy production due to irreversible processes, and $d_{eq}S$ is the entropy flow due to exchange with surrounding. For $d_{eq}S$, the formula in Equation 21 holds. The key to solving problems of evolution, stability, and stationarity of biological systems is the determination of Equation 23. The necessary methods can be found in works of Gyarmati,[39] Prigogine,[35] and Dvořák.[34]

Models of evolution of thermodynamic systems often used the Prigogine criterion of minimum entropy production[33,34] to determine the direction of evolution of systems

$$\frac{d}{dt}\left[\frac{1}{m}\frac{d_{ir}S}{dt}\right] < 0 \tag{24}$$

where m is the mass of system.

For equilibrium state of systems the condition holds

$$\Delta_{ir}S < 0 \tag{25}$$

where $\Delta_{ir}S$ is the entropy fluctuation. Equation 24 holds for linear nenoquilibrium systems. For dissipative structures, the Prigogoine criterion has another form.[34]

The theoretical background of thermodynamics is remarkably complicated. On the other hand, the necessary experimental data is usually easy to obtain. Often the calorimetric measuring and the data on concentrations, pressure, and energy exchanges are sufficient.

The thermodynamic approach is frequently used for modeling processes on molecular and subcellular levels.[40] Nevertheless, the thermodynamic approach can also be useful for modeling on a multicellular level.[41]

VII. STATE-SPACE MODELS

The state-space models are built on the idea of state-space principle. This method was developed in physical sciences to represent temporal evolution of multivariable systems. The state-space is defined by a set of state-variables and t sets of their values. The dynamics of the system is described by mapping the state-space onto itself in the deterministic case. In the simplest stochastic case, the state transitions can be described as a Markov chain by the matrix of transition probabilities of the size $k^n \times k^n$ (provided there are *n* variables with *k* values each). It is obvious that only "global" properties of the system can be theoretically analyzed in tasks of such huge dimensions.[16,42,43]

The choice and completeness of state variables are important problems in developing state-space models. For example, the diastolic blood pressure and the serum cholesterol level are two of the putative state variables in modeling of the aging process. There also exists problems of a methodological nature, for example, difficulties in measurement of many physiological quantities and obstacles in collecting and processing data from longitudinal observations. Nevertheless, the state-space models can be very useful in the abstract analysis of global properties of multivariable systems in biology.

VIII. CONCLUDING REMARKS

The models discussed in this chapter were chosen from a great number of existing models of growth, development, and aging. Our selection of models was governed by the following criteria:

1)   Topical or potential contribution to understanding the nature of problems under study.
2)   The system approach to modeling.
3)   The possibility of unambigous experimental verification.

Thus, for instance, the so-called vitality models were not included in the survey, and similarly some models of aging on the molecular level were found unverifiable. Nevertheless, a number of genuinely useful models were impossible to include, simply because of limited space of this chapter.

REFERENCES

1. **Bertalanffy, L.,** *General System Theory,* George Braziller, New York, 1968.
2. **Mesarovič, M.D. and Takahava, Y.,** *General System Theory,* Academic Press, New York, 1975.
3. **Padulo, L. and Arbib, M.,** *System Theory,* W.B. Saunders, London, 1974.
4. **Ziegler, B. P.,** *Theory of Modelling and Simulation,* John Wiley & Sons, New York, 1976.
5. **Klir, G. J.,** *An Approach to General Systems Theory,* D. Van Nostrand, New York, 1970.
6. **Bertalanffy, L.,** Quantitative laws in metabolism and growth, *Q. Rev. Biol.,* 32, 217, 1957.
7. **Richards, F. J.,** A flexible growth function for empirical use, *J. Exp. Bot.,* 10, 290, 1957.
8. **Nelder, J. A.,** The fitting of generalization of the logistic curve, *Biometrics,* 17, 89, 1961.
9. **Turner, M. E.,** A theory of growth, *Math. Biosci.,* 29, 367, 1976.
10. **Dixon, J. W.,** *Introduction to Statistical Analysis,* McGraw-Hill, New York, 1969.
11. **Strehler, B. L.,** *Time, Cells and Aging,* Academic Press, New York, 1977.
12. **Starý, I.,** *Theory of Reliability,* Technical University, Prague, 1973 (in Czech).
13. **Gompertz, B.,** On the nature of the function expressive of the law of human mortality and on a new mode of determining life contingencies, *Phil. Trans. R. Soc. London Ser. A,* 115, 513, 1825.
14. **Henderson, R.,** *Mortality Laws and Statistics,* John Wiley & Sons, New York, 1915.
15. **Ďoubal, S.,** Theory of reliability biological systems and aging, *Mech. Ageing Dev.,* 18, 339, 1982.
16. **Strehler, B. L. and Mildvan, A. S.,** General theory of mortality and aging, Science, 132, 14, 1973.
17. **Comfort, A.,** *The Biology of Senescence,* Holt, Rinehart & Winston, New York, 1956.
18. **Pakin, J. V. and Hrisanov, S. M.,** Critical analysis of applicability of Gompertz-Makeham law in human population, *Gerontology,* 30, 8, 1984.
19. **Barlow, R. E. and Porschan, F.,** *Mathematical Theory of Reliability,* John Wiley & Sons, New York, 1965.
20. **Kim, H. Y., Case, K. E., and Ghare, P. M.,** A method for computing complex system reliability, *IEEE Trans. Reliab.,* R-21, 215, 1972.
21. **Witten, M.,** A return to time, cells, systems and aging, *Mech. Ageing Dev.,* 21, 69, 1983.
22. **Abernethy, J. D.,** The exponential increase in mortality rate, *J. Theor. Biol.,* 80, 333, 1979.
23. **Milsum, J. H.,** Biological systems analysis and control theory, in *Biomedical Engineering Systems,* Clynes, M. and Milsum, J. H., Eds., McGraw-Hill, New York, 1970, chap. 6.
24. **Bligh, J.,** *Temperature Regulation in Mammals and Other Vertebrates,* North-Holland, Amsterdam, 1973.
25. **Milsum, J. H.,** *Biological Control Systems Analysis,* McGraw-Hill, New York, 1966.
26. **Kalmus, H.,** *Regulation and Control in Living Systems,* John Wiley & Sons, New York, 1966.
27. **Grodins, F. S.,** *Control Theory and Biological Systems,* Columbia University, Press, New York, 1963.
28. **Cruz, J. B.,** *Feedback Systems,* McGraw-Hill, New York, 1972.
29. **Nixon, F. E.,** *Principles of Automatic Controls,* Prentice-Hall, Englewood Cliffs, N. J., 1960.
30. **Švec, J. and Kotek, Z.,** *Theory of Automated Management,* State Publishing, House of Technical Literature, Prague, 1969 (in Czech).
31. **Eykhoff, P.,** *System Identification,* John Wiley & Sons, New York, 1974.
32. **Onsager, L.,** Reciprocal relation in irreversible processes. I, *Phys. Rev.,* 37, 405, 1931.
33. **Onsager, L.,** Reciprocal relation in irreversible processes. II, *Phys. Rev.,* 38, 2265, 1931.
34. **Dvořák, I.,** *Biothermodynamics,* Academia, Prague, 1982 (in Czech).
35. **Prigogine, I.,** *Non-equilibrium Statistical Mechanics,* John Wiley & Sons, New York, 1962.
36. **Katchalsky, A. and Curran, P. F.,** *Non-equilibrium Thermodynamics in Biophysics,* Harvard University Press, Cambridge, 1967.
37. **Schanakenberg, J.,** *Thermodynamical Network Analysis of Biological Systems,* Springer-Verlag, Berlin, 1977.
38. **Peusner, L.,** Hierarchies of irreversible energy conversion, *J. Theor. Biol.,* 102, 7, 1983.
39. **Gyarmati, I.,** *Non-equilibrium Thermodynamics,* Springer-Verlag, Berlin, 1970.
40. **Eigen, M.,** Selforganisation of matter and the evolution of biological macromolecules, *Naturwissenschaften,* 58, 465, 1971.

41. **Lurie, D.,** Non-equilibrium thermodynamics and biological growth and development, *J. Theor. Biol.,* 78, 241, 1979.
42. **Woodbury, M. A. and Manton, K. G.,** A mathematical model of the physiological dynamics of aging, *J. Gerontol.,* 38, 4, 398, 1983.
43. **Sacher, G. A. and Truco, E.,** The stochastic theory of mortality, *Ann. N.Y. Acad. Sci.,* 96, 985, 1962.

Chapter 15

CIRCADIAN RHYTHMS

J. Vaněček

TABLE OF CONTENTS

I. INTRODUCTION

The vast majority of biological variables shows diurnal periodic fluctuations.[1] This daily rhythmicity reflects an adaptation to 24-hr periodicity of physical environmental factors resulting from rotation of the earth, notably light-dark and temperature cycles.[2,3] However, most biological rhythms continue in the aperiodic environment, i.e., under constant lighting and temperature conditions, food and water *ad libitum,* controlled humidity, and acoustic isolation. Under constant conditions, the endogenous daily rhythms run with the period close to 24 hr and are thus termed "circadian" rhythms.[4] This finding along with other evidence led to the conclusion that circadian rhythms are driven by an endogenous biological clock.[5] Although several observations suggest that the mammalian circadian system consists of more than one circadian oscillator,[6,7] the oscillators are hierarchically coupled so that under natural conditions the system acts as if driven by a master circadian pacemaker. In mammals, the central role in the regulation of circadian rhythms is ascribed to the suprachiasmatic nucleus of the hypothalamus. Its bilateral destruction abolished the circadian rhythms in locomotory activity, drinking behavior, corticosterone secretion, pineal *N*-acetyl-transferase activity, etc.[8-10]

Daily environmental cycles entrain the circadian rhythms to exactly a 24-hr period and ensure the proper phasing of the internal program with external changes.[5] The most important entraining stimulus — Zeitgeber — is the light-dark (LD) cycle, although temperature cycle, food and water availability, and social cues are also of significant value. The light-perceptive structures and the photic pathways are thus integral parts of the circadian system.

The questions addressed by circadian research usually fall into one of these categories: (1) localization of the circadian pacemaker, the secondary rhythmic centers, and the entraining and effector pathways; (2) the search for the mechanism of the circadian pacemaker function, its entrainment, and its coupling to the overt rhythms; (3) the search for biochemical processes underlying the circadian oscillations and their entrainment; and (4) the search for the importance of a circadian system in human and animal physiology and pathology and the consequent improvement of diagnosis and therapy of those diseases. Besides, the precise description of the regulation of any biological system requires knowledge of its daily rhythm. The definition and the control of the rhythmic component of the studied system is thus necessary for total biological research. This chapter surveys the most important methods used in circadian research for measurement, recording, and manipulation of daily rhythms. It is focused on mammalian rhythms only.

II. TYPES OF METHODS USED

A. Rhythms Used in Circadian Studies

Since the vast majority of biological functions undergo daily periodical changes, almost any method used in biological research might be used for measurement of circadian rhythms when sampled around the clock. However, easily measurable and reproducible rhythms with a reasonably high amplitude and precisely defined phase-reference point are mostly studied in circadian research. The methods used for rhythm measurement are behavioral, biochemical, histochemical, physiological, electrophysiological, etc.

Behavioral methods record daily changes in animal behavior. Diurnal species rest during night and fulfill most of the locomotory, feeding, drinking, and reproductive activities during the day; in nocturnal animals the reverse is true.[11] Also learning performance and memory show daily rhythms. *The locomotory activity rhythm* (Figure 1) is one of the most studied circadian rhythms. It is recorded by means of an activity wheel[12,13] or by various switches under the floor.[14,16] In field studies, the activity is recorded as absence from nest or retreat.[17,18] *Drinking activity* is recorded in drinkometers as number of licks per time unit,[19-21] or the

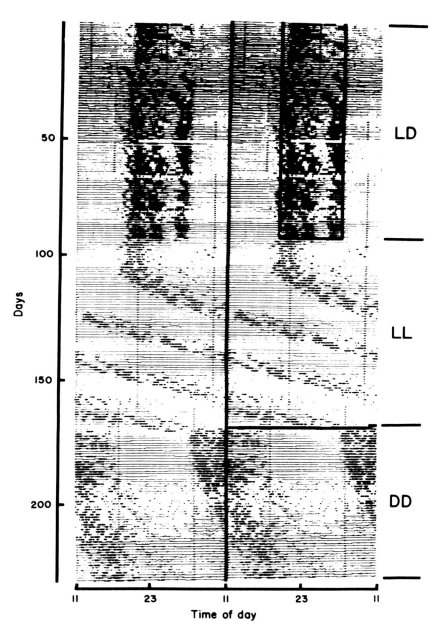

FIGURE 1. Double-plotted wheel-running activity record of a male Golden hamster synchronized in light-dark (LD) 14:10 regime and "free running" in constant light (LL) or darkness (DD). The succesive days are plotted below each other. (From Rusak, D. and Boulos, Z., *Photochem. Photobiol.*, 34, 267, 1981. With permission.)

volume of water intake is measured.[22] *Food intake* is recorded as the number of small food pellets eaten per unit of time. Every press of a lever mounted next to a small food cup delivers a food pellet, and the event is recorded.[21] Alternatively, the number of food approaches is recorded.[23]

Physiological and electrophysiological methods are used for measurement of circadian rhythms in body temperature, blood pressure, single- and multiunit neuronal activities, EEG, etc. *The body temperature rhythm* (Figure 2), although its amplitude is usually less than

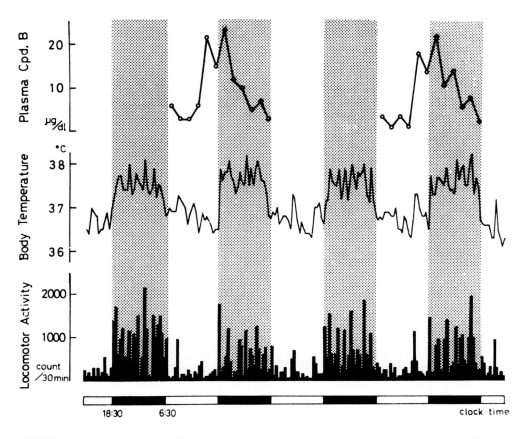

FIGURE 2.   Circadian rhythms of locomotor activity, body temperature, and plasma corticosterone (Cpd. B) determined simultaneously in a single rat kept under LD 12:12 (light from 6.30 to 18.30) (From Honma, K. I. and Hiroshige, T., *Am. J. Physiol.*, 235, R243, 1978. With permission.)

1°C, is very often studied for its good reproducibility and easy recording.[14,15,24] The thermistor probe converting the temperature changes to an electric signal is inserted into the rectum or colon, or operated under the skin. The probe is connected to a recorder by a long wire or telemetrically.[25] *The electrical activity of single neurons* and the *multiunit activity* of several brain areas show circadian rhythm. Neuronal activity of the hypothalamic suprachiasmatic nucleus (SCN), the presumed site of the circadian pacemaker, is high during the day and low at night, and the rhythm persists after the surgical isolation of the hypothalamic island containing SCN.[26,27] Multiunit activity recorded from other brain areas is high during the night, and the rhythm is abolished after SCN isolation. Neural activity is recorded by the microelectrodes stereotaxically inserted to the desired nuclei. The signals are amplified and filtered, and number of spikes per time unit is counted with the help of window discriminators. The discharge of SCN and other nuclei is also possible to record from brain slices incubated in vitro.[28] The firing rate of cells in the incubated SCN shows a circadian rhythm for at least one circadian cycle.

Biochemical and histochemical methods are used to record the circadian rhythms in enzyme activities, hormone and neurotransmitter concentrations, metabolic activity of neural tissue, plasma ion and metabolite concentrations, etc. *Hormone and neurotransmitter rhythms* are often determined by specific radioimmunoassays (RIA), by fluorometric and radioenzymatic methods, and by high-performance liquid chromatography (HPLC) with electrochemical or fluorometric detection. These are the most widely used methods for their specificity, sensitivity, and simplicity, but sometimes other methods are used. The determination of plasma, cerebrospinal fluid, or urine concentrations is advantageous, since it allows serial sampling via the permanent catheter. Cortisol and corticosterone concentrations in plasma and their metabolites in urine[7,14] show marked daily rhythms (Figure 2) and are frequently assayed by RIA or fluorometric or protein binding methods.[29-31] Melatonin rhythm was determined in pineal gland, plasma, cerebrospinal fluid, urine, and saliva.[32-35] Melatonin is assayed by RIA or HPLC with electrochemical detection.[36,37] The latter method allows simultaneous determination of various precursors and metabolites. Plasma concentrations of LH, FSH, prolactin, norepinephrine, epinephrine, and other hormones show daily rhythms as well as cerebrospinal concentrations of vasopressin and serotonin and urinary excretion of catecholamines and 17-hydroxysteroids.[38-42] *Enzyme activity rhythms* have been described in pineal *N*-acetyltransferase (NAT; EC 2.3.1.5),[43] brain monamine oxidase (EC 1.4.3.4),[44] urinary β-glucoronidase (EC 3.2.1.31),[42] and several other enzymes.[45] *Metabolic activity* of brain structures can be measured by autoradiographic technique using tracer amounts of intravenously injected 2-deoxy-*D*-(^{14}C)-glucose[46] (2-DG). Since brain structures are dependent on continuous supply of glucose for energy, the amount of glucose utilized by the neurons should reflect the functional activity of that area. Using the 2-DG method, a circadian rhythm in metabolic activity of SCN has been shown as being high during the day and low at night.[47]

B. Methods for Rhythm Manipulation

The methods of manipulation of the period, amplitude, phase, and waveform of the circadian rhythms is the specific methodology in circadian rhythm research. Various photic, temperature, olfactory, and acoustic stimuli, food restriction, various drug administration, and surgical interventions are used.

Photic stimuli represent the most important method of rhythm manipulation. LD cycle is the most important "Zeitgeber" of circadian rhythms in mammals.[2,48] In the absence of LD cycle, and in isolation from other external periodicities, the circadian rhythms start "free running" with their endogenous period. To induce free running, animals are exposed to the constant darkness (DD) or to the low-intensity constant light (LL) (Figure 1). The intensity of LL affects the endogenous period[49] and changes the internal phase-relationship among several rhythms within one organism. High-intensity LL induces splitting of rhythms into two subcomponents, their decomposition or suppression.[43,50,51]

Short light pulses lasting usually 1 to 60 min are used to induce the phase shifts of circadian rhythms[13,52] (Figure 3). Animals kept in darkness are exposed to a light pulse and released to constant darkness. Phase shifts are determined immediately or after several days when the new steady state is achieved. Light exposure during the animals' subjective day does not induce phase shifts in its circadian rhythms, while light exposure at the beginning of subjective night induces phase delays, and illumination at the end of night results in phase advances of the rhythm.[13] The direction and the size of the phase shift thus depends on circadian time of the light exposure and is also influenced by duration, intensity, and wavelength of the light.[53,54] Dark pulses are sometimes used to phase shift the circadian rhythms in animals kept in LL.[55] To be effective, dark pulses have to last several hours. Light steps might also be used for phase shifting:[56] low-intensity light is abruptly changed for higher intensity light (step up) or vice versa (step down).

LD cycles are used for entrainment of circadian rhythms, for determination of the limits of entrainment, and for manipulation of internal phase-angle relationship among several rhythms.[19,53,54] LD cycles might vary in their frequency, in the length of the light period (photoperiod), or in the relative intensity of the light and dark periods. Full photoperiods consisting of one uninterrupted light period per cycle might be simulated by skeleton photoperiods which consist of two light pulses per cycle separated by several hours of darkness.

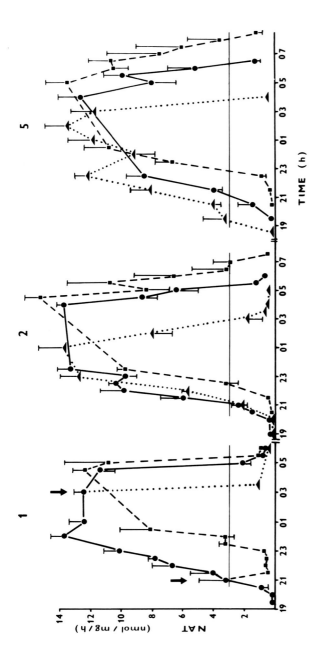

FIGURE 3.   Chronogram showing the effect of 1-min light pulse on pineal *N*-acetyltransferase (NAT) rhythm in rats. Animals adapted to LD 12:12 (light from 06.00 to 18.00) were exposed to the light pulse at 2100 hr (--■--) or at 0300 hr (. . .▲. . .), or were unpulsed (——). Arrows indicate the time of the light pulse. Rats were then kept in constant darkness and they were sacrificed the same night (1), the following night (2), or the fifth night (5). The horizontal line indicate the NAT activity of 3 nmol/mg/hr where the phase-shifts were determined. (Data from Illnerová and Vaněček.[52,92])

The first pulse then represents the beginning of the photoperiod while the second represents the end. The light-dark and dark-light transitions might be abrupt or gradual. The limits of entrainment are determined by exposing the animal to the LD cycle with a gradual decreasing or increasing period.[7,53] Many mammals entrain to cycles between 22 and 26 hr or even longer. The limits of entrainment depend on properties of the Zeitgeber and on the responsiveness of the circadian system to the entraining signal.[57] The strength of a LD cycle as an entraining agent is determined by the intensity and duration of the light period of the cycle: the cycles with very short or long photoperiods are weaker Zeitgebers. The responsiveness of the circadian system depends on species, the particular rhythm, and the previous light history.[57]

Ambient temperature cycle is a powerful Zeitgeber in heterotherms: in homoiotherms, however, it is much less effective.[57] Nevertheless, the temperature cycle was shown to entrain the pig-tailed macaque (*Maccaca nemestrina*)[58] and the pocketmouse (*Perognathus longimembris*).[59] Apart from the cycles, the temperature pulses or steps might be applied. The homeostatic regulation of the deep body temperature in homoiotherms defends their circadian pacemaker from the effect of ambient temperature. To study the temperature dependence of their oscillator, it is necessary to anesthetize the animal first: body cooling is reached by wetting the fur with ethanol and covering with ice.[60,61] The temperature dependence is then described by a temperature coefficient (Q_{10}).* Good opportunity to study temperature effects on circadian system give facultative heterotherms as bats, hamsters, pocket mice, etc.[59,60,62]

Restricted food availability is also effective as an entraining stimuli.[63] Restricted daily feeding is accomplished either by making available to the animal an unlimited amount of food for a limited time or a limited quantity of food for an unlimited duration. In the latter case, the quantity of food offered is considerably less that the daily intake of an animal fed *ad libitum,* thus ensuring that the entire ration is consumed within a few hours. Since so many daily rhythms are connected with feeding, the forced shift of the feeding time may phase shift several other rhythms, too. The feeding schedule may even override the entrainment by LD cycle. Limits of entrainment to periodic feeding may be determined by making food available at periods shorter and longer that 24 hr.[64] Although there are some similarities between the entrainment of circadian rhythms by LD cycles and by food, the central pacemaker does not appear to be entrained by food cycle.[65]

Social stimuli are a group of Zeitgebers responsible for mutual entrainment among conspecific animals or for entrainment by heterospecific predators or competitors.[11,57] The social entrainment is probably due to auditory, olfactory, or tactile information. To study this phenomenon, the various sounds and handling are tested as Zeitgebers.[7,11] Alternatively, the entrainment of the blinded animal housed together with the sighted one to the imposed LD cycle, or the mutual synchronization among animals housed together in constant conditions, is studied. Social entrainment is studied thoroughly in humans, since it seems to be one of the most important Zeitgebers in this species.[66] Gong and alarm-clock signals are used in humans, and also the effects of knowledge of time and mutual entrainment are tested.

Drug effects on circadian rhythms are studied in order to determine the biochemical nature of the circadian pacemaker and its entrainment. Drugs are administered either systemically or to the proximity of the suprachiasmatic nucleus, the presumed circadian pacemaker. Drugs are injected in a single dose, or the injections are repeated in 24-hr intervals. Alternatively, drugs might be implanted or administered by programed infusion. Several drugs have been

---

* $Q_{10} = \left(\dfrac{R_2}{R_1}\right)^{10/T_1 - T_2}$ where R_1 and R_2 are the rates of the clock at temperature T_1 and T_2.

found to affect circadian rhythms. Drugs may either affect the period of free running rhythms, e.g., lithium, D_2O, estradiol, or they may cause entrainment or induce phase shifts, e.g., carbachol, melatonin, or avian pancreatic polypeptide.[67-72] The effect of neonatal administration of various drugs on the development of daily rhythms may also be studied.[73]

Surgical interventions are used to localize the structures responsible for rhythm generation, the entraining and effector pathways. Alternatively, the surgical approach is used to exclude the defined part of the circadian system and to study the function of remaining structures. Lesions of SCN, paraventricular nucleus (PVN) and other hypothalamic nuclei, transections of photic pathways, and surgical isolation of SCN in hypothalamic islands are examples of the most common techniques.[8,10,26] Lesions are usually performed by stereotaxically implanted electrodes, transections and SCN isolations by various knives. Localization of all interventions should be verified by histological examination of the brain postmortem.

III. NECESSARY EQUIPMENT

A. Controlled-Environment Room

A controlled-environment unit is an absolute necessity for circadian rhythm research. To induce free running of circadian rhythms, the animals must be kept isolated from all external periodical stimuli. In studies directed at the mechanism of entrainment, only one type of physical stimuli is studied as a Zeitgeber, while other environmental conditions must be kept constant. For these and other reasons the animal room must be provided with control of illumination, temperature, humidity, and should be isolated from outside noise. Most important is the control of illumination. Animal room must be light attenuated, and the entrance should be from the dark vestibule to avoid the light leak during the entry. It is advantageous to have the room furnished with small light-proof boxes, each for one or a few animals, since it allows the application of different lighting conditions at once. Each animal unit must have the light source uniformly illuminating the whole place. The LD cycle is controlled by an automatic clock switch, and another timer should generate additional light exposure, pulses, etc. Alternatively, the illumination is controlled by a computer which permits the changes of the period of the LD cycle. Duration of the photoperiod must be changeable from 0 (DD) to 24 hr (LL). The regulation of the intensity of light should also be possible. Often it is necessary to have a monochromatic light source; various interference filters and additional light sources are used for this purpose.

The ambient temperature in the unit must be controlled by a thermistor switch. The animal room should be sound attenuated; background noise is sometimes provided by a white-noise generator. Fresh-air flow in each animal unit is provided by an exhaust fan, and humidity is kept at constant level. For human studies, an even more sophisticated isolation unit is necessary.[74]

B. Special Equipment for Rhythm Measurement

Registration of running activity and drinking rhythms requires special equipment.

A running wheel is a simple but widely used device for the measurement of animal locomotory rhythm. The wheel turns freely around a horizontal axle and is hollow inside. Access to the wheel is through a small tunnel in the rear of the animal cage. The animal may step in and run inside, which makes the wheel turn around. Each revolution is mechanically or electrically recorded by an event recorder.

A drinkometer may work on the principle of lick counting, or it may measure actual volume of water intake. The first type of apparatus consists of an infrared light source and a photoelectric receiver mounted across the water spout.[75] The infrared beam is interrupted by each lick, which is monitored on an event recorder. In the second type of apparatus, a water tank is connected with drinking spout through a cartridge, which makes 0.05-mℓ drops accurately.[22] When a rat drinks water, water falls drop by drop. The short circuit generated by each drop activates an electromagnetic counter and the drinking counts of each 1-hr period are printed out.

IV. CHOOSING THE OPTIMAL METHOD

The choice of the optimal method depends on the question addressed by the experiment, and must involve the choice of the animal or in vitro model, the choice of the rhythm to be studied, and the method of its manipulation.

A. Choice of the Animal and/or In Vitro Model

It is possible to study many circadian rhythms in various animals. For example, drinking, feeding, locomotory activities, and body temperature show daily rhythm in various rodents, dogs, rabbits, sheep, monkeys, or humans.[11,57] The experimental animal should be chosen according to the quality and reproducibility of the rhythm in a given species, its amplitude, phase-reference point, and the absence of interfering events. For example, the hamsters feed regularly during both the "active" and "rest" phase of their daily cycles which makes the feeding rhythm in this species much less clear than in rats.[21,76] *N*-acetyltransferase activity rhythm has much higher amplitude in rats than in golden hamsters.[77,78] Another important criterion is the size of the animal. Small animals as rodents are often used in circadian research since their size permits keeping a large number of individuals in small cabinets. However, the use of a bigger animal such as a rabbit or sheep might sometimes be an advantage. The repeated blood withdrawal is then enhanced, which makes it possible to study the daily rhythm of various hormones in plasma in a single individual. Some rhythms, e.g., the neuronal electric activity, persist after isolation of the tissue, and it is possible to study them in vitro.[28] These models have an advantage in their simplicity, in that the interference from other regulatory systems is excluded. However, in vitro models show only a few daily cycles at most, and the short time series represents a certain difficulty for interpretation and evaluation. Moreover, in vitro models do not permit study of the circadian system in its whole complexity, and in vitro techniques might introduce some artifacts.

B. Choice of the Rhythm and the Method of its Measurement

In research devoted to the study of the mechanism of the circadian clock function and its entrainment, it is advantageous to directly follow the rhythm of the central pacemaker. Peripheral rhythms are driven by the circadian pacemaker, but they do not necessarily reflect all changes of the pacemaker, and not all changes of the overt rhythms are due to those of the pacemaker. Peripheral rhythms are subject to other regulations, e.g., homeostatic, since they fulfill other roles apart from having rhythmic characteristics. The oscillatory component of peripheral rhythms thus might be obscured by the nonoscillatory one. Nevertheless, the peripheral rhythms are widely used because of their simple measurement allowing long-time series and because there have been no methods for measuring the rhythm of the pacemaker until recently. The behavior of the pacemaker is inferred from the response of the overt rhythm to the various stimuli. However, we must be aware that the peripheral rhythms serve as good markers of the circadian pacemaker state only if the whole system is in a steady state.

Before selecting the rhythm to be studied, the following should be considered:

1.   The higher the amplitude the more accurately the reference phase may be usually determined, which in turn allows the precise determination of phase shifts, changes of the period, entrainment, etc. Moreover, in high-amplitude rhythms, the background noise interferes less with the measurement.
2.   Simple measurement of the rhythm allows frequent determinations and hence long-time series. Both are very important for the description of the course of the rhythm and its changes.
3.   Longitudinal following of the course of daily rhythm in a single individual has several advantages as compared with transversal determination of average rhythm value. Transversal method calculates the course of the rhythm from average values from several animals for each time point and uses different groups of animals for different time points. The interpretation of the results may be more complicated and often more than one explanation is possible. The methods that require sacrificing the animal for the rhythm determination are thus disadvantageous.

C. Choice of the Method of Rhythm Manipulation

Method of manipulation depends completely on the aim of research. For entrainment of mammalian circadian rhythms, the LD cycle is of major importance.[13,57] In addition, access to food and social stimuli are also important Zeitgebers in mammals, with the latter being especially important in humans.[57,66] In research devoted to the mechanism of entrainment in mammals, the photic and social stimuli and restricted feeding are thus preferably used. In studies designed to localize the pacemaker or the entrainig pathways, brain stereotaxic microsurgery and electric stimulation should be employed. In studies of the biochemical mechanisms of the circadian function, the pharmacological approach is used, while in clinical research the internal phase relationship of several rhythms and its pathological changes are often studied.

V. ADVANTAGES AND DRAWBACKS OF PARTICULAR METHODS

The metabolic and electric activity rhythms of hypothalamic SCN in mammals have several advantages as compared with other circadian rhythms. The mammalian central pacemaker appears to be located in these nuclei.[26] The SCN rhythms should closely reflect the state of the central pacemaker and should be free of interference from noncircadian regulatory centers. The electrical neural activity of SCN in both diurnal and nocturnal mammals is high during the day and low at night.[26,79] SCN metabolic activity in diurnal and nocturnal mammals measured as 2-DG uptake is also higher during the day than at night.[57,80,81] These rhythms continue in constant darkness and respond instantly to light exposure.[82,83] The 2-DG uptake rhythm was detected in fetal rats before birth, earlier than any other circadian rhythm.[84] However, the 2-DG method does not permit the longitudinal observation since it is necessary to kill the animal for each determination. The method is very expensive and time consuming, and the quantitative evaluation of autoradiographs requires a densitometer coupled to the image-processing system. The recording of electrical neural activity of SCN is a much less expensive method and permits longitudinal studies. A daily rhythm of SCN multiunit activity has been recorded even for periods of several days, which makes this method widely applicable.[65]

Although the pineal gland appears to be the integral part of the avian and reptilian circadian pacemaker,[3,85] its role in the mammalian circadian system is not clear. Nevertheless, the rhythm of its hormone, melatonin, has several advantages and has similar properties as the rhythms of SCN. In both diurnal and nocturnal species, the melatonin concentration is high at night and low during the day and the rhythm is instantly affected by light stimuli, while the effect of other non photic stimuli is negligible.[32,33,86,87] The melatonin rhythm has a high amplitude and two well-defined phase-reference points: the evening increase and the morning decrease. The rhythm in pineal melatonin is reflected in plasma, cerebrospinal fluid, saliva, and urinary concentrations, which allows continuous sampling and longitudinal studies.[32-35] The rhythm in the activity of the melatonin-forming enzyme, the N-acetyltransferase (NAT) is also used in circadian studies.[52] In rats, the NAT rhythm shows very high amplitude and the assay is simple enough to allow a large number of determinations. However, its main disadvantage is the necessity of killing the animals for NAT determination which makes longitudinal studies impossible.

Several other hormonal rhythms are studied in circadian research. Concentrations are, however, subjected to homeostatic regulations apart from a control by the circadian pacemaker, and they may be affected by stress, food intake, presence of the sexual partner, etc. These rhythms thus include the nonrhythmic component more or less pronounced, which might interfere with the rhythmic component and obscure its course. Body temperature, locomotory activity, and drinking rhythms are probably the most studied rhythms. Their detection and recording is very simple, usually noninvasive, and permits longitudinal studies in very long time series. The amplitude is reasonably high and the reference phases well-defined. In locomotory activity and drinking rhythms, the time of the beginning of the activity is used since the end of activity is more variable. In body temperature rhythm the times of the peak (acrophase) or trough (bathyphase) are used as reference phases. However, these rhythms have a common disadvantage. They are not part of the circadian pacemaker system and the functions serve other purposes and are thus also regulated from non oscillatory homeostatic centers. Therefore the body temperature, locomotory, or drinking activities are affected by several internal and external changes: stress, access to food, disease, presence of other individual, ambient temperature, etc.

VI. RESULT INTERPRETATION AND STATISTICAL EVALUATION

Daily rhythms are often presented as chronograms or actograms, both showing the course of the function through a 24-hr period. Consecutive days are usually depicted in consecutive lower lines to enhance the orientation and reading of the phase shifts (Figure 1). The dependence of the phase shift on the time of the exposure to the resetting stimuli might be summarized in phase-response or -resetting curves. The phase-response curve depicts a phase-shift as a function of the phase at which the stimulus is applied; the phase-resetting curve depicts the new phase as the function of the old at which the light exposure occurs.

Several mathematical methods have been developed for statistical evaluation of time-series data.[88,89] Only the two most often used methods will be discussed here. Power spectral analysis is based on the assumption of a time series representing the sum of an infinite number of sinusoidal functions with different periods. The variance spectrum or so-called "power spectrum" shows the contribution of oscillations with various periods to the total variance. This method detects rhythms and their frequency in noisy time series and evaluates their statistical significance. It is especially valuable if a biological variable oscillates with several frequencies.[90] This technique requires sampling at regular intervals over a relatively long time span.

The cosinor method is especially appropriate in analyzing short synchronous time series.[91] The first step consists of fitting a cosine function by the least squares method:

$$y(t) = M + A \cdot \cos(\omega t + \phi)$$

where M is mesor (overall 24-hr mean), A is amplitude (one half of the total cosine excursion), ω is angular velocity (360°/τ), τ is period, t is time, and φ is the acrophase of the rhythm. The acrophase represents the crest of the fitted curve in relation to an arbitrarily selected reference point (usually midnight) along the 24-hr time scale. In the second step, a probability or P value of the fit of the cosine curve to the data is tested by Fisher's F-test. Cosinor quantified data are often displayed on a polar coordinate system. In such a display, time is expressed in degrees (a 24-hr time scale is equated to a 360° circle). The amplitude is presented as a line starting from the center and directed to the point on periphery. The vector of this line represents the computed time of acrophase. The ellipse at the end of the vector represents 95% confidence limits on amplitude and acrophase.

VII. TYPICAL EXAMPLES OF THE METHODS USED IN CIRCADIAN RESEARCH

The effects of short light pulses on the rhythm of pineal NAT were studied.[52,92] Animals adapted to LD 12:12 were exposed to 1-min light pulse at various night times and then were kept in constant darkness until decapitated. The NAT activity was determined in animals sacrificed the same night, the following night, or the fifth night after the light exposure. The phase shifts were determined by comparison with NAT rhythm in unpulsed control animals. Two phases of the NAT rhythm were used as reference points — NAT increase and decrease. Generally, the light pulses in the early night induced delays and the pulses at late night induced advances of the rhythm (Figure 3). However, the phase shifts of NAT increase and decrease were not always parallel. For example, the light pulse at 2100 hr delayed the NAT increase by about 1.7 hr, regardless of whether the NAT rhythm was studied the second or the fifth day after the light exposure. The NAT decrease, however, was delayed the first night after this pulse by 0.6 hr, the following night by 1.3 hr, and only the fifth night by about 1.8 hr. On the other hand, the light pulse at 0300 hr immediately induced the large advance of NAT decrease, while NAT increase was advanced only after 4 more days. The size and direction of the phase shift thus depends on the timing of light exposure as well as on the phase of the rhythm used as the reference point and on the interval between the light pulse and the determination of the phase shift. The dependence of the phase shift on the timing of the light exposure is summarized as the phase-response curve (Figure 4).

The experiment described is more or less typical for studies directed at the mechanism of entrainment of circadian rhythms. This experiment strongly supports the hypothesis that NAT rhythm is driven by two coupled pacemakers[52] as proposed originally for the locomotory activity rhythm of rodents.[54] The first oscillator controls the evening increase and the other the morning decrease of NAT activity.

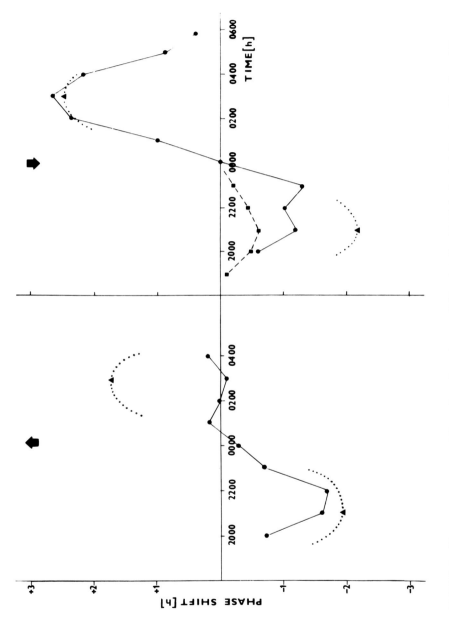

FIGURE 4. Phase-response curves of *N*-acetyltransferase increase (left) and decline (right) for 1-min light pulses. The phase-shifts were determined the same night (—■—), the following night (——●——), and the fifth night (...▲...) after the light pulse as the time difference between the time when NAT reached the value of 3 nmol/mg/hr during its rise (↑) or decline (↓) in pulsed and control rats. (Data from Illnerová and Vaněček.[52,92])

REFERENCES

1. **Bünning, E.**, *The Physiological Clocks*, Springer-Verlag, New York, 1973.
2. **Pittendrigh, C. S.**, Circadian clocks: what are they?, in *The Molecular Basis of Circadian Rhythms*, Hastings, J. W. and Schweiger, H. G., Eds., Dahlem Workshop, Berlin, 1975, 11.
3. **Rusak, B. and Zucker, I.**, Neural regulation of circadian rhythms, *Physiol. Rev.*, 59, 449, 1979.
4. **Halberg, F.**, Temporal coordination of physiologic function, *Cold Spring Harbor Symp. Quant. Biol.*, 14, 289, 1960.
5. **Pittendrigh, C. S.**, Circadian systems: general perspective in *Handbook of Behavioral Neurobiology*, Vol. 4, Aschoff, J., Ed., Plenum Press, New York, 1981, 57.
6. **Moore-Ede, M. C.**, The circadian timing system in mammals: two pacemakers preside over many secondary oscillators, *Fed. Proc. Fed. Am. Soc. Exp. Biol.*, 42, 2802, 1983.
7. **Aschoff, J. and Wever, R.**, The circadian system of man, in *Handbook of Behavioral Neurobiology*, Vol. 4, Aschoff, J., Ed., Plenum Press, New York, 1981, 311.
8. **Moore, R. Y. and Eichler, V. B.**, Loss of a circadian adrenal corticosterone rhythm following suprachiasmatic lesions in the rat, *Brain Res.*, 42, 201, 1972.
9. **Moore, R. Y. and Klein, D. C.**, Visual pathways and the central neural control of a circadian rhythm in pineal serotonin N-acetyltransferase activity, *Brain Res.*, 71, 17, 1974.
10. **Stephan, F. K. and Zucker, I.**, Circadian rhythms in drinking behavior and locomotor activity of rats are eliminated by hypothalamic lesions, *Proc. Natl. Acad. Sci. U.S.A.*, 69, 1583, 1972.
11. **Rusak, B.**, Vertebrate behavioral rhythms, in *Handbook of Behavioral Neurology*, Vol. 4, Achoff, J., Ed., Plenum Press, New York, 1981, 183.
12. **Bruce, V. G.**, Environmental entrainment of circadian rhythms, *Cold Spring Harbor Symp. Quant. Biol.*, 14, 29, 1960.
13. **De Coursey, P. J.**, Phase control of activity in a rodent, *Cold Spring Harbor Symp. Quant. Biol.*, 15, 49, 1960.
14. **Honma, K. I. and Hiroshige, T.**, Internal synchronization among several circadian rhythms in rats under constant light, *Am. J. Physiol.*, 235, R243, 1978.
15. **Wever, R. A.**, Fractional desynchronization of human circadian rhythms. A method for evaluating entrainment limits and functional interdependencies, *Pfleugers Arch.*, 396, 128, 1983.
16. **Tokura, H. and Hirari, N.**, Circadian locomotor rhythms of the night monkey (*Aotus frivirgatus*) under constant light conditions, *Folia Primatol.*, 33, 15, 1980.
17. **Quanstrom, W. R.**, Behaviour of Richardson's ground squirel *Spermophilus Richardsonii Richardsonii*, *Anim. Behav.*, 19, 646, 1971.
18. **Bovet, J. and Oertli, E.**, Free running circadian activity rhythms in free-living beaver *(Castor Canadensis)*, *J. Comp. Physiol.*, 92, 1, 1974.
19. **Stephan, F. K.**, Circadian rhythms in the rat: constant darkness, entrainment to T cycles and to skeleton photoperiods, *Physiol. Behav.*, 30, 451, 1983.
20. **Albers, H. E., Lydic, R., and Moore-Ede, M. C.**, Entrainment and masking of circadian drinking rhythms in primates: influence of light intensity, *Physiol. Behav.*, 28, 205, 1982.
21. **Boulos, Z. and Terman, M.**, Splitting of circadian rhythms in the rat, *J. Comp. Physiol.*, 134, 75, 1979.
22. **Kuribara, H., Hayashi, T., Alam, M. R., Tadokoro, S., and Miura, T.**, Automatic measurement of drinking in rats: effects of hypophysectomy, *Pharmacol. Biochem. Behav.*, 9, 697, 1978.
23. **Rietveld, W. J. and Gross, G. A.**, The role of suprachiasmatic nucleus afferents in the central regulation of circadian rhythms, in *Biological Rhythms in Structure and Function*, Von Mayersbach, H., Schening, L. E., and Pauly, J. E., Eds., Alan R. Liss, New York, 1981, 205.
24. **Fuller, C. A., Lydic, R., Sulzman, F. M., Albers, H. E., Tepper, B., and Moore-Ede, M. C.**, Circadian temperature of body temperature persists after suprachiasmatic lesions in the squirrel monkey, *Am. J. Physiol.*, 241, R385, 1981.
25. **Silverman, R. W. and Lomax, P.**, Techniques for the measurement of temperature in the biological range — a review, in *Environment, Drugs and Thermoregulation*, Lomax, P., Ed., S. Karger, Basel, 1983, 5.
26. **Inouye, S. T. and Kawamura, H.**, Persistence of circadian rhythmicity in hypothalamic island containing the suprachiasmatic nucleus, *Proc. Natl. Acad. Sci. U.S.A.*, 76, 5962, 1979.
27. **Nishino, H., Koizumi, K., and Brooks, C. M.**, The role of suprachiasmatic nuclei of the hypothalamus in the production of circadian rhythm, *Brain Res.*, 112, 45, 1976.
28. **Groos, G. and Hendriks, J.**, Circadian rhythm in electrical discharge of rat suprachiasmatic neurones recorded in vitro, *Neurosci. Lett.*, 34, 283, 1982.
29. **Gomez-Sanchez, C., Murry, B. A., Kem, D. C., and Kaplan, N. M.**, A direct radioimmunoassay of corticosterone in rat serum, *Endocrinology*, 96, 796, 1975.
30. **Glick, D., von Redlich, D., and Levine, S.**, Fluorometric determination of corticosterone and cortisol in 0.02—0.05 milliliters of plasma or submiligram sample of adrenal tissue, *Endocrinology*, 74, 653, 1964.

31. **Murphy, B. E. P.,** Some studies of the protein-binding of steroids and their application to the routine micro and ultramicromeasurement of various steroids in body fluids by competitive protein-binding radioassay, *J. Clin. Endocirnol. Metab.,* 27, 973, 1967.

32. **Wilkinson, M., Arendt, J., Bradtke, J., and de Ziegler, D.,** Determination of a dark-induced increase in pineal *N*-acetyl-transferase activity and simultaneous radioimmunoassay of melatonin in pineal, serum and pituitary tissue of the male rat, *J. Endocrinol.,* 72, 243, 1977.

33. **Hedlund, L., Lischko, M. M., Rollag, M. D., and Niswender, G. D.,** Melatonin: daily cycle in plasma and cerebrospinal fluid of calves, *Science,* 195, 686, 1977.

34. **Rivest, R. W. and Wurtman, R. J.,** Relationship between light intensity and the melatonin and drinking rhythms of rats, *Neuroendocrinology,* 37, 155, 1983.

35. **Vakkuri, O.,** Diurnal rhythm of melatonin in human saliva, *Acta Physiol. Scand.,* 24, 409, 1985.

36. **Arendt, J., Paunier, L., and Schizonenko, P. D.,** Melatonin radioimmunoassay, *J. Clin. Endocrinol., Metab.,* 40, 347, 1975.

37. **Mefford, I. N., Chang, P., Klein, D. C., Namboodiri, M. A. A., Sugden, D., and Barchas, J.,** Reciprocal day/night relationship between serotonin oxidation and *N*-acetylation products in the rat pineal gland, *Endocrinology,* 113, 1582, 1983.

38. **Berndtson, W. E. and Desjardins, C.,** Circulating LH and FSH levels and testicular function of hamsters during light deprivation and subsequent photoperiodic stimulation, *Endocrinology,* 95, 195, 1974.

39. **Mattheij, J. A. M. and Swarts, J. J. M.,** Circadian variations in the plasma concentrations of prolactin in the adult male rat, *J. Endocrinol.,* 79, 85, 1978.

40. **Linsell, C. R., Lightman, S. L., Mullen, P. E., Brown, M. J., and Causon, R. C.,** Circadian rhythms of epinephrine and norepinephrine in man, *J. Clin. Endocrinol. Metab.,* 60, 1210, 1985.

41. **Reppert, S. M., Artman, H. G., Swaminathan, S., and Fisher, D. A.,** Vasopressin exhibits a rhythmic daily pattern in cerebrospinal fluid but not in blood, *Science,* 213, 1256, 1981.

42. **Wisser, H. and Breuer, H.,** Circadian changes of clinical chemical and endocrinological parameters, *J. Clin. Chem. Clin. Biochem.,* 19, 323, 1981.

43. **Klein, D. C. and Weller, J. L.,** Indole metabolism in the pineal gland. A circadian rhythm in *N*-acetyltransferase, *Science,* 169, 1093, 1970.

44. **Chevillard, C., Barden, N., and Saavedra, J. M.,** Twenty-four hour rhythm in monoamine oxidase activity in specific areas of the rat brain stem, *Brain Res.,* 223, 205, 1981.

45. **North, C., Feuers, R. J., Scheving, L. E., Pauly, J. E., Tsai, T. H., and Cosciano, D. A.,** Circadian organization of thirteen liver and six brain enzymes of the mouse, *Am. J. Anat.,* 162, 183, 1981.

46. **Sokoloff, L.,** Relation between physiological function and energy metabolism in the central nervous system, *J. Neurochem.,* 29, 13, 1977.

47. **Schwartz, W. J. and Gainer, H.,** Supracheasmatic nucleus: use of ^{14}C-labeled deoxyglucose uptake as a functional marker, *Science,* 197, 1089, 1977.

48. **Pittendrigh, C. S.,** On the mechanism of the entrainment of a circadian rhythm by light cycles, in *Circadian Clocks,* Aschoff, J., Ed., North-Holland, Amsterdam, 1965, 277.

49. **Aschoff, J.,** Exogenous and endogenous components in circadian rhythms, *Cold Spring Harbor Symp. Quant. Biol.,* 15, 11, 1960.

50. **Pittendrigh, C. S.,** Circadian rhythms and the circadian organization of living systems, *Cold Spring Harbor Symp. Quant. Biol.,* 15, 159, 1960.

51. **Honma, K. I. and Hiroshige, T.,** Endogenous ultradian rhythms in rats exposed to prolonged continuous light, *Am. J. Physiol.,* 235, R250, 1978.

52. **Illnerová, H. and Vaněček, J.,** Two-oscillator structure of the pacemaker controlling the circadian rhythm of *N*-acetyl-transferase in the rat pineal gland, *J. Comp. Physiol.,* 145, 539, 1982.

53. **Pittendrigh, C. S. and Daan, S.,** A functional analysis of circadian pacemakers in nocturnal rodents. IV. Entrainment: pacemaker as clock, *J. Comp. Physiol.,* 106, 291, 1976.

54. **Pittendrigh, C. S. and Daan, S.,** A functional analysis of circadian pacemakers in nocturnal rodents. V. Pacemaker structure: a clock for all seasons, *J. Comp. Physiol.,* 106, 333, 1976.

55. **Ellis, G. B., McKeveen, R. E., and Turek, F. W.,** Dark pulses affect the circadian rhythm of activity in hamsters kept in constant light, *Am. J. Physiol.,* 242, R44, 1982.

56. **Aschoff, J.,** The phase angle difference in circadian periodicity, in *Circadian clocks,* Aschoff, J., Ed., North-Holland, Amsterdam, 1965, 262.

57. **Aschoff, J.,** Freerunning and entrained circadian rhythms, in *Handbook of Behavioral Neurobiology,* Vol. 4, Aschoff, J., Ed., Plenum Press, New York, 1981, 81.

58. **Aschoff, J.,** Circadian rhythms: general features and endocrinological aspects, in *Endocrine Rhythms,* Krieger, D., Ed., New York Raven Press, 1979, 1.

59. **Lindberg, R. G. and Hayden, P.,** Thermoperiodic entrainment of arousal from torpor in the little pocket mouse, *Perognatus longimembris, Chronobiologia,* 1, 356, 1974.

60. **Rawson, K. S.,** Effects of tissue temperature of mammalian activity rhythms, *Cold Spring Harbor Symp. Quant. Biol.,* 15, 105, 1960.

61. **Gibbs, F. P.**, Temperature dependence of rat circadian pacemaker, *Am. J. Physiol.*, 241, R17, 1981.
62. **Vaněček, J., Janský, L., Illnerová, H., and Hoffmann, K.**, Arrest of the circadian pacemaker driving the pineal melatonin rhythm in hibernating golden hamsters, *Mesocricetus auratus, Comp. Biochem. Physiol.*, 80A, 21, 1985.
63. **Boulos, Z. and Terman, M.**, Food availability and daily biological rhythms, *Neurosci. Biobehav. Rev.*, 4, 119, 1980.
64. **Stephan, F. K.**, Limits of entrainment to periodic feeding in rats with suprachiasmatic lesions, *J. Comp. Physiol.*, 143, 401, 1981.
65. **Inouye, S. I. T.**, Restricted daily feeding does not entrain circadian rhythm of the suprachiasmatic nucleus in the rat, *Brain. Res.*, 232, 194, 1982.
66. **Aschoff, J., Fatranská, M., and Giedke, H.**, Human circadian rhythms in continuous darkness: entrainment by social cues, *Science,* 171, 213, 1971.
67. **Kripke, D. F. and Wyborney, V. G.**, Lithium slows rat circadian activity rhythms, *Life Sci.*, 26, 1319, 1980.
68. **Daan, S. and Pittendrigh, C. S.**, A functional analysis of circadian pacemakers in nocturnal rodents. III. Heavy water and constant light: homeostasis of frequency?, *J. Comp. Physiol.*, 106, 267, 1976.
69. **Zatz, M. and Herkenham, M. A.**, Intraventricular carbachol mimics the phase-shifting effect of light on the circadian rhythm of wheel-running activity, *Brain Res.*, 212, 234, 1981.
70. **Redman, J., Armstrong, S., and Ng, K. T.**, Free-running activity rhythms in the rat: entrainment by melatonin, *Science,* 219, 1089, 1983.
71. **Morin, L. P., Fitzgerald, K. M., and Zucker, I.**, Estradiol shortens the period of hamster circadian rhythms, *Science,* 196, 305, 1977.
72. **Albers, H. E., Ferris, C. F., Leeman, S. E., and Boldman, B. D.**, Avian pancreatic polypeptide phase shifts hamster circadian rhythms when microinjected into the suprachiasmatic region, *Science,* 223, 833, 1984.
73. **Taylor, A. N. and Lengvári, I.**, Effect of combined perinatal thyroxine and corticosterone treatment on the development of the diurnal pituitary-adrenal rhythm, *Neuroendocrinology,* 24, 74, 1977.
74. **Elliott, A. L., Mills, J. N., Minors, D. S., and Waterhouse, J. M.**, The effect of real and simulated time zone shifts upon the circadian rhythms of body temperature, plasma 11-hydroxycorticosteroids, and renal excretion in human subjects, *J. Physiol.*, 221, 227, 1972.
75. **Murakami, H. and Imai, H.**, A digital auto-recorder for measuring the drinking frequency of mice, *Lab. Anim. Sci.*, 25, 634, 1975.
76. **Zucker, I. and Stephan, F. K.**, Light-dark rhythms in hamster eating, drinking and locomotor behaviors, *Physiol. Behav.*, 11, 239, 1973.
77. **Illnerová, H. and Vaněček, J.**, Pineal rhythm in *N*-acetyl-transferase activity in rats under different artificial photoperiods and in natural daylight in the course of a year, *Neuroendocrinology,* 31, 321, 1980.
78. **Vaněček, J. and Illnerová, H.**, Effect of light at night on the pineal rhythm in *N*-acetyltransferase activity in the Syrian hamster, *Masocricetus auratus, Experientia,* 38, 513, 1982.
79. **Sato, T. and Kawamura, H.**, Circadian rhythms in multiple unit activity inside and outside the suprachiasmatic nucleus in the diurnal chipmunk *(Eutamias Sibiricus), Neurosci. Res.*, 1, 45, 1984.
80. **Schwartz, W. J., Reppert, S. M., Eagan, S. M., and Moore-Ede, M. C.**, In vivo metabolic activity of the suprachiasmatic nuclei: a comparative study, *Brain Res.*, 274, 184, 1983.
81. **Flood, D. G. and Gibbs, F. P.**, Species difference in circadian $[^{14}C]$ 2-deoxyglucose uptake by suprachiasmatic nuclei, *Brain Res.*, 232, 200, 1982.
82. **Inouye, S. T.**, Light responsiveness of the suprachiasmatic nucleus within the island with the retinohypothalamic tract spared, *Brain Res.*, 294, 263, 1984.
83. **Schwartz, W. J., Smith C. B., and Davidsen, L. C.**, In vivo glucose utilization of the suprachiasmatic nucleus, in *Biological Rhythms and their Central Mechanism,* Suda, M., Hayaishi, O., and Nakagawa, H., Eds., Elsevier, Amsterdam, 1979, 355.
84. **Reppert, S. M. and Schwartz, W. J.**, Maternal coordination of the fetal biological clock in utero, *Science,* 220, 969, 1983.
85. **Menaker, M.**, The search for principles of physiological organization in vertebrate circadian systems, in *Vertebrate Circadian System,* Aschoff, J., Daan, S., and Groos, G. A., Eds., Springer-Verlag, Berlin, 1982, 1.
86. **Tamarkin, L., Reppert, S. M., and Klein, D. C.**, Regulation of pineal melatonin in the Syrian hamster, *Endocrinology,* 104, 385, 1979.
87. **Illnerová, H., Backström, M., Sääf, J., Wetterberg, L., and Vangbo, B.**, Melatonin in rat pineal gland and serum; rapid paralled decline after light exposure at night, *Neurosci. Lett.*, 9, 189, 1976.
88. **Mercer, D. M. A.**, Analytical methods for the study of periodic phenomena obscured by random fluctuations, *Cold Spring Harbor Symp. Quant. Biol.*, 15, 73, 1960.

89. **Scheving, L. E.,** Chronobiology, a new perspective for biology and medicine, in *Proc. 11th Collegium Interantionale Neuro-Psychopharmacologicum Congress,* Saletu, B., Ed., Pergamon Press, Oxford, 1978, 629.

90. **Halberg, F., Panofsky, H., and Mantis, H.,** Human thermo-variance spectra, *Ann. N. Y. Acad. Sci.,* 117, 254, 1964.

91. **Halberg, F., Tong, Y. L., and Johnson, E. A.,** Circadian system phase — an aspect of temporal morphology, procedures and illustrative examples, in *The Cellular Aspects of Biorhythms,* von Mayersbach, H., Ed., Springer-Verlag, Berlin, 1967, 20.

92. **Illnerová, H. and Vaněček, J.,** Extension of the rat pineal N-acetyltransferase rhythm in continuous darkness and on short photoperiod, *Brain Res.,* 261, 176, 1983.

93. **Rusak, B. and Boulos, Z.,** Pathways for photic entrainment of mammalian circadian rhythms, *Photochem. Photobiol.,* 34, 267, 1981.

Chapter 16

HORMONAL REGULATIONS

D. Ježová and P. Tatár

TABLE OF CONTENTS

INTRODUCTION

A. The Neuroendocrine System

Hormonal regulations may be characterized as a complex system of humoral, neural, and metabolic events directed to the maintenance of the homeostasis.[1] The activity of endocrine glands is controlled by both neural and humoral factors.

The regulatory influences of the central nervous system (CNS) on endocrine gland function are mediated through the hypothalamo-pituitary unit. In the hypothalamus, the neural control is integrated and transduced into hormonal signals via hypothalamic-releasing and -inhibiting hormones produced by neurosecretory cells. The axons of such neurons are located in the external zone of median eminence, where hypothalamic hormones enter the hypophyseal portal circulation to reach the anterior pituitary. In contrast, the hormones of posterior pituitary are produced by hypothalamic magnocellular nuclei and reach the gland by axonal transport. The release of hypothalamic hormones may be stimulated or inhibited by different neurotransmitters and neuromodulators.[2]

The humoral control of endocrine gland function is achieved by several feedback systems which may be negative or positive. The very simple feedback interrelationships are between the hormone production and metabolic change evoked by the same hormone. For example, the increase in blood glucose concentration stimulates and its decrease inhibits insulin secretion. The most complicated regulatory system involves feedbacks at several levels of the hypothalamo-pituitary-peripheral endocrine gland unit (Figure 1). In this system, hypothalamic-releasing hormone (liberin) induces appropriate pituitary hormone release which stimulates the hormone production in peripheral endocrine glands. Increased concentration of peripheral hormone in blood results in negative feedback inhibition at the level of both pituitary and hypothalamus. Appropriate hypothalamic inhibiting hormone (statin) may be influenced in the opposite way.

This short description of neuroendocrine regulations has to be considered only as a very simplified view. In fact, neuroendocrine regulations are formed by a high number of simultaneously occurring processes and their mutual connections. The activity and control of endocrine system function cannot be analyzed or understood in its whole complexity. However, we have to keep this fact in mind when studying any interrelations of hormonal systems.

B. Possible Levels of Interference with Hormonal Regulations

Certain endocrine functions may be influenced at different levels of the neuroendocrine system (Figure 2). Generally, there are three most important levels at which the hormonal regulations may be studied and modified having also their specific methodological approaches:

1. Central regulatory system. With the use of a combination of histochemical fluorescence techniques, biochemical methods, and selective brain lesions, the main monaminergic and peptidergic systems have been recognized.[3] For investigation of their role in the control of various endocrine functions, the pharmacological and electrochemical methodologies are very useful (see Sections II.A and B).

2. Hormone biosynthesis and secretion. For evaluation of changes in hormone secretion, the biochemical determination of their concentrations in biological fluids is the most widely used approach. Highly sensitive techniques developed in recent years, especially radioenzymatic and radioimmunoassays, have markedly contributed to our knowledge on hormone regulations.[1,4,5] However, the availability of sensitive, reproducible biochemical techniques for hormone analysis is not the only inevitable assumption for obtaining reliable results. The experimental events preceeding the analysis of hormone concentration are of the same importance and are often underestimated (see Section III.A).

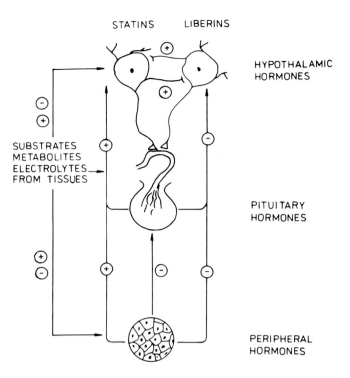

STATINS   LIBERINS

HYPOTHALAMIC
HORMONES

SUBSTRATES
METABOLITES
ELECTROLYTES
FROM TISSUES

PITUITARY
HORMONES

PERIPHERAL
HORMONES

FIGURE 1.   Scheme of neuroendocrine feedbacks.

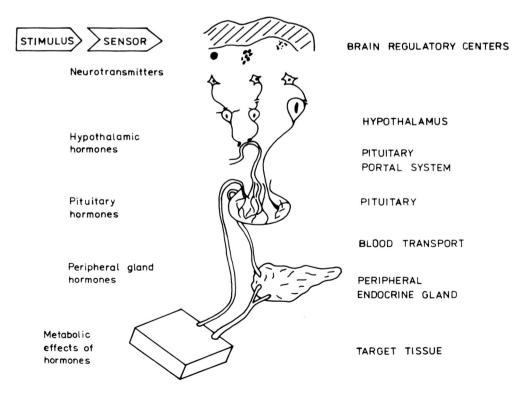

STIMULUS   SENSOR

BRAIN REGULATORY CENTERS

Neurotransmitters

HYPOTHALAMUS

Hypothalamic
hormones

PITUITARY
PORTAL SYSTEM

Pituitary
hormones

PITUITARY

BLOOD TRANSPORT

Peripheral gland
hormones

PERIPHERAL
ENDOCRINE GLAND

Metabolic
effects of
hormones

TARGET TISSUE

FIGURE 2.   Neuroendocrine regulations.

3.   Hormone action. The investigation of the mechanisms of hormone action on metabolic events is a very important field of research. Besides the studies on hormone receptors, the immunological and genetic approaches are becoming more and more significant. These studies utilize mostly in vitro techniques and fit into the field of molecular endocrinology, genetics, and biochemistry.[6-8]

It is impossible to briefly describe all models and methods used in hormonal regulation studies. In the following review the attention is given only to selected methodologies of general importance, and some errors in methodological approaches are mentioned which might remain unrecognized and complicate the comparison of results obtained by different laboratories. An important part of our knowledge of hormonal regulations is based upon studies in rats, and therefore the attention is focused on this animal species. A lot of information on hormone regulations has been obtained by in vitro studies; however, the methods used in in vitro experiments were not included in this review.[9]

II. METHODS FOR EVALUATION AND MODIFICATION OF CENTRAL REGULATORY MECHANISMS

A. Models of Pharmacological Modification of Central Neurotransmitters

The neural control of hypothalamic regulatory hormones by stimuli from extra- and intrahypothalamic regions is mediated by neurotransmitters. The most important neurotransmitters involved in the control are noradrenaline, dopamine, serotonine, histamine, acetylcholine, gamma-aminobutyric acid, endogenous opioids, and other neuroactive peptides.[1,2,10,11]

Neurohumoral functions may be influenced by pharmacological interaction at different stages in the process of biosynthesis, receptor binding, and metabolism of neurotransmitters.[12] An increased effect of neurotransmitter may be achieved by:

1.   Administration of the neurotransmitter or its agonist
2.   Increased biosynthesis of the neurotransmitter induced by administration of its precursor
3.   Lowering the reuptake of the neurotransmitter into the nerve endings
4.   Decrease of the neurotransmitter degradation

The opposite effect may be evoked by administration of the neurotransmitter receptor blockers, by administration of false precursor, etc.[12]

Recent progress in psychopharmacology resulted in a large variety of drugs affecting the CNS which may be used in neuroendocrine studies. However, the use of many different drugs and treatment schedules has led not only to the better understanding of central regulation of hormone secretion but also to some contradictory conclusions and to difficulties in comparison of results.

It is not possible to list here the drugs used for pharmacological manipulations together with their doses and main effects because of too many possibilities for various modifications in the widespread field of animal endocrinology. Such reviews are valuable if only a certain pharmacological group or certain endocrine function is concerned.[13-15] Even these are of relatively short-time value as new drugs are developed very quickly. Therefore, instead of reviewing the action of individual drugs, we will focus on some general considerations with regard to our own experience (Table 1).

From the methodological problems, which are at least partially responsible for variability in results, the first the problem of drug specificity has to be mentioned. All drugs available are more or less unspecific. Next to the main effect on certain neurotransmitter systems, they partially affect other neurotransmitters as well as other physiological functions. Unspecific effects increase with increasing dose.[10] Even if high specificity of a drug is indicated

Table 1
SOME METHODOLOGICAL PROBLEMS IN STUDIES USING
PHARMACOLOGICAL MODIFICATION OF ENDOCRINE FUNCTIONS

Problem	How to minimize its interference
Unspecificity of drugs and unwished side effects (increasing with the dose)	Treatment with minimal effective doses
	Comparison of the effects of several drugs with the same main action
Central vs. peripheral drug effects (these cannot be distinguished following systemic drug administration)	Use of central drug administration to investigate central action only
	Administration of drugs which do not cross the blood/brain barrier to investigate peripheral effects only
Time schedule of drug administration	Administration of drugs always at the same time of the day
	Measurement of the drug-induced effects at more than one time interval

by a pharmaceutical company or by the experimental data available, some unspecific effects have to be expected with higher doses. The administration of minimal effective doses of all drugs in any experimental protocol may be recommended.

Although it is clear that only the effects within the CNS are important for studies of central regulatory mechanisms, the other drug effects are often neglected. In the case of drugs which cross the blood-brain barrier, systemic administration may be chosen. This route is simple to perform and convenient from several points of view. The results are also better comparable with human studies in which central drug administration is hardly possible. However, following systemic administration, the peripheral metabolic, hemodynamic, and other functions may be affected and this in turn modifies the CNS and/or endocrine function studied. This disadvantage may be overcome by low-dose drug injections into the cerebral ventricles or cisterna magna.[16] Some drugs may be implanted into the chosen brain region in the form of pellets. The place of central drug administration need not to be the place of its action, and it cannot be determined always which brain structures were reached by the drug and which were not. Even centrally administered drugs may induce undesirable side effects, e.g., microconvulsions.

The next important question is the time schedule of drug administration and the evaluation of the endocrine changes under study. The secretion of hormones is not constant, with defined circadian rhythms and episodic hormone release being well known.[17] Also the drug-induced endocrine changes may differ with respect to the time of drug administration. Similarly, it is not reliable to measure the drug-induced effects in single time intervals after the drug administration, as many valuable experimental data may be omitted.

Summarizing, the results of experiments with pharmacological manipulations of central neurotransmitters have to be interpreted with caution. It seems useful to test several drugs, as well as dosage and time schedules and, if possible, to strengthen the data using other methods. The experiments with pharmacological approaches if performed and evaluated properly have already led and will further lead to valuable information on central regulatory mechanisms of hormone secretion.

B. Models for Selective Modification of Central Regulatory Centers by Neurophysiological Methods

The main purpose for the utilization of neurophysiological methods in neuroendocrine studies is to evaluate the significance of individual brain structures in the regulation of hormone secretion. In principle, two approaches are available — the surgical or electrical lesion of the brain structure in question, or the opposite way, that is, electrical stimulation of discrete brain areas.

Many interesting data on the hypothalamic control of pituitary hormone secretion have been obtained in rats with various hypothalamic deafferentations performed by a special "Halász-type" knife.[18] Three types of deafferentation are usually performed: (1) anterolateral cuts around the medial basal hypothalamus (MBH); (2) posterior cuts around the MBH; and (3) complete surgical ablation of the MBH.

The crucial importance of good methodology and proper histological control may be documented on the development of research aimed to find out the location of nerve cells producing corticoliberin (CRH). On the basis of deafferentation studies, it was believed for a long time that the cells producing CRH were located within the MBH. The substantial evidence for this hypothesis was that the rats with complete MBH deafferentation retained full pituitary-adrenocortical response to stress stimuli.[19] Later on, it gradually became evident that CRH activity was present outside the MBH. The earlier deafferentation studies were reinvestigated in an elegant series of experiments by Makara et al.[20-22] It has been pointed out that the stress-induced corticosterone release was retained when the basal region of the lateral retrochiasmatic area was not completely transected. Experiments with careful histological control has shown that if the cuts were complete in the anterolateral direction, there was no or only small activation of the pituitary-adrenal system following tuberal electrical stimulation,[20] ether inhalation,[22] immobilization,[23] insulin hypoglycemia,[24] and other stress situations. Nowadays, it is generally accepted that the main CRH cell bodies are actually located outside the MBH, mainly in the paraventricular nucleus. It is suggested that high-quality and thin, coronal sections with no freeze artifacts are necessary for proper histological control and that the sections taken at intervals of no more than 50 μm should be examined.[21] For these reasons, the earlier findings without appropriate control have to be considered with caution and reevaluated. For example, in some studies the rise of plasma corticosterone in response to ether stress was examined to verify the MBH lesions. Only animals which responded adequately were used in subsequent experiments.[25] This is quite uncorrect because the animals with complete MBH lesions do not respond to ether stress adequately.[22]

Electrical brain lesions applied to cortical and subcortical brain structures and often to individual hypothalamic nuclei are performed by anodal direct current guided to steel electrodes introduced stereotaxically into the corresponding area.[26-28] Some hypothalamic nuclei may be lesioned using a special quadrangular-shaped knife.[29]

Sterotaxic implantation of electrodes into the discrete brain regions makes possible their electrical stimulation followed by evaluation of endocrine changes induced.[25,29-31] The crucial point in the methodology is the current intensity, because higher intensities may evoke unspecific responses.

The perfusion of a circumscribed site in the brain of the unanesthetized animal has become a widely utilized scientific procedure in neuroscience.[32] The push-pull cannula technique, developed several years ago,[33] has been succesfully used to study in vivo release of endogenous compounds from discrete brain regions.[34,35] The entire principle upon which a push-pull perfusion is based is an exchange of constituents between perfusate and brain, which occurs presumably by diffusion. This exchange is two way. One of the advantages of the push-pull cannula technique is that the cannula can be implanted a long time prior to experimentation which may then be performed in freely moving unstressed animals. However, push-pull cannula cause some tissue damage, especially if used by an unskilled or uninformed worker. Moreover, chronically implanted cannula can induce the reaction of the glial system in the brain. In spite of its limitations, this is a useful methodology in the studies of in vivo release of neurotransmitters and neuropeptides.

III. METHODS FOR EVALUATION AND MODIFICATION OF HORMONE SECRETION

A. Hormones in Biological Fluids — How to Obtain Reliable Data

1. Introduction

With the exception of special studies on hormone secretory mechanisms, the hormone concentrations in biological fluids provide a good indicator of endocrine gland function and hormone secretion. In spite of very frequent measurements of hormone levels, the obtained or even published data are often of limited value because of unappropriate methodological approaches. The most important sources of methodological errors are the fluid sampling techniques and general conditions of experiments; however, inadequate interpretation of the results also occurs.

The biological fluids used for hormone determinations are blood, urine, saliva, and cerebrospinal fluid. The blood analysis is preferable in animal physiology. The analysis of urine is a convenient method for clinical and experimental purposes in man, while in animals it requires somewhat complicated measurement of diuresis and urine sampling. Estimation of hormone concentrations directly in endocrine glands or other tissues is also performed for certain purposes but not for evaluation of hormone secretion. Without having additional indicators, it cannot be decided whether, for example, a decrease in glandular hormone content is due to decreased hormone synthesis or to increased secretion.

As to the methodology, the crucial points for hormone data and their interpretation are (1) blood sampling techniques and time schedule used; (2) general conditions of experiments performed; and (3) analytical method for hormone determination (not included in this review; see References 1, 4, and 5).

2. Blood Sampling Techniques

The route of blood sampling may considerably influence the actual results and the approach to their evaluation. The methods of choice differ with regard to the animal species studied and the use of anesthetized vs. conscious animals.

a. Anesthetized Animals

Obviously, the blood from peripheral circulation, venous or arterial, is being analyzed in studies of hormonal regulations. Peripheral blood may be obtained by venous or cardiac puncture, decapitation, or cutting the tail. For more proper evaluation of hormone secretory patterns, blood samples from the vein from the endocrine gland in question may be taken. Thus, adrenal venous corticoid output can indicate the changes in pituitary-adrenocortical function,[36] and the concentration of testosterone in the testicular vein reflects the androgen secretory activity of the testes.[37] For determination of gastrointestinal peptides, the sampling of portal blood is an appropriate approach because of their degradation in the liver. The cannulation of vena portae may be performed also in conscious animals.[38] A special feature is the analysis of blood from the hypothalamo-hypophyseal portal system in neuroendocrine studies.[39]

The essential disadvantage of blood sampling in anesthetized animals is that anesthesia itself influences the hormonal regulations. The range of induced changes depends on the anesthetic drug, as well as on the dose and route of its administration.

To illustrate this subject, the effects of barbiturate anesthesia on some hormonal systems may be mentioned as this type of anesthesia is very frequently used in animal physiology research.

The pituitary-adrenocortical axis in rats does not seem to be influenced by barbiturate anesthesia to a great extent, as no elevated corticosterone basal levels and normal stress responses were observed.[40]

On the other hand, barbiturate anesthesia exerts a strong inhibitory effect on the activation of the sympathetic-adrenomedullary system.[41] Thus, in anesthetized animals, no stress-induced increase in plasma catecholamine levels occurs. This is of particular importance in stress research; however, the lack of sympathetic-adrenomedullary activation and of its metabolic consequences may influence many other physiological functions.

Barbiturate anesthesia in rats results in a continuous decrease of plasma thyroxine level for at least 8 hr[42] which appears to be due to a displacement of this hormone from plasma protein binding and thus an increased disposal rate.[43] Even the earlier report on the suppression of thyroid secretion by the same barbiturate[44] may apparently be explained by such a mechanism.

Furthermore, anesthesia may modify the circadian rhythms of hormone secretion. For example, the normal pulsative release of growth hormone in rats is prevented by anesthesia and so the studies on growth hormone regulation in this animal species have to be performed under a conscious state.[45]

As has been recently reviewed,[46] urethane, another widely used anesthetic in animal experimentation, produces a variety of potentially disturbing side effects at the endocrine level.

Summarizing, anesthesia affects the secretion of several hormones and influences the reactivity of the organism. The conclusions made in studies in anesthetized animals[36] may differ from those obtained in experiments with conscious animals.[47] These limitations have to be taken into account in sudies in which the use of anesthesia cannot be omitted.

b. Conscious Animals

The most important aspect for evaluation of blood sampling techniques in conscious animals is to what extent the sampling requires the handling and restraint of the animal. The latter procedures evoke unspecific stress effects which should be avoided.

Sampling of trunk blood after rapid decapitation of small laboratory animals is simple to perform and a relatively high volume of blood can be obtained. However, plasma levels of many hormones depend on even minor disturbing events before decapitation. Plasma corticosterone levels in eight rats succesively removed for decapitation within 15 min from one common cage are shown in Figure 3. The hormone level in the blood of the last animal is already considerably high. Therefore, the rats kept in a common cage should be decapitated by an adequate number of staff contemporally, at best within 30 sec. Decapitation is quite unadequate for catecholamine studies because it increases plasma levels of both epinephrine and norepinephrine to extremely high values within several second.[41,48]

To determine the truly basal status of various hormones in a conscious animal, the chronic cannulation is a suitable technique. Chronic cannulation is of special value for studying the regulation of endocrine system easily stimulated by handling, restraint or other disturbing events, the sympathoadrenomedullary system, hypothalamo-pituitary-adrenal system, somatotrophic and lactotrophic systems to be mentioned the first.

Several methods for implanting chronic cannulas for remote blood sampling in stress-free, conscious, and unrestrained rats have been developed. Polyethylene or silastic catheters may be put into the femoral artery,[49] jugular vein,[44] descending aorta,[50] abdominal aorta,[51] or tail artery,[52] depending on the needs of the experiments. The ends of the catheters are brought to the back of the neck in a subcutaneous tunnel and exteriorized to the cage top through stainless steel springs sewn to the skin. At least a 1- to 3-to day period before the onset of experiments is necessary to allow postsurgical recovery of baseline hormone secretion. A 3-day recovery is needed for normal body weight gain and food intake restoration.[49] Plasma levels of some hormones such as corticosterone mich be slightly higher in chronic cannulated rats compared with those measured in carefully decapitated rats,[53,54] but usually still within the limits of physiological range. The surgical procedure and recovery seem to

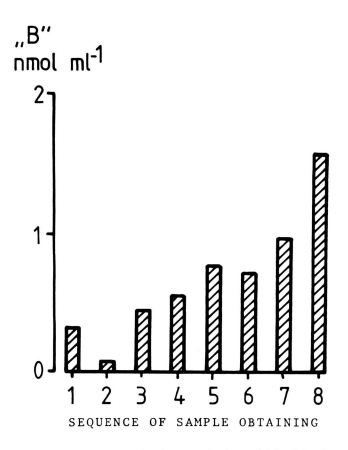

"B"
nmol ml⁻¹

FIGURE 3.   Concentration of corticosterone in plasma of eight adult male rats kept in a common cage and successively bled by decapitation during 15 min. (From Németh, S., in *Stress: The Role of Catecholamines and Other Neurotransmitters,* Gordan and Breach, New York, 1984. With permission).

be more simple for tail artery cannulation as compared to other vessels; however, blood reinfusion and drug injections are inconvenient in this model. If some drug administration is needed, the tail artery catheter model may be combined with another catheter insertion into the peritoneal cavity.[55] An indwelling catheter for infusions into subcutaneous tissue of freely moving rats was also described.[56] The advantages and disadvantages of chronic cannulation in comparison with decapitation are indicated in Table 2.

3. General Conditions of Experiments

It is generally accepted that to obtain reliable data, it is necessary to perform the experiments under defined conditions and limitation of interfering events.

In all research activities and especially in studies on hormone regulations, the possibility of distortion of results by unrecognized stress should be avoided. The fact that an increase in stress hormone release and consequent metabolic and hemodynamic changes may modify the general reactivity of the organism is often neglected. Especially in nonendocrine research, it is sometimes considered that animals were not under stress conditions because normal blood glucose levels and normal blood pressure were found. For example, this statement was applied for experiments performed about 60 min following surgery in anesthesia,[57] and it may be suggested with certainty that plasma corticosterone or other stress hormones would show a stress response.

Table 2
BLOOD SAMPLING OF CONSCIOUS RAT — COMPARISON OF
DECAPITATION WITH CHRONIC CANNULATION

Model	Advantages	Disadvantages
Decapitation	Simple, no surgery needed	Handling cannot be avoided
	Big volume of blood taken	Unsuitable for some hormone studies (catecholamines)
		Repeated sampling impossible
		Quick performance is needed after opening the cage
Chronic cannulation	No handling	Need for surgery and anesthesia beforehand
	No restraint	
		Individual housing
	Possibility of repeated blood sampling	Need for adaptation to novel conditions
	Possibility of drug administration without stress of injection	Limited volume of blood taken (increased by reinfusion of resuspended erythrocytes)

However, stimuli of a much lesser intensity than surgical trauma may induce a stress reaction. Entering the animal room, opeining the cage, mild noise and handling can be considered as disturbing events. The comparison of the effect of several stimuli on plasma corticosterone levels in rats is shown in Figure 4. A brief exposure of rats to a novel environment (transport of rats in their cages to another room) was sufficient to cause significant elevation of corticosterone release to the levels comparable to those after intensive physical activity (swimming). It is therefore undesirable to transport the animals from the housing room to another experimental room on the day of experimentation, as it is usually done in many laboratories. Instead, the animals should be adapted to the conditions of experimental chamber or laboratory for at least 24 hr.

It is also worth mentioning that even very quickly performed, systemic injections of saline in conscious animals cause adrenocortical activation (Figure 4). In experiments in which acute drug administration is necessary, the animals should be adapted to the injection procedure by repeated handling and saline injections to avoid unspecific stress effects. Another approach is to use one of the chronic cannulation techniques.

There are many other factors that might be underestimated. At least two of them should be underlined — circadian rhythms and feeding. The experiments performed during the morning hours cannot be compared to those performed in the afternoon because of daily rhythms of endocrine reactivity, hormone secretion, and other physiological functions. In many experimental procedures, the animals may change their feeding behavior, and under certain circumstances a substantial decrease of food intake may occur (following surgery, in repeatedly stressed animals, etc.). This partial fasting could influence the results of experiments. With the exception of very short experiments, the food and water intake should be measured and appropriate pair-fed animals should be studied, if necessary.[41]

B. Models of Endocrine Hypo- and Hyperfunction

The very classical methodology in endocrinology is extirpation of endocrine glands. It played a historical role in the assessment of endocrine gland function and hormone deficiency syndroms, and their metabolic consequences. The extirpation and also transplantation of endocrine glands is being used until now with somewhat better understanding of resulting events.[59] First of all, it has to be kept in mind that extirpation of certain glands results not only in the lack of corresponding endocrine function, but also in a modification of other

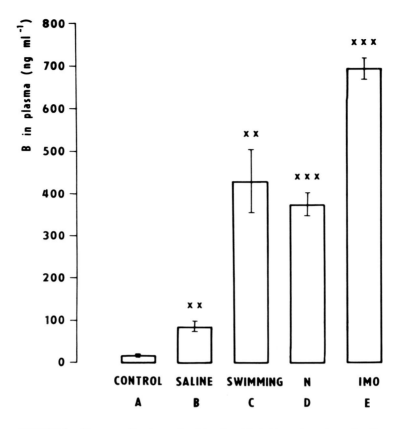

FIGURE 4.   Plasma corticosterone level in rats subjected to various stress stimuli. (A) Nonstressed control; (B) 60 min following subcutaneous saline injection; (C) 15 min of swimming; (D) 15 min following transport to novel environment; (E) restraint for 60 min. Means of five to eight values ± SE. Statistical significance against nonstressed control: xx — $p < 0.01$; xxx — $p < 0.001$. (From Ježová-Repčeková, D., Vigaš, M., and Kulifajová, A., *Endocrinol. Exp.*, 12, 3, 1978. With permission.)

neuroendocrine systems. The interglandular relationships were mentioned in Section I.B. After removal of the endocrine gland, particular substitutive hormonal treatment may be necessary. Results of all operations should be checked by hormone estimation and by histological examination after sacrifice of the animal.

Here only short notes to surgical operations on endocrine glands will be given.[60,61]

Thyroidectomy is a relatively simple operation, though in small animals (rat, mouse) it is nearly impossible to preserve intact parathyroid glands. Parathyroid glands can remain intact in dogs, when carefully prepared. In the rabbit, parathyroid glands are localized extrathyroidally. Parathyroidectomy in smaller animals leads to devastation of the thyroid gland. Autotransplantation of the thyroid gland usually is successful and transplants become functional. The success of homotransplantation depends on the enhanced secretion of TSH after previous thyroidectomy in the recipient animal. Autotransplantation of parathyroid glands is more frequently successful than homotransplantation.

Pancreatectomy is a difficult operation. A total pancreas can be removed only in young dogs and cats; the rabbit pancreas is dispersed and adhering to vessels. In rats the total pancreatectomy is almost impossible, but subtotal extirpation results in diabetes. For transplantation, the pancreatic tissue with atrophic excretory parts due to prior duct ligation can be used. It is possible to implant the islet cells themselves obtained by digestion of acinar tissue. Embryonal and neonate pancreases have given better results.[62]

Adrenalectomy is a reliable endocrine operation, providing that accesory adrenals are not present. Moreover, glomerulosa cells remaining in adrenal capsula *in situ* may grow to sufficient adrenal cortex during 1 month. By adrenomedullectomy, the adrenal cortex is preserved but its venous drainage may be harmed. Adrenocortical transplantation is successful only in adrenalectomized animals.[63]

Castration in male animals is the most simple endocrine operation. Testicular monotransplantation provides hormonal production, but spermatogenesis may occur only in scrotal sites of implantation. Castration in females is technically somewhat similar to adrenalectomy. Caution must be given not to damage the ovarian tissue, remnants of which are capable of regeneration. Even heterotransplantation of ovarian tissue to various sites usually provides full endocrine function of the graft.

Hypophysectomy, even if made by an experienced person by either the parapharyngeal or transauricular route, results only in about 80 to 90% in the total removal of pituitary without damage to adjacent tissue. Anterior pituitary-grafted animals are an excellent model of chronic hyperprolactinemia without secretion of significant amounts of other pituitary hormones from the graft.[64]

Stimulation of autonomic nerves innervating endocrine glands (e.g., n. vagus-islets of Langerhans) or denervation of endocrine glands (e.g., ovary) changes the neural control of secretion of a particular gland. However, such a procedure may also impair the blood supply, as often seen after the denervation of adrenals.

Synthesis and/or release of hormones may also be inhibited by pharmacological means. For example, synthesis of glucocorticoids can be blocked chemically (by aminogluthetimid, metopirone, or etomidate[1,65]). Certain analogs of natural steroids interfere with binding to receptors, and thus they interfere with effects of natural hormones in target tissues. These antihormones, as they are called, include antiandrogens (e.g., cyproterone acetate), antiestrogens (e.g., clomiphene, dimethylstilbestrol) and antialdosterone (spironolactones), and are available for experimental models of particular endocrine hypofunction.

The classical model for endocrine hyperfunction is the administration of exogenous hormones. Also controlled-release delivery systems (pellets, pumps, etc.) for hormones can be used to study certain physiological and pathophysiologic states.[66] Some caution is necessary with repeated hormone administration because the treatment may induce changes, e.g., in receptor sensitivity (see Section IV.A), in feedback relationships, and, in case of protein hormones, it may induce immune response. Endocrine functions can be influenced using various other substances ranging from specific effect to general toxicity, a list of which cannot be fully given here.

IV. METHODS FOR EVALUATION AND MODIFICATION OF HORMONE ACTION

A. Introduction

In the recent years, our knowledge on the mechanisms of hormone action in target tissues has increased and is still rapidly increasing. The efforts of numerous investigators have resulted in the identification of a complex array of cellular and subcellular events involved in hormone action. The primary observations are usually being made on in vitro systems, later to be applied for systems in vivo. The methods used are mostly special biochemical techniques, which are outside the scope of this review.[6-9] Therefore, only brief comments will be given to this very progressive part of endocrinology.

B. Methods for Investigation of Hormone Receptors

For estimation of hormone action in target tissue, specific receptor and postreceptor events can be measured. Specific assays for hormonal receptors have been developed using either specific ligands or monoclonal antibodies against receptors.[6,7,67] Also, raising anti-idiotypic antibodies with activity against the corresponding receptor offers a novel method in the isolation and characterization of receptors.[68]

The number and binding affinity of particular receptors are dependent on various conditions which may influence the experimental results. Action of agonists themselves lowers the number of available receptors (down regulation), and this should be considered if hormones or their agonists are administered repeatedly in an experimental protocol.[69,70] Other hormones than direct agonists may also influence the receptor number and/or binding, as it is known for thyroid hormones and adrenergic receptors.[71] The receptor status may be influenced by certain experimental conditions, e.g., fasting or overfeeding elevates or lowers, respectively, the number of insulin receptors. Moreover, the binding parameters of hormone receptors change during postnatal development.[72-74] However, it should be taken into account that changes of receptor number and affinity do not always correlate with changes of hormonal effects in tissues. To some extent this is due to the different methods used for estimation of receptors.[75]

The use of isolated physiologically responding tissues to study receptors provides an index of both affinity and intrinsic physiological efficacy.[76]

Assessing the postreceptor hormonal effects, ''second messenger'' measurement[77] is useful, in addition to the measurement of metabolic consequences of hormone action. Activation of adenylate cyclase with CAMP accumulation and enhancement of calcium uptake in the cell or inhibition of these processes are elicited by several hormones.[77] Protein phosphorylation-dephosphorylation systems represent a major mechanism of cellular regulation.[78] There are suggestions that one type of receptor in a single target cell may be linked to multiple potential effector systems. The molecular biology approach can evaluate the hormonal action on the level of regulation of gene expression.[8,80] A very simplified scheme of the possible events involved in the mechanisms of hormone action of target cells is given in Figure 5.

C. Immunological Approaches in Hormone Regulation Studies

Immunological methods in hormone regulation studies are exploiting specific antibodies to hormones and their receptors. Injection of antigen to rabbit or guinea pig is used for developing antibodies for radioimmunological methods.[5] An improved sensitivity and specificity of radioimmunoassay was achieved with the use of monoclonal antibodies.[81] New possibilities in endocrine research offer antiidiotypic antibodies, which are useful in receptor assays, and as a model of autoimmune diseases.[68] Antibodies are valuable tools in immunocytochemistry of hormonal and receptor materials in tissues.[3]

The administration of antibodies against particular hormones or receptors inactivates the physiological effect of their antigens. This approach is often used to assess hormone involvement in the regulatory events under conditions studied (e.g., anticorticoliberin antibody,[82] antisomatostatin antibody[83]). However, a hormone-potentiating effect was also seen when administering hormone-specific monoclonal antibodies, which challenges the generally held opinion that an antibody can merely neutralize hormonal functions. This may be important in animal physiology.[84]

On the other hand, immune and endocrine systems do mutually influence each other. Evidence was provided that the immune response itself can bring about neuroendocrine responses with immunoregulatory consequences.[85] The measurement of immune response to endocrine manipulations may serve as a model of a target system for hormonal regulations. Another methodological approach consists of a search for soluble messengers released in vitro by activated immune cells which, upon injection into normal animals, may influence hormone secretion.[86] Therefore, the measurement of immune response to endocrine manipulations serves as another model of target system for hormonal regulations.

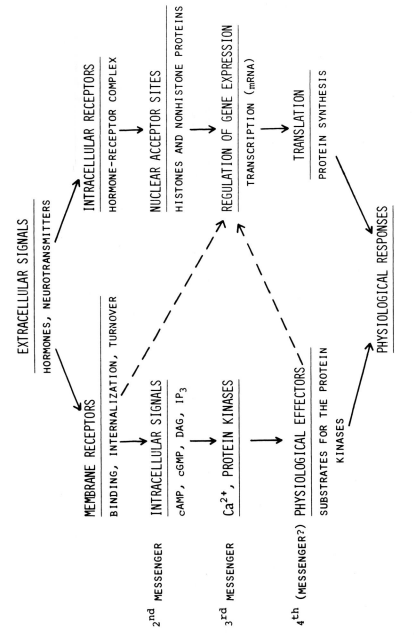

FIGURE 5. Mechanisms of hormone action (very simplified).

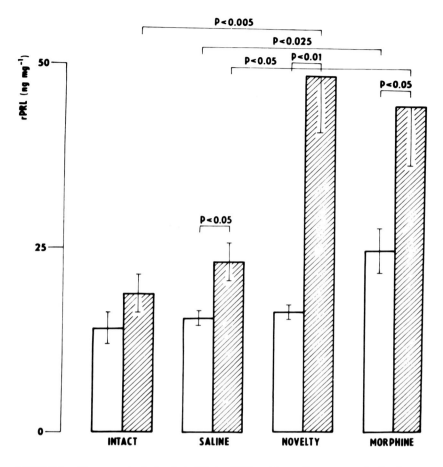

FIGURE 6.   Plasma prolactin levels in Wister-AVN and Long Evans rats under basal conditions (intact) and in response to subcutaneous injection of saline, morphine (6 mg kg^{-1}), or novelty stress for 15 min. Empty columns represent values of the Wistar-AVN; cross-hatched bars those of the Long Evans rats. Means of eight values ± SE. (From Jurčovičov, J., Vigaš, M., Klír, P., and Ježová, D., *Endocrinol. Exp.*, 18, 209, 1984. With permission.)

D. Genetic Approaches in Hormone Regulation Studies

The influence of inheritance on neuroendocrine secretory patterns is of great importance. Functioning of the neuroendocrine system is not identical in different strains of laboratory animals, even under the same circumstances (Figure 6), which may lead to different results when different strains of the same species are used.

According to the multiple factors controlling hormone synthesis, release, and action, there are numerous sites where genetic mutations could interrupt the correct flow of endocrine information. Such genetic errors as experiments of nature may provide useful insights into biochemical and physiological processes.

The most frequent genetic defects exploited in endocrine research occur in laboratory rats and mice.[88,89] Obese and/or diabetic rats and mice are useful in nutrition, obesity, and diabetes research. Several mouse mutants exhibit dwarfism and demonstrate the diversity of endocrine dysfunction associated with failure of normal growth. Mutations directly affecting reproductive function have also been described; the testicular feminization mouse and the hypogonadal mouse are useful models in understanding actions of androgens on intracellular receptors and the role of gonadotropin-releasing hormone in reproduction. The diabetes insipidus mouse is a model of nephrogenic diabetes insipidus due to a failure in its response to vasopressin. The diabetes insipidus rat is a model of central diabetes insipidus.[50] When comparing rat and mouse mutants as experimental model strains, the disadvantage of the mouse in endocrine research is the small blood volume and the difficulty in serial blood sampling.

The new way of investigating peptide-secreting cells and tissues, as well as hormone action, uses the methods and principles of molecular genetics.[8,90-92] The use of recombinant deoxyribonucleic acid (DNA) techniques enables one to determine if a certain hormone gene is expressed in a particular tissue by measuring the level of specific messenger ribonucleic acid (mRNA) in that tissue.[93] With the widespread application of molecular cloning, our concept of the mechanisms of hormone action and of regulation of hormone secretion may change tremendously.

The mechanisms of hormone action on gene expression in target tissues can be studied by measurement of mRNA accumulation. The primary structure of peptide hormones and hormonal receptors can be analyzed from the sequence of complementary DNA.[94] The methods of recombinant DNA technology make it possible to locate defects at the gene level, as has been shown by identification of the genetic defect in diabetes insipidus rats.[95]

It may be claimed without doubt that the molecular biology and genetic methods represent one of the most progressive methodological approaches in hormone studies.

ACKNOWLEDGMENTS

The authors thank Drs. Milan Vigaš, Štefan Németh, Marta Dobrakovová, Pavol Langer, and Štefan Zórad for their contribution to the data on which this review is based. The secretarial help of Mrs. M. Bajerlová, D. Kubíková, and Mr. T. Dobias is gratefully acknowledged.

REFERENCES

1. **Gray, C. H. and James, V. H. T., Eds.,** *Hormones in Blood,* Vol. 1 to 3, Academic Press, New York, 1979.
2. **Weiner, R. I. and Ganong, W. F.,** Role of Brain monoamines and histamine in regulation of anterior pituitary secretion, *Physiol. Rev.,* 58, 905, 1978.
3. **Hökfelt, T., Elde, R., Johansson, O., Ljungdahl, A., Goldstein, M., Luft, R., Efendic, S., Nilsson, G., Terenius, L., Ganten, D., Jeffcoate, S. L., Rehfeld, J., Said, S., Perez de la Mora, M., Rossani, L., Tapia, R., Teran, L., and Palacios R.,** Aminergic and peptidergic pathways in the nervous system with special reference to the hypothalamus, in *The Hypothalamus,* Reichlin, S., Baldessarini, R. J., and Martin, J. B., Eds., Raven Press, New York, 1978, 434.
4. **Frieden, E. H.,** *Chemical Endocrinology,* Academic Press, New York, 1976.
5. **Renter, A. M., Ketelslegers, J. M., Hendrick, J. C., and Franchimont, P.,** Radioimmunoassay of protein hormones: principles and methodology, *Horm. Res.,* 9, 404, 1978.
6. **Blecher, M., Ed.,** *Methods in Receptor Research,* Marcel Dekker, New York, 1976.
7. **Barne, M. and Hennessen, W., Eds.,** *Developments in Biological Standardization. Monoclonal Antibodies: Standardization of Their Characterization and Use,* Vol. 57, S. Karger, Basel, 1984.
8. **Maniatis, T., Fritsch, E. F., and Sambrook, J.,** *Molecular Cloning: A Laboratory Manual,* Cold Spring Harbor Laboratory, Cold Spring Harbor, N. Y., 1984.
9. **Barnes, D. W., Sirbasku, D. A., and Sato, G. H., Eds.,** *Methods for Serum-free Culture of Cells of the Endocrine System. Cell Culture Methods for Molecular and Cell Biology,* Vol. 2, Alan R. Liss, New York, 1984.
10. **Krieger, D. T.,** Neurotransmitter regulation of ACTH release, *Mt. Sinai J. Med. N.Y.,* 40, 302, 1973.
11. **Palkovits, M.,** Effect of stress on catecholamine- and neuro-peptide-containing neurons in the central nervous system, in *Stress: The Role of Catecholamines and Other Neurotransmitters,* Usdin, E., Kvetňanský, R., and Axelrod, J., Eds., Gordon and Breach, New York, 1984, 75.

12. **Brown, G. M., Friend, W. C., and Chambers, J. W.,** Neuropharmacology of hypothalamic-pituitary regulation, in *Chemical Neuroendocrinology. A Pathophysiological Approach,* Tolis, O., Ed., Raven Press, New York, 1979, 47.

13. **Preziosi, P. and Nistico, G.,** Psychotropic drugs: mechanism of action at the neurotransmitter level, *Int. J. Clin. Pharmacol.,* 15, 497, 1977.

14. **Müller, E. E., Cella, S., Locatelli, V., Peñalva, A., and Cocchi, D.,** Drugs affecting prolactin secretion, in *Prolactin and Prolactinomas,* Tolis, G. et al., Eds., Raven Press, New York, 1983, 83.

15. **Nathan, R. S. and Van Kammen, D. P.,** Neuroendocrine effects of antipsychotic drugs, in *Antipsychotics,* Vol. 3, Burrows, G. D., Norman, T. R., and Davies, B., Eds., Elsevier, Amsterdam, 1985, 11.

16. **Myers, R. D.,** Chronic methods: intraventricular infusion, cerebrospinal fluid sampling, and push-pull perfusion, in *Methods in Psychobiology,* Vol. 3, Myers, R. D., Ed., Academic Press, New York, 1978, 281.

17. **Krieger, D. T., Ed.,** *Endocrine Rhythms,* Raven Press, New York, 1979.

18. **Halász, B. and Gorski, R. A.,** Gonadotropic hormone secretion in female rats after partial or total interruption of neural afferents to the medial basal hypothalamus, *Endocrinology,* 80, 608, 1967.

19. **Halász, B., Slusher, M., and Gorski, R.,** Adrenocorticotrophic hormone secretion in rats after partial or total deafferentation of the medial basal hypothalamus, *Neuroendocrinology,* 2, 43, 1967.

20. **Makara, G. B., Stark, E., and Palkovits, M.,** ACTH release after tuberal electrical stimulation in rats with various cuts around the medial basal hypothalamus, *Neuroendocrinology,* 27, 109, 1978.

21. **Makara, G. B., Stark, E., Rappay, G., Kárteszi, M., and Palkovits, M.,** Changes in corticotrophin releasing factor of the stalk median eminence in rats with various cuts around the medial basal hypothalamus, *J. Endocrinol.,* 83, 165, 1979.

22. **Makara, G. B., Stark, E., and Palkovits, M.,** Reevaluation of the pituitary-adrenal response to ether in rats with various cuts around the medial basal hypothalamus, *Neuroendocrinology,* 30, 38, 1980.

23. **Makara, G. B., Kvetňanský, R., Ježová, D., Jindra, A., Kakucska, I., and Opršalová, Z.,** Plasma catecholamines do not participate in pituitary-adrenal activation by immobilization stress in rats with transection of nerve fibers to the median eminence, *Endocrinology,* 119, 1757, 1986.

24. **Ježová, D., Kvetňanský, R., Kovács, K., Opršalová, Z., Vigaš, M., and Makara, G. B.,** Insulin-induced hypoglycemia activates the release of adrenocorticotropin predominantly via central and propranolol insensitive mechanisms, *Endocrinology,* 120, 409, 1987.

25. **Feldman, S., Conforti, N., and Siegel, R. A.,** Adrenocortical responses following limbic stimulation in rats with hypothalamic deafferentations, *Neuroendocrinology,* 35, 205, 1982.

26. **Myers, R. D., Ed.,** *Methods in Psychobiology,* Vol. 2, Academic Press, New York, 1972.

27. **Adler, M. W.,** The effect of single and multiple lesions of the limbic system on cerebral excitability, *Psychopharmacologia,* 24, 216, 1972.

28. **Filaretov, A. A. and Filaretova, L. P.,** Role of the paraventricular and ventromedial hypothalamic nuclear areas in the regulation of the pituitary-adrenocortical system, *Brain Res.,* 342, 135, 1985.

29. **Lengvári, I., Kovács, M., Liposits, Z., and Szelier, M.,** The lack of effect of electrochemical stimulation of the hypothalamus and median eminence on the plasma corticosterone level after lesion of the paraventricular nuclei, *Exp. Clin. Endocrinol.,* 85, 314, 1985.

30. **Bureš, J., Petráň, M., and Zachar, J.,** *Electrophysiological Methods in Biological Research,* Academia, Prague, 1967.

31. **Dobrakovová, M., Opršalová, Z., and Kvetňanský, R.,** Plasma catecholamines in rats electrostimulated in different brain areas, in *Stress: The Role of Catecholamines and Other Neurotransmitters,* Usdin, E., Kvetňandký, R., and Axelrod, J., Eds., Gordon and Breach, New York, 1984, 649.

32. **Myers, R. D.,** Methods, for perfusing different structures of the brain, in *Methods in Psychobiology,* Vol. 2, Myers, R. D., Ed., Academic Press, New York, 1972, 169.

33. **Geddum, J. H.,** Push-pull cannulae, *J. Physiol.,* 155, 1, 1961.

34. **Levine, J. E. and Ramirez, V. D.,** In vivo release of luteinizing hormone-releasing hormone estimated with push-pull cannulae from the mediobasal hypothalami of ovariectomized, steriod-primed rats, *Endocrinology,* 107, 1782, 1980.

35. **Wuttke, W., Demling, J., Roosen-Runge, G., Siegel, R., Fuchs, E., and Düker, E.,** In vivo release of catecholamines and amino acid neurotransmitters, in *Stress: The Role of Catecholamines and Other Neurotransmitters,* Usdin, E., Kvetňanský, R., and Axelrod, J., Eds., Gordon and Breach, New York, 1984, 93.

36. **Ganong, W. F., Kramer, N., Salmon, J., Reid, I. A., Lovinger, R., Scapagnini, U., Boryczka, A. T., and Shackelford, R.,** Pharmacological evidence for inhibition of ACTH secretion by a central adrenergic system in the dog, *Neuroscience,* 1, 167, 1976.

37. **Eik-Nes, K. B.,** An effect of isoproterenol on rates of synthesis and secretion of testosterone, *Am. J. Physiol.,* 217, 1764, 1969.

38. **Sirek, A. and Sirek, O. V.,** A new technique for hepatic portal sampling in the conscious dog, *Proc. Soc. Exp. Biol. Med.,* 172, 397, 1983.

39. **Porter, J. C. and Smith, K. R.,** Collection of hypophysial stalk blood in rats, *Endocrinology,* 81, 1182, 1967.

40. **Ježová, D. and Németh, Š.,** unpublished data.

41. **Németh, Š.,** Possible distorting of experimental results by stress, in *Stress: The Role of Catecholamines and Other Neurotransmitters,* Usdin, E., Kvetňanský, R., and Axelrod, J., Eds., Gordon and Breach, New York, 1984, 59.

42. **Langer, P., Földes, O., Brozmanová, H., and Gschwendtová K.,** Studies on the effect of pentobarbiturate, ether and heparin on plasma thyroxine level in rats, *Endocrinologie,* 76, 309, 1980.

43. **Langer, P., Kokešová, H., and Gschwendtová, K.,** Acute redistribution of thyroxine after the administration of univalent anions, salicylate, theophylline and barbiturates in rats, *Acta Endocrinol.* (Copenhagen), 81, 516, 1976.

44. **Brown, M. R. and Hedge, G. A.,** Thyroid secretion in the unanesthetized, stress-free rat and its suppression by pentobarbital, *Endocrinology,* 9, 158, 1972.

45. **Martin, J. B.,** Studies on the mechanism of pentobarbital induced GH release in the rat, *Neuroendocrinology,* 13, 339, 1974.

46. **Maggi, C. A. and Meli, A.,** Suitability of urethane anesthesia for physiopharmacological investigations. III. Other systems and conclusions, *Experientia,* 42, 531, 1986.

47. **Holland, F. J., Richards, G. E., Kaplan, S. L., Ganong, W. F., and Grumbach, M. M.,** The role of biogenic amines in the regulation of growth hormone and corticotropin secretion in the trained conscious dog, *Endocrinology,* 102, 1452, 1978.

48. **Popper, C. W., Chiueh, C. C., and Kopin, I. J.,** Plasma catecholamine concentrations in unanesthetized rats during sleep, wakefulness, immobilization and after decapitation, *J. Pharmacol. Exp. Ther.,* 202, 144, 1977.

49. **Raff, H. and Fagin, K. D.,** Measurement of hormones and blood gases during hypoxia in conscious cannulated rats, *J. Appl. Physiol.,* 56, 1426, 1984.

50. **Valtin, H.,** Experimental models for the study of brain peptides, in *Neuroendocrinology of Vasopressin, Corticoliberin, and Opiomelanocortins,* Baertsch, A. J. and Dreifuss, J. J., Eds., Academic Press, London, 1982, 307.

51. **Engberg, A.,** A technique for repeated renal clearance measurements in undisturbed rats, *Acta Physiol. Scand.,* 75, 170, 1969.

52. **Chiueh, C. C. and Kopin, I. J.,** Hyperresponsivity of spontaneously hypertensive rat to indirect measurement of blood pressure, *Am. J. Physiol.,* 234, H690, 1978.

53. **Fagin, K. D., Shinsako, J., and Dallman, M. F.,** Effects of housing and chronic cannulation on plasma ACTH and corticosterone in the rat, *Am. J. Physiol.,* 245, E515, 1983.

54. **Ježová, D.,** Stimulation of ACTH release by naloxone: central or peripheral action? *Life Sci.,* 37, 1007, 1985.

55. **Ježová, D., Jurčovičová, J., Vigaš, M., Murgaš, K., and Labrie, F.,** Increase in plasma ACTH after dopaminergic stimulation in rats, *Psychopharmacologia,* 85, 201, 1985.

56. **Mucha, R. F.,** Indwelling catheter for infusions into subcutaneous tissue of freely-moving rats, *Physiol. Behav.,* 24, 425, 1980.

57. **Hardebo, E. and Andersson, J.,** Neutralization of endothelial surface charge and blood-brain barrier function, *J. Cereb. Blood Flow Metab.,* Suppl. 1, 407, 1983.

58. **Ježová-Repčeková, D., Vigaš, M., and Kulifajová, A.,** Effects of phentolamine and clonidine on pituitary-adrenocortical axis during stress in conscious rats, *Endocrinol. Exp.,* 12, 3, 1978.

59. **Simeonovic, C. J., Dhall, D. P., Wilson, J. D., and Lafferty, K. J.,** A comparative study of transplant sites for endocrine tissue transplantation in the pig, *Aust. J. Exp. Biol. Med. Sci.,* 64, 37, 1986.

60. **DeBoer, J., Archibald, J., and Downie, H. G., Eds.,** *An Introduction to Experimental Surgery,* Elsevier, New York, 1975.

61. **Waynforth, H. B.,** *Experimental and Surgical Technique in the Rat,* Academic Press, New York, 1980.

62. **Korec, R.,** Treatment of alloxan and streptozotocin diabetes in rats by intrafamiliar homo (allo) transplantation of neonatal pancreases, *Endocrinol. Exp.,* 14, 191, 1980.

63. **Engeland, W. C.,** Pituitary-adrenal function after transplantation in rats: dependence on age of the adrenal graft, *Am. J. Physiol.,* 250, E87, 1986.

64. **Adler, R. A., Herzberg, V. L., and Sokol, H. W.,** Studies of anterior pituitary-grafted rats. II. Normal growth hormone secretion, *Life Sci.,* 32, 2957, 1983.

65. **Dörr, H. G., Kuhnle, V., Bocthausen, H., Bidlingmaier, F., and Knorr, D.,** Etomidate: a selective adrenocortical 11-β-hydroxylase inhibitor, *Klin. Wochenschr.,* 62, 1011, 1984.

66. **Beck, L. B. and Pope, V. Z.,** Controlled-release delivery systems for hormones. A review of their properties and current therapeutic use, *Drugs,* 27, 528, 1984.

67. **Kahn, C. R., Flier, J. S., Jarret, D., and Roth, J.,** Use of antireceptor antibodies as probes of receptor structure and function, in *Horomonal Receptors in Digestive Tract Physiology,* Boufils, S., Fromagest, P., and Rosselin, G., Eds., North-Holland, Amsterdam, 1977, 69.

68. **Farid, N. R., Briones-Urbina, R., and Nazrul-Islam, M.,** Biologic activity of anti-thyrotropin antiidiotypic antibody, *J. Cell Biochem.,* 19, 305, 1982.
69. **Kolena, J. and Šeböková, E.,** Hormonal regulation of testicular LH/hCG receptors in rat, *Exp. Clin. Endocrinol.,* 82, 1, 1983.
70. **Harden, T. K.,** Agonist-induced desenzitization of the beta-adrenergic receptor-linked adenylate cyclase, *Pharm. Rev.,* 35, 5, 1983.
71. **Clur, A.,** Idothyronines and iodotyrosines as hypothetical receptor for catecholamines and opiates, *Hypotheses Med.* 16, 97, 1985.
72. **Ficková, M. and Macho, L.,** Insulin receptors in isolated adipocytes from rats with different neonatal nutrition, *Endocrinol. Exp.,* 15, 259, 1981.
73. **Brtko, J. and Knopp, J.,** Nuclear receptors for thyroxine in rat liver during postantal development in relation to DNA-dependent RNA polymerase activity, *Biol. Neonat.,* 43, 245, 1983.
74. **Taylor, R.,** Insulin receptor assays-clinical applications and limitations, *Diabete. Med.,* 1, 181, 1984.
75. **Zorad, Š., Ficková, M., Klimeš, I., and Macho, L.,** Comparison of goodness of fit of two mathematical models for the estimation of insulin binding, *Diab. Croat,* 15, 183, 1986.
76. **Ruffolo, R. R.,** Use of isolated, physiologically responding tissue to investigate neurotransmitter receptors, *Neurol. Sci.,* 10, 53, 1984.
77. **Greengard, P.,** Intracellular signals in the brain, *Harwey Lect.,* 75, 277, 1981.
78. **Krebs, E. G.,** The phosphorylation of proteins: a major mechanism for biological regulation, *Biochem. Soc. Trans.,* 13, 813, 1985.
79. **Cheng, K. and Larner, J.,** Intracellular mediators of insulin action, *Annu. Rev. Physiol.,* 47, 405, 1985.
80. **Seo, H.,** Growth hormone and prolactin: chemistry, gene organization, biosyntesis, and regulation of gene expression, in *The Pituitary Gland,* Imura, H., Ed., Raven Press, New York, 1985, 57.
81. **Aston, R. and Ivanyi, J.,** Monoclonal antibodies to growth hormone and prolactin, *Pharmacol. Ther.,* 27, 403, 1985.
82. **Linton, E. A., Tilders, F. J. H., Hodgkinson, S., Berkenbosch, F., Vermes, I., and Lowry, P. J.,** Stress-induced secretion of adrenocorticotropin in rats is inhibited by administration of ovine corticotropin-releasing factor and vasopressin, *Endocrinology,* 116, 966, 1985.
83. **Štrbák, V., Jurčovičová, J., and Vigaš, M.,** Maturation of the inhibitory response of growth hormone secretion to ether stress in postnatal rat, *Neuroendocrinology,* 40, 377, 1985.
84. **Holder, A. T., Aston, R., Preece, M. A., and Ivanyi, J.,** Monoclonal antibody-mediated enhancement of growth hormone activity in vivo, *J. Endocrinol.,* 107, R9, 1985.
85. **Besedovsky, H. G., del Rey, A., and Sorkin, E.,** Interaction of activated immune cell products in immune-endocrine feed-back circuits, in *Leucocytes and Host Defense,* Oppenheim, F., Ed., Alan R. Liss, New York, 1986.
86. **Besedovsky, H. O., del Rey, A., Sorkin, E., Lotz, W., and Schwulera, U.,** Lymphoid cells produce an immunoregulatory glucocorticoid increasing factor (GIF) acting through the pituitary gland, *Clin. Exp. Immunol.,* 59, 622, 1985.
87. **Jurčovičová, J., Vigaš, M., Klír, P., and Ježová, D.,** Response of prolactin, growth hormone and corticosterone secretion to morphine administration or stress exposure in Wistar-AVN and Long Evans rats, *Endocrinol. Exp.,* 18, 209, 1984.
88. **Herberg, L. and Coleman, D. L.,** Laboratory animals exhibiting obesity mad diabetic syndromes, *Metabolism,* 26, 59, 1977.
89. **Charlton, H. M.,** Mouse mutants as models in endocrine research, *J. Exp. Physiol.,* 69, 655, 1984.
90. **Maurer, R. A.,** Transcriptional regulation of the prolactin gene by ergocryptine and cyclic AMP, *Nature,* 294, 94, 1981.
91. **Barinaga, M., Yamonoto, G., Rivier, C., Vale, W., Evans, R., and Rosenfeld, M. G.,** Transcriptional regulation of growth hormone gene expression by growth hormone-releasing factor, *Nature,* 306, 84, 1983.
92. **Douglass, J. O., Civelli, O., Birnberg, N., Comb, M., Uhler, M., Lissitzky, J. C., and Herbert, E.,** Regulation of expression of opioid peptide genes, *Ann. Neurol.,* 16, S22, 1984.
93. **Chan, L.,** Hormonal control of gene expression, in *The Liver: Biology and Pathology,* Arias, I., Popper, H., Schachter, D., and Shafritz, D. A., Eds., Raven Press, New York, 1982.
94. **Numa, S.,** Structures, expression and products of neuropeptide and receptor genes, in *Endocrinology, Proc. 7th Int. Endocrinology Congr., Quebec City, 1984,* Labrie F., Proulx, L., Eds., Excerpta Medica, Amsterdam, 1984, 49.
95. **Richter, D. and Schmale, H.,** Vasopressin — expression in normal and diabetes insipidus (Brattleboro) rats, *Trends Neurosci.,* 7, 317, 1984.

Chapter 17

DEPOSITORY PROCESSES

Z. Deyl and M. Adam

TABLE OF CONTENTS

I. INTRODUCTION

Depository and accumulatory processes in tissues have a definite relation to a number of pathological situations and aging. They are based on chemical interactions occurring between metabolites or exogenous compounds and protein constituents. In short-lived, rapidly metabolized proteins, such processes are difficult to see as long as the products of the reaction are rapidly excluded from the body. On the other hand, in long-lived, slowly metabolized proteins, they are fairly common. Posttranslational or genetic alterations of enzymes may result in the inability to metabolize compounds normally occurring in the body, which consequently leads to their accumulation (storage diseases).

The very depository and accumulatory processes may be categorized according to what is deposited. Basically both low- and high-molecular entities may be involved; proteins and lipids and their oxidation products represent typical examples. Regarding the chemistry of depository processes, better understanding is available of the reactions involving low-molecular components. From the physiological point of view, it has to be emphasized that a primary reaction of a low-molecular compound with, for example, a protein or a lipid may initiate changes in interactions with cells or cell metabolism leading to serious consequences for the living individual.[1-3]

Such processes can be explained, for example, by metabolic disturbances in diabetes. Here, excessive attachment of glucose to extravascular matrix proteins with a slow metabolic rate could be the biochemical link between persistent hyperglycemia, excessive high binding of plasma constituents, and immunologically mediated tissue damage initiated by the *in situ* formation of immune complexes.[4] The initial, rapidly formed ketoamine products of nonenzymatic protein glycosylatron, which have been well characterized in a number of diabetic tissue proteins, are in the course of time subjected to a number of rearrangements (degradation, dehydration, Amadori rearrangement), leading to brown pigments possessing reactive carbonyl groups.[5,6] Because of these carbonyls, late glycosylation products have the ability of reacting with additional amino groups of other proteins to form intermolecular crosslinks, immobilize such proteins, and form depots. Trapping of plasma constituents by such reactive products of nonenzymatic glycosylation thus leads to the extravascular diabetic protein accumulation. In the next step, some of these highly bound proteins can function as planted antibodies or antigens in the *in situ* formation of immune complexes.[1]

II. POSTTRANSLATIONAL REACTIONS IN PROTEINS

A. Introduction

Clearly, the first step in such a chain of events is the chemical modification of the protein polypeptide chain, i.e., its nonenzymatic posttranslational modification. This, until recently a rather neglected area, was excellently reviewed by Harding.[1] The first questions to ask are what are the categories of reactions involved, what types of products are formed, and how can their products be identified. It is beyond the scope of this review to go into details, but the selection of available literature in this area is quite large.

B. Nonenzymatic Glycosylation

Glycosylation of proteins by a variety of sugars, especially glucose, has been firmly proven as a common consequence of the simultaneous occurrence of glucose and free protein amino groups in tissues. The initial reaction consists of a nucleophilic attack by the amino groups on the open-chain form of glucose (or another reducing sugar) acting as an aldehyde. The resulting product is an aldimine (a Schiff base) which is rearranged to an Amadori product. In Figure 1, the course of reactions involved in glycosylation of the N-terminal valine residue in hemoglobin is depicted.[7,8]

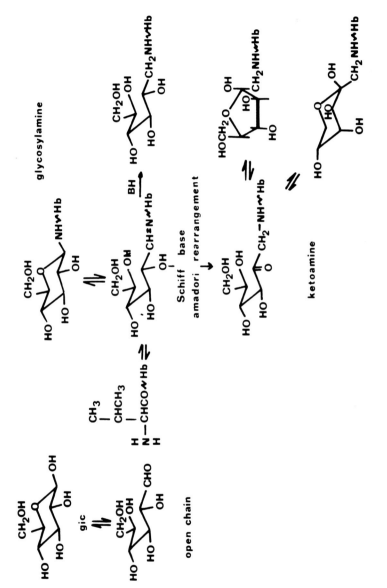

FIGURE 1.   Glycosylation scheme of N-terminal valine in hemoglobin

Table 1.
SURVEY OF NONENZYMATIC
GLYCOSYLATION OF PROTEINS

Protein investigated	Method of estimation
Hemoglobin	Electrophoresis
	Ion-exchange chromatography
Collagen	
Aorta	Colorimetric
Human tendon	HPLC of furosine
Tail tendon	Tritiated hexosyllysine
Glomerular basement	Tritiated hexosyllysine
membrane	Colorimetric
Placental	Tritiated hexosyllysine
Arteries (human)	HPLC of furosine
Serum albumin	Colorimetric
Low-density	HPLC of furosine,
lipoprotein	tritiated hexosyllysine
Erythrocyte membrane	Tritium incorporated
protein	from NaB^3H_4,
	HPLC of furosine
Renal glomerulus protein	Colorimetric, HPLC of furosine
Myelin	Glycosylated amino acids on affinity
	chromatography
Peripheral nerve	HPLC of furosine
Lens crystallins	Tritiated hexosyllysine, colorimetric
Lens capsule	Colorimetric

Usually this reaction does not go further than the product of Amadori rearrangement. However, in tissues constituted of proteins with a low metabolic turn-over, the Amadori product may move on through an entire panoply of browning products, which as a matter of fact, are close if not identical with those occurring during food storage and processing.[9]

It has to be stressed that nonenzymatic glycosylation (at least as far as the Amadori product is concerned) is a general reaction in animals and occurs, though to a different extent, with all proteins and with all reducing sugars (Table 1). Hemoglobin, collagen, erythrocyte membrane proteins, serum proteins, lens proteins, renal glomerular proteins, myelin, and ribonuclease A were the proteins studied from this point of view; diabetes and aging were the typical situations in which the nonenzymatic glycosylation was shown to occur at high intensity.[10-12] The biological consequences are believed to result primarily from altered protein conformation. Three ingredients have to be found in the tissues with a high rate of glycosylation, particularly long-lived proteins, elevated level of glucose, and time. This, in diabetes, leads to the well known sequelae, i.e., cardiovascular disease, microangiopathy, retinopathy, peripheral neuropathy, and cataract.[10,11]

C. Reactions of Carbonyls: Aldehydes and Ketones

In the preceding paragraphs we have shown the potential of glycosylation reactions into which reducing sugars (glucose) enter with a free aldehydic group. Evidence is now accumulating that a variety of other aldehydes and ketones may react with a protein amino group under Schiff-base formation in vivo.[13,14] This binding may have deleterious effects as described above, but some of these reactions are essential for proper physiological functions of the proteins involved. Thus, for example, binding and the release of retinal from opsin is essential for vision; formation of Schiff base-based cross-links in extracellular matrix proteins like elastin and collagen is necessary to ensure the physical properties, i.e., strength and elasticity of tissues.[15] Of the deleterious reactions, one can mention the five adducts arising during the reaction of acetaldehyde with hemoglobin (in alcoholics). In other proteins, reaction with acetaldehyde may lead to nonsulfidic covalent cross-links, as in the case of lens crystallins. Such reactions are naturally intensified in those individuals who continue to drink alcohol while medicated with aldehyde dehydrogenase inhibitors.[14,16]

Pyridoxal and pyridoxal phosphate represent another category of aldehydes involved in those reactions. Pyridoxal, unlike pyridoxal phosphate, exists in solution at neutral pH predominantly in the hemiacetal form. Therefore, only a small proportion is available for the reaction as an aldehyde. Nevertheless, the identification of a minor fraction of human hemoglobins as a pyridoxylated protein (one pyridoxal residue per hemoglobin tetramer) demonstrated that such reactions indeed occur in vivo. More recent studies indicate the possibility of nonspecific binding of pyridoxal phosphate with different proteins and different tissues with unknown involvement in neoplastic and developmental processes. Covalent binding seems to account for the trapping of pyridoxal within the cells.[17,18]

Glucocorticoids, i.e., steroids with a keto group adjacent to a hydroxyl group, are also involved in covalent binding and posttranslational protein modifications. Thus, for example, cataract is a typical side effect of the administration of glucocorticoids, especially prednisolone, to patients with rheumatoid diseases or asthma;[16] increased 16α-hydroxylation of estradiol was shown to occur in lupus erythematosus.[19] Erythrocytes and plasma membrane proteins are also capable of binding these steriods. All these events clearly involve reaction with proteins, and the respective adducts were identified. Studies about the binding of 16α-hydroxyestrone and cortisone to serum albumin have shown that the steroids are bound in the first step via a Schiff base; the sequence reactions is shown in Figure 2.[20]

Of the other carbonyl compounds, one has to keep in mind reactions with formaldehyde (see reactions with lens proteins in vivo[14]. This aldehyde yields probably two bound hydroxymethyl groups in the first step of the reaction (it does not produce a Schiff base). In the organism it can be produced as a result of N-demethylation reactions, or may result from the metabolism of certain antitumor agents.

A number of aldehydes and ketones are produced during lipid peroxidation; they react primarily with lysine and tyrosine residues in proteins. Malonaldehyde, the main product of lipid peroxidation, is a potential cross-linking agent, and its reactions may mimic protein alterations seen in aging.[21,22]

D. Reaction of Quinones

Another category of reactive compounds occuring in the body that are involved in posttranslational reactions with proteins are quinones. Thus, for example, hardening of the insect cuticle or the formation of melanin are essentially protein-quinone reactions.[23,24] In sclerotization of the insect cuticle, diphenols are oxidized enzymatically to 1,2-quinones which react with free amino terminals and free lysine amino groups in proteins. Another metabolic pathway used for the same purpose is the conversion of N-acetyldopamine to acetylarteronone which reacts as a ketone with cuticle proteins. Finally, tyrosine-rich proteins are oxidized at their phenolic side chains to epoxides or quinones which cross-link the proteins in the next step.[25]

Melanin is produced by the reaction of proteins with a variety of quinones derived from dopaquinone (also indole-5,6-quinone). The reaction starts, at the thiol groups of cysteinyl residues; however, further covalent modifications take place, notably with free amino groups. Cross-linking and metal binding increases the complexity of the arising products.[24]

Within the category of quinones, one can find a number of xenobiotics, some of which are quite common, and through their ability to react with proteins may be accumulated in some tissues. Thus, for example, naphtoquinone, arising from naphthalene metabolism, reacts with lens protein thiol and amino groups in vivo (and causes cataract).[26] Acetaminophen

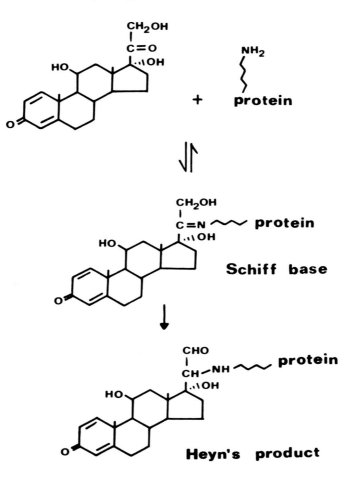

FIGURE 2. Reaction of prednisolone with the ε-amino group of protein-bound lysine. (From Manabe, S., Bucala, R., and Cerami, A., *J. Clin. Invest.*, 74, 1803, 1984. With permission.)

(paracetamol, a widely used analgesic-antipyretic drug) is converted by the action of cytochrome P-450 into a reactive quinone-like product that depletes glutathione, binds covalently to proteins, and causes necroses in the liver and kidney.[27] Butylated hydroxytoluene, a common antioxidant in processed foods causes hemorrhagic death and severe lung injuries in mice;[28] both these effects are believed to arise from the conversion of the parent compound through cytochrome P-450-linked monooxygenase to an alcohol which dehydrates to quinone methide. This product binds to cysteinyl residues in microsomal proteins. The list is far from complete (see Table 2). 17α-Ethinylestradiol, the active substance of a number of contraceptive pills, is converted in the liver to a reactive metabolite that binds covalently to proteins.[29] Binding as a quinone distinguishes this steroid from corticosteroids that bind as ketones. (Readers looking for more detailed information in this respect are referred to Table 2 and the review by Harding.[1])

E. Miscellaneous Xenobiotics and Posttranslational Modifications of Proteins

In order to give the reader an image about what categories of compounds and reactions may be involved in posttranslational protein modifications, an assortment of xenobiotics that, mostly through microsomal oxidation, are converted to reactive intermediates that bind with proteins is presented in Table 3.

Table 2
SURVEY OF ALDEHYDES AND KETONES CAPABLE OF PROTEIN MODIFICATIONS IN VIVO

Compound	Protein involved	Type of the product	Note
Acetaldehyde	Hemoglobin lens proteins	Acetaldehyde-val, -lys, and -tyr; Schiff-base type adducts; reaction with tyrosine is also possible	Five adducts with hemoglobin; formation of cross-links in lens proteins
Cross-links of collagen and elastin	Collagen (different types), elastin	Lysinonorleucine dehydrolysinonorleucine (also glycosylated forms), pyridinoline, desmosines; multiple-step biosynthesis involving Amadori rearrangement, formation blocked by β-aminopropionitrile, penicillamine, homocysteine	Arise from Schiff-base adducts in a series of reactions; in the first step lysine is oxidized to α-aminoadipic acid α-semialdehyde that forms a Schiff base with unmodified lysinal; also aldol condensation of α-semialdehydes occurs in vivo
Formaldehyde	Lens proteins	Formaldehyde reacts after deionization yielding two hydroxymethyl groups	Reacts also with protein thiols; probably cross-links proteins to DNA
Glucocorticoids	Serum albumin, erythrocyte membrane proteins, lens proteins	Schiff-base-type adducts; Heyn's rearrangement is expected with prednisolone	
2,5-Hexanedione, methyl-2,5-hexanedione, 3,4-dimethyl-2,5-hexanedione	Proteins of neurofilaments, spectrin	Pyrrole adducts (Schiff-base-type adducts formed first) that autooxidize to pigmented components	Causes protein cross-linking (by pigmented components)
Lipid peroxidation products (malonaldehyde, 4,5-dihydroxy-2,3-decanal, 4-hydroxynonenal	Thiol enzymes (inhibition of glucose-6-phosphate dehydrogenase, aminopyrine demethylase), hemoglobin	Irreversible modification of lysine and tyrosine; unsaturated aldehydes form thio ethers	Mimics aging processes
Pyridoxal	Hemoglobin	Schiff-base-type adducts	Binding to amino-termini of α-chains; formation of salt bridges to arginine; hydrogen bonding with serine and threonine
Pyridoxal phosphate	Hemoglobin, different enzymes	Schiff-base-type adducts	Binds to α-amino groups in β-chains of hemoglobin; formation of salt bridges and histidine cross-linking of β-chains
Retinal	Opsin	Schiff-base-type adducts	Basically physiological reaction of opsin which at its N-terminus undergoes enzymatically controlled acetylation, phosphorylation, and glycosylation

F. Models of Nonenzymatic Glycosylation of Proteins

Brownlee and co-workers[30] have developed a methodology to test nonenzymatic glycosylation of proteins as well as the further affinity of the glycosylated products to bind either with lysine or other proteins. The method was originally worked out for collagen, however, it can be easily used with other proteins as well. The method has the following steps:

Table 3
SURVEY OF QUINONES AND SOME XENOBIOTICS CAPABLE OF BINDING TO PROTEINS[a]

Compound	Protein involved	Type of the product	Note
Naphthoquinone and 1,2-naphthalene oxide	Leus proteins, glutathione, hepatic proteins	Reacts with thiol andamino groups (forming brown pigments); thioesthers and thiazolidene ring compounds arise	
1,2-Quinones and diphenols	Arthropod cuticle proteins	Amino terminals are involved; reaction products not identified; the pathways similar to elastin and collagen cross-link formation	Physiological sclerotization of arthropod cuticle; alternative pathways based in the oxidation of *N*-acetyldopamine or in the oxidation of tyrosine-rich proteins
Dopaquinone, indole-5,6-quinone	Various proteins	Thiol group (cysteinyl residues), amino groups, metal binding is possible; thiol groups and other nucleophiles	
Acetaminophen (paracetamol)	Glutathione, liver, renal, and blood proteins		
Butylated hyroxytoluene	Liver proteins		
17 α-Ethinyl-estradiol		Binding as quinone distinguishes this compound from corticosteroids (see Table 2)	Converted in liver in a reactive intermediate that binds to proteins
4-Dimethylaminophenol	Hemoglobin	Thiol groups and histidine residues	
Epoxides (ethylene and propylene oxide)	Hemoglobin, other proteins; glutathione	Ethylhistidine, N-3′-(2-hydroxypropyl) histidine	Alkylation of cysteine, histidine, and amino terminal with other epoxides or compounds yielding epoxides in vivo
Azo dyes	Diverse proteins, proteins of liver, nuclear proteins	Reactions occur with tyrosine, methionine, unstable adducts with cysteine; involves cysteine and methionine residues in liver proteins	
2-Acetylaminofluorene	Diverse proteins, proteins of liver nuclear proteins	Methionine residues involved; reactions with tyrosine, tryptophan, and cysteine are probable	Reaction with acidic nuclear proteins is preferred to histones
Acetylhydrazine	Proteins of liver	Acylation of cysteine residues	
N-methyl-*N*′-nitro-*N*-nitrosoguanidine	Basic proteins (histones)	Amino groups	Increase in template activity of reconstituted chromatin
Methylmethane sulfonate	Hemoglobin	Methylation of cysteine and histidine residues (S-methylcysteine, N-3-methylhistidine)	
Nitrosamines	Hemoglobin, proteins of liver	3-Methylhistidine	General metabolic involvement

Table 3 (continued)
SURVEY OF QUINONES AND SOME XENOBIOTICS CAPABLE OF BINDING
TO PROTEINS[a]

Compound	Protein involved	Type of the product	Note
Halogenated hydro-carbons (carbon tet-rachloride)	Liver proteins	Formation of reactive radi-cals in liver, induction of lipid peroxidation products (see Table 2)	Direct binding of halogen-ated radicals is also possi-ble

[a]   This list is by far not exhaustive and indicates only those compounds that are in the focus of experimental work. From the others that are capable of binding with proteins in vivo it is worth mentioning β-propiolactone, oxotremorine, O,S,S,-trimethylphosphothioate, 3-methylfuran, 4-ipomeanol, misonidazole, 2,6-dinitrotoluene, 2,2′-dichlorobiphenyl, tetrachlorobiphenyls, furosemide, phenetidine, niridazole, aflatoxins, and 5,9-di-methyldibenzo [c,g] carbazole (for further information see Reference 1).

1.   Immobilization of collagen, in which soluble collagen is immobilized on Affi gel 10; this sorbent is preferred to the more commonly used CNBr activated agarose. In this step, 17 to 18 mg of protein are bound per milliliter of the sorbent.

2.   Nonenzymatic glycosylation, in which immobilized collagen (about 80 mg) is washed with phosphate buffer (0.5 mol/ℓ, pH 7.5, 50 mℓ) and then resuspended in 5 mℓ of the same buffer which was previously made 500 mmol/ℓ with respect to glucose. The reaction mixture is incubated at 37° for 10 days. Then, the gels are washed with 200 mℓ of phosphate buffered saline and can be stored at 4°C. Controls are prepared in the same way without added glucose. All solutions must contain 3 mmol/ℓ NaN_3 to avoid bacterial contamination.

3.   Proteins to be bound to glycosylated collagen have to be radioactively labeled; usually the protein in question is purified by a suitable chromatographic technique (commer-cially available proteins are frequently contaminated, and if used as such may yield erroneous results) and iodinated with NaI^{125} using the IODOGEN procedure. Specific activities of the labeled proteins should be of the order $0.5 \cdot 10^6$ cpm/nmol.

4.   Binding of labeled proteins or lysine to immobilized glycosylated collagen, in which C^{14}-lysine (about 75 cpm/nmol) or ^{125}I-labeled protein (see above) are mixed with immobilized collagen; around 10 mg of collagen are used per 1 nmol of labeled protein or 20 nmol of ^{14}C lysine. The amount of ligand bound to immobilized collagen is determined by filtering aliquots corresponding to 1 mg of collagen on Durapore hy-drophilic membrane. The filters are washed exhaustively (six times) with glycine buffer (0.2 mol/ℓ, pH 8) and (six times) with the same buffer containing 2 mol/ℓ NaCl. Samples containing ^{14}C are hydrolyzed with 6 M HCl at 110° for 12 hr and neutralized before adding hydrofluor scintillation fluid. Samples containing ^{125}I can be counted directly. Binding of lysine and proteins to agarose alone is asessed under similar conditions.

If it is desirable to test glycosylated protein for the *in situ* formation of immune complexes, the procedure can be roughly as follows: 25 nmol of IgG are mixed with 9 to 10 mg of collagen and incubated as described above. The reaction mixture after 10 days of incubation is filtered and washed, and 500 μg of ^{125}I-IgG is added to the filter and incubated at room temperature for 10 min. The procedure is repeated four times. The amount of radioactive antigen or antibody fixed to the corresponding planted antibody or antigen bound to im-mobilized collagen can be determined by counting material retained on the filter.

Instead of labeling the protein counterparts (or lysine), there are situations where it is necessary to estimate the amount of bound glucose, for example, when the level of monotopically bound glucose is to be assayed in hemoglobin. Here the Higgins' procedure is applicable[31] Red cell hemolysate is prepared and purified hemoglobin is obtained by liquid chromatography. Column fractions are pooled, pressure concentrated, and dialyzed against Krebs-Ringer phosphate buffer pH 7.3. Uniformly labeled D-^{14}C glucose is prepurified (either by preincubation with the hemolysate and subsequent column chromatography on Dowex 50WX2 or by thin layer chromatography) and incubated with hemoglobin fractions in sterile Krebs-Ringer phosphate buffer (pH 7.3, 37°). Unbound glucose is removed by rapid gel chromatography through Sephadex G25, and the amount of bound glucose is determined by measuring the hemoglobin concentration and radioactivity.

III. LIPOPIGMENTS

A. Introduction

Progressive accumulation of autofluorescent lipopigments in the cytoplasm of long-lived cells is the oldest and perhaps the most universal manifestation of cell aging. Lipopigments are mainly corpuscular substances characterized by a variable degree of natural color, autofluorescence, apolarity, and higher molecular mass, and are extractable with organic solvents only to a limited extent. They are considerably heterogenous in their ultrastructure; their intracellular localization relates to the lysosomal system.[32] It is assumed that lipopigments take their origin from unsaturated lipids through peroxidation and copolymerization mainly with proteins. It is the preoxidation step in which the biochemistry of lipopigments is linked to the existence of free radicals. An idea how such compounds can originate in the body emerged from the work of Chio and Tappel.[33] Malondialdehyde, a product of lipid peroxidation, is assumed to react with an amino group containing compounds to form conjugated Schiff bases (aldimines). These compounds are highly fluorescent with spectral characteristics virtually identical to the chloroform-methanol extracts of purified lipofuscin granules. Many experiments show that autooxidation of polyunsaturated fatty acids or tissue extracts in the presence of amino-group-containing compounds ranging from proteins to phosphatidyl ethanolamine produce substances that exhibit conjugated Schiff base-like fluorescence. Schiff bases formed between malondialdehyde and amino compounds can be quantitatively determined by the thiobarbituric acid test because the acidic conditions of the test promote hydrolysis of the Schiff-base linkages, thus making the malondialdehyde available for thiobarbituric acid.[34-36]

The name lipofuscin, was introduced at the beginning of this century and was derived from the Greek ''lipo'' meaning fat and ''fuscus'' meaning dark. Though most of the findings regarding lipofuscin were obtained with nerve tissue, there are a number of other tissues that show the accumulation of lipofuscin (Table 4). Early views as to the origin of these lipopigments held that the pigments are formed in relation to the activity of nuclear material in the cell. Certainly, the pigment accumulates in proximity to the nucleus. A large number of investigators have supported the hypothesis of a mitochondrial origin of the pigment.[32] At least it would be feasible to speculate in this way, keeping in mind the involvement of free radicals in the lipid peroxidation step. On the other hand, there is another large group of investigators who have postulated that lipfuscin arises from lysosomes.[37] Here the rationale is seen in the possible release of hydrolytic enzymes from damaged lysosomes that may be detrimental to a wide range of metabolic processes in the cell resulting in incomplete lysis or degradation of certain products of metabolism related to energy production and synthesis of structural elements in the cytoplasm. There are also other concepts regarding lipofuscin origin; particularly, it was suggested to originate in neurofibrils or endoplasmic reticulum and play a role as a neurosecretory carrier.[38-40] Beyond lipofuscin, there are in the body also related (more complex) pigment bodies. Neuromelanin may serve here as a typical example; in contrast to the lipofuscin pigments, neuromelanin was shown to decline on aging in the

64. **Robinson, W. G., Kuwabara, T., and Bieri, J. G.,** Vitamin E deficiency and the retina: photoreceptor and pigment epithelial changes, *Invest. Ophthal. Vis. Sci.*, 17, 683, 1979.

65. **Manocha, S. L. and Sharma, S. P.,** Reversibility of lipofuscin accumulation caused by protein malnutrition in motor cortex of squirrel monkeys Saimiri sciureus, *Acta Histochem.*, 58, 219, 1977.

66. **Sharma, S. P. and Manocha, S. L.,** Lipofuscin formation in developing nervous system of squirrel monkeys consequent to maternal dietary protein deficiency during gestation, *Mech. Age. Dev.*, 6, 1, 1977.

67. **Hervonen, A., Koistinaho, J., Alho, M., Helen, P., Santer, R. M., and Rapoport, S. I.,** Age related heterogeneity of lipopigments in human sympathetic ganglia, *Mech. Age. Dev.*, 35, 17, 1986.

68. **Katz, M. L., Drea, C. M., and Robison, W. G.,** Relationship between dietary retinol and lipofuscin in the retinal pigment epithelium, *Mech. Age. Dev.*, 35, 291, 1986.

69. **Winstanley, E. K. and Pentreath, V. W.,** Lipofuscin accumulation and its prevention by vitamin E in nervous tissues quantitative analysis using snail buccal ganglia as a simple model system, *Mech. Age. Dev.*, 29, 299, 1985.

70. **Ikeda, Il., Tauchi, H., and Sato, T.,** Fine structural analysis of lipofuscin in various tissues of rats of different ages, *Mech. Age. Dev.*, 33, 77, 1985.

71. **Desnicle, R. J., Thorpe, S. R., and Fiddler, M. B.,** Toward enzyme therapy for lysosomal storage diseases, *Physiol. Rev.*, 56, 57, 1976.

72. **Farrell, D. F., Baker, H. J., Herndon, R. M., Lindsey, J. R., and McKhann, G. M.,** Feline G_{M1} gangliosidosis: histochemical and ultrastructural comparisons with the disease of man, *J. Neuropathol. Exp. Neurol.*, 32, 1, 1973.

73. **Suzuki, K., Suzuki, Y., and Fletcher, T. F.,** Further studies on galactocerebroside β-galactosidase in glotoid cell leukodystrophy, in *Sphingolipid, Sphingolipidoses and Allied Disorders*, Volle, B. W. and Aronson, S. M., Eds., Plenum Press, New York, 1972, 487.

74. **Jolly, R. D.,** Animal model of human disease: mannosidosis, *Am. J. Pathol.*, 74, 211, 1974.

75. **Hickman, S. and Neufeld, E. F.,** A hypothesis for I-cell disease: defective hydrolases that do not enter lysosomes, *Biochem. Biophys. Res. Commun.*, 49, 992, 1972.

76. **Lebovitz, B. E. and Siegel, B. V.,** Aspects of free radical reactions in biological systems: aging, *J. Gerontol.*, 35, 45, 1980.

77. **Rothstein, M.,** *Biochemical Approaches to Aging*, Academic Press, New York, 1982.

78. **Deamer, D. W. and Gonzales, J.,** Autofluorescent structures in cultured W138 cells, *Arch. Biochem. Biophys.*, 165, 421, 1974.

79. **Buchanan, J. H. and Sidhu, J.,** Autofluorescence and aging: changes in ribosome accuracy and lysosome function, *Mech. Age. Dev.*, 36, 259, 1986.

80. **Poot, M., Verkerk, A., and Jongkind, J. F.,** Accumulation of a high molecular weight glycoprotein during in vitro ageing and contact inhibition of growth, *Mech. Age. Dev.*, 34, 219, 1986.

81. **Bladen, H. A., Nylen, M. U., and Glenner, G. G.,** The ultrastructure of human amyloid as revealed by the negative staining technique, *J.Ultrastruct.Res.*, 14, 449, 1966.

82. **Manry, O. P. J.,** Serum amyloid a protein-current status, *Scand. J. Rheumatol.*, 13, 97, 1984.

83. **De Beer, F. C., Balz, M. L., Holford, S., Feirestein, A., and Pepys, M. B.,** Fibronectin and C_4-binding protein are selectively bound aggregated amyloid P component, *J. Exp. Med.*, 154, 1134, 1981.

84. **Ohkubo, I., Sahashi, W., Namikawa, C., Tsukada, K., Takeuchi, T., and Sasaki, M.,** A procedure for large scale purification of human plasma amyloid P component, *Clin. Chim. Acta.*, 157, 95, 1986.

85. **Zuckerman, S. H., Steven, H., and Supernant, Y. M.,** Simplified microelisa for the quantitation of murine amyloid A protein, *J. Immunol. Meth.*, 92, 37, 1986.

86. **Griswold, D. E., Hillegass, L., Antell, L., Shatzman, A., and Hanna, N.,** Quantitative western blot assay for measurement of the murine acute phase reactant, serum amyloid P component, *J. Immunol. Meth.*, 91, 163, 1986.

87. **Maury, C. P. J. and Teppo, A. M.,** Radioimmunoassay for urinary amyloid component, *J. Lab. Clin. Med.*, 106, 619, 1986.

88. **Kimura, K., Ogawa, K., and Sato, K.,** Changes in natural killer activities in experimental secondary amyloidosis, *Scand. J. Immunol.*, 19, 513, 1984.

89. **Adam, M., Deyl, Z., and Rosmus, J.,** *Interactions of Collagen with Metals In Vivo*, Academia, Prague, 1970.

90. **Adam, M., Bartl, P., Deyl, Z., and Rosmus, J.,** Uptake of gold by collagen in gold therapy, *Ann. Rheum. Dis.*, 24, 578, 1965.

91. **Adam, M., Fietzek, P., Deyl, Z., Rosmus, J., and Kühn, K.,** Investigations on the reaction of metals with collagen in vivo. III. The effect of bismuth copper, a mercury component, *Eur. J. Biochem.*, 3, 415, 1968.

Chapter 18

BODY FLUIDS DISTRIBUTION AND TURNOVER

J. Kuneš

TABLE OF CONTENTS

I. INTRODUCTION

Water makes up about 45 to 70% of the total body mass in mammals and constitutes the medium in which all biochemical reactions take place.[1] The distribution and regulation of body water and consequently the distribution of substances for which it acts as a solvent are of such importance to the comprehension of the function of the organism as to warrant a separate field of study.

The total volume of body water is divided into two major compartments, i.e., intra- and extracellular water. Aside from these two, there is a much smaller third compartment, referred to as transcellular water (Figure 1). This last compartment represents only about 2.5% of total body water. The fluid of this compartment comprises the rest of the extracellular fluids which are not just simple transudate and which have the common property of being secreted by various epithelial membranes in the body.[2] The largest collection of transcellular liquids is located within the tracheobronchial tree, the excretory system of the kidneys and glands, cerebrospinal fluid, and aqueous humor of the eyes.

The intracellular compartment represents the largest portion of body water. It accounts for more than one half of the total body water and makes about 30 to 40% of the total body mass. This heterogeneous compartment is made up of the sum of the fluid of all body cells. Because different types of cells vary greatly in their water content as well as in their chemical composition, it is difficult to provide a simplified representation of this compartment. The volume of intracellular water cannot be measured directly and is usually determined by subtracting the value for extracellular water from that for the total body water.

The extracellular compartment represents all fluids that exist outside of cells, with the exception of some special fluids (i.e., the transcellular fluid and fluid present in connective tissues). This compartment is further subdivided into two components: the plasma component and the interstitial-lymph component. Plasma accounts for 25% of the extracellular fluid (ECF). The interstitial fluid makes up the remaining 75% of ECF. Plasma circulates rapidly in the vascular bed, whereas interstitial fluid seeps more slowly through tissue interstices. Plasma proteins which leak through the capillary wall together with an excess interstitial fluid are returned to the plasma through lymphatic vessels. Lymph flow is sluggish, and the volume of the lymphatic compartment is only a rough estimate.

The rest of the water, which is not located in either the intra- and extracellular compartments, is in the dense connective tissue (cartilages and bones). The fluid present in these tissues is difficult to measure because of their complex physicochemical nature and slow equilibration processes taking place within these specialized tissue. The interpretation of these "inaccessible" areas has been the subject of much controversy and speculation. Edelman and Leibman[3] estimated that these areas make up 75% of the ECF of dense connective tissue and cartilage and 90% of bone.

II. EXCHANGE OF WATER BETWEEN THE ORGANISM AND ITS EXTERNAL ENVIRONMENT

The body liquids, their volume, and total electrolyte concentrations are maintained and regulated to a considerable degree in various exchanges with the external environment. The body water, in general, has three sources: (1) oral liquids, (2) water content of food, and (3) metabolic water. In the lack of food intake, water can be derived first of all by oxidation of body fat and protein.

The loss of body water is normally materialized through four channels: expired air, sweat, gastrointestinal discharges, and urine. Losses of water by vaporization from lungs, skin, and in stool are obligatory, and these losses are maintained without regard to the intake. The balance between output and input of water is maintained mainly by daily changes in

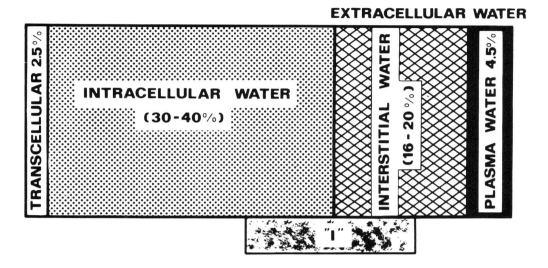

FIGURE 1. Body fluid compartments. Volumes are given as percentages of body weight. "I" indicates water of "inaccessible" areas as dense connective tissue, cartilages, and bones.

the urine volume. There are some limitations of this mechanism under conditions of diminished or absent water and salt intake. Urea excretion in mammals requires a certain minimum of urine; the rate at which renal water loss occurs is dependent on the rate of protein catabolism.

On the other hand, thirst regulates the intake of water very effectively. The sensation of thirst is probably a consequence of cellular dehydration detected by osmoreceptors. These receptors lie in the paraventricular and supraoptic nuclei and are likely to be identical with the neurons responsible for the secretion of antidiuretic hormone.[4] While the loss of water from the body is a continuous process, its replacement by drinking takes place only intermittently and matches the water deficit with differing degrees of accuracy in different animals.

III. EXCHANGE OF WATER BETWEEN BODY FLUID COMPARTMENTS

Body fluids provide a typical example of a steady-state system with a tight relationship existing between principal body fluid compartments.

A. Intracellular and Extracellular Fluid Volumes

Exchange of water and electrolytes between intra- and extracellular fluids involves both the solvent and solutes and is accomplished through the active and passive transport. There are some main principles for water and electrolyte exchange between both of these fluids which can be summarized as follows.

1. The active or passive transport of ions produces an osmotic gradient which causes net movement of water.
2. Changes in the amount of organic metabolites in cells tend to produce permanent or transient shifts. These exchanges are related to metabolism and can simulate active

transport of water. However, it represents a strictly passive osmotic transfer which is actually secondary to changes in the activity of water produced by solutes.

3.   Alterations in extracellular osmolality also provoke a passive osmotic water transfer.

Water distributes rapidly across cell membranes to establish essential equality of the osmotic concentrations of intra- and extracellular fluids. Relative intra- and extracellular volumes depend on the quantities of ions contained within these compartments. In extracellular fluids, 90 to 95% osmotically active components are represented by sodium as the major cation, and chloride and bicarbonate as the major anions. On the other hand, in intracellular fluids, potassium and magnesium are the major cations and organic phosphate compounds and proteins are the major anions.

B. Plasma and Interstitial Fluid Volumes

The distribution of fluid between plasma and interstitial compartments is determined by the balance of the hydrostatic pressure, the colloid osmotic pressure, and the tissue-turgor pressure operative across the capillary endothelium. The relative duration of the dilator vs. the constrictor phases of vasomotion, the permeability of the capillaries, the distensibility of the tissues, and the adequacy of lymphatic drainage all play supplementary roles.

In general, the interstitial fluid is determined by the capillary filtration and reabsorption. Of great importance is the permeability of capillaries for proteins and the rate at which proteins are returned to plasma through the lymph. The regulation of lymph flow is still unclear, but it is evident that the increased lymph flow has two important effects. The part of the excess capillary filtrate is returned to plasma through increased lymph flow, and thus the expansion of interstitial fluid volume will be less (edema preventing mechanism[5]). The second effect of increased lymph flow is to reduce an interstitial protein mass by transferring a part of proteins to plasma.

The important role of protein content of the interstitial fluid in the determination of the interstitial fluid volume as well as of the effective colloid osmotic pressure across the capillary wall is well recognized. The mechanisms which regulate interstitial fluid transfer through the capillary wall have been the objects of intensive study.[6,7]

A very important parameter in transcapillary fluid balance and interstitial compliance is the interstitial fluid pressure or "fluid equilibration pressure".[5] Interstitial fluid pressure is a somewhat suspect term, because under normal conditions, essentially all of the interstitial fluid exists as fluid imbibed in a gel-like ground substance.[5] However, the interstitial fluid pressure is measured, namely by recording the pressure in a small volume of saline brought into contact with the interstitium through a catheter, a wick, a perforated capsule, or a micropipette (for comparison of these methods see Reference 8). This pressure will be equal to that of any free fluid in the interstitium and will also be in equilibrium with a gel phase. Most studies[5,9,10] have found a subatmospheric interstitial fluid pressure in most normal tissues, but its absolute magnitude has varied from -1 to -7 mmHg, though it rose to positive values in edema.[5] These discrepancies may represent true differences between species and sites of measurement and/or methodological differences.

IV. MEASUREMENT OF BODY FLUIDS

A. Introduction

The principles of measurement of all body fluid volumes are the same. They can be easily determined by the indicator-dilution method. This method depends upon the relationship between the quantity of a given test substance and its concentration in a distribution space. The formula for calculating distribution volume is

$$V = \frac{I}{C}$$

where I is the amount of injected substance and C is the concentration in plasma. In other words, if one wants to determine a particular fraction of body fluid volume, the easiest way is to inject a known quantity (I) of appropriate substance and after a sufficient equilibrium time a sample of plasma is taken for the determination of concentration (C) of the injected substance. From these two quantities, the respective volume (V) can be calculated.

It is important to emphasize that the indicator should possess the following properties:

1. The distribution of the indicator must be limited to the compartment being measured and must be uniformly distributed throughout that compartment.
2. The indicator must not be either metabolized nor synthesized in the body and it must not be injurious.
3. If the indicator is excreted from the organism, the amount excreted must be easily detected.

For the indicator that is excreted, a correction of the formula for calculating its volume distribution must be made, i.e.,

$$V = \frac{I - E}{C}$$

where E is the quantity of the indicator excreted up to the time when the plasma sample was taken.

A wide variety of indicators have been used for the measurement of body fluid volumes. These indicators can be commonly divided into two major groups: radioisotopes and chemical substances. For any indicator, it is best that the concentration in the samples and in the standards are about the same and that both lie in the optimum range for determination. For chemical substances, the optimum optical density for spectrophotometric determination is usually from 0.2 to 0.7. For radioactivity counting, high activity increases statistical accuracy. Alternative techniques and indicators for the determination of particular body fluid volumes will be specified in details in the respective sections.

The test substance is usually administered either by means of the infusion or as a single injection (Figure 2). If the continuous infusion is used (Figure 2A), a constant concentration in plasma is reached after a time period which varies with the test substances used. After the achievement of a steady value, the infusion is stopped and the decrement with time of plasma concentration is observed. When the single injection method is used (Figure 2B), a number of plasma samples are taken and the falling concentration of a substance is plotted against time in a semilogarithmic scale. The linear part of the curve is extrapolated to zero time (dotted line) to give the concentration (C_o) which would be obtained if instantaneous mixing and no excretion of the indicator had occurred. The distribution volume is then calculated as the injected amount of the indicator divided by C_o.

B. Total Body Water

Though the first estimates of body water content were made by desiccation of cadavers, indirect methods for the measurement of total body water also exist. The dilution of urea, antipyrin, and N-acetyl-4-aminoantipyrin can be used for the measurement of total body water. Antipyrin is particularly useful. This drug is distributed uniformly throughout all the body water. It diffuses rapidly across cell membranes, and is not specifically bound to any cellular or extracellular component, but it is slowly excreted and also slowly metabolized.

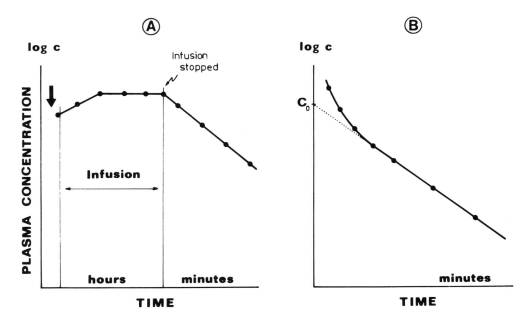

FIGURE 2.   (A) Constant-infusion technique for the determination of the indicator distribution space. Distribution volume is calculated from the disappearance rate of this indicator from plasma. Arrow represents a priming dose of the indicator. (B) Single-injection technique for the estimation of the distribution space. C_o, indicator concentration obtained after the extrapolation to zero time. There is a semilogarithmic plot of plasma indicator concentration (C) vs. time in both parts of figure.

The two isotopes of water, deuterium oxide (D_2O) and tritium oxide (HTO), can be also used for this purpose. In body fluids, D_2O is quantified by exploiting its effect on the specific gravity of water; the method applied uses the rate of fall of a precisely measured drop through an organic solvent. The sensitivity of this approach is quite high. On the other hand, HTO is quantified by measuring the radioactivity of tritium by liquid scintillation methods. Both isotopes behave in organisms exactly as in water. They are excreted in urine, feces, expired air, and sweat. Deuterium and tritium also replace hydrogen in certain compounds which contain exchangeable hydrogen atoms. Both isotopes are incorporated into newly synthesized body constituents. However, over the interval required for the equilibration of D_2O or HTO with body water, these errors are insignificant. The results obtained by measuring body water by desiccation, and with the D_2O or HTO dilution methods[1] as well as the simultaneous measurements using D_2O and antipyrin[2] were found to be in close agreement.

C. Extracellular Fluid Volume

Before we go to more detailed consideration of the problems of extracellular fluid volume (ECFV) measurement, it is necessary to emphasize the prerequisites for indicators that can be used. These prerequisites may be summarized as follows: (1) ready and uniform distribution throughout the whole anatomical extracellular space (ECS); (2) no penetration into the cells and no effect on the size of the ECS; (3) any easy and accurate estimation at low indicator concentrations (recently the radioactively labeled compounds were used for this purpose); and (4) any tendency for the indicator to be lost, e.g., by urinary elimination following whole body infusion must be considered. In experimental animals the urinary loss may be prevented by a bilateral nephrectomy.

There are many indicators that have been used to measure the volume of the extracellular water. These can be divided into two categories: (1) ions, which resemble sodium and/or

Table 1
LIST OF MAIN INDICATORS WHICH ARE USED FOR EXTRACELLULAR FLUID VOLUME MEASUREMENT

Electrolytes	Ref.	Nonelectrolytes	Ref.
^{22}Na, ^{24}Na	21	^{3}H, ^{14}C-inulin	23
Chloride	12	Sucrose	23 — 25
Bromide	13	Mannitol	24
Thiosulfate	18	^{3}H-raffinose	23
Thiocyanate	17	^{14}C-sorbitol	22
^{35}S-sulfate	14	Polyfructosan S	28
^{51}Cr, ^{60}Co-EDTA	21,22		
Sodium ferrocyanide	19		
Poly-L-glutamate	20		

Note: For complete literature see Reference 11.

chloride in distribution and (2) saccharides, which readily cross the capillary walls but not the cell membranes (Table 1).

The estimation of ECFV by sodium and/or chloride is not precise because both are also present behind the boundaries of ECS. The earliest attempts to measure of ECFV by chloride were based upon the erroneous assumption that tissue chloride is entirely extracellular. The calculation of ECS on this basis results in grossly misleading values not only for whole body or for tissues with low intracellular chloride concentration, but also for tissues with a high intracellular concentration of chloride.[11,12] The bromide space[13] is even by 2 to 7% higher than the chloride space. Sulfate[14] has been widely used because its distribution volume was about 20 to 40% lower than chloride space, but a considerable disadvantage of this substance is its elimination into the gastrointestinal tract,[15] and the accumulation in bones and in red blood cells.[16] Thiosulfate and thiocyanate, used in several earlier studies, have some disadvantages, too. Thiocyanate binds with blood albumin,[17] and accumulates in blood cells and in the gastrointestinal tract. Thiosulfate is also accumulated in blood cells and metabolized in tissues.[18]

Other ECS indicators including sodium ferrocyanide,[19] poly-L-glutamate,[20] and cyanocobalamine[21] have been used (for additional information see Table 1 and Reference 11).

In recent years, ethylendiaminetetraacetic acid (EDTA) labeled either with chromium[21] or cobalt[22] is regarded as a reliable ECS indicator.

Also, the application of saccharides for measurement of ECFV has some disadvantages. Widely used insulin,[23] with a relatively great molecular mass, has a long time of distribution within the ECS. It penetrates some parts of the ECS only after 15 hr.[21] Cotlove[24] has studied the leak of indicators into the ECS. He suggested that this leak is dependent on the length of the diffusion path, on the speed of the blood flow throughout the tissues, and on the permeability of the vascular walls for the indicator's molecules. The equilibrium is achieved faster for the indicators with smaller molecular mass.[25]

Labeled and unlabeled forms of other saccharides have also been extensively used. Mannitol, sucrose, as well as raffinose are not suitable indicators for ECFV estimation because they are eliminated into the gastrointestinal tract[25] and they are metabolized in tissues.[23,26]

A number of workers have made simultaneous measurements of the distribution volumes of several extracellular indicators in a variety of tissues. These results were summarized by Barr and Malvin,[27] who themselves made in vitro measurements on intestinal smooth muscle with seven different indicators the molecular mass of which varied from 60,000 for albumin to 60 for urea. The distribution volumes varied with this molecular mass.

It is evident that an ideal ECFV indicator does not exist. Many indicators penetrate behind the boundaries of the ECS and thus overestimate the ECFV values. Manery[12] suggested indicating these volumes as "distribution spaces" of each indicator.

It is important to note that recently a new inulin-like substance, Polyfructosan S (INU-TEST, Laevosan, Linz, Austria) was prepared. It is a starch-like polymer with molecular mass approximately 3500. In comparison with inulin, it has a number of advantages: (1) it is totally soluble in cold water, (2) it is alkali stable, and (3) weak acids hydrolyze it only half as fast as inulin. This circumstance is desirable insofar as inulin sometimes slightly hydrolyzes during the dissolution in boiling water. Mertz[28] has shown in humans that both the renal clearance and the distribution space of Polyfructosan S are in good agreement with the corresponding values of inulin. The calculated volume of distribution ratio for polyfructosan/inulin was 1.005 ± 0.093. Polyfructosan is nontoxic and produces no pyrogenic action in all experiments.

D. Plasma and Blood Volume

Evans blue (T-1824) is widely used for the measurement of plasma volume (PV). Therefore, considerable attention has been given to the exact details of methodology, improvements in optical measurements of this dye, and clinical and experimental interpretation.

The spectral absorption curves of Evans blue in the plasma obtained from various animal species are different.[29] For rat plasma, the peak absorption is at 605 mm.[30] The optical density values observed at the peak absorption are also different when Evans blue is added to different fluids. Therefore, the standard solutions of Evans blue should be prepared by using homologous plasma. Evans blue binds tightly to plasma albumin. Using a model system consisting of bovine albumin and Evans blue in the concentrations found in animals in which the PV is measured, it was possible to show that about 99.9% of the dye molecules are bound by albumin. Plasma volume will be overestimated if albumin escapes from the vascular bed or is metabolized during the course of equilibration. This is also true for radioiodinated human serum albumin (^{131}I-RISA), which is another substance commonly used today. Results with Evans blue and RISA are comparable.[31,32] A choice between both these substances depends solely on personal preference and the availability of the measuring equipment.

Blood volume (BV) is a sum of (PV) and cell volume (CV). The greatest part of published results on BV are derived from the determination of either PV or CV alone. From these quantities and the hematocrit value (Hct), the BV may be calculated from the following equations:

$$BV = \frac{PV \times 100}{100 - Hct}$$

or

$$BV = \frac{CV \times 100}{Hct}$$

The hematocrit value is usually derived from the observed venous hematocrit; this value is not identical with the volume percentage of cells in the blood sample.[33] In order to obtain the true hematocrit value, a correction must be made for the plasma trapped in packed cell column. This so-called "plasma trapping factor" has been the subject of numerous investigations. Lesson and Reeve[34] found that in normal human blood about 5% of the packed cell column consisted of plasma, and in normal rabbit blood about 5.5% of the cell column was plasma. Gregersen and Rawson[33] found that in normal dog blood 3.5 to 4% packed cell volume was plasma. These findings have also been supported in rats,[30] where the "plasma trapping factor" was found to be 0.95.

It has been recognized for many years that the concentration of red cells is not the same in all parts of the circulatory system. Thus, the cell/plasma ratio is lower in the capillaries and small vessels than in the larger arteries and veins. Consequently, one would expect the average or overall percentage to be somewhat lower than the venous cell percentage. This, in fact, is the situation in several species, including man. It has been suggested that the plasma-hematocrit method of blood volume measurement overestimates the BV values, while the cell-hematocrit method underestimates the BV values by a constant amount (9 to 10%). For reliable BV studies it would therefore be desirable that both red blood cell (RBC) and plasma volume are measured simultaneously. The RBC volume is estimated with red blood cells labeled with chromium (^{51}Cr), technetium (Tc-99m), phosphorus (^{32}P), or iron (^{59}Fe). Lee and Blaufox[35] summarize several studies of RBC volume, PV, and BV in rats done by various investigators. From simultaneous determinations of PV and RBC volume, the total BV is calculated as the sum of these measurements. From these data the ratio of overall cell percentage/arterial cell percentage (F_{cells} factor) can also be calculated. This F_{cells} factor should be used in calculating total BV in those experiments where only PV was measured. On the basis of both correction factors, the total blood volume can be calculated from PV values according to the following formula:

$$BV = \frac{PV \times 100}{100 - Hct \times 0.95 \times F_{cells}}$$

E. Interstitial Fluid Volume and Plasma Volume to Interstitial Fluid Volume Ratio

The interstitial fluid volume (IFV) cannot be measured directly because no techniques are available for its measurement. The difference between the extracellular fluid volume and the plasma volume gives the extravascular part of the ECFV which is composed largely, but not solely, of the interstitial fluid. As mentioned earlier, no substance has yet been found which will remain in the extracellular space and which will distribute itself rapidly and evenly throughout the interstitial space. Therefore, calculated IFV involves errors in the determination of ECFV.

The values of IFV and PV give the information about the distribution of extracellular water between its extra- and intravascular spaces. Thus the PV/IFV ratio is a very important parameter[36] that has the normal value about 0.33. The changes of this value can warn of possible changes in the capacity of the interstitial space. More detailed discussions and further references about these problems may be found in several recent reviews.[5,9,37]

V. APPENDIX

Although a great number of indicators of different chemical compositions, electrical charges, and molecular masses have been used to estimate the extracellular fluid volume (ECFV), there is no general consensus among workers in this field on the identity of the indicator which accurately measures the total ECFV. On the other hand, the measurement of plasma volume is much easier and precise. A simultaneous measurement of both PV and ECFV is important in a variety of clinical and experimental conditions.

A. Simple Method for the Measurement of Plasma Volume and Extracellular Fluid Volume in the Rats

1. Application of Indicators

For PV determination 100 $\mu\ell$/100 g of body mass (b.m.) of a 0.5% solution (w/v) of Evans blue in physiological saline are injected. For ECFV determination, polyfructosan S

(25% solution w/v, INUTEST, Laevosan) is injected in an amount of 20 $\mu\ell$/100 g of b.m. To determine the exact amount of indicator applied in the injection syringe, the gravimetric technique is preferred. The specific weight of polyfructosan is 1.039.

2. Procedure

Both indicators are applied into the exposed jugular vein. Five minutes after the application of Evans blue, about 0.2 mℓ of blood is taken from the tail artery into the heparinized microhematocrit tube to determine the dye concentration and hematocrit value. At this time the dye escaped from the circulation is not yet demonstrable.[38] Then bilateral nephrectomy is carried out to prevent elimination of polyfructosan. Eighty minutes after the polyfructosan injection blood is taken from the carotid artery for ECFV estimation.

The amount of ECFV removed with the kidneys may be calculated from renal mass. In vitro measurements using inulin space have shown that renal ECFV is of the order of 25% of kidney mass.[39] ECFV loss caused by bilateral nephrectomy was found to be less than 2% of distribution space.

3. Analytical Methods

The plasma level of Evans blue is determined in diluted plasma samples (50 $\mu\ell$ of plasma in 1 mℓ of water) against standards. Standards should be made in 20 times diluted plasma, serving also as the blank value. The optical density of the dye is read at 605 nm.

Polyfructosan is determined by the method used for inulin[40] with slight modification. Plasma is deproteinated (50 $\mu\ell$ plasma + 100 $\mu\ell$ 5% trichloroacetic acid + 100 $\mu\ell$ water) and centrifuged. Clear supernatant (100 $\mu\ell$) is incubated with 20 $\mu\ell$ of β-indolylacetic acid (0.5% solution in 96% ethanol) and 800 $\mu\ell$ concentrated hydrochloric acid for 80 min at 37°C together with standards (from 0.01 to 0.06 mg of polyfructosan in 1 mℓ). After cooling, the concentration of the dye is read in a spectrophotometer at 535 nm.

REFERENCES

1. **Steele, J. M., Berger, E. Y., Dunning, M. F., and Brodie, B. B.,** Total body water in man, *Am. J. Physiol.,* 162, 313, 1950.
2. **Edelman, I. S.,** Exchange of water between blood tissues. Characteristics of deuterium oxide equilibration in body water, *Am. J. Physiol.,* 171, 279, 1952.
3. **Edelman, I. S. and Leibman, J.,** Anatomy of body water and electrolytes, *Am. J. Med.,* 27, 256, 1959.
4. **Robertson, G. L.,** Control of posterior pituitary and antidiuretic hormone secretion, *Contr. Nephrol.,* 21, 33, 1980.
5. **Guyton, A. C., Taylor, A. E., and Granger, H. J.,** *Circulatory Physiology II: Dynamics and Control of the Body Fluids,* W. B. Saunders, Philadelphia, 1975, chap. 1—7.
6. **Wiederhielm, C. A.,** Dynamics of transcapillary fluid exchange, *J. Gen. Physiol.,* 52, 29S, 1968.
7. **Mellander, S. and Johansson, B.,** Control of resistance, exchange and capacitance functions in the peripheral circulation, *Pharmacol. Rev.,* 20, 117, 1968.
8. **Wiig, H.,** Comparison of methods for measurement of interstitial fluid pressure in cat skin, subcutis and muscle, *Am. J. Physiol.,* 249, H929, 1985.
9. **Aukland, K. and Nicolaysen, G.,** Interstitial fluid volume: local regulatory mechanisms, *Physiol. Rev.,* 61, 556, 1981.
10. **Brace, R. A.,** Progress toward resolving the controversy of positive or negative interstitial fluid pressure, *Circ. Res.,* 49, 281, 1981.
11. **Law, R. O.,** Techniques and applications of extracellular space determination in mammalian tissues, *Experientia,* 38, 411, 1982.
12. **Manery, J. F.,** Water and electrolyte metabolism, *Physiol. Rev.,* 34, 334, 1954.
13. **Brodie, B. B., Brand, E., and Leskin, S.,** The use of bromide as a measure of extracellular fluid, *J. Biol. Chem.,* 130, 555, 1939.

14. **Walser, M., Seldin, D. W., and Grollman, A.,** Radiosulfate space of muscle, *Am. J. Physiol.,* 176, 322, 1954.
15. **Everett, N. B. and Simmons, B. S.,** The distribution and excretion of S^{35} sodium sulfate in the albino rat, *Arch. Biochem. Biophys.,* 35, 152, 1952.
16. **Dziewiatkowski, D. D.,** One the utilization of exogenous sulfate sulfur by the rat in the formation of etheral sulfates as indicated by the use of sodium sulfate labeled with radioactive sulfur, *J. Biol. Chem.,* 178, 389, 1949.
17. **Scheinberg, I. H. and Kowalski, H. J.,** The binding of thiocyanate to albumin in normal human serum and defibrinated blood with reference to the determination of ''thiocyanate space'', *J. Clin. Invest.,* 29, 475, 1950.
18. **Nichols, G., Nichols, N., Weil, W. B., and Wallace, W. M.,** The direct measurement of the extracellular phase of tissues, *J. Clin. Invest.,* 32, 1299, 1953.
19. **Kuneš, J. and Jelínek, J.,** Extracellular fluid distribution in rats with chronic one- and two-kidney Goldblatt hypertension, *Clin. Exp. Pharmacol. Physiol.,* 6, 507, 1979.
20. **Ling, G. N. and Kromash, M. H.,** The extracellular space of voluntary muscle tissues, *J. Gen. Physiol.,* 50, 677, 1967.
21. **Mann, G. E.,** Alterations of myocardial capillary permeability by albumin in the isolated, perfused rabbit heart, *J. Physiol.,* 319, 311, 1981.
22. **Brading, A. F. and Jones, A. W.,** Distribution and kinetics of Co EDTA in smooth muscle, and its use as an extracellular marker, *J. Physiol.,* 200, 387, 1969.
23. **Esposito, G. and Csáky, T. Z.,** Extracellular space in the epithelium of rat's small intestine, *Am. J. Physiol.,* 226, 50, 1974.
24. **Cotlove, E.,** Mechanism and extent of distribution of inulin and sucrose in chloride space of tissues, *Am. J. Physiol.,* 176, 396, 1954.
25. **Swann, R. C., Madiseo, H., and Pitts, R. F.,** Measurement of extracellular fluid volume in nephrectomized dog, *J. Clin. Invest.,* 33, 1447, 1954.
26. **Pierson, R. N., Price, D. C., Wang, J., and Jain, R. K.,** Extracellular water measurements: organ tracer kinetics of bromide and sucrose in rats and man, *Am. J. Physiol.,* 235, F254, 1978.
27. **Barr, L. and Malvin, R. L.,** Estimation of extracellular spaces of smooth muscle using different sized molecules, *Am. J. Physiol.,* 208, 1042, 1965.
28. **Mertz, D. P.,** Observations on the renal clearance and the volume of distribution of Polyfructosan-S, a new inulin-like substance, *Experientia,* 19, 248, 1963.
29. **Chien, S. and Gregersen, M. I.,** Determination of body fluid volumes, in *Physical Techniques in Biological Research,* Vol. 4, Nastuk, W. L., Eds., Academic Press, New York, 1962, chap. 1.
30. **Wang, L.,** Plasma volume, cell volume, total blood volume and F_{cells} factor in the normal and splenectomized Sherman rat, *Am. J. Physiol.,* 196, 188, 1959.
31. **Schultz, A. L., Haumersten, J. F., Heller, B. I., and Ebert, R. V.,** A critical comparison of the T-1824 dye and iodinated albumin methods for plasma volume measurement, *J. Clin. Invest.,* 32, 107, 1953.
32. **Huang, K.-CH. and Bondurant, J. H.,** Simultaneous estimation of plasma volume, red cell volume and thiocyanate space in unanesthetized normal and slpenectomized rats, *Am. J. Physiol.,* 185, 441, 1956.
33. **Gregersen, M. I. and Rawson, R. A.,** Blood volume, *Physiol. Rev.,* 39, 307, 1959.
34. **Leeson, D. and Reeve, E. B.,** The plasma in the packed cell column of the haematocrit, *J. Physiol.,* 115, 129, 1951.
35. **Lee, H. B. and Blaufox, M. D.,** Blood volume in the rat, *J. Nucl. Med.,* 25, 72, 1985.
36. **Gauer, O. H.,** Introduction. Proc. Symp. on capillary exchange and the interstitial space, *Pflügers Arch.,* 336(Suppl.), 3, 1972.
37. **Guyton, C. A., Taylor, A. E., and Granger, H. J.,** *Circulatory Physiology II: Dynamics and Control of the Body Fluids,* W. B. Saunders, Philadelphia, 1975, 10.
38. **Belcher, E. H. and Harris, E. B.,** Studies on plasma volume, red cell volume and total blood volume in young growing rats, *J. Physiol.,* 139, 64, 1957.
39. **Whittam, R.,** The permeability of kidney cortex to chloride, *J. Physiol.,* 131, 542, 1956.
40. **Heyrovsky, A.,** A new method for the determination of inulin in plasma and urine, *Clin. Chim. Acta,* 1, 470, 1956.

Chapter 19

EXPERIMENTAL ORGAN TRANSPLANTATION

P. Pavel, V. Kočandrle, J. Černý, M. Nožičková, P. Šebesta, and J. Hökl

TABLE OF CONTENTS

I. INTRODUCTION

Organ and tissue transplantation constitutes an integral part of the physiological research and has been done, on an experimental basis, ever since the early 20th century. Let us recall the classic experiments by Lee and Schraut[52] demonstrating that diabetic vascular lesions are secondary complications of diabetes. Transplantation of endocrine pancreas to a diabetic animal resulted in the halting, or regression, of microangiopathy.

Surgical techniques of transplantation of all organs were originally developed in animals. Essentially, transplantation involves the establishment of a connection between the retrieved organ and the recipient's blood circulation.[1-3] Employing the techniques of vascular surgery, it is possible to transplant such organs as the heart, liver, and kidney to a new site.

If the organ is transplanted to the same animal (autotransplantation), it survives indefinitely. If the organ is transplanted to another animal of the same species (allotransplantation), it survives and functions for about 5 to 7 days, then it stops functioning. If the organ is transplanted to an animal of a different species (xenotransplantation), it either functions for a couple of minutes, or fails to start functioning altogether.

When designing the experimental protocol, the investigator is faced with several limitations:

1. Preservation techniques, the preservation time of an organ
2. Immunological incompatibility between the graft and the recipient animal (pair of animals)
3. The surgical technique, and/or particular properties of a specific organ

Those carrying out transplantation must not fail to keep in mind that the graft is completely denervated, and its lymphatic drainage discontinued.

II. TYPES OF MODELS AND METHODS

A. Introduction

Depending on the question that the experiment is designed to explore, the investigator selects either allotransplantation or autotransplantation. In allotransplantation, there are no time limits imposed on the investigator. He can retrieve the organ from the donor animal at any time and transplant it to an experimental animal. A distinct disadvantage is the immune response leading to graft rejection.[4,5] However, this drawback can be eliminated by the selection of immunologically identical inbred animals.

In autotransplantation, the researcher is limited by a certain period of time over which the graft is preserved (several hours to days). It is only over this specific period of time that the control animal can live without its own organ. If only very short-term preservation is possible, the experimental protocol can be largely simplified. The organ is left *in situ* in the experimental animal's body, and dissected in such a way that it is connected with the body only via its artery and vein. The vessels can be clamped, and the graft preserved in the animal's body. After a certain period of time, the clamps are released and blood circulation restored. This is a most simple technique requiring no vascular (or other) anastomoses.

B. Organ Removal

The technique of organ removal depends on whether auto- or allotransplantation is to be performed. Regardless of the method chosen, it is always crucial to do one's best to protect the graft from damage on retrieval. Dissection must be done very gently to avoid damage to vascular supply, and contusion of the parenchyma. In autotransplantation, the situation is even more complicated. Maximum caution must be exercised to spare, in addition to the graft, the experimental animal. In some organs, anatomy renders autotransplantation virtually impossible or, if accomplished, the procedure carries a high postoperative mortality (heart, pancreas).

To achieve immediate posttransplantation development of graft function, the retrieved organ must be preserved. It is vital to perfuse the organ with cold-storage solution as soon as possible after retrieval. Every effort should be made to cut down the time of warm ischemia. However, the organ can be perfused *in situ*, thus minimizing the interval of warm ischemia, or eliminating it altogether. The organ is not removed from the donor's body until it has been perfused.

C. Organ Preservation

1. Introduction

Metabolism can be reduced by metabolic inhibitors, hypothermia, and freezing of the organ.[6,7] There are several alternatives to achieve organ preservation. While simple cooling storage is the most frequently used technique, pulsatile or nonpulsatile perfusion and organ freezing are the other options for the transplantation surgeon.

Using these techniques, the graft is invariably first perfused with an appropriate storage solution. The aim of initial perfusion is to cool down the organ and wash out blood. The perfusate temperature is 0 to 4°C, and the pressure 30 to 60 cm H_2O.

2. Simple Cooling Storage

Once initial perfusion is over, the organ is stored in a container at 4°C.

3. Perfusion Methods

The retrieved organ is connected to a perfusion machine designed to perfuse the graft employing the pulsatile, nonpulsatile, or intermittent technique. These are relatively complicated methods featuring no distinct advantages compared with simple cooling storage.

4. Organ Freezing

If the organ is frozen to -196°C in liquid nitrogen, metabolism in the graft is completely abolished.

Because of the technical problems, the results are not satisfactory at this time.

D. Graft Rejection

A graft transplanted to an animal of the same species (allotransplantation) survives for about several days, after which time it stops functioning. The cause of graft failure is the host immune response to the graft (rejection). During rejection, the graft becomes congested, shows cloudy swelling and necroses, and becomes infiltrated with mononuclear elements.

Reaction of inbred animals is similar to that in autotransplantation.

The pair of animals (donor and recipient) is currently selected by a variety of methods. Of these, the most prospective one is the method of tissue typing according to the antigenic properties of lymphocytes.

To inhibit graft rejection, various methods of immunosuppression are employed. Azathioprine and glucocorticoids are the most widely used drugs. Cyclosporin A has been introduced in experimental and clinical practice. Antilymphocyte globulin, which has an effect on lymphoid cells, is also used.

Currently, monoclonal antibodies are yet another favorite on the transplantation scene, both experimental and clinical.

III. NECESSARY EQUIPMENT, INSTRUMENTATION, AND CONDITIONS FOR ORGAN TRANSPLANTATION

Technically, organ transplantation is the establishment of an anastomoses between the donor's organ and the recipient's blood circulation. Unless transplantation of minute fragments are involved, connection with the recipient's blood circulation must be effected by means of a vascular anastomoses of the graft's artery and vein with the recipient's vessels. Thus, essentially, the procedure is one of vascular surgery. To perform actual transplantation, instrumentation for vascular surgery, and/or possibly microsurgery must be used. The vessels are sutured together with an unabsorbable suture material (Prolene, Ethibond). Heart transplantation requires extracorporeal circulation and, besides the organ donor, also blood donors. To preserve organs by means of continuous perfusion, a perfusion machine is necessary. Cryopreservation of organs, islet or bone marrow isolation, and transplantation can be carried out only in special laboratories specifically equipped for these procedures.

If the animal is to survive, the availability of good postoperative care after transplantation is a prerequisite.

To perform allotransplantation, an immunological laboratory must be available to monitor immunosuppressive therapy, and/or typing, or cross-match. The arsenal of drugs, apart from the standard ones, must include immunosuppressive agents (Imuran, corticoids, and/or cyclosporine, antilymphocyte globulin).

IV. KIDNEY TRANSPLANTATION

A. Introduction

Kidney transplantation involves the connection of renal vessels with the appropriate vessels of the recipient. The next step is to implant the ureter to the urinary bladder or elsewhere, or to perform a nephrostomy or an ureterostomy, depending on the nature and aims of the experiment.[8-12]

Kidney retrieval is carried out in median laparotomy. Provided the donor need not be spared (in allotransplantation), the kidneys can be perfused via the abdominal aorta that is dissected above and below the renal arteries. Following administration of heparin, a cannula is inserted into the aorta and once clamps have been placed above the origins of renal arteries, perfusion with cold perfusate is started. To drain the perfusate from the renal veins, a cannula can be inserted into the inferior vena cava, or the vein can be cut open to draw blood. After a 5-min perfusion, the kidney can be removed from the donor's body.

In case autotransplantation is carried out, the first step is to release the kidney from its surrounding fatty tissue, exposing first the anterior, then its posterior surface. Next, it is necessary to dissect the renal artery and vein. To achieve this, we gently pull the kidney. The vessels are dissected at the site of their origin behind the aorta, and/or the vena cava. The periportal fatty tissue is left at the hilum to minimize the risk of ureteral ischemization. We first locate the ureter and sufficiently mobilize it, then vascular loops are passed under the artery, vein, and ureter. Next we administer heparin (not necessarily, though), and place clamps on the renal vessels that are subsequently transected. The ureter is also transected. The kidney is submerged in ice-cool saline and perfused with cold solution (Collins', Sacks', Ringer's lactate) for a period of 5 min at a pressure of 60 cm H_2O. The kidney can be placed in a container with crushed ice and stored for 24 to 48 hr. When using Ringer's lactate, the kidney should be transplanted within 24 hr.

Provided that the kidney has been preserved by means of a perfusion machine, it can be used as late as 72 hr after removal.

B. Transplantation Technique

The kidney can be transplanted extraperitoneally or intraperitoneally to the pelvic position

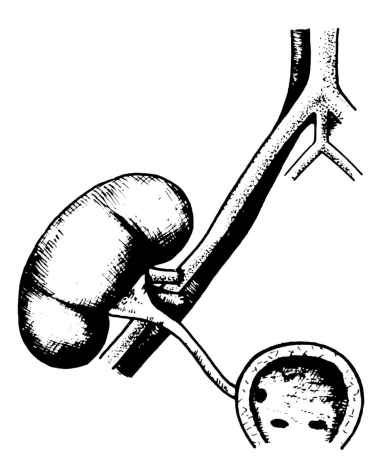

FIGURE 1.   Technique of kidney transplantation to pelvis. The renal artery and vein are anastomosed end-to-side into iliac vessels. The ureter is implanted to the bladder.

(with the ureter implanted to the urinary bladder (Figure 1), or to its original site (a more complicated procedure since the ureter cannot be reimplanted to the bladder), subcutaneously to the neck of an experimental animal with a nephrostomy or ureterostomy.[12,13]

In transplantation to the pelvis, we perform median lower laparotomy to open the abdominal cavity. Further, we dissect the external iliac vessels. Vascular anastomoses are established: first venous, end-to-side to the iliac vein, then the renal artery end-to-side to the iliac artery (Figure 1). Once the anastomoses have been completed, clamps are released. Within seconds, the kidney's homogeneous coloration is restored.

The same technique is employed to perform renal allotransplantation. If, however, a pair of mix-bred animals is used, diuresis can be expected to last (without immunosuppression) for a couple of days only.

In an *in situ* experiment, the kidney is completely isolated so that it is connected with the animal's body only via the hilar vessels and ureter. A needle, connected with a perfusion set, is inserted into the renal artery to perfuse the kidney during clamping of renal vessels. While a hyperkalemic solution must be discharged through a tiny opening in the renal vein, nonhyperkalemic solutions can be drained into circulation. After a 5-min perfusion, the kidney can be further cooled by the application of crushed ice. After some time, vascular clamps are released and the kidney is reperfused with blood. Should the need arise, we surgically close the site of needle insertion into the renal artery and/or the opening into the renal vein. While an advantage of this technique is that there is no need for vascular and ureteral anastomoses, preservation time is limited by the duration of animal anesthesia.

C. Ureter Implantation

According to the nature of the experiment, the ureter can be implanted to the urinary bladder, to the venous system (vena cava, iliac vein), or to the intestine (rectosigmoideum).[14]

V. PANCREAS TRANSPLANTATION

A. Introduction

There are several techniques available to carry out pancreas transplantation for physiological research: total pancreas transplantation, pancreas segment transplantation, transplantation of isolated islets, or of pancreatic fragments.

Ever since the first experiments, the question has remained how to handle external secretion of the pancreas. Many techniques[15,16] have been developed to eliminate or derive external pancreatic secretion while maintaining endocrine secretion (transplantation of the pancreatoduodenal block, duct implantation to the bowel, urinary bladder, ureter, duct ligature, duct obliteration, open-duct technique, and a number of modifications).

B. Total Pancreas Transplantation

Autotransplantation is almost never used, in view of pancreatico-duodenal vascular supply. However, it has been reported that the pancreas can be removed even without ischemic involvement of the duodenum or liver. Still, the mortality rate is rather high.

The technique of pancreas removal for allo- or autotransplantation is identical, the only difference being that, in autotransplantation, maximum caution must be exercised to spare the duodenum. The pancreatoduodenal vessels must be left with the graft. During removal, the uncinate process is separated from the duodenum, the spleen is released, and vessels of the gastrolienal ligament ligated. Next, the gastric artery and left gastric vein are ligated. The celiac trunk is then dissected, and the proper hepatic artery behind the origin of the upper pancreaticoduodenal artery is ligated. As the next step, the portal, splenic, and mesenteric vessels are dissected. Arterial supply is considerably variable. Following removal of the pancreas, the graft is perfused with storage solution (Ringer's lactate, Collins', Sacks', and many others); for preservation over a period of 24 to 48 hr, it can be perfused using a perfusion machine. The pancreas is transplanted to the peritoneal cavity, to the neck, or to the groin. Here, it is crucial to connect to the celiac trunk, with or without patch, and the portal vein with the recipient's vessels. The pancreatic duct is then implanted to the intestine, ureter, or the urinary bladder, or an external pancreatic fistula is constructed. Other alternatives are to leave the duct open into the peritoneal cavity, or to obliterate the duct with an appropriate material (Prolamine, Neoprene, etc.). In allotransplantation models, total pancreatectomy of one's own pancreas can be performed (Figure 2). Diabetes may be induced by streptozotocine or alloxan either before or after transplantation.

C. Pancreas Segment Transplantation

The left lobe of the pancreas supplied with blood from the branches of the splenic artery and draining into the splenic vein is used.[17] Still, due to frequent anomalies in blood supply (from the mesenteric artery), it is not always possible to use the left lobe for transplantation.

In an *in situ* experiment, the left lobe is completely isolated, the spleen exposed and vessels in the gastrolienal ligament are ligated. Once the gastric artery and vein have been ligated, the pancreas in dissected up to the celiac trunk. The splenic vein is dissected at its junction into the portal vein. The left tail of the pancreas is transected and its proximal

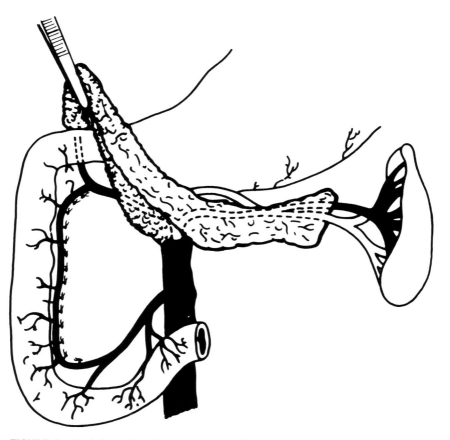

FIGURE 2. Technique of total pancreatectomy. The pancreas is carefully removed to avoid damaging pancreaticoduodenal vessels.

stump ligated (alternatively, pancreatectomy of the residual pancreas is performed). After full isolation of the pancreatic parenchyma has been achieved, the segment can be perfused with cold solution (Ringer's lactate, Collins', Sacks') via a cannula inserted retrogradely into the splenic artery, and the same technique can be used to discharge the solution from the splenic vein). After perfusion, the pancreas can be cooled locally for a specified period of time. when performing *in situ* experiments, the duct is left open into the peritoneal cavity, or is obliterated with an appropriate polymer, and/or implanted in to the stomach, duodenum, or jejunum.

In autotransplantation, the segment is dissected using the same technique (Figure 3), to be retrieved after perfusion from the donor's body and stored in a cold solution, and/or connected with a perfusion machine. In this way, it can be preserved for a period of 24 to 48 hr.[18]

The techniques of auto- and allotransplantation are identical, both involving venous and arterial anastomoses to the recipient's vessels (iliac vessels, carotid artery-jugular vein, femoral vessels (Figures 4 and 5). The duct can be left open (in the peritoneal cavity) or implanted to the intestine (Figure 6), stomach, urinary bladder, or obliterated with an appropriate polymerizing substance.[15,19,20-24]

VI. LANGERHANS' ISLETS TRANSPLANTATION

A. Introduction
The advantages of islet transplantation include a relatively easy technique, absence of

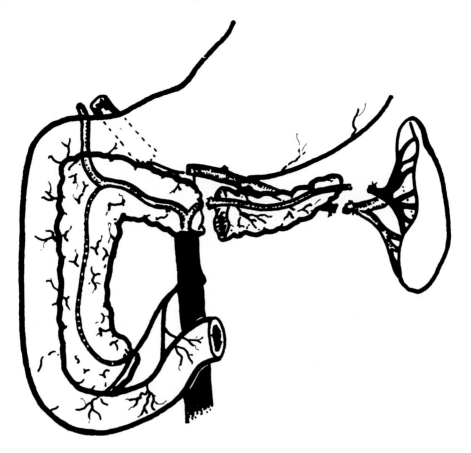

FIGURE 3.   Left pancreatic lobe (pancreatic segment) and its blood supply. Anatomic variations in blood supply are frequent. Scheme of removal.

exocrine tissue, possibility of preservation for later use, possibility of immunological regulation, and the possible use of xenografts.

The disadvantages include a low islet procurement rate and, consequently, the need for several donors, vulnerability of the tissue in the interval between implantation and engraftment, and possible destruction of islet cells by the primary process, which has been the cause of diabetes.

Sources of endocrine tissue include adult, neonatal, or fetal pancreas.

To date, transplantation of isolated "adult" islets has been successful in rodents only. In higher mammals, they are soon rejected. As a rule, immunosuppressive therapy is ineffective.[25-28]

Somewhat better results are reported when neonatal tissue is used, with the main advantage being easier isolation of the islets.

The technique regarded as most advantageous is transplantation of fetal fragments carried out at a time when the exocrine tissue has not yet been sufficiently differentiated and immunogenity is lower than in the "adult" tissue.[29]

B. Isolation Method

The basic principle of isolation is organ destruction and removal of the connective stroma and exocrine tissue. Presently, the modification by Lacy et al.[30] is used. In the removed pancreas, edema is produced by instillation of Hanks solution into the pancreatic duct or

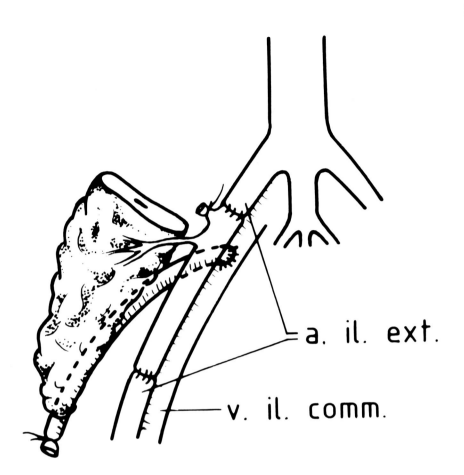

FIGURE 4. Pancreas segment transplantation into peritoneal cavity. The splenic vein is anastomosed end-to-side into the iliac vein. The splenic artery interposed into the iliac artery. The duct is left open into the peritoneum.

the pancreatic vein. The organ is then trimmed free of the connective capsule and septa. The tissue is cut into small parts 1- to 1.6-mm in diameter. The fragments are then incubated with collagenase at 37°C for about 15 min. As the next step, the tissue must be well flushed with Hanks solution. For implantation, either pancreatic fragments or isolated islets are used.[25,30-35]

Actual isolation can be done using one of the following techniques: (1) sedimentation method, (2) centrifugation method, or (3) division by means of the Ficoll gradient.

The sedimentation method is used most frequently. On repeated sedimentation, intact islets fall to the bottom of the test tube and can be retrieved mechanically from the sediment using a preparation microscope. The islets appear as graying or brown-red round or ovoid structures. Light microscopy has revealed that islets prepared by this method are neither damaged nor changed. Their insulin response to a glucose stimulus is good. When using the centrifugation method, the insulin response to glucose is not sufficient enough. This explains why these methods are not suitable for metabolic studies.

C. Islet Preservation and Implantation Methods

In the interval between isolation and implantation, the islets can be preserved in a viable condition by the following techniques:

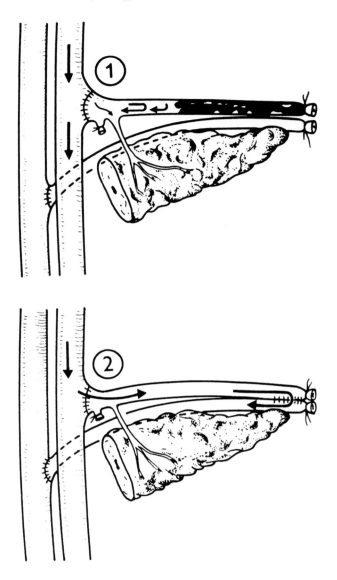

FIGURE 5.   Other techniques of arterial anastomoses. (1) End-to-side
with ligature of distal end of splenic artery and (2) end-to-side with
arteriovenous fistula.

1.   Simple hypothermia. When storing islets at 4°C, metabolism, and oxygen demand in
      particular, are kept at a minimum.[7,36,37]
2.   Cultivation in tissue cultures. The islets, or fragments of the pancreatic tissue, are
      placed in an appropriate medium at 37°C in an atmosphere containing 5% CO_2 and
      95% O_2 where they grow and multiply. The insulin response to glucose starts within
      8 hr following administration of glucose and lasts for 6 days. During this time, however,
      the content of insulin in the medium keeps on decreasing. After implantation of the
      cultivated tissue to the recipient, B-cells start filling with granulae, and insulin pro-
      duction more than doubles within 10 days.[37]
3.   Cryopreservation. Islet viability and restored function have been proven after 17 days
      of freezing. While dimethylsulfoxide seems to be the best cryoprotective agent, liquid
      nitrogen at $-196°C$ makes an ideal medium for freezing.[36-39]

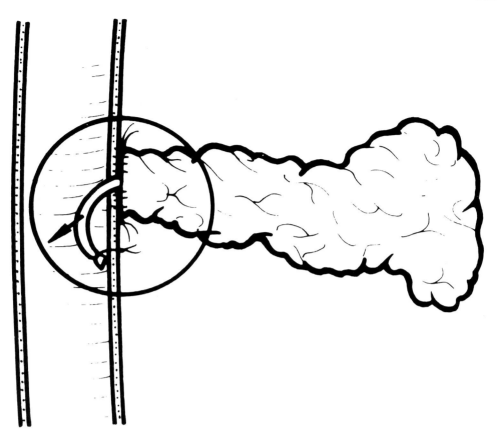

FIGURE 6.   Implantation of the duct to the small bowel (most easy and reliable technique).

The viable and functioning tissue prepared by the above techniques can be used for implantation to the recipient's body both for auto- or allotransplantation. In rodents, the islets are transplanted most often subcutaneously, intramuscularly, or intraperitoneally. In larger laboratory animals (pigs, dogs), endocrine tissue is applied to the spleen retogradely through the splenic vein or in the liver through the portal vein. Implantation beneath the renal capsule seems to be another technique offering good results. In this tissue, morphologic examination has proved undamaged islets able to maintain normoglycemia in previously diabetic animals. At the same time, the function and metabolism of the organ with implanted endocrine pancreatic tissue have been preserved.[25,27,40,41]

In cases in which the donor and recipient were histocompatible, islet allografts normalized the blood sugar level for 2 or 3 days only. After this time, hyperglycemia returned due to islet cell destruction by rejection.[25,42,43]

D. Experimental Islet Transplantation

The first animals to be used in experiments were rats and mice. The pancreas of these species is soft, with little connective tissue, rendering islet isolation, with the disruption of acinar tissue and the effect of collagenase, relatively easy. While the pancreas of a rodent may yield between 150 and 450 intact cells, a successful transplantation requires 600 to 1500 cells, i.e., islets from two to six donors. Here, another advantage in rodents is the possibility of using inbred strains and transplanting syngeneic grafts that are not rejected. The results of transplantations of adequate numbers of syngeneic islets in rodents are excellent.

VII. HEART TRANSPLANTATION

Heart transplantation is most difficult to perform. During heart removal, maximum care must be taken not to damage the conductive system. The heart must be removed with its whole right atrium and parts of the superior and inferior vena cava, which explains why, technically, autotransplantation is a most demanding procedure.[44,45]

To a certain extent, autotransplantation can be substituted by an *in situ* experiment with an attempt at periaortic denervation. In this type of operation, however, we never succeed in completely interrupting the vascular anastomoses between the coronary and bronchial circulation systems, and extracorporeal circulation must be established. First, cannulae are inserted into the superior vena cava (through the femoral and right jugular veins) and, next, an arterial cannula via the femoral artery. As the next step, periaortic denervation is performed by removing adventitia from the aortic root and from the root of the pulmonary artery. We must prevent the blood from filling the right-side chambers of the heart by placing clamps or tourniquets on the caval veins. Then, the aorta is clamped, and cool cardiopledgic solution is administered into the aortic root. The perfusate is drained through an opening in the right atrium. The coronary bed is perfused for 3 min, with the heart cooled by the application of ice-cold solution. The period of heart preservation is limited by the period of extracorporeal circulation, i.e., it amounts to several hours according to the type of the oxygenator used. Warming of the heart by blood from bronchial collateral circulation cannot be prevented altogether, and its adverse effect can only be suppressed by repeated cardioplegic perfusion.

After some time, the aortic clamp is released, the opening caused by the cardioplegic needle sutured, tourniquets are released from the caval veins, and the opening in the right atrium is sutured. Extracorporeal circulation is gradually discontinued.

A. Heart Allotransplantation

The heart can be transplanted to its natural position (orthotopic transplantation), or as an auxiliary heart to the thoracic or abdominal cavities (heterotopic allotransplantation). (For the purpose of short-term experiments, the heart can be perfused with the blood of an *ex vivo* perfusor animal.)

The following describes heart removal in the dog, approaching through a median sternotomy. The pericardial cavity is opened, and the v. azygos divided. While avoiding the superior and inferior vena cava and the pulmonary artery, a needle for the administration of cardioplegic solution is inserted into the ascending aorta. After ligating the inferior and superior vena cava and clamping of the aorta, cardioplegic solution is administered. Both the inferior and superior vena cava are transected, and the perfusate is sucked off. Following a 5-min perfusion, the aorta, pulmonary veins, and the pulmonary artery are transected. The atria are prepared to carry out anastomoses (Figure 8).

In orthotopic transplantation, with extracorporeal circulation established as previously described, the recipient's heart is removed while leaving the posterior parts of the atria and the orifice of the vena cava and pulmonary veins (Figure 7). The graft, constantly cooled with crushed ice, is sutured first to the atria (left, right atrium), then aortic anastomosis is completed, followed by pulmonary artery anastomosis (Figures 8 and 9). Next, air evacuation from the heart must be performed: a needle for active air suction is inserted into the ascending aorta, the aortic clamp released, and air repeatedly evacuated from the left atrium and ventricle.[46]

The caval veins are released and extracorporeal circulation is discontinued gradually.

In heterotopic auxiliary allotransplantation, the heart can be transplanted to the thoracic cavity without using extracorporeal circulation (Figure 10). The heart works as a bypass of the right and/or left ventricle(s). Similarly, the heart can be transplanted to the abdominal cavity.

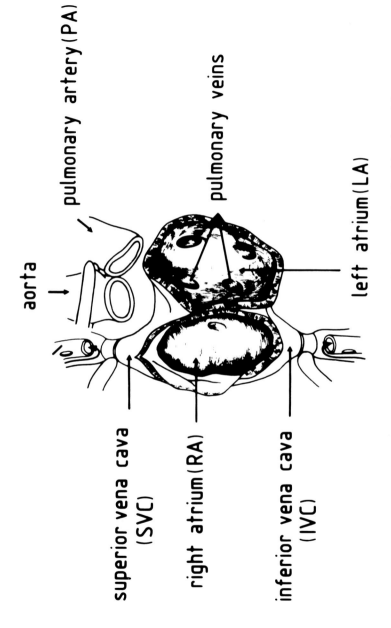

superior vena cava
(SVC)

right atrium (RA)

inferior vena cava
(IVC)

aorta

pulmonary artery (PA)

pulmonary veins

left atrium (LA)

FIGURE 7. Recipient operation. The own heart is removed, so that the back walls of the atria are left *in situ*.

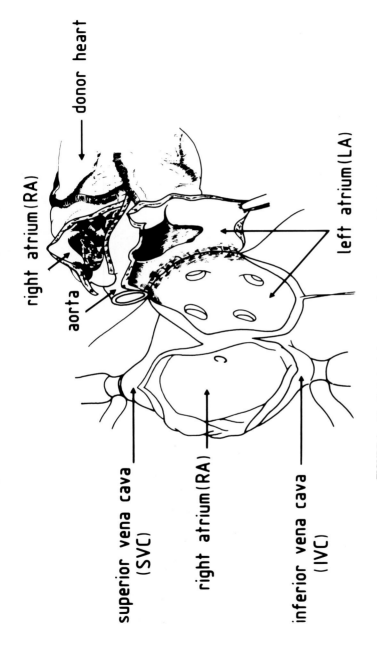

FIGURE 8.   The first step in heart transplantation: sutures of the left atria.

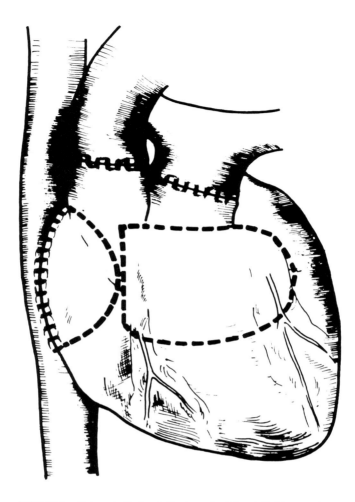

FIGURE 9. Sutures in the heart after transplantation: left atrium, septum, right atrium, aorta, pulmonary artery.

VIII. LUNG TRANSPLANTATION

Technical problems are known to be common in experimental lung transplantation. Many studies have demonstrated that preservation methods enabling short-term storage of the heart, kidney, and pancreas are not effective in lung preservation. Lungs that are ischemic, but inflated or ventilated for 2 hr, can provide pulmonary function.

In an *in situ* model, the left pulmonary artery and veins are clamped for 2 hr, and the lungs are perfused and ventilated with 40% oxygen. After 2 hr, the clamps are released. (The right lung is left in normal position, or its pulmonary artery is ligated.)

Autotransplantation, performing either a left thoracotomy, or a median sternotomy, the left lung is removed after clamping the pulmonary artery and pulmonary veins with a patch of the left atrium. The bronchus is transected and the lung is preserved in an inflated state. The preservation period could be extended by cooling the organ down to 4 to 10°C. A lung preserved for several hours after removal by simple flushing with Sacks' solution (Collins' solution) could provide pulmonary function for 10 days after transplantation.

The pulmonary arteries are sutured end-to-end, and the pulmonary veins are sutured with the left atrium. Bronchial suture must be performed very carefully.

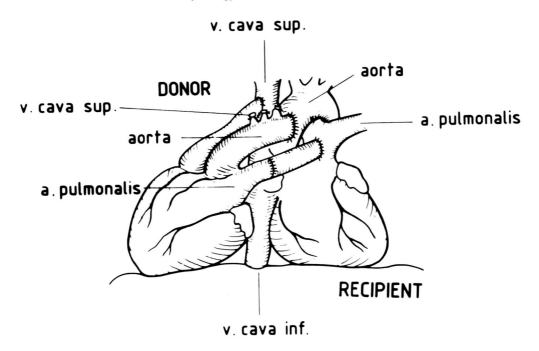

FIGURE 10.   Scheme of heterotopic heart transplantation. The heart works as a bypass of both ventricles.

For allotransplantation, lung removal, preservation, and transplantation are carried out in the same manner as previously described. A portion of the allograft left atrium and pulmonary artery can be retained to facilitate the reimplant procedure. A factor limiting donor/recipient availability is the need for the donor and recipient to have hilar structures (donor bronchus) of compatible sizes. The donor bronchus should be equal to or slightly smaller than the recipient bronchus to minimize technical problems of this critical anastomosis.

IX. HEART AND LUNG TRANSPLANTATION

Only an allotransplantation model is available. In this model, performing a median sternotomy, the heart graft is prepared as previously described, the left and right pleural cavities are opened, and the trachea is prepared. The preservation technique is the same as described above. The lungs are inflated. The heart and lungs must be removed *en bloc*. Only the aorta, caval veins, and trachea are transected.

In a recipient operation, the dorsal wall of the right atrium is left *in situ* (Figure 11). The aorta and trachea are transected and the heart and lungs are removed *en bloc*. The phrenic nerves and mediastinal structures must be spared.

The grafts are then placed in an optimal position, and the right atrium, aorta, and trachea are sutured (Figure 12). Next, the clamps are released.

X. LIVER TRANSPLANTATION

Liver transplants can be either heterotopic or orthotopic. Heterotopic transplantation can be either (1) auxiliary, i.e., the recipient's liver is left in its original site with the graft implanted to an ectopic site, or (2) nonauxiliary, i.e., the recipient's liver is removed and the graft is implanted to an ectopic site. In orthotopic transplantation, the recipient's liver is removed and replaced by the graft.

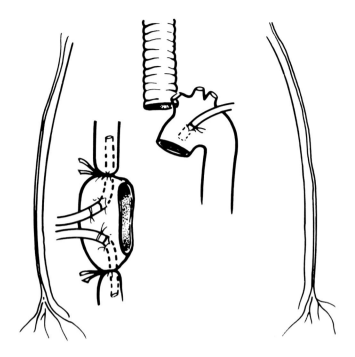

FIGURE 11. Heart and lung transplantation. Recipient operation: the own heart and lungs are removed.

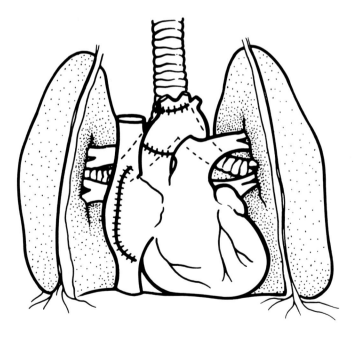

FIGURE 12. Sutures of the right atrium, aorta, and trachea.

Orthotopic transplantations are performed especially in dogs, pigs, monkeys, and sewer rats. One of the most suitable laboratory animals is the pig. In most pigs, rejection is rather slow and mild, or it does not occur at all. For long-term experiments, the minipig is a most suitable animal since it does not grow to weigh much even when mature.

The animals are fed thinned milk and steamed barley groats with grape sugar 2 to 3 days before transplantation, and left fasting for the last 12 hr.

During transplantation, combined anesthesia is employed (Thalamonal, dinitrogen monoxide with oxygen, halothane). Blood losses are compensated for by blood transfusions with simultaneous infusions of glucose and $NaHCO_3$.

The operative technique in the donor involves liver dissection while leaving behind vascular stumps which should be as long as possible, and a bile duct stump.

In preserving of the liver, acellular storage solutions as well as homologous blood can be used to perfuse the organ. Oxygenated blood with Hartman's solution in a 1:1 ratio with glucose, insulin, and antibiotics can be also used advantageously. The perfusate temperature is 4°C and pH is kept at 7.4. The liver is perfused via the portal vein, and stored in a container with saline (4°C). A list of preservation techniques in liver transplantation has been compiled by Lambotte.[47]

The operative technique in the recipient is as follows. First, structures of the liver hilus are prepared, and a temporary portojugular bypass is constructed. Before establishing the shunt, heparin must be administered. The next step is hepatectomy, with the vessels and bile duct transected as close to the liver as possible. Actual transplantation involves four vascular anastomoses and a bile duct connection. The vascular anastomoses are usually established in the following order: suprahepatic inferior vena cava, hepatic artery, portal vein, infrahepatic inferior vena cava. Once connection with the recipient's circulation has been accomplished, the venous bypass is removed. The optimal technique of bile duct reconstruction is a choledocho-choledocho-anastomosis end-to-end with cholestostomy enabling long-term flushing of the bile duct with saline supplemented with antibiotics (Figure 13). Following transplantation, the abdominal cavity is drained for 24 to 48 hr.[48]

In the early postoperative stage, care includes fusions of glucose with vitamins, hydrocortisone, antibiotics, and heparin administered according to the results of biochemical and hematologic investigations (and supplemented by blood and plasma transfusions in case of blood losses). Metabolic acidosis is controlled by an $NaHCO_3$ solution. Permanent bile duct flushing lasts for 10 to 14 days, after which time the catheter is removed from the gallbladder. On the first 2 to 3 postoperative days, the animals are fed thinned milk to be later switched over to a standard diet.

Experimental orthotopic liver transplantation makes it possible to study the physiology and pathophysiology of liver transplants, to increase the knowledge of the diverse liver functions, and to determine the origin of proteins. Liver transplantation represents a major contribution to transplantation immunology in the search for transplantation tolerance.

Heterotopic auxiliary transplantation can make a useful model in the study of liver physiology involving, e.g., liver enzyme deficiency, need for insulin, and the hepatotropic factors for optimum liver function. While an advantage of this method is that the recipient is not directly dependent on the transplant function, a disadvantage is the small space available for an auxiliary liver.

XI. BOWEL TRANSPLANTATION

Because of a number of technical, and especially immunological, problems, bowel transplantation has remained an experimental procedure to date.[49,51]

The properties of the small intestine as a graft make it a less suitable candidate for transplantation than other organs. The main reason is the considerable amount of transplanted

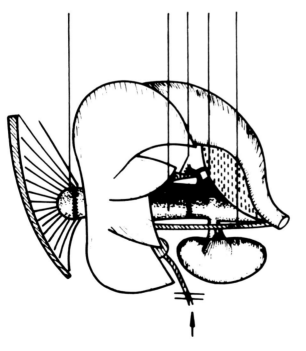

Suprahepatic inferior vena cava

Infrahepatic inferior vena cava
Hepatic artery
Portal vein
Common bile duct

FIGURE 13. Schematic representation of liver transplantation. Sutures of suprahepatic inferior vena cava, hepatic artery, portal vein, infrahepatic inferior vena cava, and choledochocholedochoanastomosis.

donor lymphoid tissue present in Peyer's plaques and in the lymph nodes in the mesenterium; their immune properties are responsible, as in bone marrow transplantation, for graft vs. host disease (GVHD).[52] Transection of lymphatic ducts in bowel transplantation causes a considerable, although temporary, reduction in the rate of resorption of nutrients, particularly fats and fat-soluble substances. Transection of afferent nerve pathways results in autonomy of the section transplanted.

Bacterial colonization, intraperitoneal positioning, and drainage of the portal vein may lead to septic and technical complications. This is especially so when maintaining the continuity of the intestinal tract, i.e., if orthotopic transplantation is attempted. Heterotopic transplantation involves an excluded intestine or its segment with both ends opened through the abdominal wall as an analogy to terminal enterostomy. Both extreme variants can be combined. From the functional point of view, a segment measuring a fourth to a third of the total length of the intestine, i.e., a segmental graft, is fully sufficient, while it significantly reduces the amount of transplanted immunocompetent tissue, it ensures survival of the recipient in terms of nutrition.

In animal experiments, the majority of studies are conducted in immunologically defined rat strains and in large animals (dogs and pigs), and concentrate on graft removal and preservation, on the suppression of its immune properties, and on the technical modifications of actual auto- and allotransplantation.[53,58] Interest in this method has been rekindled by the discovery of cyclosporin A which, compared with other conventional immunosuppressive agents, largely increases the survival rate of the bowel graft, and prevents GVHD. Other methods successful to a certain extent include those of direct control of the immunocompetent graft cells such as lymphadenectomy, and radiation therapy of the entire donor's body of the transplant *ex vivo*.[50,52,55,56]

In autotransplantation in the dog, to achieve access to the abdominal cavity, a median laparotomy is performed. The small intestine is sufficiently eviscerated while every effort is made not to twist the mesenterium. The upper mesenteric vessel in the mesenterial radix are dissected and the required length of the graft is marked by clamps placed in such a manner so as not to impair the viability of either end of the graft. With the arcades ligated, the mesenterium is transected up to the exposed vascular radix. The graft is perfused either *in situ* with Collins' solution or Ringer's lactate (4°C, 1 ℓ supplemented with 1000 units/ 10 mg heparin), or with the graft placed *ex situ* in saline at 4°C, and the pressure of the perfusate beyond the cannula not exceeding 35 cm H_2O. At the same time, the intestinal lumen is flushed isoperistaltically with the same solution supplemented with antibiotics while utmost attention is paid to maintain sterility of the graft. Following perfusion, the intraluminal temperature should be 8 to 10°C, and the period of cold ischemia should not exceed 3 to 4 hr. A well-flushed intestine is diffusely discolored, without any mesenterial swelling. In the case of orthotopic autotransplantation, vascular anastomoses are sutured to the vascular stumps left behind after removal, or end-to-side to the aorta or the inferior vena cava. To do this, atraumatic monophylament unabsorbable material, most often Prolene 6/0, is used.

After enterectomy of the recipient's intestine, the procedure for allotransplantation is carried out without any major differences. Continuity of the intestinal tract is restored by enteroanastomosis. The surgeon must check the vascular anastomosis, mesenterium pulsation, and the state of nutrition of the entire intestine, especially in the area of anastomoses with the recipient's loop. The abdominal cavity is not drained.

In the postoperative stage, the parameters that must be monitored include the recipient's overall condition, electrolyte and water balance, frequency and volume of stool, fat excretion in the stool, total serum protein, and albumin.[59] To be able to recognize safely a rejection episode, histologic examination of an intestinal mucosal biopsy specimen is necessary.

XII. CONCLUSIONS

Experimental organ transplantation as described in this chapter includes transplantation of the kidney, pancreas or islets of Langerhans, heart, lungs, heart and lungs, liver, and bowel. Essentially two model, autotransplantation (rejection-free, used to master transplantation techniques, to explore the influence of preservation, ischemia, etc.) or allotransplantation (includes rejection whose suppression can be studied) are available to the investigator. The chapter deals briefly with the specific issue of transplantation of individual organs and surveys the techniques of their preservation.

REFERENCES

1. **Hardy, J. D.,** The transplantation of organs, *Surgery,* 56, 685, 1964.
2. **Largiadér, F.,** *Organ Transplantation,* Georg Thieme Verlag, Stuttgart, 1966.
3. **Woodruf, M. F. A.,** *The Transplantation of Tissue and Organs,* Charles C Thomas, Springfield and Ryerson Press, Toronto, 1960.
4. **Calne, R. Y.,** Allografting in the pig, in *Immunological Aspects of Transplantation Surgery,* Calne, R. Y., Ed., John Wiley & Sons, New York, 1973, 296.
5. **Najarian, J. S. and Simmons, R. L.,** *Transplantation,* Urban-Schwarzenberg, Munich, 1972.
6. **Calne, R. Y.,** Comparison of various perfusion fluids used in cold storage, *Transplant. Proc.,* 9, 1541, 1977.
7. **Toledo-Pereyra, L. H.,** *Basic Concepts in Organ Procurement, Perfusion and Preservation for Transplantation,* Academic Press, London, 1982.
8. **Belzer, F. O., Ashby, B. S., and Dunphy, J. E.,** 24-Hours and 72-hours preservation of canine kidneys, *Lancet,* II, 536, 1967.
9. **Chatterjee, S. N.,** *Manual of Renal Transplantation,* Springer Verlag, New York, 1979.
10. **Dempster, W. J.,** Kidney homotransplantation, *Br. J. Surg.,* 40, 447, 1953.
11. **Hamburger, J., Crosnier, J., Dormont, J., and Bach, J. F.,** *Renal Transplantation. Theory and Practice,* Williams & Wilkins, Baltimore, 1972.
12. **Mosley, J. G. and Castro, J. E.,** Arterial anastomoses in renal transplantation, *Br. J. Surg.,* 65, 60, 1978.
13. **Stewart, B. H.,** The surgery of renal transplantation, *Surg. Clin. North Am.,* 51, 1123, 1971.
14. **Markowitz, J., Archibald, J., and Downie, H. G.,** *Experimental Surgery,* Williams & Wilkins, Baltimore, 1964.
15. **Agnes, S., Castagneto, M., and Castiglioni, C. C.,** Segmental pancreatic transplantation in the pig: comparative study of different techniques, *Transplant. Proc.,* 12(Suppl. 2), 129, 1980.
16. **Toledo-Pereyra, L. H., Castellanos, J., Lampe, E. W., Lillehei, R. C., and Najarian, J. S.,** Comparative evaluation of pancreas transplantation techniques, *Ann. Surg.,* 182, 567, 1975.
17. **Toledo-Pereyra, L. H., Castellanos, J., Manifacio, G., and Lillehei, R. C.,** Basic requirements of pancreatic mass for transplantation techniques, *Arch. Surg.,* 114, 1058, 1979.
18. **Baumgartner, D., Sutherland, D. E. R., Heil, J. E., Zweber, B., Awad, E. A., and Najarian, J. S.,** Cold storage of segmental canine pancreatic graft for 24 hours, *J. Surg. Res.,* 20, 248, 1980.
19. **Brynger, H., Mjörnstedt, L., and Ollausson, M.,** Heterotopic grafting of pancreas to the neck in the rat — an experimental model, *Transplant. Proc.,* 12(Suppl. 2), 148, 1980.
20. **Garvin, J. P., Castaneda, M. A., Niehoff, M. L., Mauller, K. A., and Brems, J. J.,** An in situ evaluation of distal splenic arteriovenous fistula on pancreas function in an isolated pancreas segment, *Arch. Surg.,* 120, 1148, 1985.
21. **Gold, M., Wittaker, J. R., Veith, F. J., and Gliedman, M. L.,** Evaluation of ureteral drainage for pancreatic exocrine secretion, *Surg. Forum,* 23, 375, 1972.
22. **Kyriakides, G. K., Nuttal, F. Q., and Miller, J.,** Segmental pancreatic transplantation in pigs, *Surgery,* 85, 154, 1979.
23. **Kyriakides, G. K., Sutherland, D. E. R., Olson, L., and Najarian, J. S.,** Segmental pancreatic transplantation in dogs, *Transplant. Proc.,* 11, 530, 1979.
24. **Martin, X., Faure, J. L., Eloy, R., Amiel, J., Margonari, J., and Dubernard, J. M.,** Systemic versus portal vein drainage of segmental pancreatic transplants in dogs, *Transplant. Proc.,* 12, (Suppl. 2), 138, 1980.

25. **Bowen, K. M., Andrus, L., and Lafferty, K. J.,** Successful allotransplantation of mouse pancreatic islets to immunosuppressed recipients, *Diabetes,* 29, 98, 1980.

26. **Lacy, P. E., Davie, J. M., and Finke, E. H.,** Prolongation of islet xenograft survival (rat to mouse), *Diabetes,* 30, 285, 1981.

27. **Leonard, R. J., Lazarow, A., and Hegre, O. D.,** Pancreatic islet transplantation in the rat, *Diabetes,* 22, 413, 1973.

28. **Mehigan, D. G., Zuidema, G. D., and Cameron, J. L.,** Pancreatic islet transplantation in dogs. Critical factors in technique, *Am. J. Surg.,* 141, 208, 1981.

29. **Mosley, J. G. and Castro, J. E.,** Arterial anastomoses in renal transplantation, *Br. J., Surg.,* 65, 60, 1978.

30. **Lacy, P. E., Kostianovsky, M., and Louis, S.,** Method for the isolation of intact islets of Langerhans from the rat pancreas, *Diabetes,* 16, 35, 1967.

31. **Bank, H. L.,** A high yield method for isolating rat islets of Langerhans using differential sensitivity to freezing, *Cryobiology,* 20, 237, 1983.

32. **Buitrago, A., Gylfe, E., Henriksson, Ch., and Pertfoft, H.,** Rapid isolation of pancreatic islets from collagenase digested pancreas by sedimentation through Percoll℠ at unit gravity, *Biochem. Biophys. Res. Commun.,* 79, 823, 1977.

33. **Kretschmer, G. J., Sutherland, D. E. R., Matas, A. J., Cain, T. L., and Najarian, J. S.,** Autotransplantation of pancreatic islets without separation of exocrine and endocrine tissue in totally pancreatectomized dogs, *Surgery,* 82, 74, 1977.

34. **Lorenz, D., Lippert, H., Tietz, W., Worm, V., Hahn, H. J., Dorn, A., Koch, G., Ziegler, M., and Rosenbaum, K. D.,** Transplantation of isolated islets of Langerhans in diabetic dogs, *J. Surg. Res.,* 27, 181, 1979.

35. **Warnock, G. L., Rajjotte, R. V., and Procyshyn, A. W.,** Normoglycemia after reflux of islet-containing pancreatic fragments into the splenic vascular bed in dogs, *Diabetes,* 32, 452, 1983.

36. **Frankel, B. J., Gylfe, E., Hellman, B., Idahl, L., Landström, U., Lovtrup, S., and Sehlin, J.,** Metabolism of cold-stored pancreatic islets, *Diabetologia,* 15, 187, 1978.

37. **Gordon, D. A., Toledo-Pereyra, L. H., and MacKenzie, G. H.,** Preservation for transplantation: a review of techniques of islet cell culture and storage, *J. Surg. Res.,* 32, 182, 1982.

38. **Bank, H. L.,** Cryobiology of isolated islets of Langerhans circa 1982, *Cryobiology,* 20, 119, 1983.

39. **Toledo-Pereyra, L. H., Gordon, D. A., and MacKenzie, G. H.,** Application of cryopreservation techniques to islet cell allotransplantation, *Cryobiology,* 20, 205, 1983.

40. **Andersson, A.,** Reversal of hyperglycemia by intrasplenic transplantation of 4-week-cultured allogeneic mouse islets, *Diabetes,* 4, 55, 1982.

41. **Bobzien, B., Yasunami, Y., Majercik, M., Lacy, P., and Davie, J. M.,** Intratesticular transplants of islet xenografts (rat to mouse), *Diabetes,* 32, 213, 1983.

42. **Barker, C., Naji, A., and Silvers, W. K.,** Immunologic problems in islet transplantation, *Diabetes,* 29, 86, 1980.

43. **Mullen, Y.,** Specific immunosuppression of fetal pancreas allografts in rats, *Diabetes,* 29, 113, 1980.

44. **Maggett, W. M., Willman, V. L., Haruda, Y., Barner, H. B., Cooper, T., and Hanlon, C. R.,** Work capacity and efficiency of the autotransplanted heart, *Surg. Forum,* 17, 222, 1966.

45. **Willman, V. L., Cooper, T., Cian, L. G., and Hanlon, C. R.,** Autotransplantation of the canine heart, *Surg. Gynecol. Obstet.,* 115, 299, 1962.

46. **Lower, R. R. and Shumway, N. E.,** Studies in orthotopic homotransplantation of the canine heart, *Surg. Forum,* 11, 18, 1960.

47. **Lambotte, L.,** Liver preservation, in *Basic Concepts of Organ Procurement, Perfusion and Preservation for Transplantation,* Toledo-Pereyra, L. H., Ed., Academic Press, New York, 1982.

48. **Hökl, J., Kořístek, V., Černý, J., Gregor, Z., Fikulka, J., and Busch, K.,** Orthotopic allotransplantation of liver in pigs with fulminant hepatic failure, *Transplantation,* 29, 424, 1980.

49. **Pritchard, T. J. and Kirkman, R. L.,** Small bowel transplantation, *World J. Surg.,* 9, 860, 1985.

50. **Raju, S., Didlake, R., Cayirli, M., Turner, M. D., Grogan, J. B., and Achord, J.,** Experimental small bowel transplantation utilizing cyclosporine, *Transplantation,* 38, 561, 1984.

51. **Wassef, R., Cohen, Z., Nordgren, S., and Langer, B.,** Cyclosporine absorption in intestinal transplantation, *Transplantation,* 39, 496, 1985.

52. **Lee, K. W. and Schraut, W. H.,** In vitro allograft irradiation prevents graft-versus-host-disease in small bowel transplantation, *J. Surg. Res.,* 38, 364, 1985.

53. **Billiar, T. R., Garberglio, C., and Schraut, W. H.,** Maltose absorption as an indicator of small-intestinal allograft rejection, *J. Surg. Res.,* 37, 75, 1984.

54. **Cohen, Z., Nordgren, S., Lossing, A., Cullen, J., Craddock, G., and Langer, B.,** Morphologic studies of intestinal allograft rejection. Immunosuppression with cyclosporine, *Dis. Colon Rectum,* 27, 228, 1984.

55. **Craddock, G. N., Nordgren, S., Reznick, R., Gilas, T., Lossing, A. G., Cohen, Z., Stiller, C. R., Cullen, J. B., and Langer, B.,** Small bowel transplantation in the dog using cyclosporine, *Transplantation,* 35, 284, 1983.

56. **Diliz-Perez, H. S., McClure, J., and Bedetti, C.,** Successful small bowel allotransplantation in dogs with cyclosporine and prednisone, *Transplantation,* 37, 126, 1984.

57. **Kočandrle, V., Houttuin, E., and Prohaska, J. V.,** Regeneration of the lymphatics after autotransplantation and homotransplantation of the entire small intestine, *Surg. Gynecol. Obstet.,* 122, 587, 1966.

58. **van Ooosterhout, J. M. A., de Boer, H. H. M., and Jerusalem, C. R.,** Small bowel transplantation in the rat: the adverse effect of increased pressure during the flushing procedure of the graft, *J. Surg. Res.,* 36, 140, 1984.

59. **Nordgren, S., Cohen, Z., and Mackenzie, R.,** Functional monitors of rejection in small intestinal transplants, *Am. J. Surg.,* 147, 152, 1984.

Selected Models of Diseased States

Chapter 20

HYPERTENSION

J. Jelínek

TABLE OF CONTENTS

I. INTRODUCTION

The study of factors and mechanisms leading to a permanent increase in arterial pressure is motivated by an effort to clarify the etiology and pathogenesis of high blood pressure which would provide a basis for its rational prevention and treatment. A retrospective analysis[1] shows that research in this area was initiated by experiments designed to test the hypothesis proposed by Traub in 1856. This hypothesis tried to explain the relationship between cardiac hypertrophy and renal disease, described 20 years earlier by Bright, by an increase in arterial pressure attributable to a decrease in blood flow through renal tissue and a decrease in diuresis resulting in fluid retention. Thus, first trials carried out in the 1880s were designed to monitor changes in the weight of the heart after ischemic renal injury, or after restricting the blood supply due to renal artery constriction. In fact, these trials preceded by some 50 years systematic research into the role played by the kidney in blood pressure regulation and in the pathogenesis of hypertension started on the basis of independent observations by Goldblatt in 1934.

The 1940s saw the first data accumulating on the hypertensive effect of deoxycorticosterone[2] and its relationship to NaCl intake.[3] This provided an impetus to carry out research into the role of excessive intake of salt and steroid hormones, and of corticoids in particular, in the pathogenesis of hypertension. Differences in the blood pressure response of experimental animals exposed to the effect of hypertensive stimuli just as in the blood pressure level in individual animals within rat colonies pointed to the participation of genetic factors in the pathogenesis of hypertension. This concept was reflected in the selection of rat strains with different sensitivity to hypertensive stimuli,[4,5] as well as strains developing hypertension spontaneously during postnatal life.[6-8]

The major role played by the central nervous system in regulation of blood pressure, in its turn, led to efforts to induce hypertension by interventions altering the activities of nervous regulatory mechanisms. While to achieve this, surgical procedures were initially employed;[9] they were later replaced by stressful stimuli.

Experimental hypertension can be divided into five categories: (1) renal hypertension caused by interventions disturbing the volumoregulatory function of the kidney, (2) salt hypertension in which the etiologic factor is a salt intake exceeding the body's physiological requirements, (3) steroid hypertension stimulated especially by corticoid administration, (4) neurogenic hypertension, and (5) genetically predetermined hypertension. This division is bound to be necessarily imprecise as, in most cases, the degree of steroid hypertension is directly related to salt intake. Steroid hypertension can therefore be regarded as a special type of salt hypertension in which the steriod effect plays a rather permissive role. A similar effect is achieved by renal tissue reduction enhancing the hypertensive action of excessive salt intake, either alone or in combination with an additional effect of corticoids. In all of these cases, each individual's genetic background plays an important role in the onset of hypertension.

II. RENAL HYPERTENSION

A. Introduction

This category includes all types of hypertension induced by restriction of renal blood supply (renovascular hypertension) or by reduction of functional renal mass. According to the results of Guyton's system of analysis of blood pressure regulation,[10] the hypertensive action of these interventions depends on to what extent they have disturbed the balance between the glomerular filtration rate and feedback reception in renal tubules of the fluid filtered. The predominance of reabsorption over filtration leads to sodium and water retention with a subsequent increase in the volume of extracellular fluid which, by a feedback mechanism, raises blood pressure to a level necessary to restore the salt and water balance.[10] In renovascular types of hypertension, a diminished blood supply to the kidney results in disturbing the glomerulotubular balance by a decrease in the glomerular filtration rate. Reduction of renal mass without disturbing the glomerulotubular balance leads to uremia rather than to hypertension. Hypertension does not develop unless sodium or water intake has increased or prepubertal lesions hindering glomerular filtration and thus disturbing glomerulotubular balance[11] are present or develop in residual renal tissue.

B. Renovascular Hypertension

1. Goldblatt's Hypertension

This type of hypertension was first induced in 1934 in the dog by constricting one or both renal arteries.[12] A silver clamp that could be mounted on an exposed segment of the renal artery was used and, by turning an inbuilt screw, the artery was constricted to the diameter required. For experiments requiring additional changes in renal artery constriction, a number of modifications have been proposed allowing changes in the kidney perfusion pressure without repeated surgical interventions. The simplest devices include a Teflon®-coated wire forming a loop around the renal artery the diameter of which (and, consequently, the degree of arterial constriction) can be changed by pulling the loose end of the wire passing through a catheter to the surface of the skin.[13]

A more sophisticated device is the occlusion cuff made of latex and filled with air or saline from the outside, with the degree of inflation determining the degree of renal artery constriction.[14] Manipulation is facilitated by a modification of Goldblatt's clamp allowing permanent access to the construction screw head via a Silastic tube. The tube is mounted on the clamp body.[15] For experiments in the rabbit, a rigid metallic clamp made of a silver $2 \times 3 \times 6$ mm quadrangular prism has been designed with a notch 5 mm deep and 0.5 mm wide cut in its base. This clamp can be used in rabbits weighing about 3 kg.[16] Rigid clamps, usually made of a band of silver sheet 0.2 mm thick, 2 mm wide, and 8 mm long, are employed for experiments in the rat. To get the required internal diameter, the clamp is shaped by turning it around a steel cuff of the appropriate size. It may be helpful to trim one end of the clamp to facilitate mounting of the clamp on the exposed renal artery. As a rule, the left renal artery is preferred since it is longer than the right one, which makes clamping easier. To reach the artery, the retroperitoneal approach is used, and the clamp is mounted as close to the aorta as possible so as to sit on the artery at the site of bending and have its open end directed caudally. Experiments carried out in the rat weighing 140 to 160 g with clamps of various diameters have shown that the most pronounced hypertensive effect response is achieved with clamps 0.2 mm in diameter.[17]

Hypertension induced by the constriction of a single renal artery is termed two-kidney, one-clamp (2K-1C) Goldblatt's hypertension. In the dog, it is not permanent, as the creation of collateral circulation improves blood supply to the kidney clamped. Removal of the contralateral unclamped kidney, however, leads to permanent hypertension,[18] referred to as one-kidney, one-clamp (1K-1C) Goldblatt's hypertension. Angiographic studies have shown[19] that collateral circulation which might improve blood supply to the clamped kidney cannot arise in the rat. The frequency and degree of 2K-1C hypertension are modulated by genetic factors.[20] The failure to induce this type of hypertension experimentally, however, may also be due to injury to the clamped kidney by ischemia.[21]

An analogy to 1K-1C hypertension is that induced by constriction of the aorta over the origin of both renal arteries which evenly restricts blood supply to all renal tissue.[22] An analogy to 2K-1C hypertension is hypertension induced, in the rat, by constriction of the aorta between the origins of both renal arteries. If this intervention restricts the blood supply to the kidney below the site of constriction to such an extent that diuresis stops without ischemic necrosis of renal tissue, a pronounced renal atrophy associated with hypertension occurs. An atrophic and, from the point of view of its excretory function, fully insufficient kidney has been termed "endocrine".[23] The intervention leading to its formation, however, often results in ischemic necrosis of the kidney without an increase in blood pressure. A similar type of hypertension can also be induced in the Sprague-Dawley rat by complete ligation of the aorta between the origins of renal arteries. This intervention, too, results in atrophy of the kidney with impaired blood supply and in hypertension. Its effectiveness is indirectly related to the animal's weight. While, in rats weighing less than 150 g, hypertension does not occur; in rats weighing over 300 g, it develops in 90% of cases.[24]

1K-1C Goldblatt's hypertension is not affected by changes in sodium intake. By contrast, 2K-1C hypertension does not occur in rats fed a sodium-free diet.[25] This observation, consistent with conclusions of Guyton's analysis of blood pressure regulation, enables the categorization of one-kidney hypertension as a diffuse-type pretubular hypertension with the glomerular filtration rate decreased evenly in all nephrons, and the categorization of two-kidney hypertension as disperse-type preglomerular hypertension in which a decrease in the glomerular filtration rate involves only part of the nephrons in the renal mass.[10]

Whereas in rats with 1K-1C hypertension, clamp removal leads to a decrease, or even normalization in blood pressure,[26] in rats with 2K-1C hypertension this effect manifests itself in the early stage of hypertension only. After a sufficiently long interval,[27] hypertension persists even after its underlying cause has ceased to exist. This residual hypertension, evident after clamp removal or, possibly, after the removal of the entire clamped kidney, is referred to as post-Goldblatt's, or metaischemic, hypertension. The maintenance of an increased arterial pressure level is attributable to hypertensive lesions that have developed in the unclamped kidney, hindering its excretory capacity.[28]

2. Hypertension Due to Renal Tissue Compression

A simple procedure inducing chronic hypertension involves compression of the kidney with a woolen thread (No. 8 gauge) in small or, with a tape, in larger laboratory animals.[29] The thread/tape forms a ligature around the kidney in the shape of an eight whose loops are placed on the renal poles (figure-eight ligature). Still, hypertension develops in a small proportion of animals (10 to 20%) within 3 to 6 months. Its development can be accelerated, and its frequency increased (up to 50%) by removing the contralateral kidney.

An effect similar to that of the figure-eight ligature is exerted by perinephritis in which a connective tissue capsule is formed. For inducing perinephritis, one or both kidneys are wrapped with cellophane. Intervention covering both kidneys is more effective and its action is further augmented by the removal of the uncompressed kidney.[30] A perirenal capsule made of colloid membrane,[31] or cellophane-acetone,[32] or latex[33] wrapping have been used as an alternative to cellophane.

Renal tissue compression is also involved in the pathogenesis of hypertension induced in the rat by a transient lack of choline that occurs during weaning.[34] This intervention results in episodes of subcapillary hemorrhage allowing formation of a fibrotic capsule compressing renal parenchyma of the growing kidney. This is evidenced by the observation that decapsulation carried out after a hemorrhagic episode prevents the onset of hypertension.[35]

C. Hypertension Due to Renal Tissue Depletion

This category includes types of hypertension induced by surgical reduction of functional renal tissue (renal ablation hypertension[36]), or by interventions damaging renal parenchyma.

1. Renal Ablation Hypertension

The onset and degree of hypertension are directly related to the degree of renal mass reduction and to the body's sodium balance. While in the rat, a 50% reduction in renal mass by unilateral nephrectomy leads to a varying (about 18%) frequency of hypertension, 75%

depletion by subtotal nephrectomy (unilateral nephrectomy combined with the removal of both poles of the residual kidney) is associated with hypertension in 60% of the cases. An increased sodium intake increases the frequency, degree, and rate of development of hypertension.[36,37] A similar effect is produced by increased protein intake.[38] When present in residual renal tissue, both these factors facilitate the onset of nephrosclerosis which further restricts the excretory capacity thus contributing to the development of hypertension. This finding is consistent with the observation that, in the rat, unilateral nephrectomy induced hypertension only on the condition that renal tissue contains nephrosclerotic foci[39] whose extent may increase in connection with the compensatory growth of the residual kidney.[40] Hypertension stimulated by subtotal nephrectomy and excessive NaCl intake is easier to develop in sexually immature rats than in adult animals.[41] This is explained by an increased susceptibility of renal tissue of young animals to the onset of hypertensive lesions since, in them, hypertension is associated with signs of impaired excretory renal function which are absent in older animals.[41]

In rats maintained in a good state by parabiotic connection with normal partners, or by peritoneal dialysis, complete reduction of renal tissue by bilateral nephrectomy leads to hypertension,[42] referred to as renoprive one. This type of hypertension can also be produced in the dog[43] and has been observed even in anephric patients.[44] Development of hypertension is potentiated by a protein-containing diet with a positive water and salt balance of the body.[45] A model of renoprive hypertension has been employed to study the hypertensive function of renal tissue independent of its excretory function.[46]

2. Hypertension Due to Renal Tissue Injury

Hypertension caused by a renal parenchyma injury can be produced by focal infarction of the kidney in the rat.[47] The infarction is induced by ligation of one of the branches of the renal artery originating from it at the hilus of the kidney. The hypertensive effect of this intervention can be enhanced by removal of the contralateral kidney and administration of 1% NaCl solution as the only drinking fluid. Hypertension develops within 4 to 8 weeks.[48] It is believed that the pathogenesis of this type of hypertension involves an autoimmune response injuring renal tissue. Hypertension does not develop in rats administered immunosuppressives,[49] and can be transferred by lymph node cells to normotensive, immunologically tolerant rats.[50] This finding is consistent with the observation that, in the rat, activation of immune systems of repeated injections of homologous renal tissue produces renal injury associated with hypertension.[49] A similar effect is exerted by the administration of antibodies reacting with basal membrane antigens. In this way, hypertension has been produced in the dog[51] as well as in uninephrectomized rats fed 1% salt solution. In both animals, hypertension can be documented as early as within 2 weeks.[52]

The autoimmune mechanism injuring the vascular system[53] could also possibly be involved in the type of hypertension which can be elicited in the rat by infusion of angiotensin (7 to 12 infusions of angiotensin II at a dose of 0.1 μg/kg/min administered daily for 8 to 10 hr). After an approximately 1 month period of latency, the infusion leads to the development of progressive hypertension associated with an injured vascular bed also involving renal tissue.[54] A similar type of hypertension has been produced by repeated subcutaneous daily injections of angiotensin (100 μg/kg) in uninephrectomized rats fed 1% salt solution. Hypertension develops within 20 to 30 days and disappears after salt intake has been decreased, after administration of spironolactone, or after transplantation of a homologous renal tissue (angiotensin-salt hypertension[55]). Transient anoxia of the kidney likewise produces hypertension. It has been described in the dog[56] and in uninephrectomized rats fed 1% salt solution, whose kidney was exposed to transient ischemia in the prepubertal period. The hypertension produced was only mild.[57] Other methods used occasionally to produce hypertension included kidney radiation with X-rays[58] or parenteral administration of macromolecular substances — methylcellulose or polyvinyl alcohol.[59]

II. SALT HYPERTENSION

After a sufficiently long period, a salt (NaCl) intake in excess of the body's physiological requirements may induce hypertension. Salt is administered to experimental animals either as a solution provided as the only drinking fluid, or in the chew. Salt hypertension was produced for the first time in chickens fed a 0.9% solution of NaCl. Hypertension developed after 3 weeks and an increase in the drinking fluid salt concentration resulted in yet another increase in blood pressure.[60] Similarly, hypertension was induced in rats fed a 2% solution of NaCl over 2 to 4 weeks.[61] This procedure, however, is not always effective.[62] This may apparently be due to the different susceptibility of each rat strain to the hypertensive effect of excessive salt.[63]

The varied susceptibility to the hypertensive effect of excessive salt manifests itself even within a single strain. Sprague-Dawley rats, fed a diet supplemented with NaCl within a range of 2.8 to 9.8%, develop hypertension whose degree is directly related to NaCl intake.[64] The length of exposure needed to develop hypertension, and its degree, however, differ in each animal. About 25% of the population do not develop hypertension altogether, and by contrast, hypertension develops rapidly, reaching the malignant stage within several months, in 2 to 3%. In the remainder, hypertension develops gradually at an uneven rate.[65] After a 1-year exposure to the salt diet, hypertension persists even after a switch back to a salt-free diet in about two thirds of animals, and is associated with an increased urea concentration in the plasma.[66] This suggests that, just as in post-Goldblatt's hypertension, hypertensive injury to the kidney is at play in the pathogenesis of this residual (post-salt) type of hypertension.

The varying hypertensive effect of excessive salt intake has pointed to the possible role of genetic factors. Accelerated development of salt hypertension by the administration of L-triiodothyronine enabled to obtain two lines of rats with minimal and maximal hypertensive response, i.e., a line of rats fully resistant (R) and completely sensitive (S) already in the third filial generation not only to salt hypertension[4] but also to other types of experimental hypertension.[67] These lines were later used to produce inbred strains referred to as R/JR of S/JR.[68] The blood pressure of S-rats is higher than that of R-rats even when fed a low-salt diet,[69] making the S-rat in this respect resemble strains with spontaneously conditioned hypertension. It is believed that sensitivity to hypertensive stimuli is predetermined by two to four genetic loci out of which only one has been identified to date. This locus determines the ratio between 11-β and 18-hydroxylation of deoxycorticosterone (DOC) in the adrenal cortex and consequently, the ratio between 18-OH-DOC and corticosterone in peripheral blood. In the R-rat, this ratio is more than twice as low as that in the S-rat and in nonselected animals.[70]

Selective cross-breeding of Sabra strain rats sensitive or resistant to hypertension induced by a combined effect of excessive salt and deoxycorticosteronacetate (DOCA) also produced a line of rats resistant (N line) and sensitive (H line) to DOCA-NaCl and to 2K-1C Goldblatt's hypertension. The varied sensitivity is explained by a genetically determined difference in catecholamine metabolism. Just as in R- and S-rats, the blood pressure of animals not affected by a hypertensive stimulus is higher in the H line than in the N line.[5]

While experiments designed to induce salt hypertension in the rabbit[71] failed, it has been produced in monkeys (*Papio hamadryas*) after a 2-year exposure to excessive salt (26 mmol/kg/day).[72]

IV. STEROID HYPERTENSION

Research into the hypertensive action of steroid hormones has shown that hypertension can be produced by administration of minor alocorticoids and glucocorticoids, and by dysfunction of the adrenal cortex induced on its regeneration or during androgen administration. Steroid hypertension can thus be subdivided into three groups.

A. Mineralocorticoid Hypertension

Mineralocorticoid hypertension depends on salt intake and can thus be regarded as a special type of salt hypertension. The most frequently used type of hypertension is that produced by administration of DOCA. While the hypertensive effect of a steroid is absent in rats fed a salt-free diet,[73] it is enhanced by unilateral nephrectomy and substitution of tap water by a 1%[74] or isotonic[75] solution of NaCl. Occasionally, the drinking fluid is supplemented with potassium to compensate for losses due to the effect of DOCA. It seems, however, that this supplement is unnecessary.[76] The hypertension (referred to as DOCA-NaCl or DOCA-salt hypertension) develops at a faster rate and achieves a higher degree in rats exposed to the effect(s) of a steroid and salt in the period of sexual maturation than in those exposed to it during adulthood.[77]

DOCA can be administered in the form of subcutaneously implanted pellets. In rats weighing 200 to 250 g, implantation of pellets containing 25 mg of DOCA, and repeated after 15 days, produces hypertension within 4 weeks.[76] The pellets can be substituted by silicone rubber bands impregnated with DOCA at a dose of 100 mg/kg of body weight.[78] This technique has also been used in the pig.[79] Most often, DOCA is administered subcutaneously or intramuscularly either in oil or in the form of a water microcrystalline suspension. The hypertensive action of both is about equal. Subcutaneous administration in sesame oil at a dose of 30 mg/kg twice a week,[80] and intramuscular administration of water suspension at a dose of 50 mg/kg once a week[81] produces hypertension within 6 weeks.

After a prolonged period of exposure (approximately 3 months) to the combined effect of DOCA and excessive salt, hypertension persists even after a switch to a standard diet.[82] This residual form is called "post-DOCA" or "metacorticoid" hypertension. In rats switched to a DOCA-NaCl regimen in the prepubertal period, this type of hypertension tends to develop at a faster rate.[77]

The pathogenesis of DOCA-NaCl and post-DOCA hypertension has not yet been fully clarified. Considering the crucial role played by the kidney in blood pressure regulation,[11] this may be due to decreased sodium excretion during development of hypertension[83] and to the functional consequences of hypertensive injury to the kidney[84] at a time when the effect of DOCA has already subsided.

Aldosterone (D-aldosteroneacetate) is used only sporadically to produce hypertension. Uninephrectomized rats fed 1% NaCl solution were subcutaneously administered aldosterone in an oil solution at 2 daily doses. Given at a total daily dose of 300 to 400 μg/100g, it produced severe hypertension after 3 weeks.[85] Hypertension could be induced, however, with substantially smaller doses (25 μg/100 g)[86] given over the same period of time. More recently, aldosterone administration has been carried out using osmotic minipumps that enable continuous administration of the steroid. Administered to rats weighing 180 g at a dose of 100 μg over 24 hr, aldosterone produced hypertension already after 10 days. By contrast, DOCA, administered at the same dose and using the same technique, failed to produce hypertension even after 3 weeks. Apparently, the hypertensive effect of aldosterone is higher than that of DOCA.[87]

B. Glucocorticoid Hypertension

The hypertensive effect of glucocorticoids, unlike mineralocorticoids, is independent of salt intake. Cortisone and cortisol produce hypertension even in rats fed a low-sodium diet,[88] and the hypertensive effect of cortisol is not enhanced by the substitution of tap water with a 1% solution of NaCl, or by unilateral nephrectomy.[89] Still, enhancement of the hypertensive effect by an increased salt intake may become manifest in glucocorticoids with a physiologically significant mineralocorticoid action, provided their plasma concentration is high enough. In adrenalectomized rats, corticosterone, with its mineralocorticoid activity, produces hypertension independent of salt intake only at a low plasma concentration.[90] The pathogenesis of this type of hypertension probably involves an increase in the volume of extracellular fluid and blood plasma, explained by fluid transport from the cellular to the extracellular compartment of tissues, and increased tubular sodium resorption inhibiting excretion of excessive fluid. Fluid transport is attributed to the glucocorticoid and its retention, and to the mineralocorticoid effect of corticosterone.[83] A similar mechanism is also assumed in the hypertension produced by 8 α-fluorocortisol[91] possessing a pronounced gluco- and mineralocorticoid action.[92]

Of glucocorticoids with negligible mineralocorticoid activity, methylprednisolone[93] and dexamethasone[94] have been used to produce hypertension. While in the rat both corticoids induce hypertension within a week, chronic infusion of methylprednisolone (2-800 mg/day over 1 to 4 weeks) failed to produce hypertension in the dog and led rather to a decrease in blood pressure explained by salt and water losses.[95] The pathogenesis of the types of hypertension produced by glucocorticoids with negligible mineralocorticoid activity involves increased sensitivity of the vascular muscular tissue to the pressoric effect of noradrenaline associated with inhibition of prostaglandin synthesis.[94] These types of hypertension are regarded as a model suitable for studying the role played by vasodilating systems (prostaglandins and kinins) in the pathogenesis of hypertension.[93]

C. Hypertension Due to Adrenal Cortical Dysfunction

There is a large body of evidence indicating that experimentally produced dysfunction of the adrenal cortex associated with hypertension goes hand in hand with increased secretion of deoxycorticosterone. Hence the types of hypertension falling into this category are close to DOCA-NaCl hypertension and can be ranked among the mineralocorticoid types of hypertension.

1. Hypertension Due to Adrenal Cortex Regeneration

In unilaterally nephrectomized rats fed a 1% solution of NaCl, regeneration of adrenal cortex produces hypertension maintained after 6 to 7 weeks even after removal of the regenerated cortex or after a switch back to a normal dietary regimen.[96] Adrenal cortex regeneration can be produced by the removal of one, and enucleation of the remaining adrenal gland. This intervention involves incision of the capsule of the gland and dislodgement of its tissue by applying gentle pressure with a bent pinsette. Unilateral nephrectomy, adrenalectomy, and enucleation are performed simultaneously and, immediately after the procedure, animals are given a 1% NaCl solution to drink.[97] Regeneration can also be induced by injuring both adrenal glands by compression or needle punctures directed longitudinally and transversally to the gland axis.[98] The hypertensive effect of a regenerating adrenal gland manifests itself in rats operated on at the age of 3 weeks. The effect is substantially less evident in adult rats.[99-101] While in female rats hypertension develops more readily than in males,[100-102] this sexual difference disappears when the animals (both males and females) are fed a moderately salt-enriched diet.[103]

Development of hypertension is associated with an increase in the DOCA concentration in peripheral blood,[104] attributable to its lowered conversion into corticosterone. This is evidenced by a decrease in 11 β-hydroxylase activity associated with a decrease of mitochondrial cytochrome P-450.[105] DOC concentration in peripheral blood can be more than three times higher than the concentration necessary to produce DOCA-NaCl hypertension.[104] The increment reaches a maximum 3 weeks after enucleation, disappearing within 4 to 7 weeks. Thus, it may play a part in the onset of hypertension, not in its maintenance. It is believed that either a similar mechanism as in post-DOCA hypertension[106] or increased production of 18-OH-DOC is involved.

2. Hypertension Induced by Androgen Administration

In uninephrectomized sexually immature female rats fed with a 1% NaCl solution, 17 α-methylandrostenediol (MAD),[18] 17 α-methytestosterone (MT),[109] and testosterone[110] produce hypertension. The hormones were administered subcutaneously in a water or oil suspension at a daily dose of 10 mg per rat, and hypertension developed within 5 weeks. Its development was associated with increased DOC production[105] which, according to some data,[110] was directly related to the degree of hypertension, Just as in hypertension due to adrenal cortex regeneration, DOC-overproduction is explained by inhibition of 11 β-hydroxylase and depletion of the mitochondrial cytochrome P-450.[105] Inhibition of 11 β-hydroxylase induced by the administration of metopirone to uninephrectomized rats fed a 1% NaCl solution indeed increases plasma DOC concentration and produces hypertension.[111]

V. NEUROGENIC HYPERTENSION

Neurogenic hypertension can be divided into two groups. The first group includes types of hypertension caused by a surgically induced disturbance in neural or neuroendocrine regulation of blood pressure. The second group includes hypertension due to stressful effects, which can be termed "psychogenic".[112]

A. Surgically-Induced Neurogenic Hypertension

The most frequently used intervention is the elimination of baroreceptor reflexes. Permanent hypertension has been described in the dog following the transection of buffer nerves.[113] Measurements carried out at a later date have shown, however, that the level of blood pressure is most unstable in this type of hypertension,[114] and its mean value of 24 hr increases very little (by about 10%).[115] According to other studies, radical denervation of sinoaortic depressor areas in the rat leads to permanent hypertension,[116] Apparently, the degree of hypertension depends on the completeness of arterial baroreceptor denervation. Carotid sinus denervation plus a section of the cervical aortic nerve increases mean pressure by a mere 11% in the dog. If combined with transection of the intrathoracic vagal branches which innervate arterial and cardiopulmonary baroreceptors, the increase in blood pressure reaches 26%.[117] Central deafferentation of baroreceptor reflexes by bilateral electrolytic lesions of the intermediate zone of the nucleus tractus solitarii increases sympathetic discharge which leads to vasoconstriction and fulminant hypertension in the rat. The increase in blood pressure which provokes heart failure within several hours requires integrity of structures lying above midbrain.[118]

Permanent hypertension has been described after extensive medial anteromedian hypothalamic destruction. Hypertension was associated with renal hypertrophy and increased corticosterone concentration in peripheral blood, and potentiated by increased NaCl intake. Apparently, a role in its pathogenesis may be played by a disturbance of the feedback mechanism controlling corticoid secretion.[119]

Hypertension can likewise be induced by interventions raising intracranial pressure[120] or restricting blood supply to the brain.[121]

B. Psychogenic Hypertension

Rats exposed daily to a combination of light, sound, and mechanical stimuli[122] developed a permanent increase in blood pressure after 3 months. A similar effect has been reported in rats kept separately since weaning until 4 to 5 months of age and then put together in common cages. Mild hypertension developed in about 30% of animals, and its degree was higher in dominant males.[123] Social interaction and competition for territory occurring once the animals have been placed in complex population cages also produces hypertension in male mice of the CBA Agouti strain.[124] Hypertension has been likewise described in female rats placed for 3 to 5 weeks in sound-isolated space (sound withdrawal hypertension).[125]

The onset of psychogenic hypertension is probably due to genetic factors. While a food-shock conflict situation produces hypertension in rats genetically sensitive to the effect of hypertensive stimuli (S-rats), rats resistant to such stimuli (R-rats) do not develop hypertension. The increase in blood pressure is, however, not permanent.[126] In primates (Rhesus monkey), experimental neurosis results in a blood pressure increase demonstrable in awake animals only.[127] Therefore, it seems that a number of the above cases of psychogenic hypertension do not involve a permanent increase in blood pressure, but rather an augmented cardiovascular response to alerting stimuli in the environment.

VI. GENETICALLY PREDETERMINED HYPERTENSION

In rat colonies, the level of blood pressure is not the same in every animal. Selective inbreeding of rats with a high blood pressure level produced several rat strains developing spontaneous hypertension in the course of postnatal development. Presently, four strains, namely, New Zealand,[6] Japanese,[8] Milan,[7] and Lyon strains, are used. Genetically predetermined types of spontaneous hypertension also occur in other species of which Schlager's strain of hypertensive mice has been recently introduced into laboratory practice.[129] Of hypertensive rat strains, the most frequently one used is the Japanese strain referred to as "spontaneously hypertensive" (SHR). This strain gave rise to two other important substrains. While the first one is characterized by a high incidence of cerebral lesions, a high-lipid and high cholesterol diet can produce marked hypercholesterolemia associated with the formation of lipid deposits in mesenteric and cerebrospinal arteries in the second one. These substrains were named "stroke-prone" and "arteriolipoidosis-prone" SHRs[130] and are referred to as SP-SHR and ALP-SHR, respectively. New Zealand rats, usually referred to as "genetically hypertensive" (GH), and Milan hypertensive strain rats (MHS) are used to a lesser extent. While to obtain controls to Japanese and New Zealand rats, normotensive animals from colonies from which the hypertensive strains were derived are used, in the Milan strain, a normotensive strain (MNS) developed simultaneously with the hypertensive strain[7] serves as the control.

In the Japanese and New Zealand strains, higher blood pressure values are demonstrable as early as the perinatal period[131,132] and their levels keep on increasing with age up to the 20th to 28th week.[133,134] In the Milan strain, hypertension develops between the 25th and 50th day of age with the level of blood pressure remaining unaltered thereafter.[7] This suggests different mechanisms involved in the pathogenesis of spontaneous hypertension.

It is believed that the pathogenesis of hypertension in rats of the Japanese and New Zealand strains is similar, with a major role attributed to adrenergic mechanisms.[133] Some data indicate that the sympathetic nervous system in the Japanese strain is hyperactive and the reason for this hyperactivity lies in abnormalities of the central nervous system.[135] It seems, however, that the neurogenic vasoconstrictive tonus is not the only factor responsible for the increase in vascular resistance and blood pressures.[136] Development of hypertension involves an increase in efferent renal activity leading to sodium retention. While renal denervation slows down the development of hypertension in SHR, it is ineffective in animals who have already developed hypertension.[137] SHR kidneys have been shown to exhibit multiplication of α_2 adrenergic receptors[138] which may hinder sodium secretion. The disposition to hypertension could thus be related to an inherent renal-neural abnormality.[135] And, indeed, SHR kidneys do respond to increased perfusion pressure by a lower increase in diuresis and natriuresis than the kidneys of control animals.[140] According to the results of the system analysis of blood pressure regulation,[10] hypertension in the Japanese strain would thus compensate for renal dysfunction, enabling the maintenance of the body's sodium balance.[111,142] A similar mechanism may also be at play in the pathogenesis of hypertension of the New Zealand rat strain in whom renal denervation, performed in the weaning period, likewise slows down development of hypertension. At the same time, sodium excretion in hypertensive rats with maintained renal innervation was comparable with sodium excretion in rats with a denervated kidney and a low blood pressure level[143] in whom the factor complicating diuresis was apparently absent.

The role of the renal component in the pathogenesis of hypertension can be regarded as proven in the Milan strain rat. Whereas no primary neural and hormonal changes have been detected, sodium retention probably due to escalated reabsorption in the proximal tubule has been proved in the development of hypertension in these animals. Cross-transplantation of the kidney between hyper- and normotensive animals decreased the value of blood pressure in hypertensive and increased it in normotensive animals.[144] The renal factor also plays a key role in the hypertensive strain of Schlager's rats in whom a decrease in the glomerular filtration rate entails an increase in the volume of extracellular fluid, consistent with Guyton's concept of the onset of hypertension.[145]

VIII. CONCLUSION

In a number of cases, attempts to induce hypertension in laboratory animals have led to the design of models of some types of hypertension known in human pathology, thus enabling research of their pathogenesis. At the same time, these models have helped expand the body of knowledge about the neural, humoral, and neurohumoral mechanisms of regulation of the circulatory system which must be viewed as a homeostatic complex maintaining the physical and chemical properties of interstitial fluid in agreement with the needs of the body's cellular mass. A key role in this complex is played by the kidney, a fact making the concept that hypertension compensates for renal dysfunction very impressive indeed. The mechanisms that are involved in the onset of this dysfunction and its compensation have not been completely elucidated as yet. Of the above types of experimental hypertension, attention has been especially focused on renal Goldblatt's hypertension, steroid-salt, and genetically predetermined types of spontaneous hypertension.

REFERENCES

1. **Gordon, B.,** Some early investigations of experimental hypertension — an historical review, *Tex. Rep. Biol. Med.,* 28, 179, 1970.
2. **Grollman, A. and Harrison, T. R.,** The effect of various sterol derivatives on the blood pressure of the rat, *J. Pharmacol. Exp. Ther.,* 69, 149, 1940.
3. **Selye, H., Hall, C. E., and Rowley, E. M.,** Malignant hypertension produced by treatment with deoxycorticosterone acetate and sodium chloride, *Can. Med. Assoc. J.,* 49, 88, 1943.
4. **Dahl, L. K., Heine, M., and Tassinari, L.,** Effects of chronic excess salt ingestion. Evidence that genetic factors play an important role in susceptibility to experimental hypertension, *J. Exp. Med.,* 115, 1173, 1962.
5. **Ben-Ishay, D., Saliternik, R., and Welner, A.,** Separation of two strains of rats with inbred dissimilar sensitivity to DOCA-salt hypertension, *Experientia,* 28, 1321, 1972.
6. **Smirk, F. H. and Hall, W. H.,** Inherited hypertension in rats, *Nature (London),* 182, 727, 1958.
7. **Bianchi, G., Fox, U., and Imbasciati, E.,** The development of a new strain of spontaneously hypertensive rats, *Life Sci.,* 14, 339, 1974.
8. **Okamoto, K. and Aoki, K.,** Development of a strain of spontaneously hypertensive rats, *Jpn. Circ. J.,* 27, 282, 1963.
9. **Koch, E.,** *Die reflektorische Selbststeuerung des Kreislaufes,* Steinkopf, Leipzig, 1931.
10. **Guyton, A. C., Manning, R. D., Jr., Hall, J. E., Norman, R. A., Jr., Young, D. B., and Yi-Jen, P.,** The pathogenetic role of the kidney, *J. Cardiovasc. Pharmacol.,* 6, S151, 1984.

11. **Guyton, A. C., Cowley, A. W., Coleman, T. G., Liard, J. F., McCaa, R. E., Manning, R. D., Norman, R. A., and Young, D. B.,** Pretubular versus tubular mechanisms of renal hypertension, in *Mechanisms of Hypertension*, M. P. Sambhi, Ed., Excerpta Medica., Amsterdam, American Elsevier, New York, 1973, 15.

12. **Goldblatt, H., Lynch, J., Hanzal, R. F., and Summerville, W. W.,** Studies on experimental hypertension. I. The production of persistent elevation of systolic blood pressure by means of renal ischemia, *J. Exp. Med.,* 59, 347, 1934.

13. **Beran, A. V., Strauss, J., Brown, C. T., and Katurich, N.,** A simple arterial occluder, *J. Appl. Physiol.,* 28, 510, 1970.

14. **Sham, G. B., White, F. C., and Bloor, C. M.,** A constructive occlusive cuff for medium and large blood vessels, *J. Appl. Physiol.,* 28, 510, 1970.

15. **Ferrario, C. M., Blumle, C., Nadzam, G. R., and McCubbin, J. W.,** An externally adjustable renal artery clamp, *J. Appl. Physiol.,* 31, 635, 1971.

16. **Brooks, B. and Muirhead, E. E.,** Rigid clip for standardized hypertension in the rabbit, *J. Appl. Physiol.,* 31, 307, 1971.

17. **Leenen, F. H. H. and Wybren de Jong,** A solid silver clip for induction of predictable hypertension in the rat, *J. Appl. Physiol.,* 31, 307, 1971.

18. **Goldblatt, H.,** Experimental hypertension induced by renal ischemia *Bull. N.Y. Acad. Med.,* 14, 523, 1938.

19. **Siegel, M. B. and Levinsky, N. G.** Collateral circulation after renal artery occlusion in the rat, *Circ. Res.,* 41, 227, 1947.

20. **Dahl, D. K., Heine, M., and Tassinari, L.,** Role of genetic factors in both DOCA-salt and renal hypertension, *J. Exp. Med.,* 118, 605, 1963.

21. **Swales, J. D. and Blake, J.,** The relation between blood pressure level and ischaemic renal contraction in the rat, *J. Pathol.,* 100, 149, 1970.

22. **Goldblatt, H., Kahn, J. R., and Hanzal, R. F.,** The effect on blood pressure of constriction of the abdominal aorta above and below the site of origin of both main renal arteries, *J. Exp. Med.,* 69, 649, 1939.

23. **Selye, H. and Stone, H.,** Pathogenesis of the cardiovascular and renal changes which usually accompany malignant hypertension, *J. Urol.,* 56, 399, 1946.

24. **Rojo-Ortega, J. M. and Genest, J.,** A method for production of experimental hypertension in rats, *Can. J. Physiol. Pharmacol.,* 46, 883, 1968.

25. **Miksche, L. W., Miksche, U., and Gross, F.,** Effect of sodium restriction on renal hypertension and on renin activity in the rat, *Circ. Res.,* 27, 973, 1970.

26. **Byrom, F. B. and Dodson, L. F.,** The mechanism of vicious circle in chronic hypertension, *Clin. Sci.,* 8, 1, 1949.

27. **Queiroz, F. P., Rojo-Ortega, J. M., and Genest, J.,** Metaischemic (post-Goldblatt) hypertensive vascular disease in rats, *Hypertension,* 3, 756, 1980.

28. **Helmchen, U., Kneissler, U., Bohle, R. M., Reher, A., and Groene, H. J.,** Adaptation and decompensation of intrarenal small arteries in experimental hypertension, *J. Cardiovasc. Pathol.,* 6, S696, 1984.

29. **Grollman, A.,** A simplified procedure for inducing chronic renal hypertension in the mammal, *Proc. Soc. Exp. Biol. (N.Y.),* 57, 102, 1944.

30. **Page, I. H.,** The production of persistent arterial hypertension by cellophane perinephritis, *JAMA,* 113, 2046, 1939.

31. **Kempf, G. F. and Page, I. H.,** Production of hypertension and the indirect determination of systolic arterial blood pressure in rats, *J. Lab. Clin. Med.,* 27, 1192, 1942.

32. **Zbinden, F.,** Neues Verfahren zur experimentellen Erzeugung des permanenten bzw. reversiblen Hochdruckes durch standartisierte Kompression der Niere, *Helv. Physiol. Pharmacol. Acta,* 5, 513, 1946.

33. **Abrams, M. and Sobin, S.,** Latex rubber capsule for producing hypertension in rats by perinephritis, *Proc. Soc. Exp. Biol. Med.,* 64, 412, 1947.

34. **Hartroft, W. S. and Best, C. H.,** Hypertension of renal origin in rats following less than one week of choline deficiency in early life, *Br. Med. J.,* 1, 423, 1949.

35. **Handler, P. and Bernheim, F.,** Effect of renal decapsulation on hypertension induced by single episode of acute choline deficiency, *Proc. Soc. Exp. Biol. Med. (N.Y.),* 76, 338, 1951.

36. **Koletsky, S. and Goodsitt, A. M.,** Natural history and pathogenesis of renal ablation hypertension, *Arch. Pathol.,* 69, 654, 1960.

37. **Koletsky, S.,** Role of salt and renal mass in experimental hypertension, *Arch. Pathol.,* 68, 11, 1959.

38. **Chanutin, A. and Ludewig, S.,** Experimental renal insufficiency produced by partial nephrectomy. V. Diets containing whole dried meat, *Arch. Int. Med.,* 58, 60, 1936.

39. **Grollman, A. and Halpert, B.,** Renal lesions in chronic hypertension induced by unilateral nephrectomy in the rat, *Proc. Soc. Exp. Biol. Med.,* 71, 394, 1949.

40. **Striker, G. E., Nagle, R. B., Kohnen, P. W., and Smuckler, E. A.,** Response to unilateral nephrectomy in old rats, *Arch. Pathol.,* 87, 439, 1969.

41. **Kuneš, J. and Jelínek, J.,** Influence of age on saline hypertension in subtotal nephrectomized rats, *Physiol. Bohemoslov.,* 32, 123, 1984.

42. **Braun-Menendéz, E. and von Euler, U. J.,** Hypertension after bilateral nephrectomy in the rat, *Nature (London),* 160, 905, 1947.

43. **Grollman, A., Muirhead, E. E., and Vanatta, J.,** Role of the kidney in pathogenesis of hypertension as determined by a study of the effects of bilateral nephrectomy and other experimental procedures on the blood pressure of the dog, *Am. J. Physiol.,* 157, 21, 1949.

44. **Dustan, H. P. and Page, I. H.,** Some factors in renal and renoprival hypertension, *J. Lab. Clin. Med.,* 64, 984, 1964.

45. **Muirhead, E. E., Stirman, J. A., and Jones, F.,** Further observations on the potentiation of postnephrectomy hypertension of the dog by dietary protein, *Circ. Res.,* 7, 68, 1959.

46. **Muirhead, E. E. and Brosius, W. L.,** Renomedullary deficiency. A permissive factor in renoprival hypertension, *Arch. Pathol.,* 95, 77, 1973.

47. **Loomis, D.,** Hypertension and necrotising arteritis in the rat following renal infarction, *Arch. Pathol.,* 41, 231, 1946.

48. **Sokabe, H. and Grollman, A.,** A study of hypertension in the rat induced by infarction of the kidney, *Tex. Rep. Biol. Med.,* 21, 93, 1963.

49. **White, F. N. and Grollman, A.,** Autoimmune factors associated with infarction of the kidney, *Nephron,* 1, 93, 1964.

50. **Okuda, T. and Grollman, A.,** Passive transfer of autoimmune induced hypertension in the rat by lymph node cells, *Tex. Rep. Biol. Med.,* 25, 257, 1967.

51. **Selkurt, E. E., Abel, F. L., Edwards, J. L., and Yum, M. N.,** Renal function in dogs with hypertension induced by immunological nephritis, *Proc. Soc. Exp. Biol. Med.,* 144, 295, 1973.

52. **Neugarten, J., Kaminetsky, B., Feiner, H., Schacht, R. G., Liu, D. T., and Baldwin, D. S.,** Nephrotoxic serum nephritis with hypertension: ameliration by antihypertensive therapy, *Kidney Int.,* 28, 135, 1985.

53. **White, F. N. and Grollman, A.,** Experimental periarteritis nodosa in the rat, *Arch. Pathol.,* 78, 31, 1964.

54. **Koletsky, S., River-Velez, J. M., and Pritchard, W. H.,** Production of hypertension and vascular disease by angiotensin, *Arch. Pathol.,* 82, 99, 1966.

55. **Muirhead, E. E., Leach, B. E., and Armstrong, B.,** Angiotensin-salt hypertension, *Clin. Sci. Mol. Med.,* 45, 257s, 1973.

56. **McCabe, R. E., Jr., Gomez, J., and Zintel, H. A.,** The production of sustained hypertension in dogs by a single transient anoxic episode, *Angiology,* 20, 237, 1969.

57. **Kuneš, J., Jelínek, J., and Zicha, J.,** Age-dependent blood pressure response to increased salt intake in rats influenced by transient renal ischemia, *Clin. Sci. Mol. Med.,* 70, 185, 1986.

58. **Fisher, E. R. and Hellstrom, H. R.,** Pathogenesis of hypertension and pathologic changes in experimental renal irradiation, *Lab. Invest.,* 19, 530, 1968.

59. **Hall, C. E. and Hall, O.,** Comparison of macromolecular hypertension due to polyvinyl alcohol and methyl cellulose, in respect to the role of sodium chloride, *Lab. Invest.,* 11, 826, 1962.

60. **Lenel, R., Katz, L. N., and Rodbard, S.,** Arterial hypertension in the chicken, *Am. J. Physiol.,* 152, 557, 1948.

61. **Sapirstein, L. A., Brandt, W. L., and Drury, D. L.,** Production of hypertension in the rat by substituting hypertonic sodium chloride solution for drinking water, *Proc. Soc. Exp. Biol. Med.,* 73, 82, 1950.

62. **Jelínek, J., Kraus, M., and Musilová, H.,** Adaptation of rats of different ages to forced intake of a 2% NaCl solution without the occurrence of salt hypertension, *Physiol. Bohemoslov.,* 15, 137, 1966.

63. **Molteni, A. and Brownie, A. C.,** Incidence of salt-induced hypertension in rats from different stocks, *J. Med.,* 3, 193, 1972.

64. **Meneely, G. R. and Ball, C. O. T.,** Experimental epidemiology of sodium chloride toxicity and the protective effect of potassium chloride, *Am. J. Med.,* 25, 713, 1958.

65. **Dahl, L. K.,** Salt and hypertension, *Am. J. Clin. Nutr.,* 25, 231, 1972.

66. **Dahl, L. K.,** Effects of chronic excess salt feeding. Induction of self-sustaining hypertension in rats, *J. Exp. Med.,* 114, 231, 1961.

67. **Rapp, J. P.,** Dahl salt-susceptible and salt-resistant rats. A review, *Hypertension,* 4, 753, 1982.

68. **Rapp, J. P. and Dene, H.,** Development and characteristics of inbred strains of Dahl salt-sensitive and salt-resistant rats, *Hypertension,* 7, 340, 1985.

69. **Bunag, R. D., Butterfield, J., and Sasaki, S.,** Hypothalamic pressor response and salt induced hypertension in Dahl rats, *Hypertension,* 5, 460, 1983.

70. **Rapp, J. P. and Dahl, L. K.,** 18-hydroxycorticosterone secretion in experimental hypertension in rats, *Circ. Res.,* 32 and 33 (Suppl. II), II-153, 1971.

71. **Goldblatt, H.,** The effect of high salt intake on the blood pressure of rabbits, *Lab. Invest.,* 21, 126, 1969.
72. **Cherchovich, G. M., Čapek, K., Jefremova, Z., Pohlová, I., and Jelínek, J.,** High salt intake and blood pressure in lower primates *(Papio Hamadryas), J. Appl. Physiol.,* 40, 601, 1976.
73. **Selye, H., Stone, H., Timiras, P. S., and Schaffenburg, C.,** Influence of sodium chloride upon the actions of desoxycorticosterone acetate, *Am. Heart J.,* 37, 1009, 1949.
74. **Selye, H. and Pentz, I.,** Pathogenetical correlations between periarteritis nodosa, renal hypertensic and rheumatic lesions, *Can. Med. Assoc. J.,* 49, 264, 1943.
75. **Greene, D. M.,** Mechanisms of desoxycorticosterone action. I. Relation of fluid intake to blood pressure, *J. Lab. Clin. Med.,* 33, 853, 1948.
76. **Tajima, Y., Ichikawa, S., Sakamaki, T., Matsuo, H., Aizava, F., Kogura, M., Yagi, S., and Murata, K.,** Body fluids distribution in the maintenance of DOCA-salt hypertension, *Am. J. Physiol.,* 244, H695, 1983.
77. **Musilová, H., Jelínek, J., and Albrecht, I.,** The age factor in experimental hypertension of DCA type in rats, *Physiol. Bohemoslov.,* 15, 525, 1966.
78. **Ormsbee, H. S. and Ryan, C. F.,** Production of hypertension with desoxycorticosterone acetate-impregnated silicon rubber implants, *J. Pharm. Sci.,* 62, 255, 1973.
79. **Miller, A. W., Bohr, D. F., Schork, A. M., and Terris, J. M.,** Hemodynamic responses to DOCA in young pigs, *Hypertension,* 1, 591, 1979.
80. **Passmore, J. C., Whitescarver, S. A., Ott, C. E., and Kotchen, T. A.,** Importance of chloride for desoxycorticosterone acetate-salt hypertension in the rat, *Hypertension,* 7(Suppl. I), I-115, 1985.
81. **Malyusz, M., Mendoza-Osorio, V., and Ochwadt, B.,** Nierenfunktion bei Ratten mit experimentellem Hochdruck, *Pfleugers Arch.,* 332, 28, 1972.
82. **Green, D. M., Saunders, F. J., Wahlgren, N., and Craig, R. L.,** Self-sustaining, post-DCA hypertensive cardiovascular disease, *Am. J. Physiol.,* 170, 94, 1952.
83. **Haack, D., Möhring, J., Möhring, B., Petri, M., and Hackenthal, E.,** Comparative study on development of corticosterone and DOCA hypertension, *Am. J. Physiol.,* 233, F403, 1977.
84. **Azar, S., Johnson, M. A., Iwai, J., Bruno, L., and Tobian, L.,** Single nephron dynamics in "post-salt" rats with chronic hypertension, *J. Lab. Clin. Med.,* 91, 156, 1978.
85. **Hall, C. E. and Hall, O.,** The comparative hypertensive activities of the acetates of D-aldosterone and deoxycorticosterone, *Acta Endocrinol.,* 54, 399, 1967.
86. **Fregly, M. J., Kim, K. J., and Hood, C. I.,** Development of hypertension in rats treated with aldosterone acetate, *Toxicol. Appl. Pharmacol.,* 15, 229, 1969.
87. **Komanicky, P. and Melby, J. C.,** Hypertensinogenic potencies of aldosterone and deoxycorticosterone in the rat, *Hypertension,* 4, 140, 1982.
88. **Knowlton, A. I., Loeb, E. N., and Soterk, H. S.,** Effect of synthetic analogues of hydrocortisone on the blood pressure of adrenalectomized rats on sodium restriction, *Endocrinology,* 60, 768, 1957.
89. **Friedman, S. M., Friedman, C. L., and Nakashima, M.,** Further observations on the hypertensive properties of compound F acetate in the rat, *Endocrinology,* 53, 633, 1953.
90. **Hyde, M. P., Grefer, C. C., and Skelton, F. R.,** Interrelationships between plasma corticosterone and dietary sodium in experimental hypertension, *Endocrinology,* 71, 549, 1962.
91. **Haack, D., Hackenthal, E., Homsy, E., Möhring, B., and Möhring, J.,** Effects of 9-alpha-fluorohydrocortisone on blood pressure, plasma volume and sodium, potassium and water balance in rats, *Acta Endocrinol.,* 76, 539, 1974.
92. **Whitworth, J. A., Butkus, A., Coghlan, J. P., Denton, D. A., Mills, E. H., Spence, C. D., and Scoggins, B. A.,** 9-α-fluorocortisol-induced hypertension: a review, *J. Hypertension,* 4, 133, 1986.
93. **Elijovich, F. and Krakoff, L. R.,** Mechanism of the response to captopril in glucocorticoid hypertension, *Clin. Exp. Hyper. Theory and Practice,* A4 (9 and 10), 1795, 1982.
94. **Handa, M., Kondo, K., Suzuki, H., and Saruta, T.,** Dexamethasone hypertension in rats: role of prostaglandins and pressor sensitivity to norepinephrine, *Hypertension,* 2, 139, 1980.
95. **Hall, J. E., Morse, C. L., Smith, F. J., Jr., Young, D. B., and Guyton, A. C.,** Control of arterial pressure and renal function during glucocorticoid excess in dogs, *Hypertension,* 2, 139, 1980.
96. **Skelton, F. R.,** A study of the natural history of adrenal regeneration hypertension, *Circ. Res.,* 7, 107, 1959.
97. **Skelton, F. R.,** Development of hypertension and cardiovascular-renal lesions during adrenal regeneration in the rat, *Proc. Soc. Exp. Biol. Med.,* 90, 343, 1955.
98. **Hall, C. E., Ayachi, S., and Hall, O.,** Hypertension in rats following multiple acupuncture of the adrenal glands, *Endocrinology,* 95, 1268, 1974.
99. **Skelton, F. R. and Guillebeau, J.,** The influence of age on the development of adrenal-regeneration hypertension, *Endocriniology,* 59, 201, 1956.
100. **Brownie, A. C., Bernardis, L. L., Niwa, T., Kamura, S., and Skelton, F. R.,** The influence of age and sex on the development of adrenal regeneration hypertension, *Lab. Invest.,* 15, 1342, 1966.

101. **Jelínek, J., Albrecht, I., and Musilová, H.,** The age factor in experimental hypertension: hypertension due to adrenal regeneration, *Physiol. Bohemoslov.*, 15, 424, 1966.

102. **Neff, A. W. and Correll, J. T.,** Influence of sex on adrenal regeneration hypertension, *Proc. Soc. Exp. Biol. Med.*, 95, 227, 1957.

103. **Jelínek, J.,** Salt intake and sexual differences in sensitivity of rats to adrenal regeneration hypertension, *Physiol. Bohemoslov.*, 16, 389, 1967.

104. **Rapp, J. P.,** Deoxycorticosterone production in adrenal regeneration hypertension. In vitro vs. in vivo comparison, *Endocrinology*, 84, 1409, 1969.

105. **Skelton, F. R., Brownie, A. C., Nickerson, P. A., Molteni, A., Gallant, S., and Colby, H. D.,** Adrenal cortical dysfunction as a basis for experimental hypertensive disease, *Circ. Res.*, 24(Suppl. I), I-35, 1969.

106. **Brown, R. D., Gaunt, R., Gisoldi, E., and Smith, N.,** The role of deoxycorticosterone in adrenal-regeneration hypertension, *Endocrinology*, 91, 921, 1972.

107. **Grekin, R. J., Dale, S. L., Gaunt, R., and Melby, J. C.,** Steroid secretion by the enucleated rat adrenal: measurements during salt retention and the development of hypertension, *Endocrinology*, 91, 1166, 1972.

108. **Skelton, F. R.,** The production of hypertension, nephrosclerosis and cardiac lesions by methylandrostendiol treatment in the rat, *Endocrinology*, 53, 492, 1953.

109. **Molteni, A., Brownie, A. C., and Skelton, F. R.,** Production of hypertensive vascular disease in the rat by methyltestosterone, *Lab. Invest.*, 21, 129, 1969.

110. **Colby, H. D., Skelton, F. R., and Brownie, A. C.,** Testosterone induced hypertension in the rat, *Endocrinology*, 86, 1093, 1970.

111. **Colby, H. D., Skelton, F. R., and Brownie, A. C.,** Metopirone induced hypertension in the rat, *Endocrinology*, 86, 620, 1970.

112. **Stanton, H. C. and Cooper, B. S.,** Comparison of effects of antihypertensive agents on normotensive rats with "metacorticoid" and adrenal regeneration hypertension, *Cardiovasc. Res. Cent. Bull. (Houston)*, 5, 16, 1966.

113. **Koch, E. and Mies, H.,** Chronischer arterieller Hochdruck durch experimentelle Dauerausschaltung der Blutdruckzugler, *Krankheitsforschung*, 7, 241, 1929.

114. **Ferrario, C. M., McCubbin, J. W., and Page, I. H.,** Hemodynamic characteristics of chronic experimental neurogenic hypertension in unanesthetized dogs, *Circ. Res.*, 24, 911, 1969.

115. **Cowley, A. W., Jr., Liard, J. F., and Guyton, A. C.,** Role of the baroreceptor reflex in daily control of arterial blood pressure and other variables in dogs, *Circ. Res.*, 32, 564, 1973.

116. **Kreiger, E. M.,** Neurogenic hypertension in the rat, *Circ. Res.*, 15, 511, 1964.

117. **Ito, C. S. and Scher, A. M.,** Hypertension following arterial baroreceptor denervation in the unanesthetized dog, *Circ. Res.*, 48, 576, 1981.

118. **Doba, N. and Reis, D. J.,** Acute fulminating hypertension produced by brainstem lesions in the rat, *Circ. Res.*, 32, 584, 1973.

119. **Nosaka, S.,** Hypertension induced by extensive medial anteromedian hypothalamic destruction in the rat, *Jpn. Circ. J.*, 30, 509, 1966.

120. **Griffith, J. Q., Jr., Jeffers, W. A., and Lindauer, M. A.,** A study of the mechanism of hypertension following intracranial kaolin injection in rats; leucocytic reaction and effect on lymphatic absorption, *Am. J. Physiol.*, 113, 285, 1935.

121. **Katsuki, S.,** Role of the brain stem in pathogenesis of hypertension, *Jpn. Cir. J.*, 30, 175, 1966.

122. **Rosekranz, J. A., Watzman, N., and Buckley, J. P.,** The production of hypertension in male albino rats subjected to experimental stress, *Biochem., Pharmacol.*, 15, 1707, 1966.

123. **Alexander, N.,** Psychosocial hypertension in members of a Wistar rat colony, *Proc. Soc. Exp. Biol. Med.*, 146, 163, 1974.

124. **Vander, A. J., Henry, J. P., Stephens, P. M., Kay, L. L., and Mouw, D. R.,** Plasma renin activity in psychosocial hypertension CBA mice, *Circ. Res.*, 42, 496, 1978.

125. **Marwood, J. F., Illet, K. F., and Lockett, M. F.,** Effects of adrenalectomy and of hypophysectomy on the development of sound-withdrawal hypertension, *J. Pharm. Pharmacol.*, 25, 96, 1973.

126. **Friedman, R. and Dahl, L. K.,** Psychic and genetic factors in the etiology of hypertension, in *Stress and the Heart*, Wheatley, D., Ed., Raven Press, New York, 1977, p. 137.

127. **Urmancheva, T. G., Fufacheva, A. A., Čapek, K., Kuneš, J., and Jelínek, J.,** Blood pressure in monkeys chronically exposed to psychoemotional stress, *Physiol. Bohemoslov.*, 34, 112, 1986.

128. **Vincent, M., Bornet, H., Berthezene, F., Dupont, J., and Sassard, J.,** Thyroid function and blood pressure in two new strains of spontaneously hypertensive and normotensive rats, *Clin. Sci. Mol. Med.*, 54, 391, 1978.

129. **Schlager, G.,** Selection for blood pressure levels in mice, *Genetics*, 76, 537, 1974.

130. **Yamamori, Y., Horie, R., Akiguchi, I., Ohtaka, M., Nara, Y., and Fukase, M.,** New models of spontaneously hypertensive rat (SHR) for studies on stroke and atherogenesis, *Clin. Exp. Pharmacol. Physiol.*, Suppl. 3, 199, 1976.

131. **Bruno, L., Azar, S., and Weller, D.,** Absence of a prehypertensive stage in post-natal Kyoto hypertensive rats, *Jpn. Heart J.,* 20(Suppl. 1), 90, 1979.

132. **Jones, D. R. and Dowd, D. A.,** Development of elevated blood pressure in young genetically hypertensive rats, *Life Sci.,* 9, 247, 1970.

133. **Trippodo, N. C. and Frohlich, E. D.,** Similarities of genetic (spontaneous) hypertension. Man and rat, *Circ. Res.,* 48, 309, 1981.

134. **Phelan, E. L., Simpson, F. O., and Smirk, F.,** Characteristics of the New Zealand strain of genetically hypertensive (GH) rats, *Clin. Exp. Pharmacol. Physiol.,* Suppl. 3, 5, 1976.

135. **Judy, W. V., Watanabe, A. M., Murphy, W. R., Aprison, B. S., and Pao-Lo Yu,** Sympathetic nerve activity and blood pressure in normotensive backcross rats genetically related to spontaneously hypertensive rats, *Hypertension,* 1, 598, 1979.

136. **Touw, K. B., Haywood, J. R., Shaffer, R. A., and Brody, M. J.,** Contribution of the sympathetic nervous system to vascular resistance in conscious young and adult spontaneously hypertensive rats, *Hypertension,* 2, 408, 1980.

137. **Winternitz, S. R., Katholi, R. E., and Oparil, S.,** Role of the renal sympathetic nerves in the development and maintenance of hypertension in the spontaneously hypertensive rat, *J. Clin. Invest.,* 66, 971, 1980.

138. **Graham, P. M., Pettinger, W. A., Sagalowsky, A., Brabson, J., and Gandler, T.,** Renal alpha-adrenergic receptor abnormality in the spontaneously hypertensive rat, *Hypertension,* 4, 881, 1982.

139. **Oparil, S.,** The sympathetic nervous system in clinical and experimental hypertension, *Kidney Int.,* 30, 437, 1986.

140. **Roman, R. J. and Cowley, A. W.,** Abnormal pressure-diuresis-natriuresis response in spontaneously hypertensive rats, *Am. J. Physiol.,* 248, F199, 1985.

141. **Coleman, T. G., Manning, R. D., Jr., Norman, R. A., and DeClue, J.,** The role of the kidney in spontaneous hypertension, *Am. Heart J.,* 89, 94, 1975.

142. **Arendhorst, W. J. and Beierwaltes, W. H.,** Renal tubular reabsorption in spontaneously hypertensive rats, *Am. J. Physiol.,* 237, F38, 1979.

143. **Diz, D. I., Nasljetti, A., and Baer, P. G.,** Renal denervation at weaning retards development of hypertension in New Zealand genetically hypertensive rats, *Hypertension,* 4, 361, 1982.

144. **Bianchi, G., Ferrari, P., Cusi, D., Guidi, E., Pati, C., Vezzoli, G., Trippodi, M. G., and Niutta, E.,** Genetic hypertension and the kidney, *J. Cardiovasc. Pharmacol.,* 6, S162, 1984.

145. **Rosenberg, W. L., Schlager, G., and Gennaro, J. F., Jr.,** Glomerular filtration and fluid balance in genetically hypertensive mice, *Proc. Soc. Exp. Biol. Med.,* 178, 629, 1985.

Chapter 21

EXPERIMENTAL CARDIAC HYPOXIA AND ISCHEMIA

B. Ošťádal and F. Kolář

TABLE OF CONTENTS

I. INTRODUCTION

Cardiovascular diseases belong undoubtedly to the most frequent diseases of modern times. Among the most dangerous (and hence the most widely studied) are hypoxic states of the cardiopulmonary system resulting from disproportion between the amount of oxygen supplied to the cell and the amount actually required by the cell. In recent years, the attention of cardiological research laboratories has been concentrated on the creation of a theoretical basis for the rational prevention and therapy of these serious states, with special reference to myocardial ischemia and chronic hypoxia leading to the development of cor pulmonale. The majority of conventional methods, which were based on clinical empiricism, were concerned mainly with the treatment of complications, i.e., of arrhythmia and heart failure, but their use was limited by inadequate knowledge of the various regulatory mechanisms controlling the relationship between oxygen supply and demand in the hypoxic myocardium.[1]

Development of knowledge in this sphere would therefore be unthinkable without a detailed analysis of the pathogenetic mechanisms involved in the origin of the disease in man. At the same time, it must be acknowledged that there is still no experimental model which adequately reproduces all the functional, structural, and metabolic changes characteristic for hypoxic states of the human cardiopulmonary system. This does not surprise us, considering the complexity of the conditions to be reproduced.[2] In addition, the possibility of generalizing the findings is limited by a number of other factors, such as the choice of a suitable experimental animal, the level of the observations (in vivo and in vitro) and, last but not least, the experience and critical approach of the experimenter.

II. CARDIAC HYPOXIA AND ISCHEMIA

As stated above, myocardial hypoxia is the result of disproportion between oxygen supply and demand (Figure 1). Owing to the high coronary arteriovenous difference, the myocardium is no longer able to bring about a substantial improvement in the oxygen supply by increased extraction of oxygen from the blood so that the only way of meeting rising demands is through an increase in the coronary blood flow. Oxygen consumption depends on the heart rate, on the contractility of the myocardium, and on the tension of the ventricular wall, which is the result of the pressure-volume relationship in the ventricle itself.

Theoretically, any of the known mechanisms leading to tissue hypoxia (Figure 2) can be responsible for a reduced oxygen supply in the myocardium, but the most common causes are undoubtedly (1) *ischemic hypoxia* induced by reduction or interruption of the coronary blood flow and (2) *hypoxic hypoxia* characterized by a drop in pO_2 in the arterial blood. For the sake of completeness we could add (3) *anemic hypoxia*, in which the arterial pO_2 is normal, but the oxygen transport capacity of the blood is decreased and (4) *histotoxic hypoxia* resulting from reduced intracellular utilization of oxygen in the presence of adequate saturation and an adequate blood flow (i.e. inhibition of oxidative enzymes in cyanide poisoning). The most frequent causes of raised oxygen consumption are increased physical or mental stress, or the administration of catecholamines or other substances with a positive inotropic and chronotropic effect. In healthy subjects these raised requirements are adequately met by an increase in the coronary blood flow.

Here it should be emphasized that the terms "hypoxia" and "ischemia" are unfortunately often used interchangeably in the literature. The consequences of the two main mechanisms at the cellular level are very different, however (Figure 3). In ischemia there is not only a drop in the supply of oxygen and substrates, there is also a significant change in the clearance of metabolites, in particular of lactic acid and hydrogen ions. Hypoxic hypoxia is usually a generalized phenomenon diffusely involving the whole myocardium, whereas ischemia is confined to the area supplied by the affected coronary artery. In this association, electric

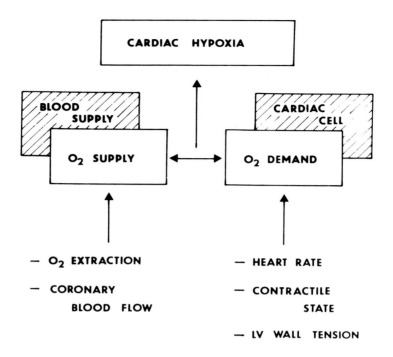

FIGURE 1. Scheme of the development of myocardial hypoxia; LV — left ventricle.

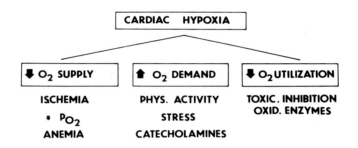

FIGURE 2. Mechanisms leading to myocardial hypoxia.

instability, arrhythmia and contractility disturbances develop, together with zones of hypo- and akinesia.[3,4]

To complete the picture by the relevant clinical syndromes, ischemic hypoxia is manifested primarily in myocardial ischemia and in its acute form, myocardial infarction, while hypoxic hypoxia is associated with chronic or pulmonale of varying origin, cyanosis due to a congenital heart disease and changes induced in the cardiopulmonary system by a temporary decrease in barometric pressure at high altitudes. In two cases, however, hypoxic hypoxia can be qualified as physiological: (1) the fetal myocardium adapted to hypoxia corresponding to an altitude of 8000 m and (2) the myocardium of subjects living permanently at high altitudes. In both these situations the myocardium is significantly more resistant to acute oxygen deficiency, but in populations living in lowlands, this property is lost soon after birth.[4,5]

From the practical aspect, the experimental models of hypoxia and ischemia of the cardiopulmonary system can be, according the induction of pathological process, divided into two large groups: in vivo and in vitro models.

	REGIONAL ISCHEMIA	GLOBAL ISCHEMIA	HYPOXIA
CORONARY FLOW	⬇	O	N
O$_2$ SUPPLY	⬇	⬇ ⬇	⬇
SUBSTRATES	⬇	⬇ ⬇	N
COLLATERALS	I + E	E	I + E
CLEARENCE OF METABOLITES	⬇	⬇ ⬇	N

FIGURE 3. Basic differences between cardiac ischemia and hypoxia. 0 — zero, N — normal, I — intracoronary, E — extracoronary.

III. IN VIVO MODELS

1. Myocardial ischemia induced by reduction or interruption of the coronary blood flow (ischemic hypoxia)
2. Myocardial hypoxia induced by a drop in pO$_2$ in the arterial blood (hypoxic hypoxia) (Table 1)

A. Experimental Ischemia

Ischemia is the most frequent and most serious cause of hypoxic lesions of the myocardial cells. It is the result of gradual (over a period of years) or sudden (within a few minutes or hours) occlusion of the coronary arteries. Gradual occlusion is due to the progressive accumulation of fibrous material in the wall of the arteries (for the chronic form of myocardial ischemia, see Section III.A.2). Acute occlusion is generally caused by thrombosis or an embolism, or by spasm of one of the branches of the coronary vessels.

The structural, functional and metabolic properties of the hypoxic myocardium depend on the intensity and duration of the hypoxic stimulus (Figure 4). In general, it can be said that the onset of ischemia is followed immediately by a decrease in contractility, by reduced aerobic metabolism and increased anaerobic metabolism, by inhibition of ATP production and fatty acid utilization and by the release of catecholamines. Early changes take place within minutes after initiation of the process; the main ones are an increase in glycogenolysis, exhaustion of intracellular ATP stores, the development of acidosis, and ion transport disturbances, in particular the increased uptake of extracellular calcium by the cells (a "calcium overload"). Structural changes are characterized chiefly by glycogen depletion and by edema of the mitochrondria and T tubules. If the myocardium is not given protective treatment, the process continues with loss of mitochondrial respiratory function, lysosomal changes together with activation of hydrolases, disruption of the sarcolemma, and finally cell death and tissue necrosis, i.e. myocardial infarction.[36]

If the blood flow is renewed when the myocytes and the terminal vascular bed are still in the reversible phase, the cells do not die and gradually regain their integrity. If reperfusion is initiated when the myocytes are already irreversibly damaged, there is no improvement. On the contrary, the disintegration process is sharply accelerated and edema of the myocytes, massive intracellular accumulation of calcium, and disruption of the contractile apparatus are the results. This leaves the question of whether myocytes still in the reversible phase are also liable to reperfusion injury.[6] The fact that the administration of scavengers affords protection against both ischemic and reperfusion/damage indicates that oxygen radicals play a role in the development of myocardial cell lesions.[7]

As mentioned in the introduction, none of the models employed hitherto are ideal ones,

Table 1

THE MOST FREQUENTLY USED EXPERIMENTAL MODELS OF CARDIAC ISCHEMIA AND HYPOXIA

In Vitro			In Vivo		
Hypoxia	Ischemia		Hypoxia	Ischemia	
Acute	**Global**	**Regional**	**Acute and Chronic**	**Chronic**	**Acute**
Lowering of PO$_2$ in medium / isolated perfused heart — isolated papillary muscle \ isolated myocytes	Lowering or cessation of coronary flow Isolated perfused heart	Coronary artery ligature	Normobaric hypoxia Hypobaric hypoxia	Atherosclerosis Ameroid	Coronary artery ligature Balloon cathether Electrolysis Arterial thrombosis Coronary spasm

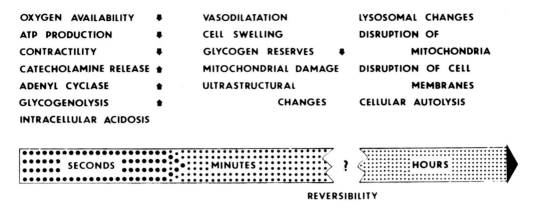

FIGURE 4.   Time course of some functional, structural and metabolic changes after occlusion of coronary artery. (Adapted from Hearse, D. J., *J. Physiol. (Paris)*, 76, 1980, 751. With permission.)

since they all differ more or less from the human disease. The basic problem is that in most experimental models we are working with an intact animal in which the systems responsible for the supply and consumption of oxygen are not afflicted by any pathological process. The methods inducing myocardial ischemia in vivo can be divided, according ot the rate of onset of the changes, into acute and chronic and, according to the way in which ischemia is induced, into "open chest", "closed chest", and "open" followed by "closed chest" methods (for a review see References 2 and 8).

1. Acute Ischemia

The most frequently used method is unquestionably *acute ligation* of one of the branches of the coronary vessels in an anaesthetized, open chest animal. Its undoubted advantage is that it allows exact determination of the onset of occlusion and the duration of ischemia and ensures total preclusion of antegrade perfusion. The exact degree of reduction of the coronary blood flow can be set by means of a flowmeter and when the ligature is released the effects of reperfusion can be studied. Another great advantage of this method is that it allows the use of small experimental animals, in particular of laboratory rats (important advantage: standardization of the experiment, the requirements of statistical evaluation, and of course, the question of costs). This method was already used in rats in 1946 by Heimburger[9] and by Johns and Olson[10] in artificially ventilated animals. Selye et al.[11] described a similar technique in spontaneously breathing laboratory rats. This model was used most extensively by McLean et al.[12] who studied possibilities of pharmacological protection of the ischemic myocardium. Evans et al.[13] described the (so far) simplest modification of this method and drew attention to the possibilities of its utilization in pharmacological screening studies. It requires far less time than work with larger experimental animals and it can be employed with success for testing various cardiotoxic and cardioprotective substances.

The disadvantage of acute ligation is that a previously healthy myocardium becomes ischemic suddenly, without any preadaptation such as gradual narrowing of the afferent vessels; the unavoidable surgical trama likewise cannot be ignored. The use of a ligature in conscious animals prevents the suppression of cardiovascular reflexes and the often negative inotropic effect of anesthesia, but the degree of reduction cannot be regulated.[8] A *balloon catheter* with regulatable inflation is evidently more satisfactory in this respect, but it can be used only in large laboratory animals. Furthermore, the size of the balloon may partly alter and the position in which it is fixed, which depends to some extent on the anatomy of the coronary vessels, is not always the same. This method can therefore be employed only by an experienced worker.[8] The main difference between acute occlusion induced by *elec-*

trolysis and classic ligation is that the former is slower and that is accompanied by the release of various factors which may potentiate or, conversely, antagonize intravascular coagulation. Electrolysis can also cause intravascular thrombosis and extensively damage the intima, however.[2]

A relatively large number of methods for producing *arterial thrombosis* in laboratory animals has been described (for reviews see References 14 and 15), but only a few are suitable for inducing acute myocardial ischemia. They use either intracoronary thrombogenic devices, or produce controlled lesions of the coronary artery wall (by means of an electric current, laser, or freezing). For instance, a coil of wire introduced into the lumen of the coronary artery induces progressive formation of a thrombus and gradual occlusion of the artery in a few hours.[46] Another model uses a teflon-coated electrode, which is introduced into the coronary artery under fluoroscopic control and, with a current of 100 to 900 microamperes, induces thrombotic occlusion in 18 to 93 min.[17] The advantage of this method is its reproducibility, its disadvantage the need for fluoroscopy and the danger of ischemia and serious arrhythmia induced by the guiding catheter. One drawback of all the above methods for inducing coronary thrombosis is that coronary spasm cannot be ruled out as a cause of ischemia.

Attempts to induce *standard coronary spasm* are only a few years old[18] and it is still difficult to evaluate them. It was found that ergonovine-induced coronary spasm in dogs, but only if experimental atherosclerotic lesions were present. The spasm was very small (only 20%) and was not manifested in ischemic changes in the ECG. The same authors induced coronary spasms (up to 75%) in minipigs by injecting histamine into sclerotic arteries stripped of the endothelium.[19]

In dogs, pigs, and rats, the incidence of myocardial ischemia after acute occlusion is almost 100% and mortality is correspondingly high.[13,20] According to Evans et al.[13] after excluding dead animals and those in which occlusion was technically imperfect, 49% of the rats can be used for further analysis (this figure is based on an evaluation of 3000 operations performed over 3 years). Mortality was higher among old animals than among young rats. The size of the ischemic zone varied from 60 to 80% of the perfused area after occlusion of the descending branch of the left coronary artery in dogs and from 90 to 100% in pigs; the difference is due chiefly to the much larger collateral circulation of dogs.[21]

2. Chronic Ischemia

The chief method for inducing chronic myocardial ischemia is *atherosclerosis*. The experimental models used to investigate this serious process have quite recently been discussed in several extensive reviews (e.g., References 22 and 23) and they will be evaluated in another chapter of this monograph. We shall therefore confine ourselves here to just a few remarks necessary for a complete survey of the problems of experimental chronic myocardial ischemia. In the first place, it must be emphasized that there are species in which atherosclerosis, with all its consequences (i.e., acute and chronic ischemia), occurs spontaneously. They include several birds (e.g., pigeons) and mammals, particularly pigs (in which the changes in the aorta and coronary vessels largely resemble the lesions in man) and certain monkeys (*Macaca mulatta*, *Macaca fascicularis*). In all these species, the degree of involvement can be increased by the administration of a high cholesterol diet. The second group comprises species in which atherosclerosis does not occur spontaneously, but can be induced experimentally relatively easily by the administration of a special high cholesterol diet. They include the rabbit (incidentally, the first laboratory animal used in the study of arteriosclerosis by Ignatowski in 1908[24]), the dog (in which hypothyroidism must also be present, however) and various monkeys (e.g., *Macaca arctoides* and *Cercopithecus aethiops*). Lastly, there is a group of species resistant, or only slightly sensitive, to an atherosclerotic diet. Unfortunately, it includes the most widely used biological experimental model in cardiological research, i.e., the laboratory rat.

Special mention should be made of methods in which coronary artery stenosis is not manifested in the resting state, but only under conditions of increased workload. *A small plastic cone* with a narrow central lumen is introduced into a coronary artery of a large experimental animal. Under basal conditions, this model assures normal perfusion distally to the stenosis, but flow rate and function are proportionally impaired when the heart rate is increased.[25,26] With this model, however, it is likewise often very difficult to ensure exact localization of the occlusion. This technique allows to study the development of a collateral circulation and the effect of long-term pharmacological treatment under chronic conditions. An *aneroid* has been used for the same reasons;[21] a cylindrical, hygroscopic plastic ring applied to a coronary artery gradually tightens around it and in 2 or 3 weeks, 50% of such arteries are completely occluded. The advantage of this method is that it stimulates the growth of collaterals in dogs and pigs. The disadvantage of the last two approaches is that it is almost impossible to determine whether the required reduction of the coronary flow has been achieved.

The advantages of methods used for inducing myocardial ischemia "in vivo" have been summed by Oliver.[2] The main merit of the "open chest" technique is that a ligature or some other constrictors can be applied exactly, under visual control, to a particular spot. In most laboratory animals, a surface electrode can be used for mapping the focus and the electrophysiological and metabolic properties of the ischemic and nonischemic part of the myocardium can be studied. Biopsy samples can be taken from exactly defined areas of the myocardium. The possibilities furnished by the "open chest" method in small laboratory animals, especially in pharmacological screening studies, have already been stressed. Their basic drawback is surgical trauma, which can lead to serious functional and metabolic disturbances. However, experience has shown that the situation is still relatively stable weeks after the operation. The advantages of the "closed chest" method are evident from the preceding paragraph, i.e., minimum trauma with minimum hemodynamic and metabolic consequences. There is no damage to the epicardium and no simultaneous ligation of small branches of the sympathetic nerve, with subsequent release of catecholamines. The possibility of induction of ischemia even a few weeks after the operation, allows the experimenter to work with conscious animals with already developing collateral circulation.

The results can be, however, influenced significantly by the choice of the biological experimental model. In dogs, for example, it is difficult to assure genetic standardization, as well as the same sex and age of the experimental animals. There are also important differences between the canine and the human myocardium: the most important include differences in the anatomy of the main coronary vessels and the much greater development of the collateral bed in the dog. The consequences of ligation are therefore never so severe as they otherwise would be in relation to the size of the area supplied by the ligated artery. Cardiologists are consequently turning to animals whose cardiovascular responses are closer to those of healthy man, such as pigs (including minipigs) and various species of monkeys.[27] The wider use of these species is still hampered, however, by complications associated with the breeding and handling of them.

Furthermore, the development of myocardial ischemia can likewise be influenced by the type, dose, and duration of anesthesia, the manner of occlusion, ventilation, heart rate, and the level of the blood pressure at the time of the operation. The only way of guaranteeing an objective comparison is therefore by keeping a detailed experimental record.

B. Experimental Hypoxic Hypoxia

Hypoxic hypoxia of the cardiopulmonary system is the result of a drop in pO_2 in the arterial blood. The coronary blood flow is usually normal and the supply of substrates is likewise unchanged; as distinct from ischemia, metabolite release is unaffected. In experimental studies of the effect of hypoxic hypoxia on the cardiopulmonary system, we most often encounter chronic hypoxic hypoxia induced by a decrease in partial oxygen pressure in a normo- or hypobaric chamber. This standard situation is very popular for investigation of the pathogenetic mechanisms involved in (1) hypoxia induced by a chronic pulmonary disease,[28] (2) high altitude hypoxia,[5,19] (3) chronic ischemic heart disease (with reference to the difficulties with the chronic model of ischemia, see above) (4) congenital cyanotic heart disease, and (5) adaptation of the cardiopulmonary system to chronic hypoxia and its raised resistance to acute ischemia.[4]

In a critical evaluation of an experimental model, it must be borne in mind that the different types of hypoxic situations which occur in man can be accompanied by a whole series of other factors (in altitude hypoxia, for instance, cold, reduced humidity, and radiation). Furthermore, various other organs and tissues, in addition to the cardiopulmonary system, are deprived of oxygen.[5,29,30] Chronic hypoxia is not always permanent; it is often of an intermittent nature, e.g., in exacerbation of a chronic obstructive pulmonary disease during an acute respiratory infection, or in repeated ascents in mountains. Hypoxia is likewise not continual in myocardial ischemia, when it depends on the instantaneous coronary blood flow. We therefore elaborated a model of chronic, but intermittent hypoxic hypoxia, in which the animals are exposed to a hypoxic environment for only a given part of the day. The intensity and duration of the hypoxic stimulus are chosen in accordance with the needs of the particular experiment.[31-33]

Models of permanent and intermittent hypoxic hypoxia lead to the development of a series of functional, metabolic and structural changes in the cardiovascular system (for a review see References 4, 5, 28, and 33 to 35). Pulmonary hypertension, hypertensive changes in the pulmonary bed and right ventricular hypertrophy develop in a relatively short time. Furthermore, polycythemia appears, cardiac output rises, and the myocardial blood flow, together with the blood flow in other zones of the peripheral circulation, increases. In both ventricles, capacity for glucose and lactate utilization rises, while the ability to break down fatty acids falls. The collagen concentration in the myocardium rises together with the changes of structural and enzymatic properties of the contractile proteins. Focal necrotic lesions, localized chiefly in the right ventricle, appear in the myocardium at the onset of the acclimatization process. It is interesting to note that primarily damaged myocardium is significantly more resistant to acute hypoxic injury. The given models permit to study the development of all the above changes, the possibility of their spontaneous reversibility when the animals are removed from the hypoxic atmosphere and/or the protective pharmacological attenuation of unwanted manifestations of acclimatization.

This leaves us with the question of suitable species of experimental animals. Sensitivity to hypoxic hypoxia is characterized by marked interspecies differences. Cattle and pigs are among the most sensitive animals; sheep and dogs seem less liable to develop hypoxic pulmonary hypertension and right ventricular hypertrophy, while the laboratory rat and the rabbit come between these two groups.[36] Many valuable findings have been made precisely in the laboratory rat; their significance for clinical practice depends on the extent to which changes observed in this species are comparable to findings in patients.[37] Pulmonary hypertension, right ventricular hypertrophy, and muscularization of the pulmonary arterioles and the enlargement of the carotid body develop in both rat and man. Similarly, the development of their ventilatory adaptation to chronic hypoxia is comparable. On the other hand, polycythemia, which is always found in rats, occurs in only some patients with a chronic obstructive pulmonary disease. Here again, for the transposal of experimental results to clinical practice, it is necessary to choose an adequate procedure and a suitable experimental animal for every question. Even so, the results should then be subjected to a critical evaluation.

The determination of the extent of ischemic (hypoxic) injury of the myocardium in ''in vivo'' experiments has not yet been resolved satisfactorily. This, however, is an essential prerequisite for evaluation of the effect of protective pharmacological treatment. Classic approach includes indirect electrocardiographic methods (for a survey see Reference 38) based chiefly on evaluation of elevation of the ST segment. Unfortunately, this is not a specific indicator of myocardial ischemia, but can also be produced by local changes in the electrolyte concentration; pericarditis and the administration of glycosides can lead to similar changes. A further possibility is to measure the height of the epicardial R wave, whose voltage is in inverse proportion to the regional coronary blood flow in the ischemic zone. Enzymatic methods are based on the finding that the increase in some serum enzyme levels is the converse of their decrease in the injured tissue. This relationship has been studied in detail in the case of creatine kinase, for example.[39] Maroko et al.[38] used the determination of creatine kinase in the myocardial tissue, combined with planimetry of hematoxylin- and eosin-stained histological sections. A further modification of this morphological method is to visualize the infarcted area by incubating sections in triphenyltetrazolium salts and then to determine its weight.[38]

Radionuclide methods offer a wide range of possibilities. "Cold spot" isotopes, which accumulate only in intact cells (e.g., ^{86}RB, ^{131}Cs, and ^{201}Tl) and "hot spot" radionuclides, which accumulate in necrotic tissue (^{203}Hg and ^{99}Tc) have been used. Gamma cameras allow spatial visualization of the damaged focus.[40] Some authors found a good correlation between the extent of cardiac lesions determined morphologically, enzymatically, and by means of radionuclides.[41-43] The choice of a suitable method is individual and depends on the facilities of the laboratory, the experimental model, the species of the experimental animal, the extent of the lesion, and the interval between forming of the lesion and the time of measurement.

IV. IN VITRO MODELS

A whole series of experimental models, with various laboratory animals, is used for inducing acute myocardial ischemia and hypoxia in vitro. The individual models employ a number of standard preparations requiring completely different methodical approaches. They can be generally divided into three groups: (1) preparations in which both the oxygen and substrate supply and the clearance of metabolites are assured by perfusion of the coronary bed, (2) nonperfused preparations in which metabolic exchange between tissue and medium is due to diffusion, and (3) isolated myocytes. It must likewise be borne in mind that all these preparations are denervated and are mostly without the influence of hormonal factors present in blood, unless, of course, the perfusion fluid is blood. This nevertheless does not detract from the fact that the models in vitro represent an important experimental contribution to the study of many questions associated with myocardial ischemia and hypoxia.

A. Experimental Ischemia

For studying acute myocardial ischemia in vitro, it is obvious that the only possible models are those from the first group, i.e., perfused preparations. Most workers who, from various aspects, study ischemic and reperfusion changes of the myocardium and, in particular, protection against it, used the isolated perfused heart — most frequently the heart of the laboratory rat (for a survey see Reference 44), less often of the rabbit,[45] the cat,[46] the dog,[47] or some other animals. The perfusion techniques most extensively used today are retrograde perfusion via the aorta (classic Langendorff preparation) and the working heart according to Neely et al.[48]

The *perfusion technique* already introduced by Langendorff[49] in 1895 consists in retrograde perfusion of the coronary bed of the isolated heart via the aorta. Since then, many different modifications of this method have been used, but the basic principle has remained the same (for methodical review see Reference 50). The most widely used perfusion medium is warm (37°C) Krebs-Henseleit bicarbonate buffer continually saturated with a mixture of 95% O_2

and 5% CO_2 (pH = 7.4). More complex media are used less often, their purpose usually being to increase oxygen transport capacity (e.g., perfusion of the heart with a perfluoro-chemical emulsion,[51] an erythrocyte suspension,[52,53] or blood[46]).

Langendorff preparation can be perfused in two ways. In perfusion under constant flow, the perfusion pressure is determined by the rate of the pump and the resistance of the coronary bed. The second, more usual, alternative is perfusion at a constant pressure determined by a hydrostatic pressure of perfusate in a reservoir connected to an aortic cannula. The perfusion pressure used by different authors for the adult rat heart varies from 60 to 120 cm H_2O.

Langendorff preparation is commonly assumed to be a model of a nonworking heart, but many authors point out that even in this experimental system the perfusion solution fills the ventricle during diastole as a result of an incompetent aortic valve and is ejected during systole (e.g. Reference 55). The atrially perfused working heart according to Neely et al.[48] differs from the classic preparation in respect of its much higher performance and its metabolic requirements, which approach those of the intact heart in vivo. This is an advantage in experiments in which postischemic functional recovery is studied, for instance. Irreversible damage occurs after a much shorter period of ischemia than in the retrogradely perfused heart. Unlike Langendorff preparation, working heart allows functional parameters which react sensitively to oxygen deficiency (e.g., cardiac output) to be recorded.

The *working heart* according to Neely et al.[48] is principally based on the classic Langendorff preparation. During retrograde perfusion, a cannula connected to a reservoir situated 10 to 20 cm above the heart (filling pressure) is introduced into the left atrium. When aortic inflow is clamped and atrial inflow is opened, the heart begins to work; the perfusate flows from the atrium into the left ventricle and is pumped out through a branch of the aortic cannula against a set hydrostatic pressure representing systemic peripheral resistance (usually 80 to 140 cm H_2O). The aortic tract contains a bubble trap chamber which simulates the elasticity of the blood vessel wall and thus improves the stability of the preparation.

Isolated perfused heart preparations allow the induction of regional or global ischemia. *Complete global ischemia* can be induced either by clamping aortic inflow in the Langendorff preparation, or atrial inflow, together with the aortic tract, in the working heart (e.g. Reference 56). Some authors prefer to use *partial global ischemia* models of varying degrees. A decrease in the coronary flow in the retrogradely perfused heart can easily be achieved by reducing the perfusion pressure,[57] for example, or by switching the perfusion pump to a lower rate.[58,59] Similarly, partial ischemia can be achieved in the working heart by replacing atrial perfusion by low flow retrograde perfusion via the aorta.[44] Another possibility is low flow atrial perfusion maintained by a pump in the presence of a closed aortic tract when the coronary flow is identical to the rate of the pump.[60] The latter authors describe yet another way of inducing ischemia in the working rat heart. It is based on the function of a one-way valve inserted in an aortic cannula just above the orifices of the coronary arteries. This does not influence aortic output or ventricular afterload, but the valve prevents the perfusate from flowing retrogradely into the aorta during diastole; the result being a drop in the coronary perfusion pressure and reduction of the coronary blood flow. In both types of isolated perfused heart preparations, *regional ischemia* can be induced by ligating the coronary artery; the study of Kannengiesser et al.[61] serves as an example for the rat heart. All the above operations used for inducing acute ischemia in the isolated perfused heart permit the renewal of normoxic conditions (reperfusion) after a chosen time, by a return to a normal perfusion pressure or perfusion rate, or by loosening the ligature on the coronary artery.

The most sensitive way of studying the effect of O_2 deficiency on the myocardium in vitro is to measure its functional performance. In Langendorff preparation, the usual way is to record the intraventricular pressure curve, measured isovolumically by means of a fluid-filled balloon introduced into the ventricle through the left atrium and connected to a pressure transducer and a recording apparatus.[52,62]

This technique cannot be employed with the working heart and the intraventricular pressure is therefore transmitted to the transducer by means of a hypodermic needle inserted into the ventricle through the apex.[48] From the pressure curve or its first derivative we can read the heart rate, systolic and diastolic pressure, or the maximum rate of pressure development, $(dP/dt)_{max}$, which is the most frequently used heart contractility index. Trouve and Nahas[55] recommend "noninvasive" measurement of pressure in the branch of the aortic cannula (supravalvular aortic pressure) and consider pressure amplitude to be a suitable parameter of left ventricular function in the Langendorff preparation. Another form of functional evaluation is to measure the contractile force with a strain gauge attached to the apex of the ventricle.[63]

Both preparations allow continual measurement of coronary flow by collecting of fluid ejected by the right ventricle (e.g., Reference 48) or using all kinds of flowmeters.[50] In Langendorff preparation it is more satisfactory to study the flow rate by means of an electronically controlled pump which maintains a constant filling pressure of perfusate in the aortic reservoir.[55]

The most sensitive indicator of the functional state of the myocardium is aortic flow,[64] which is easily measured in the working heart preparation. Aortic and coronary flow give the cardiac output value, which allows stroke volume at the given heart rate to be computed. The external (pressure-volume) work of the perfused heart is then usually expressed as the product of cardiac output and the pressure gradient (the difference between aortic pressure and left atrial filling pressure).[53,65] Oxygen deficiency is also manifested in characteristic ECG changes which can be recorded from the surface of the isolated perfused heart in the routine manner.[59,66] More detailed information about the electrophysiological consequences of ischemia can be obtained using intracellular potential recordings.[67,68]

The isolated perfused heart is a satisfactory model for a biochemical analysis of ischemic changes in the myocardium. It not only permits tissue metabolite levels to be studied, it also allows to detect myocardial injury from the composition of the coronary effluent. The basic biochemical approach for studying myocardial ischemia includes the determination of tissue high energy phosphate (phosphocreatine, ATP, ADP, AMP), glycogen, and lactate levels by standard enzymatic methods. High-performance liquid chromatography[69,70] or nuclear magnetic resonance[71,72] represent a modern alternative for high energy phosphate level determination. The last of these methods also furnishes information on the intracellular pH (ischemia leads to acidosis) and it allows continual observation throughout the experiment.

In the isolated heart, the coronary effluent is used for the detection of certain metabolites (lactate) and particularly enzymes, whose release is a sign of impaired integrity of the cellular membranes; creatine kinase, malate dehydrogenase, and lactate dehydrogenase, for example, are determined for this purpose.

Morphological detection of myocardial injury is also possible in isolated perfused heart preparations. Whenever, in the course of the experiment, the heart can be fixed, the usual way is to replace the normal perfusion solution by a fixing solution with a suitable composition (e.g., Reference 59) or to take tissue biopsies.

B. Experimental Hypoxia

Acute hypoxic states of the myocardium can be studied "in vitro" in both perfused and nonperfused preparations, including isolated myocytes; but the most widely used model is probably the isolated perfused heart. With both types of perfusion (aortic and atrial), hypoxia of the isolated heart can be induced by using nitrogen instead of oxygen in the mixture with which the perfusate is saturated.

Isolated nonperfused preparations are used rather less often than the perfused heart for studying myocardial hypoxia; the relatively frequent preparations of this type are the trabeculae and (particularly) the papillary muscles, whose structure and muscle fiber orientation are suitable for testing the mechanical function or the myocardium. Other parts of the heart, e.g., strips from the various parts or whole atria or ventricles, are now seldom used. The preparation is fixed in a chamber filled with buffered physiological saline at a temperature of less than 37°C. Contractions in response to electric stimulation of the muscle are recorded in an isotonic or isometric arrangement. Transmembrane potential changes can be recorded simultaneously with intracellular microelectrodes. Hypoxia is induced in the same way as in the perfused heart, i.e., by substituting nitrogen for oxygen in the gas mixture with which the medium is saturated (for detailed descriptions of these techniques and their various modifications and of methods for evaluating the electrical and mechanical activity of the hypoxic and reoxygenated myocardium see References 31 and 74 to 77).

The above "in vitro" models of myocardial ischemia and hypoxia can be of use only if adequate tissue oxygenation under normoxic conditions can be assured. Since the isolated heart is usually perfused with medium without erythrocytes, the deficiency must be compensated by raising the pO_2 in the perfusion solution. The coronary flow also helps to keep the tissue adequately supplied with oxygen, since in a heart perfused with a solution of a low viscosity it is several times greater than in vivo. Some of the experimental data show that myocardial oxygenation, under normal conditions, is adequate not only in Langendorff preparation, but also in the working heart, where metabolic requirements are much greater.[44,78]

In nonperfused preparations the situation is quite different, since metabolic exchange with the surrounding medium is dependent entirely on diffusion, and oxygenation of the tissue is adequate only to a given depth from the surface. That means that the preparation has a critical diameter and that if this exceeded, mechanical function per cross section area unit diminishes. The critical diameter depends on the experimental conditions; the interspecies differences may exist as well. The values for the myocardium of different mammals vary from 0.5 to 1.2 mm (for survey see Reference 79), but the value determined by Schouten and Keurs[80] for the rat myocardium is far smaller (0.2 mm). This factor severely limits the choice of a suitable preparation of this type for studying hypoxia; many of the models used hitherto were evidently more or less hypoxic even under normal conditions.

The mechanisms of cellular injury are being studied increasingly in *isolated myocytes* (for a survey see References 81 and 82). The main advantages of this attractive approach is that it permits work with a homogeneous cell population. In addition to the effect of acute and chronic oxygen deprivation, the direct effect of exogenous metabolic inhibitors, or of endogenous catabolites formed during ischemic damage in vivo, can also be studied. It is likewise possible to test the effect of cardiotoxic substances whose in vivo study is impossible owing to their systemic toxicity. As distinct from in vivo conditions, hypoxia of the isolated myocyte is not accompanied by disruption of the plasmalemma, evidently owing to the smaller intracellular osmotic load (the greater extracellular space prevents the accumulation of a large amount of catabolites). The drawback of the isolated myocyte is that it is cut off from all the types of cells with which it is normally in constant contact in vivo. It has no blood supply and no innervation and its properties may likewise be modified by the isolation procedure (hyaluronidase, collagenase, and calcium-free medium). It should not be forgotten that in vivo ischemia or hypoxia comprises a series of pathogenetic mechanisms as a consequence of the mutual interaction of the cellular elements and the extracellular environment. The direct application of results obtained with isolated myocytes is, therefore, not simple.

In conclusion, it can be assumed that we do not yet possess an experimental model simulating human myocardial hypoxia and ischemia in every respect. All the methods now employed nevertheless have their justification for exactly defined situations, when seeking answers to clearly defined experimental questions. Despite sepsis and criticism, nobody today is any longer in doubt that experimental studies have contributed an infinite amount of important information on the pathogenesis of these serious cardiovascular diseases.

REFERENCES

1. **Nayler, W. G.**, The pharmacology of calcium antagonist drugs, in *Colloquium on the Use of Calcium Antagonists in Heart Disease,* Knoll, A. G., Ed., Ludwigshafen, 1978, 9.
2. **Oliver, M. F.**, *Methods of Inducing Myocardial Ischemia,* Oliver, M. F., Julian, D. G., and Donald, K. W., Eds., Churchill Livingstone, Edinburgh, 1972, 11.
3. **Hearse, D. J.**, Cardioplegia: the protection of the myocardium during open heart surgery: a review, *J. Physiol. (Paris),* 76, 1980, 751.
4. **Moret, P. R.**, Hypoxia and the heart, in *Hearts and Heart-Like Organs,* Vol. 2, Bourne, G. H., Ed., Academic Press, New York, 1980, 33.
5. **Heath, D. and Williams, D. R.**, *Man at High Altitude,* Churchill Livingstone, Edinburgh, 1981.
6. **Jennings, R. B. and Steenbergen, Ch., Jr.**, Nucleotide metabolism and cellular damage in myocardial ischemia, *Annu. Rev. Physiol.,* 47, 1985, 727.
7. **Hammond, B. and Hess, M. L.**, The oxygen free radical system: potential mediator of myocardial injury, *J. Am. Cell Cardiol.,* 6, 1985, 215.
8. **van der Giessen, W. J.**, Experimental models of myocardial ischemia, *Prog. Pharmacol.,* 5, 1985, 47.
9. **Heimburger, R. F.**, Injection into pericardial sac and ligation of coronary artery of the rat, *Arch. Surg.,* 52, 1946, 677.
10. **Johns, N. P. and Olson, B. J.**, Experimental myocardial infarction. I, A method of coronary occlusion in small animals, *Ann. Surg.,* 140, 1954, 675.
11. **Selye, H., Bajusz, E., Grasso, S., and Medell, P.**, Simple techniques for the surgical occlusion of coronary vessels in the rat, *Angiology,* 11, 1960, 398.
12. **MacLean, D., Fishbein, M. C., Braunwald, F. E., and Maroko, P. R.**, Long-term preservation of ischemic myocardium after experimental coronary artery occlusion, *J. Clin. Invest.,* 61, 1978, 541.
13. **Evans, R. G., Val Mejias, J. E., Kulevich, J., Fischer, V. W., and Mueller, H. S.**, Evaluation of a rat model for assessing interventions to salvage ischaemic myocardium: effects of ibuprofen and verapamil, *Cardiovasc. Res.,* 19, 1985, 132.
14. **Henry, R. L.**, Methods for inducing experimental thrombosis, *Angiology,* 13, 1962, 554.
15. **Philip, R. B.**, Experimental models of thrombosis and the effect of platelet inhibiting drugs, in *Methods of Testing Proposed Antithrombotic Drugs,* Philip, R. B., Ed., CRC Press, Boca Raton, 1981, 171.
16. **Kordenat, R. K., Kedzie, P., and Stanley, E. L.**, A new catheter technique for producing experimental coronary thrombosis and selective coronary visualization, *Am. Heart J.,* 83, 1972, 360.
17. **Salazar, A. E.**, Experimental myocardial infarction. Induction of coronary thrombosis in the intact closed-chest dog, *Circ. Res.,* 9, 1961, 1351.
18. **Nakamura, M., Kawachi, Y., and Tomoike, H.**, Augmented responses to ergonovine of the atherosclerotic canine coronary artery, *Circulation,* 64 (Suppl. IV), 1981, IV-21.
19. **Shimokawa, H., Tomoike, H., Nabeyama, S., Yamamoto, H., and Araki, H.**, Coronary artery spasm induced in atherosclerotic miniature swine, *Science,* 223, 1983, 560.
20. **Pelouch, V., Oštádal, B., and First, T.**, Structural and enzymatic properties of cardiac myosin in ischemic and non-ischemic regions of the rat myocardium, *Pflugers Arch.,* 364, 1976, 1.
21. **Schaper, W.**, *The Collateral Circulation of the Heart,* North Holland, Amsterdam, 1971.
22. **Jokinen, M. P., Clarkson, T. B., and Prichard, R. W.**, Recent advances in molecular pathology, animal models in atherosclerosis research, *Exp. Mol. Pathol.,* 42, 1985, 1.
23. **Comai, K., Feldman, D. L., Goldstein, A. L., and Hamilton, J. G.**, Atherosclerosis: an overview, *Drug Dev. Res.,* 6, 1985, 113.
24. **Ignatowski, A. C.**, Influence of animal food on the organism of rabbits, *S. Peterburg Izv. Imp. Voen. Med. Akad.,* 16, 1908, 54; cited according to Reference 22.
25. **Millard, R. W.**, Induction of functional coronary collaterals in the swine heart, *Basic Res. Cardiol.,* 76, 1981, 468.
26. **Gewirtz, H. and Most, A. S.**, Production of a critical coronary stenosis in closed chest laboratory animals, *Am. J. Cardiol.,* 47, 1981, 589.
27. **Bruyneel, K. and Opie, L. H.**, The baboon as an experimental animal for the production of myocardial infarction. Comparison with the mongrel dog, in *Effect of Acute Ischemia on Myocardial Function,* Oliver M. F., Julian, D. G., and Donald, K. W., Eds., Churchill Livingstone, Edinburgh, 1972, 18.
28. **Herget, J. and Peleček, F.**, Experimental chronic pulmonary hypertension, *Int. Rev. Exp. Pathol.,* 18, 1978, 347.
29. **Bouverot, P.**, Adaptation to altitude hypoxia in vertebrates, *Zoophysiology,* 16, 1985, 1.
30. **Trojan, S.**, Adaptation of the central nervous system to oxygen deficiency during ontogenesis, *Acta Univ. Carol. (Med. Monogr.) Praha,* 85, 1978, 5.
31. **McGrath, J. J., Procházka, J., Pelouch, V., and Oštádal, B.**, Physiological responses of rats to intermittent high-altitude stress: effects of age, *J. Appl. Physiol.,* 34, 1973, 289.

32. **Widimský, J., Urbanová, D., Ressl, J., Ošťádal, B., Pelouch, V., and Procházka, J.,** Effect of intermittent altitude hypoxia on the myocardium and lesser circulation in the rat, *Cardiovasc. Res.,* 7, 1973, 798.

33. **Ošťádal, B. and Widimský, J.,** Intermittent hypoxia and cardiopulmonary system, *Rozpravy ČSAV,* Vol. 95, Academia, Prague, 1985.

34. **Moret, P. R.,** Myocardial metabolism: acute and chronic adaptation to hypoxia, *Med. Sci. Sports Exerc.,* 19, 1985, 48.

35. **Widimský, J., Ošťádal, B., Urbanová, D., Ressl, J., Procházka, J., and Pelouch, V.,** Intermittent altitude hypoxia, *Chest,* 77, 1980, 383.

36. **Tucker, A., McMurtry, I. F., Reeves, J. T., Alexander, A. F., Will, D. H., and Grover, R. F.,** Lung vescular smooth muscle as a determinant of pulmonary hypertension at high altitude, *Am. J. Physiol.,* 228, 1975, 762.

37. **Kay, J. M., Suyama, K. L., and Klane, P. M.,** Effect of intermittent normoxia on muscularization of pulmonary arterioles induced by chronic hypoxia in rats, *Am. Rev. Respir. Dis.,* 123, 1981, 454.

38. **Maroko, P. R., Deboer, L. W. V., and Davis, R. F.,** Infarct size reduction: a critical review, *Adv. Cardiol.,* 27, 1972, 282.

39. **Kjekshus, J. K. and Sobel, B. E.,** Depressed myocardial creatine phosphokinase activity following experimental myocardial infarction in rabbit, *Circ. Res.,* 27, 1970, 403.

40. **Keyes, J., Leonard, P. F., and Brody, S. L.,** Myocardial infarct quantification in the dog by single photon emission computed tomography, *Circulation,* 58, 1978, 227.

41. **Botvinick, E. H., Shames, D., Lappin, H., Tyberg, J. V., Townsend, R., and Parmley, W. W.,** Noninvasive quantitation of myocardial infarction with technetium-99m pyrophosphate, *Circulation,* 52, 1975, 909.

42. **Poliner, L. R., Buja, L. M., Parkey, R. W., Stokely, E. M., Stone, M. J., Harris, R., Saffer, S., Templeton, G. H., Bonte, F. J., and Willerson, J. T.,** Comparison of different non-invasive methods of infarct sizing during experimental myocardial infarction, *J. Nucl. Med.,* 18, 1977, 517.

43. **Faltová, E., Mráz, M., Konrád, L., Protivová, L., and Šedivý, J.,** Studies on isoprenaline induced myocardial lesions. Quantitative evaluation by Mercurascan uptake, *Basic Res. Cardiol.,* 72, 1977, 454.

44. **de Leiris, J., Harding, D. P., and Pestre, S.,** The isolated perfused rat heart: a model for studying myocardial hypoxia or ischaemia, *Basic Res. Cardiol.,* 79, 1984, 313.

45. **Michio, F., Hideo, I., and Tetsuya, A.,** In vitro assessment of myocardial function using a working rabbit heart, *J. Pharmacol. Methods,* 14, 1985, 49.

46. **Vogel, W. M. and Lucchesi, B. R.,** An isolated, blood-perfused, feline heart preparation for evaluating pharmacological interventions during myocardial ischemia, *J. Pharmacol. Methods,* 4, 1980, 291.

47. **Mulch, J., Schoper, J., Gottwik, M., and Hehrlein, F. W.,** Recovery of the heart after normothermic ischemia. Part II. Myocardial function during postischemic reperfusion, *Thorac. Cardiovasc. Surg.,* 27, 1979, 145.

48. **Neely, J. R., Liebermeister, H., Battersby, E. J., and Morgan, H. E.,** Effect of pressure development on oxygen consumption by isolated rat heart, *Am. J. Physiol.,* 212, 1967, 804.

49. **Langendorff, O.,** Untersuchungen am überlebenden Säugetierherzen, *Pflugers Arch.,* 61, 1895, 291.

50. **Döring, H. J. and Dehnert, H.,** Das isolierte perfundierte Warmblüter-Herz nach Langendorff, Biomesstechnik No. 5, Biomesstechnik-Verlag March GmbH, March (FRG), 1985.

51. **Deutschmann, W., Lindner, E., and Deutschländer, N.,** Perfluorochemical perfusion of the isolated guinea pig heart, *Pharmacology,* 28, 1984, 336.

52. **Bergnam, S. R., Clark, R. E., and Sobel, B. E.,** An improved isolated heart preparation for external assessment of myocardial metabolism, *Am. J. Physiol.,* 236, 1979, H644.

53. **Gauduel, Y., Martin, J. L., Teisseire, B., Duruble, M., and Duvelleroy, M.,** Hemodynamic and metabolic responses of the working heart in relation to the oxygen carrying capacity of the perfusion medium, *Gen. Physiol. Biophys.,* 4, 1985, 573.

54. **Katz, G., Paine, W. G., and Tiller, P. M.,** A new method for coronary perfusion of the mammalian heart, *Arch. Int. Pharmacodyn. Ther.,* 61, 1939, 109.

55. **Trouve, R. and Nahas, G.,** Cardiac dynamics of the Langendorff perfused heart., *Proc. Soc. Exp. Biol. Med.,* 180, 1985, 303.

56. **Taegtmeyer, H., Roberts, A. F. C., and Raine, A. E. G.,** Energy metabolism in reperfused heart muscle: metabolic correlates to return of function, *J. Am. Coll. Cardiol.,* 6, 1985, 864.

57. **Weisfeldt, M. L. and Shock, N. W.,** Effect of perfusion pressure on coronary flow and oxygen usage of nonworking heart, *Am. J. Physiol.,* 218, 1970, 95.

58. **Kligfield, P., Horner, H., and Brachfeld, N.,** A model of graded ischemia in the isolated perfused rat heart, *J. Appl. Physiol.,* 40, 1976, 1004.

59. **Manning, A. S., Hearse, D. J., Dennis, S. C., Bullock, G. R., and Coltart, D. J.,** Myocardial ischaemia: an isolated, globally perfused rat heart model for metabolic and pharmacological studies, *Eur. J. Cardiol.,* 11, 1980, 1.

60. **Neely, J. R., Rovetto, M. J., Whitmer, J. T., and Morgan, H. E.,** Effects of ischemia on function and metabolism of the isolated working rat heart, *Am. J. Physiol.,* 225, 1973, 651.

61. **Kannengiesser, G. J., Lubbe, W. F., and Opie, L. H.,** Experimental myocardial infarction with left ventricular failure in the isolated perfused rat heart. Effects of isoproterenol and pacing, *J. Mol. Cell. Cardiol.,* 7, 1975, 135.

62. **Fallen, E. L., Elliot, W. C., and Gorlin, R.,** Apparatus for study of ventricular function and metabolism in the isolated perfused rat heart, *J. Appl. Physiol.,* 22, 1967, 836.

63. **Cotten, M. V.,** Direct measurement of changes in cardiac contractile force, *Am. J. Physiol.,* 174, 1953, 365.

64. **Manning, A. S., Keogh, J. M., Coltart, D. J., and Hearse, D. J.,** Propranolol in the ischemic, reperfused, working rat heart: association between beta-adrenergic blocking activity and protective effect, *J. Mol. Cell, Cardiol.,* 13, 1981, 1077.

65. **Bünger, R., Sommer, O., Walter, G., Stiegler, H., and Gerlach, E.,** Functional and metabolic features of an isolated perfused guinea pig heart performing pressure-volume work, *Pflugers Arch.,* 380, 1979, 259.

66. **Sahyoun, H. A. and Hicks, R.,** Electrocardiographic recording of normal and infarct bearing rat hearts in a perfused isolated preparation, *J. Pharmacol. Methods,* 1, 1978, 351.

67. **Kléber, A. G.,** Resting membrane potential, extracellular potassium activity, and intracellular sodium activity during acute global ischemia in isolated perfused guinea pig hearts, *Circ. Res.,* 52, 1983, 442.

68. **Inoue, F., MacLeod, B. A., and Walker, M. J. A.,** Intracellular potential changes following coronary occlusion in isolated perfused rat hearts, *Can. J. Physiol. Pharmacol.,* 62, 1984, 658.

69. **Bedford, G. K. and Chiong, M. A.,** A high performance liquid chromatographic method for the simultaneous determination of myocardial creatine phosphate and adenosine nucleotides, *J. Chromatogr.,* 305, 1984, 183.

70. **Sellevold, O. F. M., Jynge, P., and Aarstad, K.,** High performance liquid chromatography: a rapid isocratic method for determination of creatine compounds and adenine nucleotides in myocardial tissue, *J. Mol. Cell. Cardiol.,* 18, 1986, 517.

71. **Bailey, I. A., Seymour, A. M. L., and Radda, G. K.,** A ^{31}P NMR study of the effects of reflow on the ischemic rat heart, *Biochim. Biophys. Acta,* 637, 1981, 1.

72. **Lavanchy, N., Martin, J., and Rossi, A.,** Graded global ischemia and reperfusion of the isolated perfused rat heart: characterization by ^{31}PNMR spectroscopy of the extent of energy metabolism damage, *Cardiovasc. Res.,* 18, 1984, 573.

73. **Fuchs, J., Veit, P., and Zimmer, G.,** Uncoupler- and hypoxia-induced damage in the working rat heart and its treatment. II. Hypoxic reduction of aortic flow and its reversal, *Basic Res. Cardiol.,* 80, 1985, 231.

74. **Winbury, M. W.,** Influence of glucose on contractile activity of papillary muscle during and after anoxia, *Am. J. Physiol.,* 187, 1956, 135.

75. **Tyberg, J. V., Yeatman, L. A., Parmley, W. W., Urschel, C. W., and, Sonnenblick, E. H.,** Effects of hypoxia on mechanics of cardiac contraction, *Am. J. Physiol.,* 218, 1970, 1780.

76. **Shigenobu, K., Asano, T., and Kasuya, Y.,** Effects of hypoxia and metabolic inhibitors on the electrical and mechanical activities of isolated guinea pig papillary muscles, *J. Pharmacobiodyn,* 4, 1981, 317.

77. **Asano, T., Shigenobu, K., and Ksuya, Y.,** Effects of hypoxia and reoxygenation on tissue ATP level and electrical and mechanical function of isolated guinea pig ventricular muscles, *J. Pharmacobiodyn.,* 7, 1984, 63.

78. **Opie, L. H., Mansford, K. R. L., and Owen, P.,** Effect of increased heart work on glycolysis and adenine nucleotide in the perfused heart of normal and diabetic rats, *Biochem. J.,* 124, 1971, 475.

79. **Paradise, N. F., Schmitter, J. L., and Surmitis, J. M.,** Criteria for adequate oxygenation of isometric kitten papillary muscle, *Am. J. Physiol.,* 241, 1981, H348.

80. **Schouten, V. J. A. and ter Keurs, E. D. J.,** The force-frequency relationship in rat myocardium. The influence of muscle dimensions, *Pflugers Arch.,* 407, 1986, 14.

81. **Reimer, K. A., Steenbergen, Ch., and Jennings, R. B.,** Isolated cardiac myocytes: is their response to injury relevant to our understanding of ischemic injury, in vivo?, *Lab. Invest.,* 53, 1985, 369.

82. **Wollenberger, A.,** Isolated heart cells as a model of the myocardium, *Basic Res. Cardiol.,* 80 (Suppl. 2), 1985, 9.

Chapter 22

HYPERLIPOPROTEINEMIA AND EXPERIMENTAL ATHEROSCLEROSIS

R. Poledne and A. Vrána

TABLE OF CONTENTS

I. INTRODUCTION

Hyperlipoproteinemia has been recognized as the main disease-promoting and disease-inducing factor of atherosclerosis, the quantitatively most important disease. In the industrialized world, every second man dies of clinical complications of atherosclerosis.[1] Atherosclerotic lesion is characterized by accumulation of lipids in the intra- and extracellular space of the intima together with cellular proliferation which leads to lumen narrowing.

The increased mortality from atherosclerotic complications stimulated interest in lipoprotein metabolism after World War II not only in epidemiological and clinical studies but also in experimental studies. Experimental models of hyperlipoproteinemia furnished the body of information necessary to understand the physiology and pathophysiology of lipoproteins and currently serve as perfect scientific tools to study atherosclerosis as well as different hypolipoproteinemic and antiatherosclerotic effects.

II. LIPOPROTEIN METABOLISM

Lipids are water-insoluble substances transported by the blood stream in the form of a polydisperse collection of particles of a high molecular weight, lipoproteins. In addition to lipids (triglycerides, cholesterol, and phospholipids), lipoproteins contain one or more specific proteins called apolipoproteins, which play a solubilizing and structural role. Moreover, some of them are cofactors in different reactions of lipoprotein metabolism or a binding part to specific receptors of plasmatic membrane of all tissues.

According to hydrated density, lipoproteins get separated into four major classes:

1. Chylomicrons (density 0.95 g/mℓ)
2. Very low density lipoproteins (VLDL) (density 1.006 g/mℓ)
3. Low density lipoproteins (LDL) (density 1.063 g/mℓ)
4. High density lipoproteins (HDL) (density 1.21 g/mℓ)

Chylomicrons are formed in the intestine after dietary lipid absorption. VLDL are secreted from the liver and transport the bulk of endogenously synthesized triglycerides, and are metabolized in the plasma. Triglycerides are hydrolized by enzyme lipoprotein lipase (LPL), attached to the vascular endothelium. Due to triglyceride hydrolysis, their particles become denser and form an intermediay product (intermediary density lipoprotein — IDL) and are finally changed to LDL. During gradual lipolysis, VLDL particles lose part of apolipoproteins and consequently, LDL particles contain only apolipoprotein B (apoB).

HDL are produced by the intestine and liver in the form of precursor-nascent discoidal HDL and they are further developed to particles circulating in the plasma. Two subfractions of HDL — less dense (HDL_2) and denser (HDL_3) — have been separated and their interconversion likewise takes place in the blood.

Increased concentration of chylomicrons frequently observed in patients with lipoprotein lipase deficiency does not represent an increased risk for premature atherosclerosis. There are partly inconsistent results concerning the increased risk of elevated endegenously synthesized triglyceride-rich particles, VLDL. Recent findings of the relationship of hypertriglyceridemia to cardiovascular mortality in epidemiological studies[2] and the prevalence of coronary arterial stenosis in a clinical trial increased interest in models of hypertriglyceridemia.[3]

The correlations between atherosclerosis and its clinical complications on the one hand and a high concentration of LDL particles carrying cholesterol on the other are so close that they are sometimes considered as a proof of causality. Therefore it is not surprising that models of hypercholesterolemia are the most frequent models of hyperlipoproteinemia.

Table 1
TOTAL SERUM CHOLESTEROL
CONCENTRATIONS BEFORE
AND AFTER INCREASE OF
ALIMENTARY CHOLESTEROL
INTAKE FOR 2 WEEKS

	N	Control diet	Cholesterol diet
Rat	20	1.45 ± 0.32	1.88 ± 0.52
Swine	12	3.81 ± 0.38	9.71 ± 1.24
Rabbit	10	2.05 ± 0.28	125.4 ± 21.50

Note: Each diet adequate to species was supplemented with 2% of crystalline cholesterol and 5% of beef lard.

Increased concentration of high density lipoproteins represents a ''negative'' risk factor of atherosclerosis as the negative correlation between HDL cholesterol concentration and atherosclerosis has been repeatedly documented.

Increased concentration of LDL cholesterol has been shown to participate in atherogenesis by:

1. Induction of endothelial cell damage leading to an increase in cholesterol inflow into the arterial wall
2. Stimulation of smooth muscle proliferation
3. Cholesterol accumulation in the cell

The exact mechanism of participation of triglyceride-rich particles in atherogenesis has not been convincingly demonstrated yet, but the atherogenic effect of free fatty acids liberated during lipolysis on the surface of the arterial wall by lipoprotein lipase and a direct influence of remnant particles of chylomicrons and VLDL must be considered.

III. MODELS OF HYPERCHOLESTEROLEMIA

A. Effect of High Cholesterol Intake

Diets high in saturated fat and cholesterol cause alterations in plasma lipoproteins, and these alterations stimulate the delivery of cholesterol to the arterial wall. An increase in alimentary cholesterol intake raises the total cholesterol concentration and changes specific lipoprotein fractions but, quantitatively, these changes are more dramatic in one species than in another (Table 1). Qualitatively, similar changes can be observed in a variety of species.[5] An increase of alimentary cholesterol intake decreases the level of typical HDL (without apoprotein E) and increases the level of HDL-containing apoprotein E, referred to as HDL_c.[6] Compared with typical HDL, HDL_c are less dense (in ultracentrifugal separation they overlap the density of LDL), migrate slowly on agarose electrophoresis as a distinct α-2-migrating band and are enriched in cholesterol esters (up to 50% of the chemical composition of HDL_c may be cholesterol esters). It is important to note that HDL_c appear regardless of whether or not total cholesterolemia increases and they are formed also in the rat with no or a minimal effect of alimentary cholesterol on cholesterolemia.[7] HDL_c are formed in the plasma compartment after enrichment of typical HDL with cholesterol taken up from peripheral tissues, and apo E from other plasma lipoproteins.

Cholesterol-induced hypercholesterolemia is also associated with the production of β-

VLDL.[5] Unlike normal VLDL which migrate in agarose electrophoresis as a pre-β band distinct from the β band of LDL, β-VLDL particles of animals fed cholesterol have β electrophoretic mobility even though they float at the same density as typical VLDL in ultracentrifugation. Although β-VLDL isolated from various cholesterol-fed species are very similar in terms of physical and chemical properties, they may be of a different origin. In cholesterol-fed rabbits, β VLDL arise as a remnant of chylomicrons and accumulate in the plasma because of saturated hepatic apoE uptake system;[8] however, apoE receptors are not repressed in the liver of a cholesterol-fed dogs and canine β VLDL are produced by the liver.[7] Irrespective of their origin, β VLDL cause marked cholesterol accumulation in smooth muscle cells or macrophages.[9] It is reasonable to speculate that after cholesterol feeding, atherosclerosis develops when inflow of cholesterol from VLDL and LDL into the arterial wall exceeds outflow of cholesterol, as typical HDL concentration drops at the same time.

Diets high in cholesterol and fat also cause an elevation of LDL in most animals. In addition, high cholesterol diet-induced LDL particles are larger in comparision with LDL from the same species on a low-cholesterol diet. The mechanism by which elevated cholesterolemia causes accelerated atherosclerosis is still partly speculative. LDL serve as the main source of cholesterol for all extrahepatic tissues. This cholesterol, required for the structure of all plasma and intracellular membranes, enters the intracellular space via the high-affinity (apoB, E receptors) pathway and down regulates endogenous cholesterol synthesis and new receptor formation.[10] In addition, the low-affinity pathway which does not possess this intracellular regulation has been demonstrated on the plasma membrane. The low-affinity pathway does not play an important role in cholesterol transport to the cell in normocholesterolemia.[11] It is supposed that in a high LDL concentration, when the receptor-mediated pathway is close to saturation, the significance of the low-affinity pathway increases. Because cholesterol entering the cell in this way does not possess down regulation in de novo cholesterol synthesis and receptor production, cholesterol starts to accumulate in the arterial wall cells and monocytes.

B. Diets Producing Hyperlipoproteinemia

All diets used for the stimulation of cholesterol concentration in blood are high in saturated fat and are almost invariably supplemented with crystalline cholesterol. The fat content varies from 2 to 30%, but a minimal content of fat in the diet is necessary not only as an adequate source of essential fatty acids, but also to solubilize crystalline cholesterol. Absorption of cholesterol varies between 20 and 65% in different animal species and individuals of the same species, but it drops to 5 to 20% when cholesterol is supplied to the diet in the form of solid crystals. Crystalline cholesterol must be first dissolved in a dietary fat and then mixed with the rest of diet. Attention must be paid to the temperature during melting of lard and cholesterol which should be as low as possible to prevent oxidation of solubilized cholesterol. A relatively low-fat diet is used in species susceptible to hypercholesterolemia (birds,[12] rabbits,[13] and some primates[14]) but diets high in saturated fat are necessary for the induction of hypercholesterolemia in the swine[4] and the dog.[7]

The cholesterol content of diets ranges between 0.2 and 6%. Like the saturated fat content, diets with only 0.2 to 1% of crystalline cholesterol produce hypercholesterolemia in rabbits,[14] cockerels,[15] and rhesus monkeys.[17] In the swine,[4,18] a 2 to 4% diet is usually used[4] and a higher cholesterol content is recommended for the dog.[7]

In animal species resistant to a high-fat, high-cholesterol diet, some additional metabolic affections are used. The method of choice is surgical thyroidectomy[7] or the pharmacological effect on the thyroid gland by propythiouracyl,[6] combined with the supplementation of a diet with bile acid salts.[19] Although this type of diet is able to produce hypercholesterolemia also in the dog and in the rat, it is necessary to stress that it is hepatotoxic and always

Table 2
EFFECT OF ALIMENTARY CHOLESTEROL, TYPE OF DIETARY FAT, SUCROSE AND DIETARY FIBRES ON CHOLESTEROLEMIA IN THE PRAGUE HEREDITARY HYPERCHOLESTROLEMIC RAT

Type of diet supplementation	Cholesterolemia
0.5% cholesterol + 5% sunflower oil	2.61 ± 0.37
0.5% cholesterol + 5% beef lard	2.91 ± 0.34
0.5% cholesterol + 5% beef lard + 10% sucrose	4.06 ± 0.67
1.0% cholesterol + 5% sunflower oil + 5% beef lard	6.44 ± 0.57
1.0% cholesterol + beef lard + 10% sucrose	8.12 ± 1.87
1.0% cholesterol + 5% beef lard + 10% sucrose + 5% citrus pectin	4.78 ± 1.02

Note: Standard laboratory diet was supplemented. Six animals in each group ± SD.

produces liver steatosis. The hepatotoxicity of such a diet together with thyroid gland affection makes all these models a far cry from the pathology of human hyperlipoproteinemia which should be modeled. Although animal models utilizing these toxic diets and above-mentioned endocrine affections are not useful for the study of the pathology of human hypercholesterolemia, they can be used for the induction of experimental atherosclerosis, as the final effect of artificial hyperlipoproteinemia on the development of atherosclerotic changes is proportional to the total serum cholesterol level and the lesions resemble human atherosclerosis.[5]

On the contrary, hypercholesterolemia can be initiated in very sensitive species or individuals without cholesterol supplementation only by the administration of a semisynthetic diet high in saturated fat and sucrose and low in dietary fibers.[20] The effect of saturated and unsaturated fatty acids, fiber and sucrose is documented in Table 2, in which stimulation of hypercholesterolemia with four different diets in a new strain of hypercholesterolemic rat is presented.

C. Interspecies Differences

It is understandable that sensitivity to alimentary cholesterol plays a very important role in selecting a suitable model of hypercholesterolemia. The most commonly used animal is the rabbit, not only because it has been used from the beginning of the century but also since it is inexpensive, sensitive to alimentary cholesterol, and very easy to handle.[21] As has been said previously, although the changes in lipoproteins after a high-cholesterol diet are qualitatively very similar in all animal species, the quantitative differences are tremendous. The example of hypercholesterolemia produced by a diet with 1 to 2% in the rabbit compared with medium susceptible (swine) and very resistant (rat) species is shown in Table 3. Although hypercholesterolemia in the swine and in the rat was produced with a doubled cholesterol content in the diet, it was several times lower compared with the rabbit. The serum cholesterol concentration in the rabbit increases immediately after switching from control to cholesterol diet and is stable in time. A significant increase in cholesterolemia in the rabbit already after 1-day-feeding of a semisynthetic diet without crystalline cholesterol was described by Terpstra.[20] In the swine, the increase in total serum cholesterol is slower, reaching maximal values around the second month of cholesterol feeding followed by a

Table 3
TIME COURSE OF TOTAL SERUM CHOLESTEROL AFTER A HIGH-CHOLESTEROL DIET

	N	% of cholesterol in diet	Days 0	5	10	20	100
Rat	8	2	1.38	1.52	1.82	1.49	—
Rabbit	17	0.5	1.85	10.44	28.51	27.90	30.22
Swine	6	4	2.55	5.98	15.25	16.43	12.11

decrease in almost all individuals. In the rat, the cholesterol concentration in the blood rises only by about 30% and then decreases almost to the starting level.

Birds are very sensitive to alimentary cholesterol and they were used for induction of atherosclerosis frequently in the 1950s and 1960s.[12] The relatively small similarity to human anatomy, physiology, and pathology of atherosclerosis is the main reason for a decrease in the popularity of birds in cardiovascular disease modeling.

By contrast, the swine is in many metabolic respects very similar to man; it is true also in the case of lipoprotein metabolism.[22] As in man, a high-fat diet produces no change in the blood triglyceride concentration, but when cholesterol is supplemented to the diet, hypercholesterolemia appears. Also the qualitative changes of plasma lipoproteins after a high-fat, high-cholesterol diet fairly resemble those observed in man.[5] Together with the appearance of two specific lipoproteins, HDL_c and β VLDL, the concentration of LDL cholesterol increases.[4] The rather complicated handling of the swine is a certain disadvantage of this animal model. Repeated blood sampling for cholesterol and lipoprotein determination during chronic experiment is possible only from the vena cava superior, requiring three skilled persons to handle even the minature swine, the most useful strain of a pig for experimental work.

Numerous reports have reviewed the response of various species of monkeys to diets with fat and cholesterol supplementation. Semisynthetic diets containing 15 to 25% of animal fat and 0.5 - 4% of cholesterol are usually used.[16] The most sensitive rhesus monkey develops hypercholesterolemia when fed a diet formulated from human table food.

D. Individual Differences

Individual differences in the sensitivity to alimentary cholesterol were found in several species in the 1970s.[7,14,23] Although differences in alimentary cholesterol absorption were suspected to be responsible for the observed variation in plasma cholesterol concentration in low and high responders.[14] This possibility has been excluded recently.

A defect of functional apoB, E receptors,[8] a shift in feedback regulation of cholesterol synthesis, cystein-arginine interchange in apoE[24] and the difference in specific lipoprotein production[25] has been shown to differentiate hyperresponders from hyporesponders. Hyper- and hyporesponders to a cholesterol diet were described in the rat,[23] mouse,[26] rabbit,[13] Japanese quail,[15] dog,[7] and primates.[14] The interest concerning the individual responsiveness in experiments was stimulated by observation of similar differences also in man.[27]

The differences in the sensitivity of rats to a high-cholesterol diet supplemented with cholic acids were used for inbreeding of hyperresponders by Imai.[19,23] Similar selection and inbreeding were also used for the Japanese quail resulting in a line of birds sensitive to experimental atherosclerosis.[15] A new line of rat, the Prague hereditary hypercholesterolemic rat (PHHC), was obtained after selection and inbreeding of individuals most sensitive to diet supplemented with cholesterol only. The distribution of basal cholesterolemia in the population is characterized by the Gaussian distribution curve showing polygenic control.

When animals were fed a diet for 2 weeks containing 2% cholesterol, a slight shift of cholesterolemia appeared both in males and females with a sign of splitting to two subpopulations. Four males and ten females with the highest stimulation of cholesterolemia were selected as parent generation. Selection and inbreeding was continued for ten generations. Cholesterolemia increased gradually up to the fourth generation and then was stabilized. A dramatic jump appeared in the eighth and nineth generations caused probably by a recombination of genes controlling cholesterolemia. In the twentieth generation, the PHHC rat displays doubled basal cholesterolemia compared to Wistar rats and fivefold cholesterolemia after a cholesterol diet. Certain differences in cholesterolemia between different inbred strains of the mouse and rat already used have been documented.[28]

E. Models of Spontaneous Hypercholesterolemia

Polygenic control of spontaneous (basal) cholesterolemia has been described in the preceding paragraph. When individuals with highest cholesterol concentration are used for selective inbreeding, mean cholesterolemia of the offspring of selected parent groups has an increasing tendency.[29] This approach was used for the selection of hypercholesterolemic lines of rat in two laboratories.[29,30] Unfortunately, fertility was lost in the twelfth generation in one line,[29] and the other line named RICO (rat with increased cholesterol) displayed more than 90% of serum cholesterol in HDL, making this line not a very suitable model of the pathology of atherogenic hyperlipoproteinemia in man.[30]

On the contrary, a similar approach to select spontaneously hypercholesterolemic rabbits was most successful. Hyperlipoproteinemic mutants appeared during selection, and their homozygous offspring obtained by brother-sister, or back-breeding display a high LDL cholesterol concentration,[31] xanthomas,[32] and atherosclerosis resembling human pathology.[33] This animal, the Watanabe hereditary hyperlipemic (WHHL) rabbit, is the best experimental model of human familial hypercholesterolemia. Like human disease, hypercholesterolemia is caused by a single gene defect in apoB, E receptors.[34] A combination of LDL overproduction[34] due to enhanced conversion of VLDL to LDL,[34] as well as its clearance,[9] leads to hypercholesterolemia. Recently, this combination of defects has been shown as the only cause of hypercholesterolemia in the WHHL rabbit,[34] as production of apoB of VLDL is not increased and no lipoproteins other than VLDL are produced in the liver.[8]

Spontaneously hypercholesterolemic animals are also more sensitive to high-cholesterol and high-saturated fat diets[29,32] and vice versa. Inbred strains of rat sensitive to alimentary cholesterol (the PHHC rat) display at the end of selection a doubled spontaneous serum cholesterol level,[35] although basal cholesterolemia was not considered a criterion in selection.

F. Models of Hypertriglyceridemia

Increased concentration of serum triglycerides can be produced by an increased rate of triglyceride-rich lipoprotein synthesis in the liver, by a decrease in their utilization or by a combination of these two processes.

Increased intake of carbohydrates usually results in an elevated concentration of triglycerides of plasma VLDL both in experimental animals and in man. The effect is particularly pronounced in susceptible individuals. Moreover, numerous experiments have revealed that, in addition to the amount of dietary carbohydrates, also their type may play a role. Disaccharide intake, and especially sucrose intake, as well as monosaccharide intake is more efficient as regards to the induction of hypertriglyceridemia than the intake of the same proportion of starch. Quantitative differences were observed also between individual hexoses, the intake of fructose (a constituent monosaccharide of sucrose) usually has a more marked hypertriglyceridemic effect than consumption of the same proportion of glucose in the diet.

Hypertriglyceridemia can be easily induced by sucrose or fructose administration either in drinking water (5 to 10%) or in the diet (30 to 70% of energy) in a number of species.

However, the rat is most frequently used for modeling of this disorder. Already after a few days of administration of sugars, triglyceridemia in the rat rises to values which are two to three times higher as compared with starch- or glucose-fed controls.

Triglycerides concentration in serum culminates after 1 or 2 weeks of feeding the sucrose diet due to the rapid conversion of fructose to precursors for synthesis of VLDL triglycerides, fatty acids and α-glycerophosphate. Activities of numerous hepatic enzymes of fructose metabolism, NADPH generation and fatty acid synthesis were found to be consistently increased by fructose or sucrose feeding. In addition, the catabolism of VLDL triglycerides, due to elevated lipolytic activity and increased oxidation of fatty acid in muscle, is increased in this model. Nevertheless, the increase of VLDL utilization in tissues is not sufficient to compensate their overproduction in the liver. The model of sucrose-induced hypertriglyceridemia is useful for studies of pathophysiological mechanisms of hypertriglyceridemia as well as treatment of this disorder.[36,37]

Rats with increased hepatic VLDL production[38] and mice with lipase deficiency are the models of choice for hypertriglyceridemia.[39]

IV. EXPERIMENTAL ATHEROSCLEROSIS

A. Rat

The rat is resistant to atherosclerosis not only because it is very resistant to a high-fat, high-cholesterol diet, as has been mentioned previously, but an additional reason makes the induction of experimentl atherosclerosis impossible unless extreme conditions are used. No sign of atherosclerosis is found when similar stimuli producing atherosclerosis in other species are used. A five- to tenfold increase in cholesterolemia in a more sensitive line of rat produces only limited lipid deposition in the aorta without smooth muscle cell proliferation and fibrous lesions.[19,35,40] When extreme doses of vitamin D are used, degenerative changes occur, but it is the alteration of calcium metabolism that is actually responsible for the effect.

B. Mouse

As in the rat, lipid cumulation in the aorta of the mouse is the only process resembling atherosclerotic changes.[41] These changes are more frequent in some inbred strains of mouse which are more sensitive to alimentary cholesterol.[24] Nevertheless, no atherosclerotic lesions have been induced in the mouse which could be used as an experimental model of human atherosclerosis.

C. Dog

The dog, too, is resistant to the development of hypercholesterolemia and does not display naturally occurring atherosclerosis. As the dog is generally less sensitive to a high-cholesterol diet, propylthiouracyl or thyroidectomy are used to increase cholesterolemia and to accelerate atherogenesis.[6] A high variation of individual responsiveness to alimentary cholesterol leads to a high variation in experimental atherosclerotic production.[7] Canine atherosclerosis after thyroidectomy involves the thoracic aorta and coronary intracranial arteries.[25] The most severe lesions are present in the arch and in the abdominal aorta at the site of origin of the intercostal branches. The nature of lesions is predominantly in extracellular lipid deposition surrounding foam cells without affecting the main coronary arteries blood flow producing large myocardial ischemia. Peripheral small arteries involvement is more severe[5] than the changes in the medium size arteries.[25] These are, together with the prominent medial changes, the main differences as compared with human atherosclerosis.

D. Rabbit

Although rabbit has been the most frequently used animal in experimental atherosclerosis modeling since the first Anichkov's experiments,[21] the similarity of the model with human disease is far from being ideal. The differences lie in localization (in the descending thoracic aorta in the rabbit whereas in the abdominal aorta in man) and the morphological structure (rabbit lesions are predominantly lipoid by nature with a small proportion of fibrotic lesions). The individual differences in sensitivity to alimentary cholesterol producing the variations in final cholesterolemia are the main cause of the high variability of atherosclerosis produced in experimental groups fed a high-cholesterol diet. This variability is decreased by an individualized dosage of alimentary cholesterol intake. For example, when 0.2% of cholesterol for high responders and 0.3% for low responders in diet is used, cholesterolemia varies between 15 and 20 mmol/1 in. in a long-term experiment with a small variation of the surface of atheromatous lesions within experimental groups.

The disadvantages of atherosclerosis in the rabbit in the nature of lesions are compensated for, on the other hand, by an important advantage. In this particular mammal, production of experimental atherosclerosis is far simpler and more reproducible than in any other animal species.

With experimental atherosclerosis induced by an increase of alimentary cholesterol intake alone, or by a synthetic diet,[20] the lesions are located in the arch and descending thoracic aorta. Potentiation of the atherogenic effect of a diet by immune disease or by hypertension[13] leads to more complicated fibrotic lesions with ulcerations and aneurysms located also in the abdominal aorta. Experimental atherosclerosis induced by a combination of mild hypercholesterolemia and renovascular hypertension is more similar to human atherosclerosis both in location as well as morphology.[13] Three animals out of twenty died within 4 months of atherosclerosis induction after acute myocardial failure, and frequent myomalacia were found in all animals in the experimental group. Still, the distal part of the coronary arteries was involved without frequent stenoses of the main coronary arteries.

E. Swine

The swine has been widely used in studies of experimental atherosclerosis.[22] In many aspects it appears to be the superior model as spontaneous atherosclerosis occurs in older individuals. The morphology and biochemistry of atheromatous lesions induced by a high-cholesterol diet resembles those observed in humans. Atheromatous lesions appeared after 8 to 10 months (when cholesterolemia increases to 10 mmol/ℓ) in the arch and the descending abdominal aorta, mainly at the site of origin of the intercostal and renal arteries. Lesions were also found in the coronary arteries and aortic valves. Disadvantages of experimental atherosclerosis in the swine include an extremely low reproducibility and a wide range of variations among individuals. Occasionally, we found individuals without any atherosclerosis within a group of animals with frequent and complicated lesions regardless of the degree of hypercholesterolemia.[4] The model should be improved by a combination of cholesterol feeding with renovascular hypertension or damage to the endothelial surface by balloon catheter.[18]

F. Nonhuman Primates

This animal species makes an attractive model of atherosclerosis because of its marked similarity in the morphology and distribution of lesions to humans. This is not surprising in view of the phylogenic proximity and similarities in the anatomy and physiology of the cardiovascular system.

Induction of experimental atherosclerosis in several species of *Macaques (nemestria, mulata)* is successful when a high-cholesterol diet (0.5 to 2%) is applied for 0.5 to 2 years. On the contrary, we were not able to obtain any atheromatous lesions in baboons after 1 year of cholesterol feeding as all individuals used were low responders to alimentary cholesterol. Although the high similarity of experimental atherosclerosis in nonhuman primates to human diseases has been described by several authors, the arterial changes involved are generally more lipoid, composed predominantly of macrophage-derived foam cells with extracellular lipid deposition. These changes regress with relative ease (compared with human atherosclerosis) when induced by cholesterol diet feeding for 6 months to 1 year.[42] In an experiment currently in progress in which a longer induction period is used, arterial disease rather resembles human fibrotic and complicated lesions, less easily regressive.[42] This result shows that also the most suitable model of human atherosclerosis can be further improved.

V. CONCLUSIONS

Numerous models of hyperlipoproteinemia have been described in the literature and summarized here. From all these models, those utilizing hepatotoxic substances in the diet and thyroid gland affection are not very useful tools to study the mechanism of hyperlipoproteinemia in man and different hypolipoproteinemic effects. Still, they can be valuable to study the pathogenesis of atherosclerosis as different types of elevation of lipoprotein concentration in the blood lead to similar morphological changes in the arterial wall. Spontaneous hypercholesterolemia and hereditary hypersensitivity to dietary hypercholesterolemic stimuli in animal models obtained by gene mutation or recombination are most promising tools to study the mechanism of hyperlipoproteinemia in man and different therapeutic interventions in an effort to control this disorder.

Animal species extremely sensitive to atherosclerotic stimuli produce lesions containing only lipids and differing from human disease of arteries. These models can be improved by a decrease in extreme hypercholesterolemia to a moderate value together with simultaneous intervention into endothelial wall integrity by an additional influence. Atherosclerosis produced in less sensitive species displays a high individual variability decreasing the value of these models. A decrease of high variability by careful control of atherogenous lipid concentration, together with prolongation of the induction phase period, are necessary to obtain atheromatous lesions similar to those in man which make good models for the study of atherosclerosis regression.

REFERENCES

1. **Wilsmut, R. W., Garrison, R. J., Castelli, W. P., Feinleit, M., McNamee, P. M., and Kaunel, W. B.,** Prevalence of coronary heart disease in the Framingham offspring study: Role of lipoprotein cholesterol, *Am. J. Cardiol.,* 46, 649, 1980.
2. **Pelkonen, R., Nikkilä, E. A., Koskinen, S., Penttinen, K., and Saru, S.,** Association of serum lipids and obesity with cardiovascular mortality, *Br. Med. J.,* 2, 1185, 1977.
3. **Gotto, A. M., Govry, G. A., Yeshurum, D., and DeBakey, M. E.,** Relationship between plasma lipid concentration and coronary artery lipid in 496 patients, *Circulation,* 56, 875, 1977.
4. **Poledne, R., Reiniš, Z., Lojda, Z., Hanuš, K., and Číhová, Z.,** The inflow rate of low density lipoprotein cholesterol to the arterial wall in experimental atherosclerosis, *Physical. Bohemoslov.,* 35, 313, 1986.
5. **Mahley, R. W.,** Dietary fat, cholesterol and accelerated atherosclerosis, *Atherosclerosis Rev.,* 5, 1, 1979.
6. **Mahley, R. W. and Holcombe, K. S.,** Alterations of the plasma lipoproteins and apoproteins following cholesterol feeding in rat. *J. Lipid Res.,* 18, 314, 1977.
7. **Mahley, R. W. and Weisgraber, K. H.,** Canine lipoproteins and atherosclerosis, *Circ. Res.,* 35, 713, 1974.
8. **Hornick, C., Kita, T., Hamilton, R. L., Kane, P., and Havel, R. J.,** Secretion of lipoproteins from the liver of normal and Watanabe heritable hyperlipidemic rabbits, *Proc. Natl. Acad. Sci. U.S.A.,* 80, 6096, 1983.
9. **Kovanen, P. T., Brown, M. S., Basu, S. K., Bilheimer, D. W., and Goldstein, J. L.,** Saturation and suppression of hepatic receptors: a mechanism for the hypercholesterolemia of cholesterol-fed rabbits, *Biochemistry,* 98, 1396, 1981.

10. **Goldstein, J. L. and Brown, M. S.,** The low density lipoprotein pathway and its relation to atherosclerosis, *Annu. Rev. Biochem.,* 46, 897, 1977.

11. **Steinberg, D.,** Metabolism of lipoproteins at the cellular level in relation to atherosclerosis, in *Lipoproteins, Atherosclerosis and Coronary Heart Disease,* Miller, N. E. and Lewis, B., Eds., Elsevier, Amsterdam, 1981, 31.

12. **Reiniš, Z., Duben, Z., Vaněček, R., and Kubát, K.,** The influence of nutrition and physical activity on experimental atherosclerosis in chickens, in *Proc. IV Int. Congress of Angiology,* B. Prusík, Z. Reiniš, O. Riedl, Eds., State Medical Publishing House, Prague 1962, 598.

13. **Pirk, J., Poledne, R., Urbanová, D., Komárek, P., and Konopková, M.,** Regression of experimental atherosclerosis induced by combination of hypercholestrolemia and hypertension, in *Proc. 5th Dresden Int. Lipid Symp., Advances in Lipoprotein and Atherosclerosis Research, Diagnostics and Treatment,* 1985, 700.

14. **Bhattacharya, A. K. and Eggen, D. A.,** Feedback regulation of cholesterol biosynthesis in rhesus monkeys with variable hypercholestrolemic response to dietary cholesterol, *J. Lipid Res.,* 22, 16, 1981.

15. **Shih, J. C., Pullman, P. E., and Kao, K. J.,** Genetic selection, general characterization and histology of atherosclerosis-susceptible and resistant Japanese quail, *Atherosclerosis,* 49, 41, 1983.

16. **Bhattacharya, A. K. and Eggen, D. A.,** Mechanism of the variability in plasma cholesterol response to cholesterol feeding in rhesus monkeys, *Artery,* 11, 306, 1983.

17. **Faggiotto, A., Ross, R., and Harker, L.,** Studies of hypercholesterolemia in the nonhuman primate. I. Changes that lead to fatty streak formation, *Atherosclerosis,* 4, 323, 1984.

18. **Fritz, K. E., Daud, A. S., Augustyn, S. M., and Jarmolych, J.,** Morphological and biochemical differences among grossly defined types of swine aortic atherosclerosis induced by combination of injury and atherogenic diet. *Exp. Mol. Pathol.,* 32, 61, 1980.

19. **Imai, Y., Matsumura, H., Shimo, A., Oke, K., and Sozuoki, Z.,** Induction of aortic lipid deposition in a high-response (ExHC) rat fed a diet containing cholesterol and cholic acid, *Atherosclerosis,* 28, 453, 1977.

20. **Terpstra, A. M. and Sanchez-Muniz, F. J.,** Time course of the development of hypercholesterolemia in rabbits fed semipurified diets containing casein or soybean protein, *Atherosclerosis,* 39, 217, 1981.

21. **Anitschkow, N. and Chalatow, S.,** Über experimentelle Cholesterin Steatose, ihre Bedeutung für Entstehung einiger pathologischer Prozesse, *Zentralbl. Allg. Pathol.,* 24, 1, 1913.

22. **Dods, W. J.,** The pig model for biomedical research, *Fed. Proc.,* 41, 247, 1982.

23. **Imai, Y. and Matsumura, H.,** Genetic studies on induced and spontaneous hypercholesterolemia in rats, *Atherosclerosis,* 18, 59, 1973.

24. **Weisgraber, K. H., Innerarity, T. I., and Mahley, R. W.,** Abnormal lipoprotein receptor-binding activity of the human E apoprotein to cystein-arginine interchange of a single site, *J. Biol. Chem.,* 257, 2518, 1982.

25. **Mahley, R. W.,** Atherogenic hyperlipoproteinemia, *Med. Clin. North Am.,* 66, 375, 1982.

26. **Weibust, R. S.,** Inheritance of plasma cholesterol level in mice, *Genetics,* 73, 303, 1973.

27. **Katan, B. and Beynen, A. C.,** Hyper-response to dietary cholesterol in man, *Lancet,* 28, 1213, 1983.

28. **Lusis, A. J., Taylor, B. A., Wagenstein, R. W., and LeBoef, R. C.,** Genetic control of lipid transport in mice, *J. Biol. Chem.,* 258, 5071, 1983.

29. **Boissel, J. L., Crouzet, B., Bourdillon, M. C., and Bles, N.,** Selection of a strain of rats with spontaneous cholesterolemia, *Atherosclerosis,* 39, 11, 1981.

30. **Müller, K. R., Li, J. R., Dinh, D. M., and Subbiah, M. T. R.,** The characteristics and metabolism of a genetically hypercholesterolemic strain of rats (RICO), *Biochem. Biophys. Acta,* 574, 334, 1979.

31. **Tanzawa, K., Shimada, Y., Kuroda, M., Tsujita, Y., Arai, M., and Watanabe, Y.,** WHHL-rabbit: a low density lipoprotein receptor-deficient model for familial hypercholesterolemia, *FEBS Lett.,* 118, 81, 1980.

32. **Watanabe, Y.,** Serial inbreeding of rabbits with hereditary hyperlipidemia (WHHL-rabbit), *Atherosclerosis,* 36, 261, 1980.

33. **Buja, A. M., Kita, T., Goldstein, J. L., Watanabe, Y., and Brown, M. S.,** Cellular pathology of progressive atherosclerosis in the WHHL rabbit. An animal model of familial hypercholesterolemia, *Arteriosclerosis,* 3, 87, 1983.

34. **Kita, T., Brown, S. M., Bilheimer, U. D., and Goldstein, J. L.,** Delayed clearance of very low density and intermediate density lipoprotein with enhanced conversion to low density lipoprotein in WHHL rabbits, *Proc. Natl. Acad. Sci. U.S.A.,* 79, 5693, 1982.

35. **Poledne, R., Scheithauer, E., and Reysserová, I.,** Cholesterol turnover in Prague hereditary hypercholesterolemic rat (PHHC), in *Advances Lipoprotein Atherosclerosis Research Diagnostics and Treatment, Proc. 5th Dresden Int. Lipid Symp.,* 1985, 504.

36. **Vrána, A. and Fábry, P.,** Metabolic effects of high sucrose or fructose intake, *W. Rev. Nutr. Diet,* 42, 56, 1983.

37. **Vrána, A. and Kazdová, L.,** Effects of dietary sucrose or glucose on carbohydrate and lipid metabolism. Animal studies, in *Prog. Biochem. Pharmacol.,* McDonald, I. and Vrána, A., Eds., Karger, Basel, 1986, 59.

38. **Wang, C. S., Fukuda, N., and Ontko, J. A.,** Studies on the mechanism of hypertriglyceridemia in the genetically obese Zucker rat, *J. Lip. Res.,* 25, 577, 1984.

39. **Olivecrona, T., Chernick, S. S., Bentsson-Olivecrona, G., Paternini, J. R., Jr., Brown, W. V., and Scow, R. O.,** Combined lipase deficiency (cld) in mice, *J. Biol. Chem.,* 260, 2552, 1985.

40. **Butkus, A., Tan, E., and Koletsky, S.,** Tissue distribution in genetically obese and spontaneously hypersensitive rat, *Artery,* 2, 53, 1976.

41. **Breckenridge, C. W., Roberts, A., and Kussis, A.,** Lipoprotein levels in genetically selected mice with increased susceptibility to atherosclerosis, *Atherosclerosis,* 5, 256, 1985.

Chapter 23

RENAL FAILURE

J. Heller

TABLE OF CONTENTS

I. INTRODUCTION

Renal failure could be defined as the inability of the kidney to maintain water and electrolyte balance and acid-base homeostasis. This broad definition also includes inflammatory states, immunological disorders, etc., which usually are not covered by the term "renal failure",[1-6] and hence will not be described here. Acute (ARF) and chronic (CRF) forms are distinguished clinically as well as experimentally.

II. ACUTE RENAL FAILURE

Acute renal failure (ARF) is a suddenly commencing failure. Standard clinical classification includes prerenal, renal, and postrenal forms. In the prerenal form, the cause of ARF lies outside the kidney (usually a circulatory shock with hypotension, etc.) and corresponding models are described elsewhere. The postrenal form is caused by obstruction of ureters, urinary bladder, and the urethra.

A simple scheme of renal types of ARF is shown in Figure 1. The causes of these forms of ARF are threefold: ischemia, nephrotoxic drugs, and pigments.

Ischemia of the kidney is invariably due to a mere decrease in renal blood flow (RBF): hypoxemia alone, even of the highest degree, does not damage the kidney.[7] A decrease of RBF is the result of a low perfusion pressure, renal vasoconstriction, obturation of the capillary lumen (e.g., by platelet aggregates), venous stasis, etc. Long-term ischemia leads to damage to the cells of the vascular wall as well as of the tubular cells.

Cellular damage (either of ischemic or nephrotoxic origin) is associated with five major consequences:

1.   A decrease in the glomerula filtration rate (GFR) due to the damage to the glomerular membrane and activation of the tubulo-glomerular feedback mechanism.[8]
2.   A significant decrease in GFR alone, without any tubular damage, could produce typical signs of ARF.
3.   Interstitial edema as a result of damage to the peritubular capillary wall. This edema hinders tubular reabsorption and compresses tubular cells.
4.   Leakage. The damage to tubular cells leads to their partial (brush border) or complete desquamation, the consequence of which is the leakage of the glomerular filtrate into the interstitium.
5.   Obstruction of the tubular lumen is another necessary consequence of this desquamation. Very rarely, tubular obstruction could be caused by accumulation of pigments in the lumen.

In the great majority of ARF forms, an initial and a maintenance phase could be distinguished. The latter either ends by partial or complete healing or results in the development of some type of chronic renal failure.

A. Models of Ischemic Acute Renal Failure

In most of these models, complete arrest of RBF is used which, except for renal surgery, is a rare situation in the clinic where a rather incomplete or intermittent stop usually occurs.

1. Renal Artery Clamping

Renal artery clamping is the most frequently used model at present. Metallic or plastic clamps of a different shape and size are used; snare is rarely employed with respect to the possible damage to the arterial wall. The degree of damage of the renal tissue depends on the duration of clamping. A 60-min. period of complete ischemia usually leads to reversible

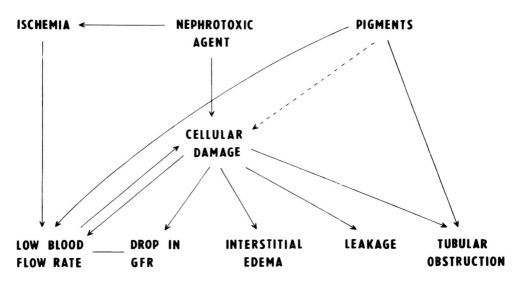

FIGURE 1. Pathogenetic events in the development of acute renal failure, with lines indicating inevitable (full line) and possible (dotted line) action.

changes even though recovery takes weeks or months and is never quite complete.[9,10] All animals survive an ischemic period of 40 min. However, care must be taken with the choice of anesthesia. As barbiturates are mainly excreted by the kidney, the danger of death from anesthesia and not from ARF is great. Short-lasting anesthetics are recommended: Ketamin (Ketalar, Parke-Davis, 100 mg/kg BW i.p.) plus Flunitrazepam (Rohypnol, Roche, 1 mg/kg BW i.p.) is a suitable combination. Whenever necessary, small doses of Flunitrazepam could be administered during the experiment to deepen the anesthesia. In larger animals, it is possible to administer either the same dose of both substances intravenously, or to give Ketamin intramuscularly and Flunitrazepam i.v. All animals awake within 1 hr after the completion of the experiment. In small laboratory animals (rats, hamsters, etc.), the danger of the "no-reflow" phenomenon must be avoided;[11] the cause of this phenomenon could be either intrarenal (and there are no means how to avoid it) or extrarenal. The latter is often caused by damage to the arterial wall when using an unsuitable clamp or snare. It is necessary to wait and watch for a change of the color of the kidney to be sure that blood flow has been really restored.

With the prolongation of the ischemic period, the causes of death are more frequent. However, dogs were described which survived even 4 hr of kidney ischemia: the temperature, humidity, and season play a role.[12]

One hour clamping of one kidney is often used in rats with the contralateral kidney left intact. This enables a comparison of the function of the damaged kidney with the intact one. This model, however, has very rarely an analogy in the clinic. Therefore, contralateral nephrectomy (UNX) is often performed immediately after the release of the clamp. Bilateral clamping is often accompanied by death of the animal: shorter ischemic times are therefore recommended (30 to 40 min). Mason and Welsch[13] pointed to another pitfall in clamping studies: a kidney prepared for micropuncture is exposed to room air so that superficial nephrons can take up some oxygen from it and, consequently, they are not completely ischemic in comparison to those located more deeply.

In larger animals, the clamp could be constricted to enable partial, instead of total, compression of the renal artery.[14] Wilson et al.[15] compared an equally long period of ischemia (2 hr) achieved either by a permanently screwed clamp or by the alternation of 15-min clamping and 3-min release: the latter type was more severe. Zager et al.[16,17] have recently developed alternative models for studying "intermittent" clamping. Rats subjected within 18 to 48 hr after the first 40-min bilateral clamping period to another ischemic insult (25 or 40 min) responded with a lesser injury than those experiencing such an insult for the first time. All these models resemble undoubtedly more clinical situations.

2. Norepinephrine Infusion

Infusion of norepinephrine (NE) into the renal artery at a dose of 0.6 to 1.0 µg/kg BW/min, results in a sudden decrease in RBF essentially to zero lasting throughout the period of infusion. With interruption, a very slow return of RBF occurs. After 3 hr following 40 min NE infusion in dogs, the RBF value amounts to approximately 40% of the initial level, and 24 hr later it is still only 75%,[18] in sharp contrast to the clamp model in which RBF is fully normalized in this period of time. A 2-hr infusion in the dog results in completely irreversible alterations making GRF unmeasurable even 8 weeks after.[19] Similar results were obtained in rats.[20] It is evident in both species that the NE model requires a significantly shorter period of time than the simple clamping to achieve the same degree of damage. The reason for this may lie in the significantly slower return of RBF to a normal value after the discontinuation of NE infusion making the actual ischemic time much longer than the very period of infusion.[21] The most important finding was made by Kleijans et al.:[22] infusion of the same amount of NE into the renal artery in conscious dogs does not induce any ARF.

3. Alternatives of Renal Ischemization

Renal ischemia occurs in a number of states associated with a decrease in cardiac output or other impairment of the cardiovascular system commonly described as "circulatory shock". These states have been modeled in animals and are described in Chapter 5.

B. Models of Nephrotoxic Acute Renal Failure

Many drugs are able to damage the kidney tissue; for modeling, there are, however, rather few really used.

1. Mercury

Mercuric chloride is used almost exclusively, at a dose of 2.5 to 15 mg/kg BW, subcutaneously. The dose mostly used is 4.7 mg/kg. The rat is the most commonly used animal although the dog has also been used.[2] Mercury typically damages the cells of the proximal convolutions; the cause of this damage, not yet fully elucidated, is the affinity of this metal (and some other heavy metals as well) to the protein sulfydryl groups. A typical feature is a much greater drop in GFR than in RBF[23] very probably due to the obstruction of tubuli accompanied by a backleak of the tubular fluid.[24,25]

2. Uranyl Nitrate

Similar damage as with mercury was observed after uranyl nitrate administration. The degree of the damage strongly depends on the dose used: 5 to 25 mg/kg BW subcutaneously was used in rats,[26] rabbits,[27] and dogs.[28] The dose up to 15 mg/kg is not accompanied by a significant tubular leakage.[29]

3. cis-Platinum (cP)

cis-Platinum (commonly used abbreviation for *cis*-diammine-dichloroplatinum) is a cytostatically active drug with considerable nephrotoxicity: 25% of patients exhibit an elevation of blood urea nitrogen concentration 1 to 2 weeks after a single dose of cP.[30] Goldstein et al.[31] studied the effect of a single dose (1.25, 2.50, 5.00, 10.00, or 15.00 mg/kg) of *cis*-Platinum after subcutaneous administration in rats. Only animals of the first two groups survived even though with severely damaged kidneys. Proximal necrosis is the typical picture;[3] tubular obstruction and mainly the backleak is considered as the main mechanism of ARF.[32]

4. Aminoglycoside Antibiotics

Widespread administration of aminoglycoside antibiotics, has been hindered by their nephrotoxicity: 10 to 26% of all patients treated with aminoglycoside antibiotics have suffered some kind of kidney impairment despite continuous monitoring of AA plasma levels. The most toxic of aminoglycoside antibiotics is gentamicin which explains why it is the most popular in animal models[33] in which mostly rats but also rabbits and dogs are used.

A remarkable feature of gentamicin caused ARF is the presence of signs of spontaneous healing despite continuous administration of the drug.[34] Another interesting observation is the typical polyuria with proteinuria;[35] oliguria typical of other types of ARF could only be seen after extremely high doses. Just as in the clinic, there is no correlation between plasma gentamicin concentration and its nephrotoxicity.[36]

An almost endless list of models of gentamicin caused ARF could be compiled. For the sake of orientation, LD_{50} in rats corresponds to a dose of 100 mg/kg BW/day for 2 weeks.[37]

The most frequently used regimens are: 20 mg/kg/day for 23 days;[38] 40 mg/kg/day for 10 days;[39] 100 mg/kg/day for 5 days.[40]

5. Cephalosporins

As soon as the nephrotoxicity of cephalosporins was recognized in the clinic, several experimental models using rats, rabbits, and guinea pigs were described.[41,42] The rabbit has turned out to be an extremely sensitive animal, and doses of 2 to 5 g/kg of cephaloridine reliably produce some renal injury.[3] The damage involves the proximal tubule and is related to the cephalosporins entry into the cell and is linked to the transport mechanism of weak organic acids.[43]

It was established that drugs transported by this mechanism compete for the carrier protein: indeed, p-aminohippuric acid or probenecid inhibit cephalosporin transport and reduce or eliminate its neprotoxicity.

6. Folic Acid

The finding[44] that folic acid produces a marked increase in renal DNA and RNA synthesis followed by intensive mitosis throughout the kidney tissue, led to the development of a model of very mild and fully reversible ARF accompanied by considerable hypertrophy of the kidney. After a single dose of 250 mg/kg BW intravenously, ARF occurs within several hours.[45]

C. Models of Acute Renal Failure Induced by Pigments

These models have been developed to mimic clinical situations in which great amounts of endogenous pigments (hemoglobin, myoglobin, and their derivatives) are released (hemolysis, crush syndrome, etc.). The original view that the pigments alone are able to obstruct the kidney tubuli, especially the thin limbs of the loop of Henle, seems to no longer be tenable: the crystals are easy to wash down.[2]

The more probable causes of nephrotoxicity of these pigments are vasoconstriction which accompanies their release or administration, and perhaps a specific toxicity of pigment molecules themselves.[46] There are two principally different types of these models.

1. Injection of Pigments

Hemoglobin was administered intravenously to dogs at a dose of 200 to 300 mg/kg;[47] in rats, methemoglobin was given at the usual dose between 0.5 and 2.0 g/kg BW.[2] A short-lasting type of anesthesia is recommended. A prerequisite is at least 24 hr, but preferably

36 hr, of dehydration,[24] otherwise ARF fails to develop. Several hours after the injection, a decrease in RBF and GFR occurs accompanied by oliguria and urine hypoosmolality.

2. Administration of Glycerol

Glycerol model is one of the most favorite models of ARF. Administration of 8 to 15 mℓ of 50% glycerol per kilogram BW i.m. induces severe necrosis at the site of injection with hemoglobin and myoglobin release from the damaged tissue and blood. These pigments and/or their metabolites are partly responsible for ARF; moreover, ischemization of the kidney due to the decrease in cardiac output and circulating volume of blood also occurs.[48] The course of ARF depends on the dose used; in every case it lasts for several days. Dehydration is an important supporting factor.[49] Since the model is painful, rats are used almost exclusively; rabbits have been also used[50] with roughly the same course.

D. Models of Postrenal Acute Renal Failure

These are models of hydronephrosis including especially those involving ureteral obstruction (UO).

1. Acute Ureteral Obstruction

An important difference exists between unilateral and bilateral UO with the latter being much less frequent; differences in the course are obvious. Modeling is rather easy: it can be materialized by ligating the ureter at a different height, by clamping, or by insertion of a polyethylene cathereter whose outer end is elevated to the height corresponding to a pressure value that is somewhat higher than glomerular capillary pressure (about 50 mmHg in the rat and 80 mmHg in the dog).[51,52] The latter alternative can be also used for modeling partial obstruction, as performed, e.g., by Blantz et al.[53] by lifting the catheter to a height equal to the pressure of 20 mmHg. In every model, when releasing the obstruction, it is imperative to check whether the release really occurred.

2. Chronic Ureteral Obstruction

Chronic UO can only be performed as the unilateral model. It is produced by ligation of the ureter at a particular height. The ureteral pressure after an initial rapid increase gradually decreases with time: Vaughan et al.[54] found it to amount to 78 mmHg after 5 hr, and 39 mmHg after 24 hr of ligation.

The consequences of UO are manyfold, depending primarily on its duration (details can be found in Reference 55).

III. CHRONIC RENAL FAILURE (CRF)

There are two types of CRF models: those which have been originally designed as CRF models, and those in which ARF evolves into a chronic stage or, more precisely, in which the animal has survived ARF. The latter actually resemble the consequences of ARF and are not discussed here.

A. Reduced Number of Nephrons

1. Unilateral Nephrectomy (UNX)

This model does not actually induce CRF. However, many symptoms accompanying reduction of renal mass are present (hypertrophy, hyperplasia, higher pressure in the glomerular capillaries possibly leading to glomerulosclerosis, etc.) and this is why it is included here. UNX is performed in anesthetized animals under sterile conditions (except for rats) from flank incision without opening the peritoneal cavity.

2. Subtotal Nephrectomy

The course of this model strongly depends on the amount of parenchyma removed.[56] Removal of about $5/_6$ of all renal tissue (i.e., UNX + $2/_3$ of the contralateral kidney) produces CRF both in the dog[57] and in the rat,[58,59] the latter animal being used more frequently. Severe CRF develops after removal of $7/_8$ of the parenchyma.[60] Various techniques to remove part of the kidney tissue have been developed: bipolar ligation of cauterization are used in rats. These methods also have been applied in the dog. Schultze et al.[61] ligated the renal arteries of the second and third order to produce infarction of 85% of the parenchyma; removal of the contralateral kidney after some time led to the picture of severe CRF. An analogous technique was employed by other investigators controlling the degree of CRF by the number of arteries ligated.[62] An aseptic procedure is imperative. Another option involves the obstruction of a large number of arterioles by the injection of large microspheres or of polymers hardening within the vasculature:[63] these techniques are rather simple but the quantification is difficult.

All these techniques provide an opportunity to carry out a two-step research project: ablation of $2/_3$ of the parenchyma of one kidney in the first phase enables one to study the changes in its remnant nephrons in an otherwise healthy nonuremic animal. In the second phase, after UNX, the same nephrons can be studied during developing uremia.

B. Other Models of Chronic Renal Failure

Christensen and Ottesen[64] observed that newborn rats fed a diet enriched with lithium (40 mmol LiCl/kg of pellets) for 8 weeks, develop renal damage associated with protein-uria, impaired concentrating ability, and later, symptoms of uremia. CRF progresses even after discontinuation of LiCl addition to the diets and exhibits a majority of properties typical of human CRF.

Several techniques have been developed in which symptoms of uremia are present without any damage to the kidney. This is achieved by returning the urine into the blood stream. McNay et al.[65] inserted a catheter (from a flank incision) into the ureter and its other end into an appropriate vein. The model can be either unilateral (with concomitant UNX) or bilateral. Emmanouel et al.[66] established a similar shunt between the urinary bladder and the caval vein in the rat. Heller et al.[67] inserted the trigonum of the urinary bladder into the peritoneal cavity of the rat. After suturing the abdominal wall, uremia develops with fluid accumulation in the peritoneal cavity.

IV. CONCLUSIONS

Resemblance of the models of renal failure to the clinical situation is different in various types of failure. The most similar are the models of nephrotoxic drug injuries: here, the way of administration, course, and therapy are often identical. The modeling of ischemic damage is more difficult: a complete absence of blood flow through the kidney, as modeled by clamping, occurs mostly iatrogenically (surgery, transplantation, etc.). Intermitent arrest appears to be closer to the clinical situation. In models of chronic renal failure, it should be remembered that, in contrast to the patients, a sudden reduction in the number of nephrons is carried out in a normal healthy kidney.

REFERENCES

1. **Brenner, B. M. and Lazarus, J. M.,** Acute Renal Failure, W. B. Saunders, Philadelphia, 1983.
2. **Olbricht, C. H. J.,** Experimental models of acute renal failure, *Contrib. Nephrol.,* 19, 110, 1980.
3. **Porter, G. A. and Bennett, W. M.,** Nephrotoxic acute renal failure due to common drugs, *Am. J. Physiol.,* 241, F1, 1981.
4. **Porter, G. A.,** Nephrotoxic Mechanisms of Drugs and Environmental Toxins, Plenum, New York, 1982.
5. **Thurau, K., Ed.,** Acute renal failure, *Kidney Int.,* 10 (suppl. 6), 1976.
6. **Wilson, C. B. and Blantz, R. C.,** Nephroimmunopathology and pathophysiology, *Am. J. Physiol.,* 248, F319, 1985.
7. **Gotshall, R. W., Mills, D. S., and Sexson, W. R.,** Renal oxygen delivery and consumption during progressive hypoxemia in the anesthetized dog, *Proc. Soc. Exp. Biol. Med.,* 174, 363, 1983.
8. **Thurau, K. and Boylan, J. W.,** Acute renal success. The unexpected logic of oliguria in acute renal failure, *Am. J. Med.,* 61, 308, 1976.
9. **Venkatachalam, M. A., Bernard, D. B., Donohoe, J. F., and Levinsky, N. G.,** Ischemic damage and repair in the rat proximal tubule: differences among the S_1, S_2 and S_3 segments, *Kidney Int.,* 14, 31, 1978.
10. **Finn, W. F. and Chevalier, R. L.,** Recovery from postischemic acute renal failure in the rat, *Kidney Int.,* 16, 113, 1979.
11. **Summers, W. K. and Jamison, R. L.,** The no-rephlow phenomenon in renal ischemia, *Lab. Invest.,* 25, 635, 1971.
12. **Bálint, P., Taraba, I., and Visy, M.,** Glomerular and tubular factors in the genesis of post-ischemic renal failure, *Acta Physiol. Hung.,* 31, 235, 1967.
13. **Mason, J. and Welsch, J.,** Glomerular filtration of the deeperlying nephrons after ischemic injury, *Nephron,* 31, 304, 1982.
14. **Daugharty, T. M., Ueki, I. P., Mercer, P. F., and Brenner, B. M.,** Dynamics of glomerular ultrafiltration in the rat. V. Response to ischemic injury, *J. Clin. Invest.,* 53, 105, 1974.
15. **Wilson, D. H., Barton, B. B., Parry, W. L., and Hinshaw, L. B.,** Effects of intermittent versus continuous renal arterial occlusion on hemodynamics and function of kidney, *Invest. Urol.,* 8, 507, 1971.
16. **Zager, R. A., Baltes, L. A., Sharma, H. M., and Jurkowitz, M. S.,** Responses of the ischemic acute renal failure kidney to additional ischemic events, *Kidney Int.,* 26, 689, 1984.
17. **Zager, R. A., Jurkowitz, M. S., and Merola, A. J.,** Responses of the normal rat kidney to sequential ischemic events, *Am. J. Physiol.,* 242, F148, 1985.
18. **Cronin, R. E., Erickson, A. M., DeTorrente, A., McDonald, K. M., and Schrier, R. W.,** Norepinephrine-induced acute renal failure: a reversible ischemic model of acute renal failure, *Kidney Int.,* 14, 187, 1978.
19. **Knapp, R., Hollenberg, N. K., Busch, G. J., and Abrams, H. L.,** Prolonged unilateral acute renal failure induced by intraarterial norepinephrine infusion in the dog, *Invest. Radiol.,* 7, 164, 1972.
20. **Conger, J. D., Robinette, J. B., and Guggenheim, S. J.,** Effect of acetylcholine on the early phase of reversible norepinphrine-induced acute renal failure, *Kidney Int.,* 19, 399, 1981.
21. **Lewis, R. M., Rice, J. H., Patton, M. K., Barnes, J. L., Nickel, A. E., Osgood, R. W., Fried, T., and Stein, J. H.,** Renal ischemic injury in the dog: characterization and effect of various pharmacologic agents, *J. Lab. Clin. Med.,* 104, 470, 1984.
22. **Kleijans, J. C. S., Smits, J. F. M., Kasbergen, C. M., Van Essen, H., and Struyker-Boudier, H. A. J.,** Evaluation of renal function during intrarenal norepinephrine infusion in conscious rats, *Renal Physiol.,* 7, 243, 1984.
23. **Vanholder, R. C., Praet, M. M., Pattyn, P. A., Leusen, I. R., and Lameire, N. H.,** Dissociation of glomerular filtration and renal blood flow in $HgCl_2$-induced acute renal failure, *Kidney Int.,* 22, 162, 1982.
24. **Mason, J., Olbricht, C., Takabatake, T., and Thurau, K.,** The early phase of experimental acute renal failure. I. Intratubular pressure and obstruction, *Pflugers Arch.,* 370, 155, 1977.
25. **Flamenbaum, W., McDonald, F. D., DiBona, G. F., and Oken, D. E.,** Micropuncture study of renal tubular factors in low dose mercury poisoning, *Nephron,* 8, 221, 1971.
26. **Flamenbaum, W., Huddleston, M. L., McNeil, J. S., and Hamburger, R. J.,** Uranyl nitrate-induced acute renal failure in the rat: Micropuncture and renal hemodynamic studies, *Kidney Int.,* 6, 408, 1974.
27. **Kobayashi, S., Nagase, M., Honda, N., and Hishida, A.,** Glomerular alterations in uranyl acetate-induced acute renal failure in rabbits, *Kidney Int.,* 26, 808, 1984.
28. **Flamenbaum, W., McNeil, J. S., Kotchen, T. A., and Saladino, A. J.,** Experimental acute renal failure induced by uranyl nitrate in the dog, *Circ. Res.,* 31, 682, 1972.
29. **Stein, J. H., Gottschall, J., Osgood, R. W., and Ferris, T. F.,** Pathophysiology of a nephrotoxic model of acute renal failure, *Kidney Int.,* 8, 27, 1975.
30. **Higby, D. J., Wallace, H. J., and Holland, J. F.,** Cis-diamminedichloroplatinum (NSC-119875): a phase I study, *Cancer Chemother. Rep.,* 59, 647, 1975.

31. **Goldstein, R. S., Noordewier, B., and Bond, J. T.**, Cis-dichlorodiammineplatinum nephrotoxicity: time course and dose response of renal functional impairment, *Toxicol. Appl. Pharmacol.*, 60, 163, 1981.

32. **Chopra, S., Kaufman, J. S., Jones, T. W., Hong, W. K., Gehr, M. K., Hamburger, R. J., Flamenbaum, W., and Trump, B. F.**, Cis-diamminedichlorplatinum-induced acute renal failure in the rat, *Kidney Int.*, 21, 54, 1982.

33. **Luft, F. C., Bloch, R., Sloan, R. S., Yum, M. N., Costello, R., and Maxwell, D. R.**, Comparative nephrotoxicity of aminoglycoside antibiotics in rats, *J. Infect. Dis.*, 238, 541, 1978.

34. **Glibert, D. N., Houghton, D. C., Bennett, W. M., Plamp, C. E., Roger, K., and Porter, G. A.**, Reversibility of gentamicin nephrotocity in rats: recovery during continuous drug administration, *Proc. Soc. Exp. Biol. Med.*, 160, 99, 1979.

35. **Cuppage, F. E., Setter, K., Sullivan, P., Reitzes, E. J., and Melnykovych, A. O.**, Gentamicin nephrotoxicity. II. Physiological, biochemical and morphological effects of prolonged administration to rats, *Virchow. Arch. (Cell Pathol.)*, 24, 121, 1977.

36. **Frame, P. T., Phair, J. P., Watanakunakorn, C., and Bannister, T. W. P.**, Pharmacologic factors associated with gentamicin nephrotoxicity in rabbits, *J. Infect. Dis.*, 135, 852, 1977.

37. **Soberon, L., Bowman, R. L., Pastoriza-Munos, E., and Kaloyanides, G. J.**, Comparative nephrotoxicities of gentamicin, netilmicin, and tobramicin in the rat, *J. Pharmacol. Exp. Ther.*, 210, 334, 1979.

38. **Barr, G. A., Mazze, R. I., Cousins, M. J., and Kosek, J. C.**, An animal model for combined methoxyflurane and gentamycin nephrotoxicity, *Br. J. Anaesth.*, 45, 306, 1973.

39. **Bennett, W. M., Hartnett, M. N., Gilbert, D., Houghton, D., and Porter, G. A.**, Effect of sodium intake on gentamicin nephrotoxicity in the rat, *Proc. Soc. Exp. Biol. Med.*, 151, 736, 1976.

40. **Heller, J.**, Effect of some simple manoeuvres on the course of acute renal failure after gentamicin treatment in rats, *Int. Urol. Nephrol.*, 16, 243, 1984.

41. **Thomas, B. L. and Faith, G. C.**, Renal tubular necrosis following cephalotin, *Nephron*, 23, 205, 1979.

42. **Atkinson, R. M., Currie, S. P., Davis, B., Pratt, P. A. H., Sharpe, H. M., and Tornick, E. G.**, Acute toxicity of cephaloridine, an antibiotic derived from cephalosporin, *Toxicol. Appl. Pharmacol.*, 40, 137, 1966.

43. **Tune, B. M. and Fernholt, M.**, Relationship between cephaloridineridine and p-aminohippurate transport in the kidney, *Am. J. Physiol.*, 225, 1114, 1973.

44. **Threlfall, G., Taylor, D. M., and Buck, A. T.**, The effect of folic acid on growth and deoxyribonucleic acid synthesis in the rat kidney, *Lab. Invest.*, 15, 1477, 1966.

45. **Schmidt, U., Tohorst, J., Huguenin, M., and Dubach, U. C.**, Acute renal failure after folate: Na-K-ATPase in isolated rat renal tubule. Ultramicrochemical and clinical studies, *Eur. J. Clin. Invest.*, 3, 169, 1973.

46. **Conn, H. L., Jr., Wood, J. C., and Rose, J. C.**, Circulatory and renal effects following transfusion of human blood and its components to dogs, *Circ. Res.*, 4, 18, 1956.

47. **Goldberg, M.**, Studies of the acute renal effects of hemolyzed red blood cells in dogs including estimations of renal blood flow with Krypton[85], *J. Clin. Invest.*, 41, 2112, 1962.

48. **Hsu, C. H., Kurz, T. W., and Waldinger, T. P.**, Cardiac output and renal blood flow in glycerol-induced acute renal failure in the rat, *Circ. Res.*, 40, 178, 1977.

49. **Thiel, G., Wilson, D. R., Arce, M. L., and Oken, D. E.**, Glycerol induced hemoglobinuric acute renal failure in the rat. II. The experimental model, predisposing factors, and pathophysiologic features, *Nephron*, 4, 276, 1967.

50. **Torres, V. E., Strong, C. G., Romero, J. C., and Wilson, D. M.**, Changes in plasma renin substrate, plasma and renal renin, and plasma osmolarity during glycerol-induces acute renal failure in rabbits, *Mayo Clin. Proc.*, 350, 111, 1975.

51. **Gottschalk, C. W. and Mylle, M.**, Micropuncture study of pressures in proximal tubules and peritubular capillaries of the rat kidney and their relation to ureteral and renal venous pressures, *Am. J. Physiol.*, 185, 430, 1956.

52. **Heller, J. and Horáček, V.**, Comparison of directly measured and calculated glomerular capillary pressure in the dog kidney at varying perfusion pressure, *Pflugers Arch.*, 385, 253, 1980.

53. **Blantz, R. C., Konnen, K., and Tucker, B. J.**, Glomerular filtration response to elevated ureteral pressure in both the hydropenic and plasma expanded rat, *Circ. Res.*, 37, 819, 1975.

54. **Vaughan, E. D., Jr., Sorenson, E. J., and Gillenwater, J. Y.**, The renal hemodynamic response to chronic unilateral complete ureteral occlusion, *Invest. Urol.*, 8, 78, 1970.

55. **Badr, K., Ichikawa, I., and Brenner, B. M.**, Renal circulatory and nephron function in experimental obstruction of the urinary tract, in *Acute Renal Failure*, Brenner, B. M. and Lazarus, J. M., Eds., W. B. Saunders, Philadelphia, 1983, 116.

56. **Sterner, G.**, Experimental chronic renal failure in rats, *Nephron*, 24, 207, 1979.

57. **Coburn, J. W., Gonick, H. C., Rubibi, M. E., and Kleeman, C. R.**, Studies of experimental renal failure in dogs. I. Effect of 5/6 nephrectomy on concentrating and diluting capacity of residual nephrons, *J. Clin. Invest.*, 44, 603, 1965.

58. **Shea, S. M.,** Chronic reduction of renal mass: glomerular morphometry by electron microscopy, *Yale J. Biol. Med.,* 51, 321, 1978.

59. **El-Nahas, A. M., Paraskevakou, H., Zoob, S., Rees, A. J., and Evans, D. J.,** Effect of dietary protein restriction on the development of renal failure after subtotal nephrectomy in rats, *Clin. Sci.,* 65, 399, 1983.

60. **Kaysen, G. A. and Watson, J. B.,** Mechanism of hypoalbuminemia in the 7/8-nephrectomized rat with chronic renal failure, *Am. J. Physiol.,* 243, F372, 1982.

61. **Schultze, R., Shapiro, H., and Bricker, N.,** Studies on the control of sodium excretion in experimental uremia, *J. Clin. Invest.,* 48, 869, 1969.

62. **Dirks, J. H. and Wong, N. L. M.,** Acute functional adaptation to nephron loss: micropuncture studies, *Yale J. Biol. Med.,* 51, 255, 1978.

63. **Peregrin, J., Kašpar, M., Haco, M., Vaněček, R., and Belán, A.,** New occlusive agent for therapeutic embolization tested in dogs, *Cardiovasc. Intervent. Radiol.,* 7, 97, 1984.

64. **Christensen, S. and Ottesen, P. D.,** Lithium-induced uremia in rats — a new model of chronic renal failure, *Pflugers Arch.,* 399, 208, 1983.

65. **McNay, J. L., Rosello, S., and Dayton, P. G.,** Effects of azotemia on renal extraction and clearance of PAH and TEA, *Am. J. Physiol.,* 230, 901, 1976.

66. **Emmanouel, D. S., Lindheimer, M. D., and Katz, A. J.,** Renal sodium handling in uremia. Studies in a rat model with structurally intact kidneys, *Miner. Electrolyte Metabl.,* 4, 288, 1980.

67. **Heller, J., Kleinová, M., Janáček, K., and Rybová, R.,** Effect of a sudden and a slow concentration increase in plasma urea on its concentration in some tissues of the dog and rat, *Physiol. Bohemoslov.,* 33, 296, 1984.

Chapter 24

RESPIRATORY DISEASES

F. Paleček

TABLE OF CONTENTS

I. INTRODUCTION

The aim of this chapter is to provide within the limited space basic information on (1) the possibilities of modeling some pulmonary diseases in animals, and (2) on the methods, which are specific for this type of animal research study.

The chapter does not contain any list of models that can be performed on the respiratory system. Such a list would be endless. There are perfect reviews both on the various models of lung diseases[1-10] and on the ways to evaluate them.[11-13] Therefore, rather a personal view is given on a few selected topics with the aim to hand over personal experience, whenever possible. It is the author's conviction that for disease modeling small experimental animals are, in general, preferable. Consequently, greater attention is given to models on rats or guinea pigs rather than dogs, cats, or cattle. Among mammals, respiratory variables are related according to certain rules. The relationships between body mass, body surface, metabolism, and breathing have been extensively studied[14-16] and provide a good basis for comparative respiratory physiology and thus also for the interpretation of models on various animal species.

A. What Disease?

The criteria used for model lung diseases are varied: according to time (acute and chronic), according to the cause (infectious, traumatic, chemical etc., including so-called ''essential'' diseases with unknown etiology and/or pathogenesis and diseases due to multifactorial causes), according to prognosis, etc. Our attention will be devoted to diseases of unknown etiology or pathogenesis, in which the model can help to elucidate its causes. This does not imply a smaller importance of studying, e.g., infectious lung diseases with known pathogenetic microorganisms for the sake of elucidating the pathogenesis and therapy. It only means that in the latter example typically rather simpler criteria are used, such as the survival of the animals. In the former group of models usually much more complex functional examinations of living organisms are necessary.

B. What Animals?

The choice of an appropriate animal species for modeling lung disease is not easy. There are several viewpoints that have to be evaluated before the decision is taken.

1.   Phylogenetic proximity of the animal species to man. In lung physiology this does not seem very important, at least within the group of mammals. The important point is that results of no animal experiments can be directly applied to man. A controlled clinical trial is always necessary and, in this way, also the possible inconsistencies between the chosen species and man should be discovered.

2.   Quantitative differences between various animals and man. There is no doubt that such differences exist and may be of major importance in the interpretation of obtained data; e.g., the effects of hydrostatic pressure on pulmonary circulation are much less in small animals than in large ones, including man. Chest wall compliance differs much between various species and also between young and old individuals. The extent of vagal control is very much different, etc. Although most research work in this area is done for medical reasons, for man, the mentioned differences need not be a basic drawback; awareness of the particular difference may help understanding, what the functional interplay would be if, e.g., the pulmonary hydrostatic pressure was less, chest wall compliance or vagal control of breathing more than in a normal man.

3.   The size of the animal may be quite important. Larger animals (e.g., dog and cat) permit a direct use of clinical equipment and methods. On the other hand, small animals (e.g., rat and guinea pig) can be maintained in large numbers for relatively

long time periods, necessary for many experiments designed to study chronic lung diseases. Thus for economic reasons, it may be more rewarding to use specialized equipment and methods designed for small animals than to keep great numbers of large animals.

4. An important variable is that of the health of the experimental animals. Animals of conventional breeding are subject to an uncontrolled and unspecified variety of infections and from that resulting functional changes of various systems of the body; e.g., spontaneous lung emphysema occurs more often in rats of conventional breeding than in the specific pathogen-free (SPF) animals.[17,18]

The idea of producing SPF animals was initiated by the aim to supply healthy and thus physiologically standard series of experimental animals. Collecting data on SPF animals indeed provides much more reliable basis for any biological research. On the other hand, the condition is to some extent artificial in the sense that, naturally, the animals (or men) are never ideally healthy. They are typically subject to a chronic disorder which may modify the responsivity and functions of some systems of the body and of the organism as the whole.

The argument is that differences between conventional and SPF animals definitely exist and that they must be borne in mind in any experiment concerned with disease modeling.

C. What Methods?

In lung disease, similarly as in disease of other organs or systems, the impaired function is of major interest. The main methodological approaches to evaluate function of the respiratory systems are physiological, biochemical, and morphological. Physiological examination provides data most pertinent to the function of the respiratory system on the whole. Analysis of biochemical changes in the various components of the system may provide insight into some of the pathogenetical mechanisms on subcellular level. The morphological examination gives only indirect indices of changed function. However, morbid anatomy (both macroscopic and microscopic) has been established in the field for the longest time and thus has become a kind of reference examination. This review will be mainly concerned with the physiological methods.

II. EXPERIMENTAL DESIGN

A. Control Group

The importance of a control group of animals cannot be stressed enough. There are no "standard" or "reference" values for animal respiratory measurements. The values depend on too many variables which are usually outside control of the experimentator. In a properly performed experiment the control group can take care of such variables as a particular breed, room temperature, level of anaesthesia, environmental influences on unanaesthetized animals, state of health, etc. Thus the control group provides a reference level of values with which to compare all the values obtained in experimental disease. With a large scatter of values within the control group more animals are needed to provide a representative sample. Here again the advantage of work with small animals is evident.

B. Ways of Administration

The material to get into contact with lung tissue can be administered either by local or general route.

1. Local Administration

For some materials this is the normal way how to get into the lung and act there. This is true, e.g., for silica dust, for toxic gases or vapors, microorganisms, etc. Experimentally it

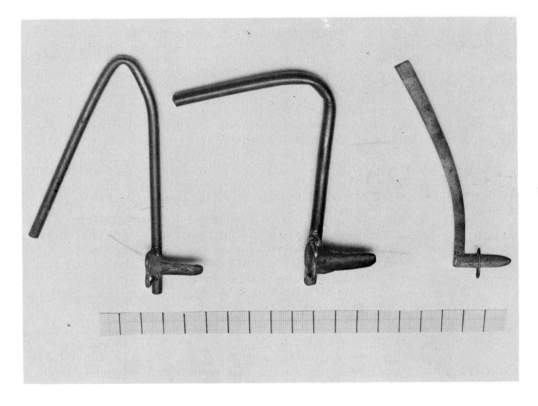

FIGURE 1.   Rat laryngoscopes of different size. The scale is in centimeters.

is usually important to quantitate the dose used. This may be difficult when using inhalation chambers or masks. Occasionally with suitable marking, the amount of inhaled and deposited material can be estimated from a postmortem analysis of the lung.

Another way of local administration is in a solution or suspension injected through the airways. In larger animals (dog, cat, rabbit) this is easily accomplished with the help of a laryngoscope, under light ether anaesthesia. For the rat, a special laryngoscope is necessary[19] (Figure 1). The volume injected into the lungs of an adult rat should not exceed 0.5 mℓ. The solution must be injected at the beginning of inspiration so that it is evenly distributed into the smaller airways. The anatomy of the pharynx of the guinea pig makes it difficult to use a laryngoscope. In this species and also in the mouse, the trachea can be punctured.

2. General Administration

Some materials have great affinity to lung tissue and affect it, when administered generally, by some of the enteral or parenteral routes; e.g., paraquat will produce lung lesions (apart from affecting some other parenchymatous organs) even after i.p. injection. But even in such a case the local administration may be more advantegous: the same local (i.e., pulmonary) effects are achieved with a dose that does not produce any marked toxic actions on other systems or organs.

C. Anesthesia

In surgical or other procedures that produce pain, local or general anesthesia is required. The problem remains when comparing the pros and cons of general anesthesia with measurements that do not require it for ethical reasons.

Naturally, values of ventilation, blood gases, etc., in unanesthetized animals should reflect more adequately the "true" values. This may be valid only if the methods used do not acutely affect the animal (plethysmograph for unanesthetized animals, preimplanted cannulas for blood sampling). Otherwise, the use of restraint devices (head masks, etc.) can affect ventilation appreciably. Some species (dogs) can be trained to tolerate such devices without much disturbance; such training is, however, tedious and time consuming.

Therefore, general anesthesia is preferred in all conditions in which it is not in direct contradiction with the design of the experiment. The choice of a suitable agent for general anesthesia is not easy either. The much favored barbiturates do not provide a steady level of anesthesia for a longer time. Additional doses of the anesthetic "as required" only mean that the depth of anesthesia is continually fluctuating. This can be much improved by a slow i.v. infusion of the anesthetic following the initial i.p. dose. However, the estimate of the supplemental dose is not easy and monitoring of the depth of anesthesia is recommended. With inhalation anesthesia and proper monitoring the fluctuations can usually be avoided. Urethane for the rat or chloralose-urethane anesthesia for the dog can provide a long lasting steady state level of anesthesia. Depression of temperature control and some other side effects must be, however, borne in mind. There is no ideal general anesthetic and the choice for each particular experiment must always be carefully considered.

III. MEASURING RESPIRATORY FUNCTIONS

Generally speaking, everything that can be measured clinically, in man can be also measured in most other mammals. The following survey will, therefore, attempt to underline the differences rather than to stress the similarities.

A. Ventilation

Minute ventilation is the product of tidal volume and frequency of breathing. The breathing frequency is easy to ascertain. In most mammals, if necessary, even aspection suffices to measure the breathing frequency. In small mammals registration is advisable; usually it is read off the volume or flow changes. Otherwise a simple device, such as a thermistor at the airway opening, is enough to provide an adequate signal.

Measuring tidal volume is more difficult. For unanesthetized animals a modification of the whole body plethysmograph can be used. This apparatus was first suggested and used with human infants[20] and later adapted for work with rats.[21] In principle, the animal moves freely inside an airtight box filled with air of defined temperature and humidity. During inspiration the inhaled air warms to body temperature and increases in volume. This increase of volume inside the box is measured as increased pressure. From this pressure change the tidal volume can be calculated. The method gives values that correlate resonably well with values obtained by direct volume measurement. In some conditions (tachypnoea, airways obstruction, small temperature difference between the box and the animal, increased activity of the animal) the method becomes much less reliable. However, for the possibility to measure tidal volume in freely moving, unrestrained animals the method is unique.

The direct methods of measuring tidal volume require some kind of bodily connection between the animal and apparatus. Therefore, the methods are best suited for anaesthetized animals, although often wake, restrained, or wake and trained animals can be used. The connections include a kind of collar for some types of body plethysmographs, a face mask or tracheal cannula. Most species can be intubated; when this is not possible, the cannula can be inserted in anaesthetized animals using tracheotomy. For intubation of small animals, such as the rat, a special laryngoscope is useful.[19,22] With advantage a transillumination of the neck region can be used (Figure 2). For larger animals a valve chamber can be used in connection with a spirograph.

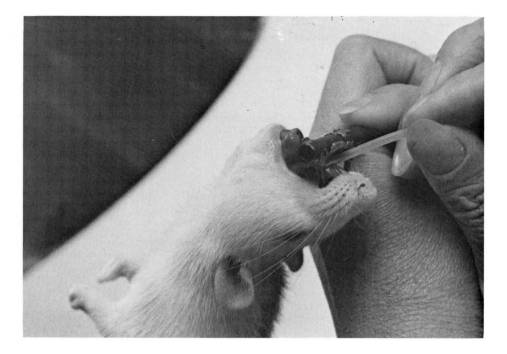

A

B

FIGURE 2. Intubation of a rat. The anaesthetized animal is held on the laryngoscope against a light source (A) and the intubation is performed under visual control (B). Alternatively, a head light can be used.

Occasionally, the air flow measurement is required. The information can be obtained by electronic derivation of the electric analogue of the volume signal, or it can be measured directly with a pneumotachograph. In this case, the information on volume can be obtained by integration of the flow signal.

In respiratory disease, the important ventilation variable is the effective alveolar ventilation. Apart from minute ventilation, arterial P_{CO_2} and carbon dioxide output must be measured. (In this case, end-tidal P_{CO_2} is not a good substitute for arterial P_{CO_2}. In lung disease typical inequalities in the ventilation-perfusion ratio exist with resulting venous admixture and alveolar dead space. Alveolar dead space cannot be derived from end-tidal CO_2 values.)

Alveolar ventilation is the basic component of pulmonary gas exchange. Much more often, however, minute ventilation is ascertained. Under standard conditions there is a relatively straightforward relationship between alveolar and minute ventilation (minute ventilation = alveolar ventilation + ventilation of the dead space). Minute ventilation is also more closely related to the respiratory center output so that the first estimate of, let us say, respiratory center depression would be based on decreased minute ventilation. Minute ventilation is followed not only as an entity, but also in its components which form a particular pattern of breathing. Thus minute ventilation is analyzed not only with regard to tidal volume and breathing frequency, but also to the duration of inspiration, duration of expiration (which may be split in the duration of "active" expiration and the expiratory pause), it may be evaluated with regard to the mean inspiratory or expiratory air flows, etc. Analysis of the pattern of breathing is of particular interest in studies concerned with the control of breathing; however, the functional basis of its changes may be in almost any functional derangement. It is good to bear in mind that although we have many refined techniques to study breathing, great discoveries were made with a primitive Marey's tambour writing on a piece of sooted paper.

B. Mechanics of the Respiratory System

1. Compliance

Compliance is the measure of elasticity of the lung, of the chest wall, or of the whole respiratory system. Apart from measurements of volume changes, also the respective pressure changes are necessary: transpulmonary pressure (i.e., interpleural vs. alveolar) for lung compliance, transmural pressure (i.e., interpleural vs. ambient) for chest wall compliance and transthoracic pressure (i.e., alveolar vs. ambient) for compliance of the respiratory system (Figure 3). Interpleural pressure is usually not measured directly; instead, intrathoracic pressure is measured, typically as esophageal pressure. Either a balloon or a water-filled cannula can be used. Generally, the agreement between interpleural and esophageal pressure is good.[23,24] Under static conditions one can assume an equal value of alveolar and tracheal pressure (or pressure at the airway opening) and, therefore, tracheal pressure can be substituted for alveolar pressure. Ambient pressure is mostly barometric, unless the experiment is performed in hyper- or hypobaric conditions.

2. Resistance

Either lung or airway resistance is measured. To measure lung resistance the pressure gradient between airway opening and interpleural (esophageal) pressure is needed. For airway resistance the pressure gradient between airway opening and alveolar pressure is necessary. In this case, under dynamic conditions, the true alveolar pressure must be measured. This is done with the use the interrupter technique or, better, of a whole body plethysmograph[24-28] (Figure 4). The other variable needed for resistance measurements is air flow. This is obtained by any of the methods mentioned under ventilation measurement.

P_{awo}

P_{tr}

P_{bs}

P_{pl}

P_{alv}

P_{oes}

P_{abd}

1

2

3

4

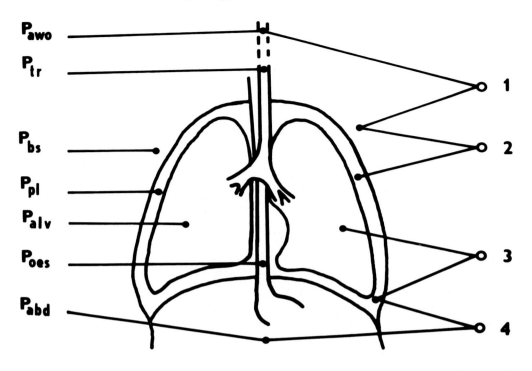

FIGURE 3.   The pressures and pressure differences most often measured in pulmonary mechanics. Pressures (P): awo — at the airway opening; tr — tracheal; bs — at the body surface; pl — pleural; alv — alveolar; oes — esophageal; abd — abdominal. Differential pressures: 1 — transthoracic; 2 — transparietal; 3 — transpulmonary; 4 — transdiaphragmatic. Under static conditions we assume that $P_{awo} = P_{tr} = P_{alv}$. For technical reasons, P_{oes} is often measured instead of P_{pl}.

3. "Loops"

A whole variety of so-called loops can be constructed by plotting air flow against volume, volume against pressure, etc. To obtain meaningful data many respiratory maneuvers may be necessary. The details can be obtained in textbooks on respiration, e.g., Reference 29. The pressure-volume loop gives the basis to calculate the work of breathing; either the work done on the lungs or, in paralyzed, artificially ventilated animal, on the whole respiratory system. In the first case the transpulmonary pressure, in the latter case the transthoracic pressure in used.

Lung mechanics is a favorite field of study in experimental lung diseases. One of the reasons probably is a relatively straightforward relationship between the morphological changes (which can be proven by independent methods) and those of the changed mechanics of the respiratory system. The methods may give useful data not only on the presence and quantity of pathological changes due to experimental disease, but also on the time course and possible therapeutic interventions.

Most of the methods are relatively simple and especially with the use of simple computer programs are easily evaluated. Breath-to-breath measurement of airways (or lung) resistance is, e.g., typical for testing bronchodilating drugs in experimentally produced models of bronchial asthma.

Recently, studies of respiratory muscle fatigue are in the focus of research interest. Measuring the work of breathing is an important component of the studies, which have to be concerned with the energetic balance of respiratory muscles, i.e., the input of energy on the one side and its output, in the form of work of breathing, on the other side.

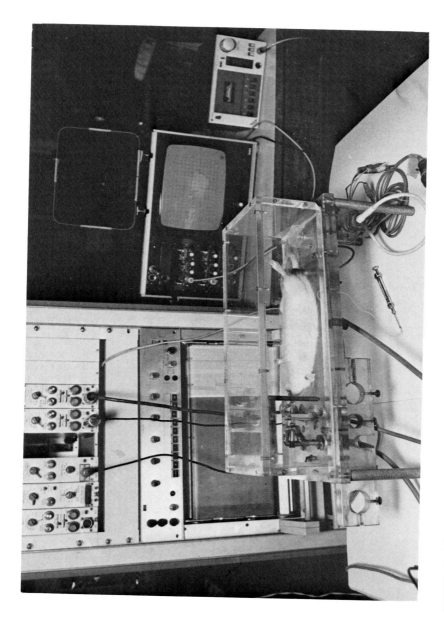

FIGURE 4. A body plethysmograph (6-ℓ volume) for small laboratory animals. The tubing permanently ventilated with breathing mixture (air) is brought inside, close to the animal so that the instrumental dead space is minimal.

C. Blood Gases

Partial pressures of oxygen (P_{O_2}) and carbon dioxide (P_{CO_2}) are important indices of overall respiratory function. The site of measuring the partial pressures is important. The most useful value is that of arterial blood. The samples of arterial blood must be obtainable unaerobically without disturbing the animal and often repeatedly. Therefore, permanent arterial cannulas help to answer these requirements, especially when implanted in a preliminary stage. The techniques of implanting chronic arterial cannulas even in small laboratory animals have been described.[30,31]

Sampling end-tidal air (to provide a sample of alveolar air) is useful for physiological studies. However, in lung disease a difference between the partial pressures of blood gases in end-tidal air and in arterial blood is typical for ventilation-perfusion inequalities. Thus end-tidal air analysis alone is not sufficient to provide adequate data in most pathological conditions of respiration.

Gas exchange between organism and the atmosphere is the vital function of the respiratory system. The function is well controlled. Values of arterial blood gas partial pressures give some insight into the control and exchange systems of respiration. Arterial hypoxemia is the most general warning that something with the system may be wrong. Hypoxemia with hypercapnia is most typical for alveolar hypoventilation, but other processes may participate. Observing a disturbance in arterial blood gases indicates, on the one hand, that the underlying disease is serious enough to break through the control mechanisms that normally maintain the respective partial pressure within narrow limits. On the other hand, such a finding says nothing about the specific cause, which may lie anywhere between a dysfunction of the control of breathing, of lung mechanics, ventilation, respiratory muscles, diffusion, and ventilation-perfusion matching.

D. Indices of Ventilation-Perfusion Inequalities

Local adjustment of alveolar ventilation and perfusion is the most important prerequisite of proper gas exchange. Its importance is in the fact, that (partly due to the enormous number of respiratory units) it is the function which is most often disturbed in any pulmonary disease. Thus, apart from areas with the "ideal" ventilation-perfusion ratio, the extent of areas with lower or higher than ideal ventilation-perfusion ratio increases. Areas with high ventilation-perfusion ratio give origin to alveolar dead space, those with a low ratio to venous admixture. Due to local and systemic control mechanisms, arterial hypoxemia alone is more often present than hypoxemia and hypercapnia. For details see textbooks on pathophysiology of respiration.[32-35]

Basically, the parameters needed to evaluate ventilation-perfusion inequalities are those of ventilation (including CO_2 output) and arterial blood gas analysis. In larger animals radioisotopic methods for regional ventilation and perfusion ascertainments can be used.

Mismatching of alveolar ventilation to perfusion on local level is the most common cause of functional disturbance of the respiratory system and is commonly found in most experimental lung diseases. Therefore, knowledge of the state of ventilation-perfusion is important. But the techniques are quite elaborate for an exact analysis. Thus, unless the experiment is primarily designed to study the ventilation-perfusion relationship under the conditions of a specific experimental model condition, the disturbance is usually assumed from the indirect indices rather than measured directly.

E. Pulmonary Circulation

1. Pulmonary Artery Blood Pressure

Pulmonary hypertension is a serious complication of some chronic pulmonary disorders. Therefore, measurement of pulmonary arterial blood pressure may be important. Again, in large animals catheterization of the pulmonary artery may be performed like in the man. In smaller animals this may be technically more difficult. Therefore, some workers used right ventricular blood pressure instead. This is not quite the same thing and it is good to know that even in the rat pulmonary artery catheterization is possible.[36] Permanent cannulation of the right ventricle[37] or pulmonary artery[38] has also been used.

2. Lung Edema

The presence of lung edema is demonstrated morphologically or by measuring lung water content. Neither method is quite reliable, especially when low degrees of lung edema are present. Morphological examination must take into account possible artifacts due to fixation and the way of standard distension of the tissue. When measuring the difference between wet and dry lung weight, standardization of blood content in the lung is difficult. Of course, more elaborate techniques with isotope labeling can be used.

F. Control of Breathing

Except for the ventilation responses to hypoxia and/or hypercapnia,[39] the testing of the regulatory systems of breathing is not much worked out. Of growing interest is the analysis of the pattern of breathing[40] and various interventions with the feed-back systems (such as vagotomy, glomectomy, etc).

A useful test of the chemical control of breathing may be the response to artificially increased dead space (''tube breathing'').[41] This includes both stimuli (hypoxia and hypercapnia); the test situation mimics that of a lung disease with alveolar dead space, and the original hypoventilation, immediately after adding the dead space, can be fully compensated with increased ventilation. (This is not possible with the classical ventilation response to CO_2, when higher concentrations of CO_2 in inhaled air must lead to increased CO_2 blood levels.)

IV. SOME SELECTED DISEASES

There is an enormous diversity of respiratory disorders studied in animal models. As examples may serve various fibrosing processes,[42-47] effects of toxic gases and vapors,[48,49] toxic effects on lungs after parenteral injections,[50] chronic bronchitis,[51] bacterial pneumonia,[52] and many others. In the following a few have been selected, briefly described, and commented.

A. Nonbacterial Pneumonia

The model may be typically used to study the time course of respiratory insufficiency, the changes in pulmonary mechanics, the respiratory control mechanisms, and the preventive or curative measures.

''Nonbacterial'' pneumonia means that no infectious agents are used primarily to produce inflammation in the lung tissue. However, secondary growth of infectious agents cannot be excluded. The best-suited materials are carragheenan or paraquat.

1. Carragheenan Pneumonia

Carragheenan is a polysaccharide from seaweeds which had been used by pharmacologists to produce standard, experimental, model subcutaneous inflammation for testing anti-inflammatory drugs.[53] It consists basically of sulfated D-polygalactose.[54] It was used in rabbits and cats to produce pneumonia by Trenchard et al.,[55] who studied the effect of the disease on pulmonary vagus nerve afferentation. Pneumonia was also produced by carragheenan in rats and mice[56] to study the changes in production and breakdown of macrophages in the course of the disease. Pneumonia after carragheenan was also studied in rats with regard to lung morphology and function.[57] The papers referred to also give details of carragheenan preparation and administration.

The morphological examination bears all the typical signs of an aspiration pneumonia: spotty or more extensive solidification of the lung parenchyma on macroscopic aspection. Microscopically, there is an infiltration of interstitial and alveolar spaces, predominantly with polymorphonuclear leukocytes and later, by macrophages.

Functionally, the typical sign of pneumonia, tachypnoea, is present. In the rat, pulmonary hypertension was observed.[58] The degree of hypertension increases with repeated carragheenan instillations in 2-week intervals. After the sixth treatment, the pulmonary hypertension is pronounced; arterial hypoxemia and hypercapnia and right heart hypertrophy are also present.

The model is reasonably well-reproducible, although the degree of scatter of pathological changes in the lungs (and corresponding functional changes) is relatively great. This is mainly due to the local action of the agent which is thus dependent on its distribution within the lungs. The inflammation is relatively mild and regresses spontaneously without treatment.

2. Paraquat Pneumonia

Paraquat is an herbicide, chemically 1,1'-dimethyl 4,4' bipyridilium dichloride. It has been widely used in agriculture and various degrees of intoxication have been described. Its affinity to lung tissue makes paraquat a suitable substance for producing model pneumonia.

Experimental lung fibrosis was described after oral or parenteral administration of paraquat.[59-62] The fibrosis is not heavy and takes several weeks to develop. The acute stage is not very pronounced, unless large doses are used, which result in mortality of the experimental animals.

To produce serious pneumonia, intratracheal administration is the best. The acute stage of this type of paraquat intoxication culminates in 2 to 3 days after the injection. In rats, a reliable sign of successful procedure is a conspicuous tachypnoea. The rate of breathing typically doubles. In functional examination, apart from tachypnoea, an increase of the functional residual lung capacity was observed, hypoxemia and occasionally hypercapnia. Histological examination revealed focal formation of hyaline membranes and edema with occasional hemorrhages and signs of inflammation.[63]

Compared to carragheenan pneumonia, intratracheally injected paraquat produces a heavy type of pneumonia which is very well reproducible. The disease is fully reversible in most animals with a suitable dose. When the animals are left to survive the acute stage, some degree of pulmonary fibrosis may be observed several weeks later.

The affinity of paraquat to parenchymatous organs differs in between various species. Apart from the lung, kidney and liver can also be damaged. In the rat the pulmonary damage is more pronounced than that of the other organs.

The rat model of paraquat pneumonia can be recommended for the pronounced signs of acute pulmonary insufficiency, good reproducibility, and spontaneous recovery of the animals.

B. Emphysema

Animal model of lung emphysema has become popular because its production by papain administration experimentally substantiated the importance of α-1-antitrypsin levels in the pathogenesis of emphysema in man. For this model several animal species have been used: rats,[64] hamsters,[65,66] rabbits,[67] and dogs.[68,69] Papain is typically administered in aerosol or by intratracheal injection. The development of the disease takes several weeks. Its ascertainment is not easy. It relies mostly on morphological picture, both macroscopic and histological.[70] Even then, methods of quantification (such as point counting or line crossing techinques)[71] are advisable. Functional measurements in vivo are useful. However, there are often reports on excised lung mechanics only.

Lung emphysema is a disease which attracts the attention of many research laboratories for its wide occurrence in human population. Attempts have been made to evaluate spontaneous incidence of the disease in various animal species.[17,72,73] The main problem seems to be that the animal models of lung emphysema in animals are not very conspicuous. The functional changes are relatively slight and the morphological distinction between experimental and control group of animals is sometimes difficult. The papain emphysema opened a path indicating that other proteolytic enzymes might participate in the pathogenesis of emphysema. This underlined the role of repeated inflammations, leukocyte disruption, and the balance of freed enzymes and antienzymes. However, the pathogenesis of emphysema is typically multifactorial and includes other effects from the outside and a complex system of internal defense mechanisms (at least partly genetically determined) which will call for other models designed specifically to elucidate the various facets of its genesis.

C. Silicosis

There are two approaches how to model the condition of lungs diseased by aggressive dusts: (1) exposure of the experimental animals to the respective dust in dust chambers,[74] and (2) injecting the dust in suspension into the lungs directly.[75,76] The former technique resembles much more the conditions of human silicosis. It is, however, time consuming and space demanding. These drawbacks are removed by the other technique; the lesions after the injection of dust suspension occur relatively fast, they are well-pronounced and no large exposure chambers are necessary. Dust breathing results in fibrotic lesions which are much more diffuse than those after dust injection. (With equivalent doses of some pollutants the final effect was similar, irrespective of the route used.[77]) Models, including lung function measurements, have been produced in various species[78,79] and often in rats.[80-83]

D. Pulmonary Hypertension

The syndrom of pulmonary hypertension has been extensively studied. Its consequence, cor pulmonale, is the fatal complication of many chronic pulmonary diseases. The pathogenesis of pulmonary hypertension can be varied.[8,9] The most outstanding feature in its pathogenesis is alveolar hypoxia. This can be produced in experimental animals with relative ease, either in hypobaric or normobaric conditions. Several animal species have been used, but again, the rat has many advantages. Three weeks in an atmosphere of 12% oxygen in nitrogen are enough to produce pronounced pulmonary hypertension. This is proven by pulmonary artery catheterization or by measurements of blood pressure in the right heart. Morphological examination shows right hear enlargement, which is usually evidenced by an increased right/left ventricle ratio. Apart from that, a typical muscularization of small pulmonary arteries is present. Important for studies is not only the time course of development of the hypertension, but also its production with intermittent hypoxia[85-87] and especially the time course of regression after withdrawal of the hypoxic stimulus.[88-92]

Hypoxia certainly is not the only mechanism in the pathogenesis of pulmonary hypertension. It is probably the most important one and the reproducibility of the model allows studies concerned with the mechanism of hypoxic vasoconstriction in the lungs. Intermittent hypoxic pulmonary hypertension or pulmonary hypertension after repeated lung inflammations[58] are models that are probably more clinically relevant than, e.g., hypobaric hypoxia. The importance of the model is especially in the perspective of finding preventive and therapeutic measures against this serious complication of chronic lung diseases.

E. Lung Edema

There are several methods in use for the production of lung edema.[93,94] The main problem, however, is not the model, but the proof and quantification of lung edema. Among the criteria used for qualitative and/or quantitative evaluation are: X-ray examination of the thorax, macroscopic examination of the lungs during autopsy, presence of fluid in the airways, relative lung weight, water content of the lungs, histological lung examination,[95] ausculation of the chest, change of electric conductance of thorax, change of the weight of perfused lungs,[93] lung/body weight ratio and dry weight of the lungs.[96]

In experiments on rabbits or rats, a reliable model of lung edema is produced by intravenous injection of a mixture of capronic and caprylic acids and olive oil.[97,98] By the morphological criteria, lung edema is present 30 min after the injection. Functional changes (e.g., lung compliance) follow the injection immediately, i.e., within 1 min. Apart from decreased lung compliance, lung edema is accompanied by tachypnoea, increased ventilation, increased pulmonary arterial blood pressure, arterial hypoxemia, and hypercapnia. The values of arterial blood gases normalize within 2 hr after the injection.

Lung edema may have different etiology. When the study is concerned with a particular type of edema, the model must be adjusted. However, the mixture of fatty acids provides a reliable method with conditions that are fully reversible.

V. ADDENDUM

A. The Rat for Modeling Respiratory Diseases

The rat is probably one of the most widely used laboratory animal. It is not suitable for all the respiratory models, e.g., its airways reactivity is very small compared with the guinea pig. On the other hand, the methods of functional examination of the respiratory system have been elaborated and now give a sound basis for also using the rat in this research area.

Some observations can be made on unanesthetized rats. The tidal volume and rate of breathing can be measured with a body plethysmograph for unanesthetized animals.[21] Even pulmonary mechanics has been measured in unanesthetized rats.[99] In general anaesthesia the body plethysmograph is used in the classical arrangement of a constant volume[24] or constant pressure box.[99-101] With the addition of pressure measurements, the system enables ascertainment of practically the whole range of mechanics of breathing[24,102-107] with the exception of volitional maneuvers. Lung mechanics was measured even on the lungs of neonatal rats.[108] The functional residual lung capacity can be obtained either by the manometric method in the body box[24,104,109] or by the gas dilution technique.[24,109,110] Lung diffusing capacity has been ascertained for oxygen[111] and carbon monoxide.[105,112] The methods of tracheal intubation[19,24] and acute or chronic cannulation of vessels[30,36,38,113] have been mentioned earlier.

Breathing in rats has been evaluated with regard to general anaesthesia;[104,114] effects of exposure to ozone were studied on the oscillatory mechanics of the respiratory system,[115] effects of oxygen breathing on ventilation-perfusion relationship.[116] Prolonged hypercapnia affects ventilation[117] and mechanical properties of rat lung.[118] Anaphylactic changes of lung mechanics were produced and prevented by dexamethasone.[119] Ventilatory response to hypoxia was followed in conscious[120] and anaesthetized[121,122] rats. Also the ventilatory response to carbon dioxide was ascertained.[24,123] Functional residual capacity increased after hypoxia[124] after a carotid body stimulant[125] and after several experimental lung diseases.[57,63,81,82,109,126-128] The rat has been widely used in models of lung emphysema.[109,126-134] Extracorporeal circulation[135] and cross-circulation[136] were used to study control mechanisms of breathing. Other studies in this area were concerned with the pattern of breathing of conscious rats,[137] effects of carotid body denervation,[138,139] etc.

This short review should indicate that the rat has been successfully used in solving most problems that otherwise would require the use of large laboratory animals.

VI. CONCLUSIONS

Respiratory diseases are commonly studied in experimental animals, either spontaneously occurring or as artificially produced models. The methods for their production as well as methods of their evaluation are numerous and complex. Therefore, the aim of research must be exactly specified first and the methods chosen accordingly. The advantage of using small laboratory animals is stressed.

REFERENCES

1. **Boren, H. G.**, Experimental emphysema. Basis, review, and critique, *Am. Rev. Respir. Dis.*, 92, 1, 1965.
2. **Eichler, O.**, *Erzeugung von Krankheitszuständen durch das Experiment*, Vol. 2, Springer-Verlag, Berlin, 1969, 308.
3. **Casarett, L. J.**, Toxicology: the respiratory tract, *Annu. Rev. Pharmacol.*, 11, 425, 1971.
4. **Staub, N. G.**, Pulmonary edema, *Physiol. Rev.*, 54, 678, 1974.
5. **Laros, C. D. and Kuyper, C. M. A.**, The pathogenesis of pulmonary emphysema (II), *Respiration*, 33, 325, 1976.
6. **Carta, P. and Sanna-Randaccio, F.**, L' emphyseme expérimental, *Bull. Bur. Physiopathol. Respir.*, 13, 561, 1977.
7. **Hugh-Jones, P. and Whimster, W.**, The etiology and management of disabling emphysema, *Am. Rev. Respir. Dis.*, 117, 343, 1978.
8. **Herget, J. and Paleček, F.**, Experimental chronic pulmonary hypertension, *Int. Rev. Exp. Pathol.*, 18, 347, 1978.
9. **Reeves, J. T. and Herget, J.**, Experimental models of pulmonary hypertension, in *Pulmonary Hypertension*, Weir, E. K. and Reeves, J. T., Ed., Futura, Mount Kisco, New York, 1984, 361.
10. **Zelter, M. and Douget, D.**, Experimental pulmonary oedemas, *Bull. Eur. Physiopathol. Respir.*, 22, 281, 1986.
11. **Barer, G. R.**, The physiology of the pulmonary circulation and methods of study, in *Respiratory Pharmacology*, Widdicombe, J., Ed., Pergamon Press, Oxford, 1981, 345.
12. **Hahn, H. L. and Nadel, J. A.**, Methods of study of airway smooth muscle and its physiology, in *Respiratory Pharmacology*, Widdicombe, J., Ed., Pergamon Press, Oxford, 1981, 501.
13. **Jennett, S.**, Methods of studying the control of breathing in experimental animals and man, in *Respiratory Pharmacology*, Widdicombe, J., Ed., Pergamon Press, Oxford, 1981, 3.
14. **Bucher, K.**, Vergleichende Charakterisierung der Lungenatmung einiger Säuger, *Helv. Physiol. Acta*, 7, 470, 1949.
15. **Drorbaugh, J. E.**, Pulmonary function in different animals, *J. Appl. Physiol.*, 15, 1069, 1960.
16. **Leith, D. E.**, Comparative mammalian respiratory mechanics, *Am. Rev. Respir. Dis.*, 128, S77, 1983.
17. **Paleček, F. and Holuša, R.**, Spontaneous occurrence of lung emphysema in laboratory rats, *Physiol. Bohemosl.*, 20, 335, 1971.
18. **Šmejkal, V., Holuša, R., Albrecht, I., and Paleček, F.**, Chronic lung diseases in SPF rats, *Physiol. Bohemoslov.*, 21, 523, 1972.
19. **Gross, P.**, A self-retaining illuminated laryngoscopic speculum for intratracheal procedures, *A. M. A. Arch. Ind. Health*, 18, 429, 1958.
20. **Drorbaugh, J. E. and Fenn, W. O.**, A barometric method for measuring ventilation in newborn infants, *Pediatrics*, 16, 81, 1955.
21. **Bartlett, D. and Tenney, S. M.**, Control of breathing in experimental anemia, *Respir. Physiol.*, 10, 384, 1970.
22. **Paleček, F.**, Tierexperimentelle Modelle chronischer Lungenkrankheiten, *Z. Erkr. Atmungsorgane*, 144, 262, 1976.
23. **Cherniack, R. M., Farhi, L. E., Armstrong, B. W., and Proctor, D. F.**, A comparison of esophageal and intrapleural pressure in man, *J. Appl. Physiol.*, 8, 203, 1956.
24. **Paleček, F.**, Measurement of ventilatory mechanics in the rat, *J. Appl. Physiol.*, 27, 149, 1969.
25. **DuBois, A. B., Botelho, S. Y., Bedell, G. N., Marshall, R., and Comroe, J. H., Jr.**, A rapid plethysmographic method for measuring thoracic gas volume: a comparison with a nitrogen washout method for measuring functional residual capacity in normal subjects, *J. Clin. Invest.*, 35, 322, 1956.
26. **Amdur, M. O. and Mead, J.**, Mechanics of respiration in unanesthetized guinea pigs, *Am. J. Physiol.*, 192, 364, 1958.
27. **Alarie, Y. and Church, F.**, Irritating properties of airborne materials to the upper respiratory tract, *Arch. Environ. Health*, 13, 433, 1966.
28. **Comroe, J. H., Forster, R. E., DuBois, A. B., Briscoe, W. A., and Carlsen, E.**, *The Lung, Clinical Physiology and Pulmonary Function Tests*, Year Book, Chicago, 1969.

29. **Fishman, A. P.**, *Pulmonary Disease and Disorders*, McGraw-Hill, New York, 1980.
30. **Popovic, V. and Popovic, P.**, Permanent cannulation of aorta and vena cava in rats and ground squirrels, *J. Appl. Physiol.*, 15, 727, 1960.
31. **Popovic, P., Sybers, H., and Popovic, V. P.**, Permanent cannulation of blood vessels in mice, *J. Appl. Physiol.*, 25, 626, 1968.
32. **Comroe, J. H.**, *Physiology of Respiration*, Year Book, Chicago, 1965.
33. **West, J. B.**, *Respiratory Physiology — the Essentials*, Williams & Wilkins, Baltimore, 1974.
34. **West, J. B.**, *Pulmonary Pathophysiology — the Essentials*, Williams & Wilkins, Baltimore, 1978.
35. **Widdicombe, J. and Davies, A.**, *Respiratory Physiology*, Edward Arnold, London, 1983.
36. **Herget, J. and Paleček, F.**, Pulmonary arterial blood pressure in closed chest rats. Changes after catecholamines, histamine and serotonin, *Arch. Int. Pharmacodyn. Ther.*, 198, 107, 1972.
37. **Popovic, V., Kent, K. M., and Popovic, P.**, Thechnique of permanent cannulation of the right ventricle in rats and ground squirrels, *Proc. Soc. Exp. Biol. Med.*, 133, 599, 1963.
38. **Rabinovitch, M., Gamble, W., Nadas, A. S., Miettinen, O. S., and Reid, L.**, Rat pulmonary circulation after chronic hypoxia: hemodynamic and structural features, *Am. J. Physiol.*, 236, H818, 1979.
39. **Severinghaus, J. W.**, Proposed standard determination of ventilatory responses to hypoxia and hypercapnia in man, *Chest*, 70, 129S, 1976.
40. **Milic-Emili, J.**, Recent advances in clinical assessment of control of breathing, *Lung*, 160, 1, 1982.
41. **Paleček, F.**, Control of breathing in diseases of the respiratory system, *Int. Rev. Physiol.*, 14, 255, 1977.
42. **Morgenroth, K.**, Experimentelle Untersuchungen zur Pathogenese der interstitiellen Lungenfibrose, *Beitr. Silikoseforsch.*, 1, 3, 1972.
43. **Ryan, S. F.**, Experimental fibrosing alveolitis, *Am. Rev. Respir. Dis.*, 105, 776, 1972.
44. **Brentjens, J. R., O'Connell, D. W., Pawlowski, I. B., Hsu, K. C., and Andres, G. A.**, Experimental immune complex disease of the lung, *J. Exp. Med.*, 140, 105, 1974.
45. **Snider, G. L., Hayes, J. A., and Korthy, A. L.**, Chronic interstitial pulmonary fibrosis produced in hamsters by endotracheal bleomycin, *Am. Rev. Respir. Dis.*, 117, 1099, 1978.
46. **Maron, Z., Weinberg, K. S., and Fanburg, B. L.**, Effect of bleomycin on collagenolytic activity of the rat alveolar macrophage, *Am. Rev. Respir. Dis.*, 121, 859, 1980.
47. **Tryka, A. F., Godleski, J. J., Skornik, W. A., and Brain, J. D.**, Progressive pulmonary fibrosis in hamsters, *Exp. Lung Res.*, 5, 155, 1983.
48. **Elmes, P. C. and Bell, D.**, The effects of chlorine gas on the lungs of rats with spontaneous pulmonary disease, *J. Pathol. Bacteriol.*, 86, 317, 1963.
49. **Bell, P. and Elmes, P. C.**, The effect of chlorine gas on the lungs of rats without spontaneous pulmonary disease, *J. Pathol. Bacteriol.*, 89, 307, 1965.
50. **Marino, A. A. and Mitchell, J. T.**, Lung damage in mice following intraperitoneal injection of butylated hydroxytoluene, *Proc. Soc. Exp. Biol. Med.*, 140, 122, 1972.
51. **Chakrin, L. W. and Saunders, L. Z.**, Experimental chronic bronchitis pathology in the dog, *Lab. Invest.*, 30, 145, 1974.
52. **Pine, J. H., Richter, W. R., and Esterly, J. R.**, Experimental lung injury, *Am. J. Pathol.*, 73, 115, 1973.
53. **Robertson, W. and Schwartz, B.**, Ascorbic acid and formation of collagen, *J. Biol. Chem.*, 201, 689, 1953.
54. **Stassen, F. L. H. and Kuyper, Ch. M.**, Connective tissue reaction to lambda carrageenan in the rat, *Exp. Mol. Pathol.*, 16, 138, 1972.
55. **Trenchard, D., Gardner, D., and Guz, A.**, Role of pulmonary vagal afferent nerve fibres in the development of rapid shallow breathing in lung inflammation, *Clin. Sci.*, 42, 251, 1972.
56. **Velo, G. P. and Spector, W. G.**, The origin and turnover of alveolar macrophages in experimental pneumonia, *J. Pathol.*, 109, 7, 1973.
57. **Wachtlová, M., Chválová, M., Holuša, R., and Paleček, F.**, Carrageenin-induced experimental pneumonia in rats, *Physiol. Bohemoslov.*, 24, 263, 1975.
58. **Herget, J., Paleček, F., Preclík, P., Čermáková, M., Vízek, M., and Petrovická, M.**, Pulmonary hypertension induced by repeated pulmonary inflammation in the rat, *J. Appl. Physiol.*, 51, 755, 1981.
59. **Conning, D. M., Fletcher, K., and Swan, A. A. B.**, Paraquat and related bipyridyls, *Br. Med. Bull.*, 25, 245, 1969.
60. **Kimbrough, R. D. and Gaines, T. B.**, Toxicity of paraquat to rats and its effect on rat lungs, *Toxicol. Appl. Pharmacol.*, 17, 679, 1970.
61. **Smith, P., Heath, D., and Kay, J. M.**, The pathogenesis and structure of paraquat-induced pulmonary fibrosis in rats, *J. Pathol.*, 114, 57, 1974.
62. **Gardiner, T. H., McAnalley, B., Heaton, J., and Reynolds, R. C.**, Changes in the pulmonary uptake and binding of drugs in an experimental model of lung fibrosis, *Toxicol. Appl. Pharmacol.*, 49, 487, 1979.
63. **Vízek, M., Holuša, R., and Paleček, F.**, Lung function in acute paraquat intoxication, *Physiol. Bohemoslov.*, 24, 559, 1975.

64. **Gross, P., Pfitzer, E. A., Tolker, E., Babyak, M. A., and Kaschak, M.,** Experimental emphysema. Its production with papain in normal and silicotic rats, *Arch. Environ. Health,* 11, 50, 1965.

65. **Goldring, I. P., Greenburg, L., and Ratner, I. M.,** On the production of emphysema in Syrian hamsters by aerosol inhalation of papain, *Arch. Environ. Health,* 16, 59, 1968.

66. **Kilburn, K. H., Dowell, A. R., and Pratt, P. C.,** Morphological and biochemical assessment of papain-induced emphysema, *Arch. Intern. Med.,* 127, 884, 1971.

67. **Caldwell, E. J.,** Physiologic and anatomic effects of papain on the rabbit lung, *J. Appl. Physiol.,* 31, 458, 1971.

68. **Pushpakom, R., Hogg, J. C., Woolcock, A. J., Angus, A. E., Macklem, P. T., and Thurlbeck, W. M.,** Experimental papain-induced emphysema in dogs, *Am. Rev. Respir. Dis.,* 102, 778, 1970.

69. **Marco, V., Meranze, D. R., Yoshide, M., and Kimbel, P.,** Papain-induced experimental emphysema in the dog, *J. Appl. Physiol.,* 33, 293, 1972.

70. **Wright, R.,** Suggested criteria for confirming pulmonary emphysema in the experimental animal, *Yale J. Biol. med.,* 40, 576, 1968.

71. **Dunnill, M. S.,** Quantitative methods in the study of pulmonary pathology, *Thorax,* 17, 320, 1962.

72. **Strawbridge, H. T. G.,** Chronic pulmonary emphysema (an experimental study). II. Spontaneous pulmonary emphysema in rabbits, *Am. J. Pathol.,* 37, 309, 1960.

73. **McLaughlin, R. F. and Edwards, D. W.,** Naturally occuring emphysema, the fine gross and histopathologic counterpart of human emphysema, *Am. Rev. Respir. Dis.,* 93, 22, 1966.

74. **Heppleston, A. G., Fletcher, K., and Wyatt, I.,** Abnormalities of lung lipids following inhalation of quartz, *Experientia,* 28, 938, 1972.

75. **Hollenbach, K., Kersten, E., Patzelt, K., and Schwesinger, G.,** Untersuchungen zur fibrogenen Wirksamkeit von Apatitnephelin- und Apatitkonzentrat-Staub in Rattenlungen nach einmaliger intratrachealer Applikation, *Int. Arch. Arbeitsmed.,* 28, 271, 1971.

76. **Morgan, A., Moores, S. R., Holmes, A., Evans, J. C., Evans, N. H., and Black, A.,** The effect of quartz, administered by intratracheal instillation, on the rat lung. I. The cellular response, *Environ. Res.,* 22, 1, 1980.

77. **Hatch, G. E., Slade, R., Boykin, E., Hu, P. C., Miller, F. J., and Gardner, D. E.,** Correlation of effects of inhaled versus intratracheally injected metals on susceptibility to respiratory infection in mice, *Am. Rev. Respir. Dis.,* 124, 167, 1981.

78. **Dale, K.,** Experimental silicosis. The relation between dose of quartz, length of action for the dust, tissue reaction and disturbance of the lung function, *Scand. J. Respir. Dis.,* 54, 306, 1973.

79. **Moorman, W. J., Lewis, R., and Wagner, W. D.,** Maximum expiratory flow-volume studies on monkeys exposed to bituminous coal dust, *J. Appl. Physiol.,* 39, 444, 1975.

80. **Verstraeten, J. M., Lacroix, E., Roels, H., and Spinoit, C.,** Les modifications de la compliance pulmonaire chez le silicotique et chez l'animal exposé à la poussière de silice, *Med. Thorac.,* 23, 160, 1966.

81. **Kuncová, M., Havránková, J., Holuša, R., and Paleček, F.,** Experimental silicosis of the rat. Correlation of functional, biochemical and histological changes, *Arch. Environ. Health,* 23, 365, 1971.

82. **Kuncová, M., Havránková, J., Kunc, L., Holuša, R., and Paleček, F.,** Experimental lung silicosis. Evolution of functional, biochemical and morphological changes in the rat, *Arch. Environ. Health,* 24, 281, 1972.

83. **Chválová, M., Kuncová, M., Havránková, J., and Paleček, F.,** Regulation of respiration in experimental silicosis, *Physiol. Bohemoslov.,* 23, 539, 1974.

84. **Ziskind, M., Jones, R. N., and Weill, H.,** Silicosis, *Am. Rev. Respir. Dis.,* 113, 643, 1976.

85. **Widimský, J., Urbanová, D., Ressl, J., Oštádal, B., Pelouch, V., and Procházka, J.,** Effect of intermittent altitude hypoxia on the myocardium and lesser circulation in the rat, *Cardiovasc. Res.,* 7, 798, 1973.

86. **Nattie, E. E., Bartlett, D., and Johnson, K.,** Pulmonary hypertension and right ventricular hypertrophy caused by intermittent hypoxia and hypercapnia in the rat, *Am. Rev. Respir. Dis.,* 118, 653, 1978.

87. **Kay, J. M., Suyama, K. L., and Keane, P. M.,** Effect of intermittent normoxia on muscularization of pulmonary arterioles induced by chronic hypoxia in rats, *Am. Rev. Respir. Dis.,* 123, 454, 1981.

88. **Grover, R. F., Vogel, J. H. K., Voigt, G. C., and Blount, S. G.,** Reversal of high altitude pulmonary hypertension, *Am. J. Cardiol.,* 18, 928, 1966.

89. **Abraham, A. S., Kay, J. M., Cole, R. B., and Pincock, A. C.,** Hemodynamic and pathological study of the effect of chronic hypoxia and subsequent recovery of the heart and pulmonary vasculature of the rat, *Cardiovasc. Res.,* 5, 25, 1971.

90. **Leach, E., Howard, P., and Barer, G. R.,** Resolution of hypoxic changes in the heart and pulmonary arterioles of rats during intermittent correction of hypoxia, *Clin. Sci. Mol. Med.,* 52, 153, 1977.

91. **Herget, J., Suggett, A. J., Leach, E., and Barer, G. R.,** Resolution of pulmonary hypertension and other features induced by chronic hypoxia in rats during complete and intermittent normoxia, *Thorax,* 33, 468, 1978.

92. **Kentera, D. and Sušič, D.**, Dynamics of regression of right ventricular hypertrophy in rats with hypoxic pulmonary hypertension, *Respiration,* 39, 272, 1980.
93. **Visscher, M. B., Haddy, F. J., and Stephens, G.**, The physiology and pharmacology of lung edema, *Pharmacol. Rev.,* 8, 389, 1956.
94. **Kartzel, K.**, Das experimentelle Lungenödem, in *Handbuch der experimentellen Pharmakologie,* Vol. VI/ 2, Springer Verlag, Berlin, 1969, 155.
95. **Lindqvist, B.**, Experimental uraemic pulmonary oedema, *Acta Med. Scand.,* 176 (Suppl. 418), 1, 1964.
96. **Poulsen, T.**, Quantitative estimation of pulmonary oedema in mice, *Acta Pharmacol.,* 10, 117, 1954.
97. **Bost, J., Sauvage, E., and Guéhemmeux, A.**, Oedéme aigu du poumon provoquée par un glycéride polyoxyéthyléne chaines courtes, *J. Physiol. (Paris),* 61, 219, 1969.
98. **Glogowska, M. and Widdicombe, J. G.**, The role of vagal reflexes in experimental lung oedema, bronchoconstriction and inhalation of halothane, *Resp. Physiol.,* 18, 116, 1973.
99. **Dorato, M. A., Carlson, K. H., and Copple, D. L.**, Pulmonary mechanics in conscious Fischer 344 rats: multiple evaluations using nonsurgical techniques, *Toxicol. Appl. Pharmacol.,* 68, 344, 1983.
100. **Paleček, F., Palečková, M., and Aviado, D. M.**, Emphysema in immature rats produced by tracheal constriction and papain, *Arch. Environ. Health,* 15, 332, 1967.
101. **Holub, D. and Frank, R.**, A system for rapid measurement of lung function in small animals, *J. Appl. Physiol.,* 46, 394, 1979.
102. **Johanson, W. G. and Pierce, A. K.**, A noninvasive technique for measurement of airway conductance in small animals, *J. Appl. Physiol.,* 30, 146, 1971.
103. **Diamond, L. and O'Donnell, M.**, Pulmonary mechanics in normal rats, *J. Appl. Physiol.,* 43, 942, 1977.
104. **Lai, Y. L. and Hildebrandt, J.**, Respiratory mechanics in the anesthetized rat, *J. Appl. Physiol.,* 45, 255, 1978.
105. **Takezawa, J., Miller, F. J., and O'Neil, J. J.**, Single-breath diffusing capacity and lung volumes in small laboratory mammals, *J. Appl. Physiol.,* 48, 1052, 1980.
106. **Jackson, A. C. and Watson, J. W.**, Oscillatory mechanics of the respiratory system in normal rats, *Respir. Physiol.,* 48, 309, 1982.
107. **Saldiva, P. H. N., Massad, E., Pires de Rio Caldeira, M., Calheiros, D. F., Saldiva, C. D., and Böhm, G. M.**, The study of mechanical properties of rat lungs by whole body plethysmography, *Acta Physiol. Pharmacol. Latinoam.,* 35, 109, 1985.
108. **Newman, L. M., Johnson, E. M., and Roth, J. M.**, Lung volume and compliance in neonatal rats, *Lab. Anim. Sci.,* 371, 1984.
109. **Oddoy, A., Eckert, H., Lachmann, B., Lux, M., Merker, G., and Vogel, J.**, Das experimentelle proteolytische Emphysem — Übersicht und eigene Ergebnisse, *Z. Erkr. Atmungsorgane,* 160, 3, 1983.
110. **King, T. K.**, Measurement of functional residual capacity in the rat, *J. Appl. Physiol.,* 21, 233, 1966.
111. **Turek, Z., Frans, A., and Kreuzer, F.**, Pulmonary diffusing capacity for O_2 in the anesthetized rat breathing spontaneously, *Respir. Physiol.,* 8, 169, 1970.
112. **Turek, Z., Frans, A., and Kreuzer, F.**, Steady-state diffusing capacity for carbon monoxide in the rat, *Respir. Physiol.,* 12, 346, 1971.
113. **Stinger, R. B., Iacopino, V. J., Alter, I., Fitzpatrick, T. M., Rose, J. C., and Kot, P. A.**, Catheterization of the pulmonary artery in the closed-chest rat, *J. Appl. Physiol.,* 51, 1047, 1981.
114. **Fukuda, Y., See, W. R., and Honda, Y.**, Effect of halothane anesthesia on end-tidal P_{CO2} and pattern of respiration in the rat, *Pflugers Arch.,* 392, 244, 1982.
115. **Kotlikoff, M. I., Jackson, A. C., and Watson, J. W.**, Oscillatory mechanics of the respiratory system in ozone-exposed rats, *J. Appl. Physiol.,* 56, 182, 1984.
116. **Truog, W. E., Hlastala, M. P., Standaert, T. A., McKenna, H. P., and Hodson, W. A.**, Oxygen-induced alternation of ventilation-perfusion relationship in rats, *J. Appl. Physiol.,* 47, 1112, 1979.
117. **Lai, Y. L., Lamm, W. J. E., and Hildebrandt, J.**, Ventilation during prolonged hypercapnia in the rat, *J. Appl. Physiol.,* 51, 78, 1981.
118. **Lai, Y. L., Lamm, W. J. E., and Hildebrandt, J.**, Mechanical properties of rat lung during prolonged hypercapnia, *J. Appl. Physiol.,* 52, 1156, 1982.
119. **Church, M. K., Collier, H. O. J., and James, G. W. L.**, The inhibition by dexamethasone and disodium cromoglycate of anaphylactic bronchoconstriction in the rat, *Br. J. Pharmacol.,* 46, 56, 1972.
120. **Favier, R. and Laciasse, A.**, Stimulus oxygène de la ventilation chez le rat éveillé, *J. Physiol. (Paris),* 74, 411, 1978.
121. **Barer, G. R., Edwards, C. W., and Jolly, A. I.**, Changes in the carotid body and the ventilatory response to hypoxia in chronically hypoxic rats, *Clin. Sci. Mol. Med.,* 50, 311, 1976.
122. **Hayashi, F., Yoshida, A., Fukuda, Y., and Honda, Y.**, The ventilatory response to hypoxia in the anaesthetized rats, *Pflugers Arch.,* 396, 121, 1983.
123. **Lai, Y. L., Tsuya, Y., and Hildebrandt, J.**, Ventilatory responses to acute CO_2 exposure in the rat, *J. Appl. Physiol.,* 45, 611, 1978.

124. **Barer, G. R., Herget, J., Sloan, P. J. M., and Sugget, A. J.,** The effect of acute and chronic hypoxia on thoracic gas volume in anaesthetized rats, *J. Physiol.,* 277, 177, 1978.
125. **Paleček, F. and Chválová, M.,** Functional residual lung capacity in rats affected by a carotid body stimulant, *Arch. Int. Pharmacodyn. Ther.,* 267, 123, 1984.
126. **Paleček, P., Palečková, M., and Aviado, D. M.,** Emphysema in immature rats produced by tracheal constriction and papain, *Arch. Environ. Health,* 15, 332, 1967.
127. **Paleček, F. and Rochová-Mikulášová, J.,** Experimental emphysema in rats; an attempt to influence its production by gestagen adminstration, *Physiol. Bohemoslov.,* 17, 445, 1968.
128. **Herget, J., Holuša, R., and Paleček, F.,** Pulmonary hypertension in rats with experimental emphysema, *Physiol. Bohemoslov.,* 23, 55, 1974.
129. **Giles, R., Finkel, M. P., and Leeds, R.,** The production of emphysema-like condition in rats by the administration of papain aerosol, *Proc. Soc. Exp. Biol. Med.,* 134, 157, 1970.
130. **Johanson, W. G., Pierce, A. K., and Reynolds, R. C.,** The evolution of papain emphysema in the rat, *J. Lab. Clin. Med.,* 78, 599, 1971.
131. **Giles, R. E., Williams, J. C., and Finkel, M. P.,** Progesterone antagonism of papain emphysema: role of sex, estrogens and serum antitrypsin, *Proc. Soc. Exp. Biol. Med.,* 144, 487, 1973.
132. **Johanson, W. G., Reynolds, R. C., Scott, T. C., and Pierce, A. K.,** Connective tissue damage in emphysema, *Am. Rev. Respir. Dis.,* 107, 595, 1973.
133. **Bell, D. P., Wade, Q. L., and Williams, T.,** Emphysema in pathogen-free and bronchitic rats: an electronic lung-scanning technique, *Am. Rev. Respir. Dis.,* 109, 297, 1974.
134. **Kobrle, V., Hurych, J., and Holuša, R.,** Changes in pulmonary connective tissue after a single intratracheal instillation of papain in the rat, *Am. Rev. Respir. Dis.,* 125, 239, 1982.
135. **Yamamoto, W. S. and McIver, W. E.,** Homeostasis of carbon doxide during intravenous infusion of carbon dioxide, *J. Appl. Physiol.,* 15, 817, 1960.
136. **Orthner, H. and Yamamoto, W. S.,** Transient respiratory response to mechanical loads at fixed blood gas levels in rats, *J. Appl. Physiol.,* 36, 280, 1974.
137. **Hanna, S. W., Drysdale, D. B., and Cragg, P. A.,** Ventilatory sensitivity and breathing patterns in conscious rats, *Proc. Aust. Physiol. Pharmacol. Soc.,* 10, 42P, 1979.
138. **Cragg, P. A. and Drysdale, D. B.,** Effect of carotid body denervation and/or vagotomy on ventilatory latencies in the anaesthetized rat, *N. Z. Med. J.,* 88, 167, 1978.
139. **Hamilton, J. H., Cragg, P. A., and Drysdale, D. B.,** The effects of carotid body denervation on ventilation in oxygen at high pressure, *Proc. Aust. Physiol. Pharmacol. Soc.,* 10, 41P, 1979.

Chapter 25

DIABETES MELLITUS AND CARBOHYDRATE METABOLISM ALTERATIONS

A. Vrána and L. Kazdová

TABLE OF CONTENTS

I. INTRODUCTION, HISTORICAL REMARKS

Diabetes mellitus is both the most frequent and probably the most dangerous disorder of carbohydrate metabolism in man, causing a number of organ complications, some of which might be fatal. Today, almost a century since diabetes was experimentally induced,[1] and more than 60 years since the discovery and isolation of the "active principle" produced by the pancreas, insulin,[2] and its clinical use, a number of questions related to the pathogenesis of diabetes and its organ complications have yet to be answered. As a result, the information essential for a rational therapy and prevention of this disease and its complications is incomplete.

As in other human metabolic disorders, an array of animal models exhibiting several, or most, of the symptoms typical of human diabetes have been developed or identified by screening over the past decades.

The development and use of experimental models of diabetes have undergone substantial changes during the past decades. Quite obviously, after the successful induction of hyperglycemia induced by extirpation of the pancreas, pancreatectomy was often employed as a reliable model to induce diabetes. Later on, it was found that hyperglycemia and other symptoms of diabetes can be elicited by chemical destruction of B cells.

Studies of nephrotoxicity of uric acid derivatives revealed in the early 1940s that alloxan destroys B cells of the islets of Langerhans and leads to permanent hyperglycemia and glycosuria. The story was repeated about 2 decades later, this time with streptozotocin. While a number of other effects were soon demonstrated for this antibiotic, it was found, among other things, that like alloxan, streptozotocin destroys B cells with the all implications. Streptozotocin has been recently preferred to alloxan because of its generally less powerful cytotoxic side effects.

Lines exhibiting metabolic and, at times, even morphological symptoms of diabetes have been identified in a number of species, and in rodents in particular over the past decades. In some cases, as in human disease, apart from genetic predisposition, also environmental factors, usually nutritional ones, are necessary for the manifestation of diabetes in these models.

Short-term hyperglycemia can also be produced by other substances, such as some analogs of glucose (2-deoxyglucose, 3-0 methylglucose), other monosacharides (mannoheptulose), and drugs interfering with intracellular metabolism of glucose. Administration of excessive doses of some hormones may elicit transient or permanent diabetes (steroid diabetes, metaglucagon diabetes, and metahypophyseal diabetes) in some species. Here we cannot fail to mention hyperglycemia induced by pricking the floor of the fourth cerebral ventricle (Claude-Bernard piqûre).

In recent years, increased attention has focused on the immunological and viral mechanisms in the pathogenesis of insulin-dependent diabetes mellitus (IDDM). This interest is due to the fact that a number of viral infections injure islets of Langerhans in animals as well as in man. Antibodies to insulin or to cellular receptors for insulin may likewise lead to manifestation of diabetes.

Because of the large number of issues involved in experimental modeling of diabetes, it is impossible to describe the individual models in detail, and we will therefore take the liberty of referring the reader to pertinent reviews whenever possible.

II. PANCREATECTOMY

As was the case with other endocrine glands, it was surgical extirpation of the pancreas which offered the first elementary information shedding some light on function of the pancreas and pathophysiology of diabetes.[1,2] It is certainly worth noting that polyuria and polydipsia observed after pancreatectomy in the dog was described by J. C. Brunner 2 centuries before that event (see Reference 3).

Diabetes induced by pancreatectomy has been studied in a number of animal species. The considerable species differences that exist in the anatomy of the pancreas not only get reflected in the varying degree of technical difficulties associated with the procedure but also in the amounts of the pancreatic mass that can be extirpated. In the rat, for example, due to the diffuse nature of the pancreas composed of the duodenal, gastric, and splenic part, it is more appropriate to apply the term subtotal pancreatectomy since total removal of the part of the tissue localized along the vena cava and the bile duct (complete pancreatectomy) is difficult to perform.

Following subtotal pancreatectomy, glycemia begins to rise within the first hours after surgery to reach about double the preoperative values and keeps on increasing. The postoperative mortality after subtotal pancreatectomy, despite the good nutritional care and administration of antibiotics, is rather high in the rat. The leading causes of death include bleeding, infection, peritonitis, and ileus.[3] Substantially lower mortality rates are reported in partial diabetogenic pancreatectomy involving removal of the gastric and lienal parts of the pancreas (about 92% of the total pancreatic mass) with the duodenal part preserved intact. Here, the development of glycemia differs from the situation observed after subtotal pancreatectomy: in the case of partial diabetogenic pancreatectomy, a marked rise of glycemia does not occur until after about 1 month after operation.[3]

Another species frequently used to induce diabetes by pancreatectomy is the dog. It was actually the dog in whom diabetes was first experimentally elicited, and the presence of insulin in the pancreas was demonstrated.[2] The canine pancreas consists of three lobes. The left lobe lies near the stomach, the right cranial lobe is adjacent to the duodenum, and the right caudal one is suspended in the mesentery. There are marked differences in the proportion of endocrine cells in the individual lobes. The highest concentration of A and B cells and a low content of D cells are present in the left and cranial right lobes, while the caudal right lobe contains few A cells, has more B cells, and the highest content of D cells.[4] The length of the interval between surgery and the onset of diabetes depends not only on the amount of tissue extirpated, but in view of the varying proportion of endocrine cells in individual parts of the gland, also on what part of the pancreas has been left intact. Even though extirpation of 80 to 90% of the pancreatic mass may lead to permanent diabetes in the dog, a reliable procedure is total pancreatectomy.[4]

An advantage of surgical pancreatectomy is that unlike "chemical pancreatectomy" induced by administration of alloxan or streptozotocin, it is not associated with toxic damage of other tissues. Drawbacks include technical difficulties of the surgical procedure, higher mortality, and the need for supplementation of digestive enzymes. Another feature making postpancreatectomy diabetes different from chemically induced, as well as from human diabetes, is the fact that not only B cells, but also other hormone-producing cells of the islets of Langerhans are eliminated.

Recently the classical partial pancreatectomy model was reinvestigated and marked regeneration of the endocrine pancreas was observed. In young rats with 90% pancreatectomy, it was found, that 8 to 10 weeks after the operation, the pancreatic remnant weight was 26% and B cells mass even 42% of values of sham-operated controls. Whereas plasma insulin in a fed state as well as after fasting was comparable in pancreatectomized and control rats, blood glucose was moderately increased in pancreatectomized rats. Insulin response to glucose in vivo and in vitro was markedly reduced, while insulin response to meal challenge or to arginin infusion was comparable with controls.[5]

III. CHEMICALLY INDUCED DESTRUCTION OF B CELLS

A number of substances impair, reversibly or irreversibly, B cells, resulting in insulin insufficiency. To induce long-lasting hyperglycemia, glycosuria, and other symptoms of diabetes, alloxan and streptozotocin are widely used. Despite their completely different structure, a different mechanism of B cells damage and their different side effects, the final metabolic effects of administration of diabetogenic doses of alloxan and streptozotocin, are similar.

Alloxan (2,4,5,6-tetraoxohexahydropyrimidine) elicits diabetes in a number of species such as dogs, sheep, rabbits, rats, mice, monkeys, birds, fish, and turtles. By contrast, the guinea pig and some subspecies of hamsters are resistant to the diabetogenic effect of alloxan. The diabetogenic dose of alloxan differs considerably in particular species; effective doses are within the range of 40 to 200 mg/kg. Intravenous administration is the most suitable way of alloxan application since it enables rapid and equal distribution of the substance. Intraperitoneal administration requires doses two to three times higher than intravenous administration.

Even though alloxan has been routinely employed for almost half a century to induce diabetes, the exact biochemical mechanism of its cytotoxic effect is not completely understood. Alloxan disturbs the integrity of cellular membrane of B cells; the mechanism that may be involved include high reactivity of alloxan with sulfhydryl (SH) groups. The possible toxic effect of free radicals cannot likewise be disregarded, since substances acting as their scavangers (e.g., superoxide dismutase) protect B cells against the effect of alloxan, and vice versa, islet superoxide dismutase is decreased after alloxan administration. It has been suggested that these two hypotheses might be linked together; there might be a combination of these two mechanisms: the interaction of free radicals with SH groups.[6]

Although the cytotoxic effect of alloxan to B cells is markedly higher than to cells of other tissues, the general cytotoxic effect of the drug cannot be ignored. One of the tissues displaying high sensitivity to the toxic effect of alloxan is the kidney. The nephrotoxic effect of alloxan can be decreased by bilateral compression of the renal arteries during administration of alloxan. This simple, but very effective maneuver protecting the kidney from the toxic effect of alloxan, can be regarded as a very good method to elicit alloxan-induced damage of B cells while minimizing its nephrotoxic effect.* Whereas alloxan nephrotoxicity cannot be completely ruled out even by this mechanical protection, animal survival rates are considerably increased by this technique.[3]

Streptozotocin (2 deoxy 2-3/methyl-3-nitrosourea/1-d-glucopyranose) a metabolite of *Streptomyces achromogenes* having tumorigenic as well as anticancerogenic properties elicits, like alloxan, diabetes in a number of rodent species, and also in dogs, pigs, and monkeys. As in the case of alloxan, the exact mechanism of its damage of B cells is not completely understood. Besides injuring the cell membrane, streptozotocin interferes with intracellular processes by, e.g., reducing the content of NAD, and in addition, like alloxan, it reacts with SH groups (for review see Reference 6).

A particular feature and an advantage of streptozotocin in the modeling of diabetes is the fact that, depending on the dose administered and on distribution of doses, its administration can produce a number of different diabetic states:

1. A single, large diabetogenic dose of the drug (see Table 1), which is most frequently used, leads to B cell necrosis, the consequences are therefore very similar to those observed after administration of alloxan (see below).

2. "Low" doses of streptozotocin, in mice, given repeatedly, after four to five admin-

* The following procedure is recommended by Korec:[3] Rats are anesthetized (e.g., with intraperitoneally given pentobarbital), placed with their abdomen down on the table, kidneys are palpated by the index and middle finger in the lumbar region and loosely supported by fingers. Both kidneys are pressed between two underlying fingers by the thumb and alloxan solution is injected intravenously within 10 to 30 sec. It is recommended to produce only moderate, 5-min lasting pressure.

Table 1
DIABETOGENIC DOSES OF
STREPTOZOTOCIN IN SEVERAL
SPECIES

Species	Multiple dose mg/kg	Days administered	Single dose (mg/kg)
Dog	15	3	50
Monkey			60
Rat			65
Mouse	40	5	100
Rabbit			300

From Preston, A. M., *Nutr. Res.*, 5, 435, 1985.

istrations (given at 1-day intervals), result in delayed but progressive diabetes which starts shortly after administration of the last dose. These changes are not observed or are attenuated by administration of antilymphocyte serum, or by irradiation, indicating that immunological mechanisms are involved.

3. In the case that approximately one half of the standard effective diabetogenic dose of streptozotocin is given to young or adult mice, diabetes also develops, even though its onset is rather delayed as compared with that after the administration of a single effective dose of the drug. The diabetic syndrome is not observed in all the treated animals and it remains to be established whether or not it is mediated by the immune mechanisms.[7]

4. Intravenous or intraperitoneal administration of streptozotocin to neonatal rats leads to hyperglycemia and glycosuria 3 to 5 days after treatment. A few days later, blood glucose is normal or moderately increased; however, glucose tolerance is impaired. Glucose-induced insulin secretion is decreased in this model, data concerning arginine-induced insulin secretion and tissue insulin sensitivity are not uniform.[8-11]

Recently it has been reported that another glucose-containing N-nitroso compound, the streptozotocin analog, chlorozotocin, may induce diabetes. Susceptibility to this agent in rodent species is completely different from that to streptozotocin. The Syrian hamster is most sensitive; less sensitive are mice, while in rats only a small increase of fasting glycemia and impairment of glucose tolerance was observed. Also the mechanism of B cell injury after chlorozotocin treatment is different from that of streptozotocin; nicotinamide pretreatment did not protect B cells from the diabetogenic effect of chlorozotocin.[12]

IV. METABOLIC AND HORMONAL CHANGES AFTER ALLOXAN OR STREPTOZOTOCIN ADMINISTRATION

Administration of alloxan or streptozotocin at effective doses exerts a pronounced effect on glucose homeostasis by interference into the spectrum of glucoregulatory hormones. The changes of glycemia have a marked three-phase character and its course is reflected in changes of plasma concentration of insulin and glucagon. Whereas the changes of plasma glucagon concentrations are similar to those of glycemia, insulinemia has an inverse course.

The first, acute hyperglycemic phase is characterized by an increase in glycemia to about double of the initial values as early as 2 hr after drug administration. The hyperglycemia is explained by an increased flux of glucose as a result of accentuated hepatic glycogenolysis.

A certain role in the mechanism of this acute hyperglycemic phase may be played by the low insulinemia, a consequence of inhibited insulin secretion. The differences between the effects of administration of diabetogenic doses of alloxan and streptozotocin are small, the onset of the above mentioned changes after streptozotocin administration, compared to that after administration of alloxan, is somewhat delayed by one to several hours.

The second, hypoglycemic phase, beginning 4 to 6 hr after administration of alloxan or streptozotocin, is the result of a massive release of insulin from destroyed B cells; in this phase, insulinemia is very high. Severe hypoglycemia can even result in death, especially when animals had fasted prior to drug administration.

The third, chronic hyperglycemic phase, begins 10 to 15 hr after the administration of alloxan or streptozotocin. At that stage, glycemia reaches values of 20 mmol and more. Due to the small number or "surviving" functioning B cells, insulinemia is very low. The glucose tolerance is markedly impaired, ketosis is more frequent in alloxan-diabetic rats.

Due to the destruction of B cells, insulinemia decreases to reach low values, and the secretory response to glucose and to other stimuli is dramatically reduced.

The metabolic situation after administration of diabetic doses of alloxan or streptoxotocin is characterized by insulinopenia, manifested by increased gluconeogenesis and glycogenesis, decreased glycolysis, increased mobilization of free fatty acids from adipose tissue, and hypertriglyceridemia. The plasma concentration of glucagon resembles the course of glycemia, being high in the first, low in the second, and increased in the third phase, especially in severely diabetic animals. A remarkable phenomenon is the paradoxical response of glucagonemia in chemically induced diabetes. Whereas in a "normal" state, glucagon secretion is usually inhibited by increased glycemia, in chemically induced diabetes the feedback mechanism is disturbed to such an extent that oral administration of glucose stimulates glucagon secretion. No matter whether glucagon secretion is increased or unaltered, chemically induced diabetes is characterized by at least a relative excess of glucagon and by a profound change of the insulin/glucagon ratio, a distinct difference as compared with the diabetic syndrome induced by pancreatectomy.

The role of another pancreatic hormone, somatostatin, in chemically induced diabetes, is less well documented. While during the first days after administration of alloxan or streptozotocin, the plasma concentration of this hormone remains unchanged; the concentration is increased and multiplication of the D cell can be observed after 2 weeks. The plasma somatostatin concentration can be decreased by insulin administration.[6]

A decrease in glycemia may occur after several months, a phenomenon probably associated with B cell regeneration, which is more frequent in the streptozotocin than in the alloxan diabetes.[6]

V. MODELS OF SPONTANEOUS DIABETES MELLITUS OR OF IMPAIRED GLUCOSE TOLERANCE IN ANIMALS

A. Rodents

Several species with spontaneous diabetes or with impaired glucose tolerance have been described and intensively studied. Most of these models are rodents, either lean or with a varying degree of obesity. The individual species, but also lines or sublines of the same species, differ apart from the presence or absence of obesity, in a number of metabolic parameters (see Table 2 and 3), as well as in the presence or absence of insulitis and of organ complications of diabetes.

Yellow mouse — This strain displays only a mild form of glucose intolerance and obesity.[13]

Ob/ob mouse — Insulinemia and glycemia in this mutant have a triphasic shape. In the first phase (at about 3 weeks of age), along with the onset of hyperphagia, insulinemia is dramatically increased (to 20 to 50 times the normal values) with a mild increase of blood

Table 2
BASAL METABOLIC CHARACTERISTICS OF SOME RODENTS WITH SPONTANEOUS DIABETES-PURE STRAINS

Animal	Obesity	Plasma glucose	Plasma insulin	Pancreatic insulin	Insulin resistance	Ketosis
Yellow mouse	+	↑	↑	↑	+	−
Obese mouse (C57BL/6J,ob/ob)	+ +	↑→N	↑→N	↑	+	−
Diabetic mouse (C57BL/KsJ,db/db)	+ +	N→↑	↑→↓	↑→↓	+	−
Zucker rat (fatty, fa/fa)	+	N or ↑	↑	↑	+	−
Chinese hamster (*Cricetulus griseus*)	=	↑	N or ↓	↓	Humoral?	rare
NZO mouse	+ +	↑	↑	↑	+	−
KK mouse	+	↑	↑	↑	+	−
T-KK mouse	+	↑	↑	↑	+	−

Note: N — normal, ↑ — high, ↑ — slightly elevated, ↓ — decreased, − — absent, + — present, and + + — prounounced.

Modified from Grodsky, G. M., et al., *Diabetes*, 31, (suppl.1), 45, 1982.

Table 3
BASAL METABOLIC CHARACTERISTICS OF SOME RODENTS WITH SPONTANEOUS DIABETES - IMPURE STRAINS (MODIFIED FROM GRODSKY ET AL. 1982)[15]

Animal	Obesity	Plasma glucose	Plasma insulin	Pancreatic insulin	Insulin resistance	Ketosis
BB rat	−	↑ ↑	↓	↓	−	+
NOD rat	−	↑	↓	↓	−	+ +
Spiny mouse (*Acomys cahirinus*)	+	↑	N → ↓	↑ ↑ a	0	Rare
Sand rat (*Psammomys obesus*)	+	↑	↑	↑ b	+	Rare
Djungarian hamster (*Phodopus sungorus*)	+	N or ↑	↑	↑ c	0	+
South African hamster (*Mystromys albicaudatus*)	−	↑	0	↓	0	−

N — normal; 0 — unknown; ↑ — high; ↑ — slightly elevated, + — present; and + + — prounouced.

a   High insulin content associated with low release.
b   Occasionally low insulin with ketosis.
c   Ketosis without marked hyperglycemia.

Modified from Grodsky, G. M. et al., *Diabetes* 31, (suppl.1), 45, 1982.

glucose concentration. Whereas in the second phase insulinemia decreases, in the third phase insulinemia and glycemia are normal. The return to normoglycemia and normalization of insulinemia is associated with regranulation, hyperplasia, and hypertrophy of B cells.[14,15]

Db/db mouse — Homozygotes of this strain are hyperphagic, obese, and after initial mild hyperglycemia, they develop severe hyperglycemia exceeding 20 mmol/ℓ. Insulinemia is increased already at 10 days of age, culminates at 2 to 3 months and then drops. The hyperinsulinemic phase is associated with hyperplasia and hypertrophy of B cells. The plasma concentration of glucagon and pancreatic glucagon content are increased. Several studies have demonstrated organ complications of diabetes in db/db mice, and findings also indicating reduced cellular immunity have been reported.[14]

Fatty (fa/fa) Zucker rat — Although the mutant is very obese, hyperphagic, hyperinsulinemic with marked resistance to insulin, and hypertriglyceridemic; glycemia is either normal or only mildly increased.[16]

Chinese hamster — The species have several strains differing in the degree of hyperglycemia and in other parameters such as insulinemia, insulin sensitivity, frequency of and onset of glycosuria (2 to 6 weeks), as well as of ketonuria (10 to 52 weeks). There are even differences within the individual substrains. Insulinemia and insulin secretion are decreased; glucagonemia and glucagon secretion are increased. The number of D cells and the content of somatostatin in the pancreas are decreased.[17]

New Zealand obese (NZO) mice — In this mutant, hyperglycemia is mild, usually not exceeding 10 to 15 mmol at 6 months of age. Insulinemia increases mildly with age; hyperplasia and islet hypertrophy are marked. There are differences among lines in the degree of glucose intolerance and insulin resistance. Antibodies against insulin receptors have been documented, and immune complexes-induced glomerulonephritis is a frequent feature.[15,16]

KK mice — The Japanese KK strain is characterized by a mild degree of hyperglycemia, slowly developing obesity, and mildly increased size and number of the islets of Langerhans. Hyperglycemia culminates between 4 to 9 months of age and then declines. Although metabolic abnormalities are normalized by 1 year of age, these animals have a shortened lifespan.[16]

Bio Breeding (BB) Wistar rats — This lean, insulinopenic animal model of diabetes was recognized in the early 1970s. Rats of both sexes are equally affected and the time of onset of hyperglycemia in susceptible animals is between 2 and 4 months of age. The onset of hyperglycemia is abrupt; prospective studies have shown that within a week, glycemia rises to a three- to a fourfold level of the initial value and, at the same time, massive glycosuria and polyuria occur. Metabolic changes are preceded by insulitis. Insulinemia is very low as a result of destruction of B cells, in the final phase, the numbers of A and D cells also decrease as does the content of glucagon and somatostatin in the pancreas. Affected animals, if not treated with insulin, die within 2 to 3 weeks. Insulitis is apparently of autoimmune etiology since administration of antilymphocyte serum, irradiation, or thymectomy prevent diabetes in susceptible animals and normalize glycemia in part of diabetics.[15,18]

The nonobese diabetic (NOD) mice — Among the mice models of diabetes, the NOD mouse is unique in that diabetes is the consequence of insulinopenia, developing on the basis of insulitis which is probably of autoimmune etiology. Interestingly, whereas the incidence of insulitis is about the same in both sexes, diabetes is approximately three times more frequent in females than in males.[16]

Apart from the above mentioned and most frequently used species and lines, manifestations of diabetes often associated with obesity can also be noted in some other species of rodents such as the spiny mouse (*Acomys cahirinus*) and the member of the same genus, *Acomys russatus*, the sand rat (*Psammomys obesus*), the Djungarian hamster (*Phodopus sungorus*), the South African hamster (*Mystromys albicaudatus*), and in some murine hybrids such as the Wellesley hybrid mouse.[6,15,19]

Recently, a line has been selected from rats of the Hebrew University strain which, when fed a standard laboratory diet, exhibits only a mild degree of the impairment of glucose tolerance and of resistance to insulin. These changes, however, markedly aggravate by the intake of a high sucrose-copper-deficient diet. During sucrose-diet feeding a majority of these animals develop organ changes; retinopathy, and diffuse intercapillary glomerulosclerosis. In hyperglycemic animals, body weight is lower as compared with that of the normoglycemic control line.[20]

B. Nonrodent Species

In addition to rodents, increased incidence of spontaneous diabetes was also observed in other species; in nonhuman primates, dogs, miniature swine, and in rabbits.

Nonhuman primates — Spontaneous diabetes has a 1 to 2 % incidence in a marjority of nonhuman primates. In some colonies of these animals, however, populations have been selected with a substantial higher incidence of the disease, especially in the *Macaca* genus whose most numerous colony and, at the same time, one with the highest incidence rate of diabetes, is the colony of *Macaca nigra* (Black celebes macaques) at the Oregon Regional Primate Research Center, Beaverton, Oregon. In this colony comprising 78 animals, overt diabetes has been demonstrated in 5, and mild diabetes in 28 animals. Overtly diabetic animals exhibit hyperglycemia and impaired glucose is deteriorated with age. The cause of diabetes is probably lesions of the islets of Langerhans, characterized by amyloid deposition. Of the organ complications of diabetes, cataracts and marked atherosclerotic alterations of the aorta, but not retinopathy, have been documented in these animals. As regarding other nonhuman primates, increased incidence of spontaneous diabetes has been reported in *Macaca mullata, Macaca radiata,* and *Macaca fascicularis.*[21]

Yucatan miniature swine — This line has a blood glucose clearance approximately three times lower than that of the control line. It appears that glucose tolerance impairment has a double cause; while decreased insulin secretion is responsible in the majority of animals investigated, in a smaller part of the population the enhanced extraction of insulin by the liver apparently plays a role. Physiological stresses, such as pregnancy, lactation, and intake of a high-fat diet further impairs glucose tolerance making the glycemic curve resembling the diabetic one. Unlike males, females become obese when fed *ad libitum.*[22]

Dogs — Spontaneous diabetes occurs in the dog with a prevalence of 1:200 to 800. Incidence is more frequent in old obese females and, at times, is due to pancreatitis. Temporary diabetes may manifest in estrum. Hereditary diabetes caused by hypoplasy of the islets of Langerhans has been reported in *keeshond dogs.* Diabetes manifests itself between 2 to 6 months of age; diabetic animals are hyperglycemic, hypoinsulinemic, hyperlipemic, and mildly ketoacidotic. Organ complications observed include cataracts, mesangial thickening in the glomeruli, but no lesions of large vessels have been noted.[23] Inherited diabetes occurring between 2 to 5 months of age was also identified in a colony of the *golden retriever.*[24]

Diabetic New Zealand white (NZW) rabbits — These animals were recognized 15 years ago, and by inbreeding, a closed colony of diabetic rabbits was established. Overt diabetes was observed in 19% of the population, abnormal glucose tolerance in 27%, and approximately half of the population did not display any significant changes in glucose utilization. In overtly diabetic animals, hyperphagia, polydipsia, mild ketonemia, and a decreased insulin response to intravenously administered glucose were observed. Since electron microscopic studies revealed hypergranulated B cells of the islets of Langerhans, probably the defect in the insulin-release may be involved. Serum triglycerides, but not serum-free fatty acids and cholesterol, were increased in overtly diabetic animals.[25]

VI. VIRUS-INDUCED DIABETES MELLITUS IN ANIMALS

Recently, increasing attention has been focused on the role of viruses in the pathogenesis of diabetes mellitus. This interest is due to the fact that the cause of insulin-dependent diabetes mellitus (IDDM) is a destruction of B cells of the islets of Langerhans, and viruses are regarded as one of the possible causes. A fact emerging from epidemiological studies is that sometimes there is a coincidence between the occurrence of some viral infections and subsequent incidence of diabetes. More direct evidence regarding the possible role of viruses in the pathogenesis of diabetes was provided by viral infections in animals. A number of viruses have been described which can produce transient, or permanent, morphological changes of B cells, or other pancreatic structures in susceptible animals. Let us briefly mention some models of diabetes induced by viral infections.

Encephalomyocarditis virus (EMC-D) — EMC-D elicits destruction of B cells in mice with subsequent insulinopenia and hyperglycemia. The consequences of infections impairment are not uniform; while in some infected animals only deteriorated glucose tolerance can be demonstrated, others display persistent hyperglycemia with or without ketosis. Morphological changes of the pancreas include lymphocyte infiltration and necrosis of B cells. The morphological changes of the islets of Langerhans are similar to insulitis observed in IDDM. There exists distinct differences in tolerance to this virus among the individual strains of mice; the ob/ob obese mice are particularly sensitive.

Coxsackie virus B4 — A repeatedly passaged virus destroys B cells in some strains of mice with subsequent insulinopenia and hyperglycemia. As in the case of encephalomyocarditis virus infection, there also exists marked differences in the severity of diabetes, with hyperglycemia being only transient in the majority of infected animals. Another analogy in the diabetic syndrome induced by this virus and that by encephalomyocarditis virus is that only some inbred strains are sensitive, with male animals being more sensitive than females. Most of the strains sensitive to encephalomyocarditis virus are also sensitive to Coxsackie virus.

Reoviruses — Due to insulitis, mice infected with reovirus develop mild transient diabetes. The mechanism may probably involve autoimmune mechanisms, since immunosuppressives are capable of preventing, or at least mitigating, the development of diabetes. Injury by reoviruses is not specific for B cells only, this infection leads to polyendocrinopathy.

In addition to these viruses, some other human and animal viruses may impair B cells in vivo and in vitro, such as the viruses of mumps, foot-and-mouth disease, rubella, and of Venezuelan equine encephalitis.[26-28]

Diabetes induced by viral infection has been reported in a colony of guinea pigs. Of morphological changes, degranulation of B cells was observed in the affected animals.[29]

VII. CONCLUSIONS

A number of animal models of surgically or chemically induced diabetes mellitus as well as of genetically transmitted or viral diabetes in rodents and in nonrodent species have been mentioned above.

Needless to say, no one of an array of animal models in question is identical with a very complex human metabolic disorder, diabetes mellitus, manifested by profound alterations of carbohydrate, lipid, protein and mineral metabolism. On the other hand, thousands of experimental studies utilizing animal models of this human disease have substantially contributed to the understanding of both insulin-dependent and -independent human diabetes. There, it is expected that both generally used as well as newly developed animal models of diabetes mellitus will be useful for further elucidation of pathophysiological mechanisms underlying the human disease and for extending the present state of the art. Such information is essential for better control of human diabetes and its organ complications and, perhaps, for the prevention of the disease itself.

REFERENCES

1. **Von Mehring, J. and Minkowski, O.,** Diabetes millitus nach Pankreasextirpation, *Naunyn-Schmiedeberg's Arch. Exp. Pathol. Pharmakol.*, 26, 371, 1890.
2. **Banting, F. G. and Best, C. H.,** The internal secretion of the pancreas, *J. Lab. Clin. Med.*, 7, 251, 1922.
3. **Korec, R.,** Alloxan diabetes, in Experimental diabetes mellitus in the rat, Korec, R., Ed., Publishing House of the Slovak Academy of Sciences, Bratislava, Czechoslovakia, 1967.

4. **Engerman, R. L. and Kramer, J. W.,** Dogs with induced or spontaneous diabetes as models for the study of human diabetes mellitus, *Diabetes,* 31, Suppl. 1, 26, 1982.

5. **Bonner-Weir, S., Trent, D. F., and Weir, G. C.,** Partial pancreatectomy in the rat and subsequent defect in glucose-induced insulin release, *J. Clin. Invest.,* 71, 1544, 1983.

6a. **Bell, R. H., Jr. and Hye, R. J.,** Animal models of diabetes mellitus: Physiology and pathology, *J. Surg. Res.,* 35, 433, 1983.

6b. **Preston, A. M.,** Modification of streptozotocin-induced diabetes by protective agents, *Nutr. Res.,* 5, 435, 1985.

7. **Grodsky, G. M., Anderson, C. E., Coleman, D. L., Craighead, J. E., Gerritsen, G. C., Hansen, C. T., Herberg, L., Howard, C. F., Lernmark, A., Matschinsky, F. M., Rayfield, E., Riley, W. J., and Rossini, A. A.,** Metabolic and underlying causes of diabetes mellitus, *Diabetes,* 31, Suppl. 1, 45, 1982.

8. **Halban, P. A., Bonner, Weir, S., and Weir, G. C.,** Elevated proinsulin synthesis in vitro from a rat model of non-insulin-dependent diabetes mellitus, *Diabetes,* 32, 277, 1983.

9. **Trent, D. F., Fletcher, D. J., May, J. M., Bonner-Weir, S., and Weir, G. C.,** Abnormal islet and adipocyte function in young B-cell-deficient rats with near-normoglycemia, *Diabetes,* 33, 170, 1984.

10. **Kergoat, M., Portha, B., and Picon, L.,** Effect of exogenous insulin on glucose kinetics in rats with noninsulin-dependent diabetes, *Metabolism,* 34, 377, 1985.

11. **Portha, B. and Kergoat, M.,** Dynamics of glucose-induced insulin release during the spontaneous remission of streptozotocin diabetes induced in the newborn rat, *Diabetes,* 34, 574, 1985.

12. **Mossman, B. T., Wilson, G. L., and Craighead, J. E.,** A diabetogenic analogue of streptozotocin with dissimilar mechanism of action on pancreatic beta cells, *Diabetes,* 34, 602, 1985.

13. **Hergerg, L. and Coleman, D. L.,** Laboratory animals exhibiting obesity and diabetes syndromes, *Metabolism,* 26, 59, 1977.

14. **Coleman, D. L.,** Diabetes-obesity syndromes in mice, *Diabetes,* 31, (Suppl. 1), 1, 1982.

15. **Grodsky, G. M., Anderson, C. E., Coleman, D. L., Craighead, J. E., Gerritsen, G. C., Hansen, C. T., Hergerg, L., Howard, C. F., Jr., and Rossini, A. A.,** Metabolic and underlying causes of diabetes mellitus, *Diabetes,* 31, (Suppl. 1), 45, 1982.

16. **Coleman, D. L.,** Other potentially useful rodents as models for the study of diabetes mellitus, *Diabetes,* 31, (Suppl. 1), 24, 1982.

17. **Gerritsen, G. C.,** The chinese hamster as a model for the study of diabetes mellitus, *Diabetes,* 31, (Suppl. 1), 13, 1982.

18. **Like, A. A., Butler, L., Williams, R. M., Appel, M. C., Weringer, E. J., and Rossini, A. A.,** Spontaneous autoimmune diabetes mellitus in the BB rat, *Diabetes,* 31, (Suppl. 1), 7, 1982.

19. **Shafrir, E. and Adler, J.,** Selective responses of spiny mice, Acomys cahirinus and Acomys russatus to affluent diets, in Proc. Int. Workshop on Lessons from Animal Diabetes, Jerusalem, Israel, 1982.

20. **Cohen, A. M., Yanko, L., and Rosenmann, E.,** Interrelationship of genetics and nutrition in the production of diabetes in animals, in *Lessons from Animal Diabetes* Shafrir, E. and Renold, A. E., Eds., John Libbey, London, 1985, 73.

21. **Howard, C. F.,** Non-human primates as models for the study of human diabetes mellitus, *Diabetes,* 31, (Suppl. 1), 37, 1982.

22. **Philips, X. R. W., Panepinto, L. M., Spangler, R., and Westmoreland, N.,** Yucatan miniature swine as a model for the study of human diabetes mellitus, *Diabetes,* 31, (Suppl. 1), 30, 1982.

23. **Engerman, R. L. and Kramer, J. W.,** Dogs with induced or spontaneous diabetes as models for the study of human diabetes mellitus, *Diabetes,* 31, (Suppl. 1), 26, 1982.

24. **Williams, M, Gregory, L., Schall, W., Gossain, V., Bull, R., and Padgett, G.,** Characterization of naturally occurring diabetes in a colony of golden retrievers, *Fed. Proc.,* 110, 740, 1981.

25. **Faas, F. H., Conaway, H. H., and Morris, M. D.,** Plasma lipid in a colony of spontaneously diabetic New Zealand white rabbits, *Biochem. Med.,* 26, 85, 1981.

26. **Rayfield, E. J. and Seto, Y.,** Viruses and the pathogenesis of diabetes mellitus, *Diabetes,* 27, 1126, 1978.

27. **Craighead, J. E.,** Viral diabetes mellitus in man and experimental animals, *Am. J. Med.,* 70, 127, 1981.

28. **Müntefering, H. and Jansen, F. K.,** Virus and experimental diabetes, in *Immunology of Clinical and Experimental Diabetes,* Gupta, S., Ed., Plenum Press, New York, 1984, 73.

28b. **Yoon, J. W. and Ray, U. R.,** Perspectives on the role of firuses in insulin-dependent diabetes, *Diabetes Care,* 8, (Suppl. 1), 39, 1985.

29. **Lang, C. M., Munger, B. L., and Rapp, F.,** The guinea pig as an animal model of diabetes mellitus, *Lab. Anim. Sci.,* 27, 789, 1977.

Chapter 26

WOUND HEALING

D. Štěpán and J. Samohýl

TABLE OF CONTENTS

I. INTRODUCTION

Successful experimental work in wound healing requires a considerable knowledge of general regularities by which the healing process is ruled. This is surveyed in the first part of this chapter. Typical examples of wound healing models are discussed in the second part of this survey.

Healing in general is a reaction of living structures towards damage and may be considered a specific type of inflammation.[1]

The course of healing is to a considerable extent determined already by the nature of the traumatic wound-causing event and by a region of lesioned tissues and disturbed cell integrity.[2] A wound is understood in this context as the contusion or disintegration of the cutaneous cover to a various depth.

II. TYPES OF WOUNDS

For the purpose of experimental work the following categories of healing wounds can be distinguished:

1. Primarily healing wounds — they represent skin defects in which the fringes can quickly integrate with minimum tissue formation (as a rule smooth cut wounds, stitched wounds)
2. Secondarily healing wounds
3. Healing wounds — epithelizing from the cutaneous adnexa
4. Epidermal wounds (combustion wounds of the second degree); only the top layer of the epidermis has been torn off and regeneration ad integrum is materialized from the basal layer

In the experiment we routinely prepare some of the above mentioned wound types. Here we attempt to assess the physiological, biophysical, and biochemical processes that may help proper healing and try to exploit them.[3]

III. PHASES OF THE HEALING PROCESS

In the process of healing we can distinguish various stages in the wound, namely the stages of alteration, exudation-infiltration, proliferation, and the stage of maturation of the scar.[4]

A. Alteration

The process of healing starts with an alteration caused by the trauma intensified by the autolytic processes in the tissues, which destroy cell membranes and circulation.[5] Bleeding contributes to washing away the impurities; activation of the hemocoagulation system prevents blood losses and covers the wound with blood coagulum. Because of the insufficient O_2 supply, metabolism follows the anaerobic pathways which consequently leads to lactic acid and fatty acid formation, overabundance of CO_2, and tissue acidosis.[6] This acidosis is in part compensated by the tissue buffering capacity and by washing off the metabolites by blood. Thus the degree of acidity can be used as a marker of the acuteness of the process. The more acute the reaction to the damage, the less it can be compensated, and the pH values found in the wound are lower.[1,7-10] Another result of the destructive metabolism is the increase of osmotic pressure that results from the decomposition of macromolecules into their individual molecular components and an increase of the extracellular concentration of potassium. Consequently the oncotic pressure inside the wound increases as well.[8]

To compensate for the concentration drop, water is absorbed and an edema arises in the vicinity of the wound. This phase of destructive metabolism of healing (including alteration and exudation — lag phase according to Dumphy[11]) continues roughly until the 4th day, when gradually the phase of constructive metabolism (proliferation and reorganization), starts to prevail. This latter phase lasts from the 2nd or 3rd day until the 14th day.[4,11]

The nervous system reacts by damaging the central stress mechanism; periferaly the tissue kinines are released and changes in the blood vessels occur.[12,13] A short arterial spasm followed by the dilatation of vessels with a turbulent-to-pendulum-like blood flow and stasis puts the transporting function of the blood out of action.

The suppression of blood circulation changes the metabolic conditions of the vessel wall, alongside with its physical and biochemical properties, particularly its permeability. Higher permeability provides for balancing the increased osmotic and oncotic pressure in the wound through the transfer of exudate, one of the causes of the inflammatory infiltration of the wound.[5]

B. Exudation (Infiltration)

Penetration of cellular elements can both be passive (erythrocytes) and active on the basis of chemotaxis with the active participation of the endothelial vessel lining.

The purpose of the inflammatory infiltration (cellulization) is to clean the wound and to establish favorable conditions for its healing.

C. Proliferation

Following the cleaning period activities of constructive anabolism, particularly cellular, proliferation prevails in the clinical picture. Proliferation of endothelial cells of the blood and lymphatic vessels (fibroblasts, myofibroblasts) start to replace the defective tissue. However, this activity is very limited and the defect is always replaced by fibrous scar tissue. The proliferating elements and the extracellular matrix synthesized by them form granulation tissue, the production of which is stimulated by anoxia, acidosis, osmotic changes in the wound, and neuro-humoral influences.[11,12,14]

Eliminating or strengthening of some of these factors during the experiment is demonstrated in the further course of healing as a change in the quality of granulation tissue formed as well as in the kinetics of granulation tissue formation.

The histological structure of the granulation tissue has no stable character. It is subject to dynamic changes in various stages of its development. Also the structure of surface layers of the focus of the growing granulation tissue differs from the basal layers and is greatly influenced by various factors acting upon the organism.[4,12]

As soon as the defective tissue is filled up and inhibition of cellular proliferation occurs, further growth stops; fibroblasts, myofibroblasts, and the fibrous tissues gradually proliferate to the detriment of venal capillaries. The granulation tissue focus changes into a rigid fibrous scar, contracting due to the action of the myofibroblasts.[15]

D. The Stage of Scar Maturation

The stage of scar maturation is characterized by the reduction of cellular activities (both metabolic and proliferative) by a gradual increase of the connective tissue proportion and last but not least also by the contraction activity of myofibroblasts. The granulation tissue gradually change into solid connective tissue with a limited blood supply system. This conversion is characterized by a collagen fiber rearrangement in the sense of a prevailing mechanical tension in the scar.

Healing of wounds interfering with the integrity of the epithelum of the body cavities and of the structures lying beneath is similar to the healing of the organ wounds, i.e., it always heals with a scar made of connective tissue with minimum proliferation of the damaged tissue. The situation in bone wounds is, however, different as under favorable conditions such wounds heal with newly formed bone tissue of full value.[1,9,10]

IV. MOST FREQUENT PROCEDURES USED IN MODELING OF WOUND HEALING

A. Anesthesia

It is imperative to prevent suffering of the experimental animals from painful procedures through anesthesia and to treat them with care during the whole experiment.

For anesthesia, a mixture of 0.01 mg Fentanil (Janssen, Belgium) and of 0.005 mg of Droperidol (Richter, Hungary) per 100 g of body weight intramuscularly can be recommended. However, application of other anesthetics will do the same job. If the animal is to be killed, it should be done in a painless way by letting it inhale diethyl ether in a lethal dose.

B. Infliction of a Standard Cut Wound

Cut wounds are inflicted with a scalpel with a marking indicating the depth of the required wound. The length of the cut should be marked beforehand.

If the experiment does not require standard wound dimensions, we can inflict it according to the anatomic structures and layers.

The process of healing is evaluated histologically, mechanically — by testing the tensile strength (the method is described below), and eventually by measuring the changes in the electrical potential in the wound.

C. Surface Wound

If the intended depth of the wound is not supposed to exceed the thickness of the epidermis. We can use a transplantation bistoury for taking skin grafts (Humby's, Watson's, Cambell's, Martin's knife or dermatomes). These tools enable us to slice off skin grafts of the required thickness so as to inflict a surface wound of the required depth. The technique of inflicting these wounds upon bigger animals is similar to taking Tiersch grafts from man. In cases when we are unable to achieve the required tension of the skin, we form a subcutaneous pocket, insert a flat implement into it (e.g., the handle of a scalpel), and by lifting it we obtain an elevated and tense area of skin ready for removal.

Deeper wounds should be cut with a scalpel and the standard quality of the wounds can be achieved by orienting the wound according to the anatomic layers and structures. The dimensions of the surface should be appropriately marked. The retraction of the margins of the wound and its premature closing can be prevented with the method using polyethylene rings as described below.

D. Following the Healing of the Open Wound with the Help of Polyethylene Rings

If we want to follow the healing of an open wound, namely the growth of granulation tissue, it is necessary to prevent the retraction of the wound. For this purpose we can use rings of inerted material stitched to the margins of the wound along its whole circumference (Figure 1).

The granulation tissue grows from the base of the defect and gradually fills up the space below the cover stitched to the upper margin of the circle. Thus the area of the wound is relatively separated from the environment, thus limiting the environmental influences and making local application of various active substances into the wound and maintaining their concentration more easy. On completing the experiment we extirpate the whole wound area. The results of the experiment can be evaluated macroscopically through visual examination by measuring the thickness of the grown granulation tissue on the cut, histologically or biochemically.[16]

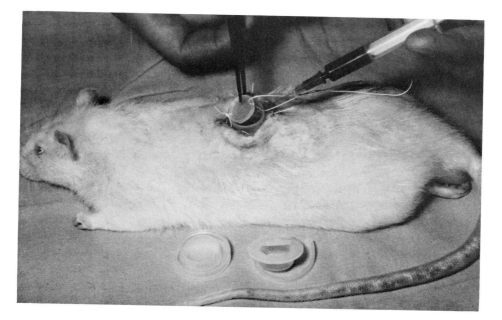

FIGURE 1.   A rat with an implanted polyethylene ring closed with a lid. The granulation tissue is growing into the space beneath the lid and the matter to be tested or the infectious agent is applied with a syringe below the surface of the granulation tissue. Alongside the rat we can see a disc ready for application. For comparison there is also the plug of a medicine bottle in the process of being adapted for use as a lid.

E. Combustion Necrosis

Standard combustion necrosis can be most easily inflicted with the help of a metal disc heated by an electric current. The shape and dimensions of the disc should correspond to the required necrosis and the temperature needed is set through the regulation of the current. The heated disc is pressed to the skin of the animal (the pressure and the time of contact. should be constant) (Figure 2). Combustion necrosis can also be inflicted by submerging part of the skin into a liquid or through radiation (infrared, ultraviolet, X-ray, or laser radiation) applied on the skin. The extent of the necrosis can be determined by shading.[17]

F. Provoking the Process of Healing through the Implantation of a Porous Material into the Subcutaneous Space

This method enables us to follow the speed and quality of infiltration of the implantate by the granulation tissue growing into it from the surrounding area. The implantate is formed by a prism cut from polyurethane foam, rubber, oxycellulose, or other porous materials inserted into the subcutaneous area (Figure 3).

The porous structure should have the character of a spatial network and not that of mutually separated cavities. This is actually the most important condition. The results of the experimental also reflect the immunity reaction towards the implanted foreign material. The implantate can be saturated prior to the implantation with a substance, the effect of which the granulation tissue formation is studied, or the substance can be injected into the implantate percutaneously in definite intervals.

A fibrous casing arises around the implantate in the subcutaneous area and a part of this casing (of varying size) must be extirpated together with the implantate. However, for purposes of biochemial and mechanical evaluation, the casing should be removed exactly along the edge of the implantate. This operation is a little tricky because with pressure it gets deformed and crushed. The best way is to freeze the whole tissue and to cut it with a

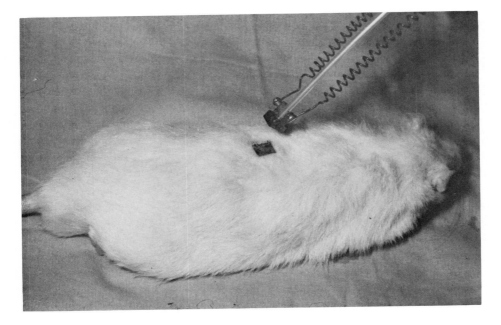

FIGURE 2.   Experimental combustion necrosis inflicted by placing on the skin a metal disk heated up by an electric current.

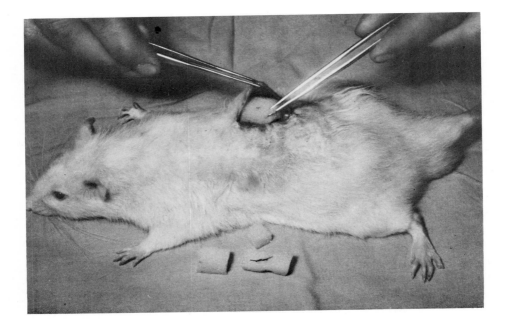

FIGURE 3.   The method of inserting the polyurethane foam prism into the subcutaneous pocket on the back of the rat and specimens of various shapes of prisms, one of them cut in two and connected with a stitch.

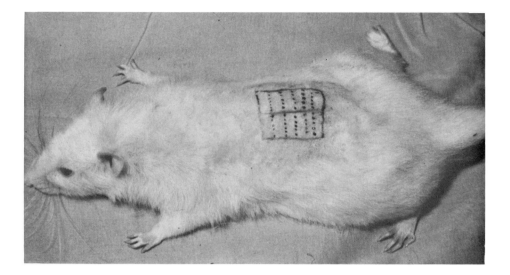

FIGURE 4.   Healing wound on the back of a rat. The marked rectangular part of the skin will be excitated and cut into strips of equal width; the strips will be tested for tensile strength.

razor blade. For mechanical tear testing the implantate should be prepared in a way similar to that used for tensile strength evaluation (see later).

After some time we extirpate the implantate and evaluated it histologically, biochemically, or according to the mechanical properties with the help of tearing machines. According to our experience with the implantation of polyurethane prisms of $7 \times 5 \times 12$ mm implanted into the subcutaneous area on the back of a rat. The time needed for a well-perceptible infiltration of the whole graft of the granulation tissue is about 21 days under physiological conditions. This period varies slightly according to the type of the implanted material. Therefore the time schedule of removing the implantate should be set during a preparatory experiment.[4]

G. The Technique of Evaluating the Healing Tissue (Scar) According to its Tensile Strength

This technique is used for judging the functional properties of the scar or granulation tissue, informing us about the amount of quality of the newly formed collagen.

We cut a strip from the scar or the entire scar so that the longitudinal axis of the scar be perpendicular to the longitudinal axis of the strip (see Figure 4). On testing the strength of the granulation tissue we implant in the above-described way a porous prism-shaped sponge, cut across the connected again with the situational stitches (Figure 3). After the desired time period the sponge is extirpated, the tissue casing is carefully cut off its surface, and the situational stitches are removed. The tissues prepared in this (or similar) way is then caught into the jaws of the teating mechanism (Figure 5) consisting of two clamps attached to the ends of the tested strip of skin or porous matter (sponge). The clamps are being pulled by a force perpendicular to the growth plane of the tissue. The pull force is represented by the weight of the vessel to which water is slowly poured. A a certain moment the two connected surfaces get torn off. The tensile strength of the wound is expressed in N/cm^2. Naturally more sophisticated tearing machines are commercially available, which as a rule are more accurate. In practice the results of individual measurements are considerably scattered (Figure 6). Therefore we have to work with a large set of data which must be statistically processed.[4]

FIGURE 5.   A detailed photograph of a polyurethane sponge infiltrated by the granulation tissue following it by cutting and fixing into the jaws for measuring tensile strength.

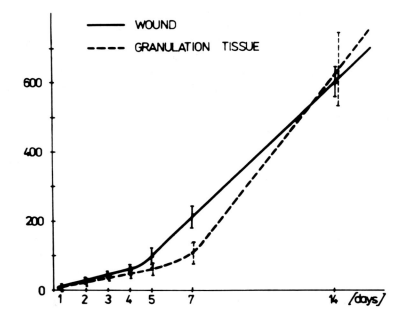

FIGURE 6.   Tensile strength of the granulation tissue infiltrating the polyurethane foam prism and the tensile strength of the healing cut wound in the skin (g/cm^2) with standard deflections.

H. Infected Wound

On modeling the healing of an infected wound it is important to have the same degree of infection in all wounds of the series. The basic principle of the method is that the given bacterial strain is cultivated to a maximum bacterial density. According to the principles of the dilution of bacterial cultures used in microbiology we obtain the bacterial suspension of the required concentration (number of bacteria per milliliter).

The infection is of clinical importance only then, when its massiveness exceeds certain limits. These limits are reached (in man) when the concentration of the bacteria exceeds $10^5/1\ cm^2$ of the wound area. With experimental animals the limit is usually higher, depending on the type of animal and on the virulence of the infection. Materialization of the viral and mycotic infections is done in a similar way.

I. Testing the Agents Locally Applied into the Wound

Besides infections, there is a number of compounds that can be experimentally applied into the wound to follow their effect upon the healing process. The method of following the healing of the open wound with the help of circles can be recommended for this purpose (Figure 1). The procedure using subcutaneous implantation of a porous material saturated with the tested substance can also be used.

J. Experimental Skin Transplantation

For elucidating the mechanisms of the healing process transplantation healing, as the most simple healing, is usually made use of. The technique of skin grafts is described here as it is also applicable for surface wound preparation.

On taking skin in its whole thickness with scalpel often parts of the subcutaneous fat layer are also taken. This layer is especially thin in rats and is very difficult to separate from the corium, but is easy to separate from deeper structures to which it is attached with very loose fibrous tissue.

After cutting out the graft, the subcutaneous fat layer must be carefully cut off, otherwise the transplantation will not heal easily. After removing the subcutaneous fat the graft is spread over a fatty tulle with the skinny surface down and with the bloody layer up. The skin will adhere to the fatty tulle, will not shrink, and can be easily handled.

The graft should be placed on clean and blood-saturated tissue, if possible on healthy fresh granulation tissue. To better show the dividing line between the graft and its surroundings during the process of its healing, turn the transplantate so that the direction of the growth of the hairs on the transplantate will be opposite to the direction of the hairs around it. We stitch the graft to the base with compression or just put it on the base and leave the fatty tulle on top of it. It is recommended transplanting the grafts to the parts of the body out of the reach of the animal's teeth or claws (on its back, near the head). It tries to get rid of the itching or aching at the graft base. In rats, the transplantation should be covered with a foil or with other material stitched to the surroundings. The animal should be separated from the others.

The graft and its surroundings are extirpated in selected intervals and are histologically evaluated (with special regards to the growth of the blood vessels, beginning to appear on the 3rd to 4th day). It is necessary to process a large number of cuts from the same preparation to check the passage of the blood vessel at the contact of the graft with the base. Further, we also evaluate the cell infiltration and the formation of the collagenous grafts and of the base. The graphic illustration of these changes in autotransplantation-isotransplantation and allotransplantation can be seen in Figures 7 and 8.

Equally important are the visual checks determining the percentage and speed of successful healing. Biochemical evaluation is also recommended.

V. METHODS OF FOLLOWING THE HEALING OF WOUNDS

A. Evaluation of the Healing of Wounds Through Visual Checks

The results of this way of evaluation can be affected by errors of the subjective attitudes of the observer. This can be limited to a certain extent through the method of a double blind experiment.

In particular we evaluate the overall condition of the animal, the surrounding of the wound, process of epithelization and retraction of the wound, properties of the exudate, fringes and the bottom of the wound, presence of necroses, proliferation of the granulation tissue, its looks, surface, and blood supply.

B. Histological Evaluation of the Healing

The histological evaluation of the samples taken from the wound in the course of healing is one of the most important and most frequently used approaches.

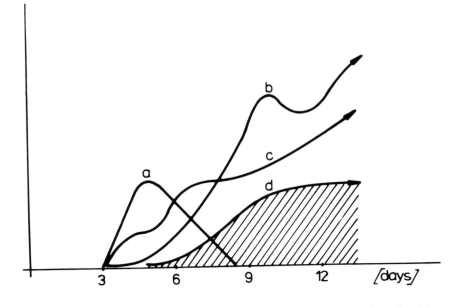

FIGURE 7.   Graphic characteristics of the processes taking place between the graft and its base during autotransplantation — isotransplanation. (a) Monocyte level; (B) degree of vascularization; (c) level of polynuclears; and (d) creation of the fibrous connection between base and graft.

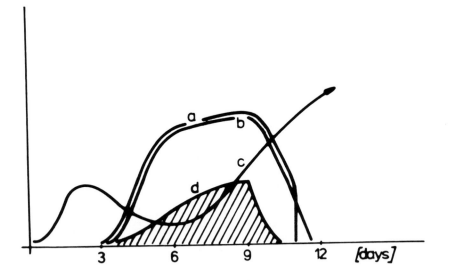

FIGURE 8.   Graphic characteristics of the processes taking place between graft and base during alotransplantation of the first set. (a) Degree of vascularization; (b) concentration of mononuclears; (c) concentration of polymorphonuclears; and (d) formation of fibrous connection between base and graft during so-called pseudo-healing.

In Table 1 we try to enumerate the most characteristic changes to be seen at individual stages of the healing process, including an approximate estimate of their duration. Their enumeration is combined with the characteristics of the measurable biochemical changes. Since the process of healing may be modified within certain limits by a series of factors the above description and its chronology are of orientational character only.

C. Skin-Window Method

This method belongs to the press-on-tape cytological methods and is very suitable for the cytological study of the healing process. Its main asset is the possibility of a recurring sample taken from the same animal in contrast to taking a histological preparate, after which the animal cannot be used for further study.

The method consists in pressing a slide onto the surface wound or excoriation and fixing it there with adhesive plaster or stitches. After some time the slide is replaced by a new one. The cellular elements present in the wound and the tissue detritus adhere to the slide. Following their fixation and dying the cytological evaluation can be carried out.

D. Measuring the Changes in the Electric Potential Around the Healing Wound

An objective evaluation of the healing process in the wound is based on the observation[18] that there is a direct relation between the electric potential changes in the wound and the character and/or intensity of the healing. Electric potential changes are measured with the help of two silver chloride electrodes. The measuring electrode is placed on the skin in the region of the wound (e.g., following of a cut wound is done at spots in the distance 5 mm from both poles of the wound jand 5 mm from the margins on both sides of the wound; the spots can be chosen at random; however, it is necessary to mark them on the skin). The second electrode is placed on the healthy skin, out of reach of local inflammatory changes in the affected area. The electrodes are connected through a millivoltmeter, on which we can read the measured difference of the electric potential between the skin in the area of healing and the skin in the physiological state. Normal variation of electric potential on various places of healthy skin is between 0 and 5 mV; the scar in an advanced stage of healing shows differences up to 10 mV as compared with normal skin. Wound healing per primam without complications shows differences up to 20 mV. Larger differences in electric potential (between 20 and 60 mV) are indicative of advanced inflammation connected as a rule with infection. These changes are observed prior to clinical manifestation of e.g., inflammation (Figure 9). The above values hold for man and for operational wounds following appendectomy.

VI. CONCLUSIONS

The above-described procedures can serve as a basis for designing a particular model experiment. In practice they should be arranged according to the objectives and purpose for which the experiment was set up. It is imperative to respect the physiological rules that govern the process of healing as overviewed briefly at the beginning of this chapter.

Table 1
CHARACTERISTIC CHANGES DURING WOUND HEALING

Time after injury	Microcirculatory changes	Light microscopy level	Electron microscopy level	Biochemical changes
0 hr	Acceleration of blood flow	Cellular debris, blood elements, blood clot	Nuclei desintegration, segregation of nucleoli, disappearance of granular endoplasmatic reticulum	Desintegration of proteins and fats
ALTERATION	Deceleration of blood flow	Spreading of alteration zone	Lysosomes desintegration	Hyperosmosis, hyperoncosis, acidosis
6 hr EXUDATION	Exudate / Swinging motion of blood flow	Tissue edema	Phagocytosis / Proteosynthetic activity of infiltrate inflammatory cells	Increase of albumins, globulins, DNA, fibrinogen, Menkin's inflammatory substances / Release of kinins
INFILTRATION 24 hr	Leukocytes located at the wall / Stasis of blood flow	Blood elements migration / Extravascular blood elements		
72 hr PROLIFERATION	Leukocyte wall adherence	Granulation tissue growing	Increase of metabolic and mitotic cell activity	Increase of collagen, glycosaminoglycans, ground substances
2—4 days WOUND CONSOLIDATION		Vascularization decay and cellulization of the granulation tissue	Fibroblasts and myofibroblasts can be distinguished	Decrease of DNA concentration
and RETRACTION		Final orientation of the collagen fibers	Inhibition of proteosynthesis	Increase of collagen concentration
1 or 2 years				

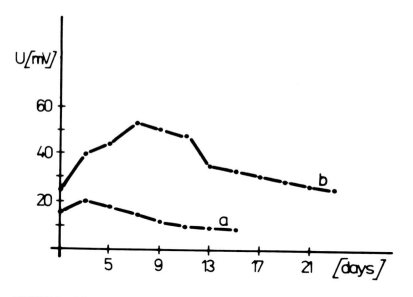

FIGURE 9. Diagram of the differences of the electric potential in cut wound in the skin in the course of healing. (a) The wound healing per primam and (b) infected wound.

REFERENCES

1. **Anderson, W. A. and Kissane, J.,** *Pathology,* Mosby, St. Louis, 1985.
2. **Kubáček, V.,** *Surgery of the Hand* (in Czech), J. E. Purkyně University, Brno, 1982.
3. **Howes, E. L. and Harwey, S. C.,** The clinical significance of experimental studies in wound healing, *Ann. Surg.,* 102, 914, 1935.
4. **Viljanto, J.,** Biochemical basis of tensile strength in wound healing, *Acta Chir. Scand.,* 333, 1, 1964.
5. **McLean, A. E. M., Ahmed, K., and Judah, J. D.,** Cellular permeability and the reaction to injury, *Ann. N.Y. Acad. Sci.,* 116, 986, 1964.
6. **Hunt, T. K., Twoney, P., Zederfeldt, B., and Dumphy, J. E.,** Respiratory gas tension and pH in healing wounds, *Am. J. Surg.,* 302, 144, 1967.
7. **Chvapil, M.,** *Physiology of Connective Tissue,* Butterworth, London, 1967.
8. **Harper, H. A.,** *Review of Physiological Chemistry,* Lange Medical, Los Altos, 1975.
9. **Peacock, E. E. and Van Winkle, X.,** *Surgery and Biology of Wound Repair,* W.B. Saunders, Philadelphia, 1970.
10. **Sedlarik, K. M.,** *WundHeilung,* VEB Gustav Fischer Verlag, Jena, 1984.
11. **Dumphy, J. E. and Udupa, K. N.,** Chemical and histochemical sequences in the normal healing of wounds, *New Engl. J. Med.,* 253, 847, 1955.
12. **Rocha e Silva, M.,** Chemical mediators of the acute inflammation reaction, *Ann. N.Y. Acad. Sci.,* 116, 899, 1964.
13. **Selye, H.,** *The Stress of Life,* McGraw-Hill, Montreal, 1956.
14. **Johnson, F. R. and McMinn, R. M.,** The cytology of wound healing of body surfaces in mammals, *Biol. Rev.,* 35, 364, 1962.
15. **Gabbiani, G., Ryan, G. B., and Majno, G.,** Presence of modified fibroblasts in granulation tissue and their possible role in wound contraction, *Experientia,* 27, 549, 1971.
16. **Štěpán, D., Pivec, P., Dvořáková, H., and Dvořák, V.,** Effect of local application of adenine on the growth of granulation tissue of rats, *Scr. Med. Fac. Med. Univ. Purkynianae Brunensis,* 58, 495, 1985.
17. **Štěpán, D., Bureš, P., Slotová, J., and Pivec, P.,** The effect of salicylic acid and benzoic acid on the debrisement combustion necroses, *Scr. Med. Fac. Med. Univ. Purkynianae Brunensis,* 58, 21, 1985.
18. **Saposnikov, J. G.,** The dynamics of electropotential changes in post-surgery wounds, *Voen. Med. Zh.,* 8, 30, 1980.

Chapter 27

CIRCULATORY, TOXIC, AND TRAUMATIC SHOCK

J. Heller

TABLE OF CONTENTS

Some of the experimental approaches described within this chapter make animals suffer. Certainly they do not conform with the ethics of animal experimentation. The Editors of this volume are of the opinion that painful animal experiments are sometimes necessary for the benefit of man; on the other hand such experiments have to be limited in the number of animals used or the researchers are urged to switch to alternative methodology (e.g., cell or tissue cultures) when possible. In any case the readers are urged to get familiar with the ethical and legislative problems of animal experiments before designing any experiment and those in stress research in particular. Fundamental information in this respect can be obtained from Thelestam, M. and Gunnarsson, A., Eds., The Ethics of Animal Experimentation, Proceedings of the Second CFN Symposium, Stockholm, April 12 to 14, 1985; *Acta Physiol. Scand.*, 128, Suppl. 554, 1986.

I. INTRODUCTION

There is not a fixed definition of circulatory shock. The following examples demonstrate this situation: "shock is a condition in which acute circulatory failure occurs in man because of derangement of circulatory control or loss of circulating fluid".[1] "Shock is a clinical term signifying a progressive cardiovascular collapse with failure of tissue perfusion leading to a profound cellular hypoxia and dysfunction".[2] "Shock is a syndrome resulting from a depression of many functions but in which reduction of the effective circulating blood volume is of basic importance, and in which impairment of the circulation steadily progresses until it eventuate in a state of irreversible circulatory failure".[3] "Shock is a significant and sustained loss of effective circulating blood volume. It eventuates in the hypoperfusion of peripheral tissues, and leads to a deficit in transcapillary exchange function."[4]

Similarly, the classification of circulatory shock is not uniform either. However, four main types can be distinguished:

1. Cardiogenic shock: accompanies cardiac dysfunction, e.g., myocardial infarction
2. Hypovolemic shock: hemorrhagia, severe dehydration, etc.
3. Endotoxic shock: accompanies septic states, etc.
4. Traumatic shock: severe damage to tissues, e.g., "crush syndrome"

In clinical practice, a combination of two, three or all four types is often seen.

The signs of shock very often accompany various situations such as potassium depletion, tetanus, etc. Corresponding models have been developed but not as models of a shock and are therefore not described here.

Numerous earlier models of shock are just on the limit (or even beyond it) of compatibility with ethics of the experimentator-animal relationship. There is no question that in many cases these have largely contributed to our understanding of the mechanisms of changes and have given a clue to adequate therapy. When modified appropriately and provided the animals are spared cruel treatment, some of these models could be helpful even today.

II. MODELS OF CARDIOGENIC SHOCK

The following criteria of cardiogenic shock have been recently suggested by Feola and Stanton:[5] (1) a decrease in mean arterial pressure (MAP) of 30% or more; (2) duration of this decrease for at least 30 min; (3) changes in ECG which are consistent with myocardial infarction; (4) absence of arrhythmias that could account for arterial hypotention; (5) decrease in cardiac output of 50% or more; (6) increase in left ventricular end-diastolic pressure; and (7) elevation of total body lactate. To fulfill all these criteria in an experimental model is difficult; three ways are used, namely ligation of coronary arteries, coronary embolization, and direct damage to the myocardium.

The latter models are rarely used as dogs are usually employed and open-chest surgery is necessary. The following procedures were used (see Reference 5 for review): application of formaldehyde, freezing by ethylchloride, cauterization, application of radom seeds, local cooling with ice, resection of a part of the left ventricle, and replacing by patches of Dacron. In the majority of cases, no shock state fulfilling the above described criteria was achieved. In rats, however, Koo et al.[6] working with electrocoagulation of an area of approximately 5 mm² were completely successful.

In dogs, Salazar[7] introduced without opening of the chest a stainless steel canulla into the lumen of the left coronary or some of its branches. After application of low-grade electric current, the picture of a shock developed. Murphy et al.[8] slightly modified this technique and used dogs treated with bretylium tosylate for blockade of norepinephrine release from neural endings. Typical signs of cardiogenic shock were observed.

III. MODELS OF HYPOVOLEMIC SHOCK

Some types of hypovolemia also occur in shock states accompanying toxic states, trauma etc.; models of these situations will be described later.

A. Models of Hemorrhagic Shock with Constant Blood Pressure ("Pressure Bottle" Model)

Although the "pressure bottle" model was described for the first time by Lamson and DeTurk in 1945,[9] it is firmly connected with the name of Carl J. Wiggers who elaborated it in full detail.[3] In principle, a larger artery (usually the femoral artery) is connected with a bottle placed at a height which corresponds to the level of the MAP chosen. As long as the MAP of the animal is higher than the desired level, blood begins to flow into the bottle; if the MAP decreases below this level, blood is "infused" into the blood stream and the decrease in the MAP is quickly compensated. Heparinization is necessary, anesthesia is not. The majority of dogs survives hemorrhagic hypotension (HH) of 50 to 60 mm Hg for several hours and complete recovery is observed after the retransfusion of the shed blood. With the MAP values of 30 to 40 mm Hg, many animals die despite complete retransfusion. As demonstrated by Wiggers[3] there is a great variability in the survival rate. The description of this "classical" Wiggers[3] model (still very often used) runs as follows: in pentobarbital anesthesia, a canulla is introduced into the femoral artery and the dog is bled by approximately 50 mℓ/ min until the MAP level of 50 mm Hg is achieved. This level is then maintained for 90 min and followed by a further decrease to 30 mm Hg maintained in the dog for another 45 min. During this latter period, blood is usually flowing from the bottle into the animal. If the dog shows some signs of approaching death (sudden decrease of the MAP, respiratory depression, etc.) it is recommended to transfuse quickly the remnant shed blood intravenously. If the course is not complicated by such events, the blood is retransfused after finishing the last 45-min period, slowly (50 mℓ/min). The MAP is normal or almost normal 30 to 60 min after retransfusion; then after, it declines quickly or more gradually until death occurs. This happens in 82% of dogs within 6 hr after retransfusion. Many variants of this "classical" procedure do exist; these variants, however, differ in a quantitative manner.

B. Models Involving Withdrawal of a Constant Volume of Blood

The amount of blood withdrawn is expressed either as a percentage of calculated blood volume or body weight. Even with these techniques, Wiggers[3] pioneered the research work. He observed that a healthy animal in every case survives the withdrawal of 30 to 40% of the calculated blood volume. If the amount withdrawn exceeds 40%, the survival rate depends on the deepness of hypotension and its duration, just as with the "pressure bottle" technique.

Another important fact is that blood withdrawal until 15/mℓ/kg body mass is usually not accompanied with a MAP drop;[10] higher amounts are usually accompanied with haemorrhagic hypotension. The degree of this hypotension depends on the state of the animal: in conscious animals it is smaller than in anesthetized ones, in dogs with chloralose smaller than with pentobarbital anesthesia;[10,12] the combination of morphin-pentobarbital is accompanied by profound HH even after withdrawal of 20 mℓ blood/kg body mass.[13] The velocity of withdrawal does not play any important role.[14]

IV. MODELS OF TOXIC SHOCK

In these models, two types are often distinguished: septic shock in which living microorganisms are injected, and endotoxic shock which is introduced by the administration of bacterial endotoxins. As endotoxins are the cause of shock in both cases, the mechanism of the shock state is practically the same. However, Hinshaw[15] comparing the effects of the injection of both living *E. coli* and their endotoxins concluded that the former technique is more relevant to the clinical situation as the toxins are released much more slowly. A simple endotoxin injection lacks the so-called "warm phase" in which the blood flow to many organs is even higher than normal. Postel et al.[16] showed that the plasma endotoxin levels must be three orders higher after simple injection than in animals with septicemia, an important quantitative difference.

Experimental models of toxic shock are associated with the work of Weil and co-workers.[17] Since their time, an explosion of various types of these models has occurred. In principle, almost all possible toxins have been injected into every laboratory animal (for review see References 15 and 18).

For the sake of approximation to the clinical situation, some additional models have been developed. Richardson et al.[19] ligated the colon cecum in dogs; later, spontaneous perforation with septical peritonitis occurred. Stone et al.[20] perforated artificially this part of the gut with a similar consequence. Nichols et al.[21] administered pieces of human feces into the peritoneal cavity.

V. MODELS OF TRAUMATIC SHOCK

There are many types of these models and only the most important will be described here. For an excellent review the reader is directed to Reference 22.

Most models discussed in this section are outright incompatible with ethics governing handling of animals since they involve atrocities which necessarily are a thing of the past. The models are briefly mentioned to provide background information.

A. Models Involving Mechanic Trauma of Bones and Muscles

Many of these models are today very rarely used. They could be found in Reference 3. A very brief description of the Best and Solandt "hammer" model from 1940[23] is as follows: a padded hammer of 800-g weight is falling 1200 times on the flexors of every hind limb of the dog. The method could be quantificated by the weight of the hammer, height from which it is falling, and the number of falls. The size and weight of dogs must be held constant. Shock state develops rather slowly but constantly even if the muscles are not damaged so far that they are not available for movements. In rats, Cuthbertson[24] constructed a device which falls from a constant height on the extremities. Shock state develops sooner in dogs and two phases could be distinguished during its course. A phase of a worsened circulatory function ("ebb" phase), and a recovery period ("flow" phase) accompanied with higher oxygen consumption and metabolic activity. To produce a comparatively mild form of traumatic shock in rats, the fracture of both femurs is a sufficient act.[25]

B. Tumbling Shock

This is a very well known technique named "Noble-Collip-drum"-method. Its first description in 1942 was designed for mice,[26] soon it was modified to also be used in rats and guinea pigs.[27] It is a closed rotating wheel similar to the wheel of fortune; the animals trying in vain to keep on the sides are repeatedly falling down. The method is therefore based on repeated falls from a constant height; quantification is thus very simple. The cause of shock is a multiorgan trauma combined with hemorrhagia and later even with sepsis. From the very beginning, Noble and Collip[26,27] regularly observed development of tolerance which is based on the "training" of abdominal muscles; tolerance to other types of trauma does not exist.[28] Anesthetized animals have a better survival rate than the conscious ones.[29]

C. Models of "Crush Syndrome"

Duncan and Blalock[30] successfully simulated the clinical picture of the "crush syndrome" by loading an anesthetized dog with a weight of 227 kg for 5 hr. The method has been modified by an increase in weight and prolongation of duration of loading; the extremities were evidently not perfused. After cessation, a picture of a heavy shock state developed with edema of extremities but without hemorrhage.[31]

D. "Tourniquette Shock"

This was also a very popular model based on ischemization of extremities with a tourniquette-like bandage,[32] tubber band,[32,33] string,[34] or mechanical clamps around a limb.[35,36] All laboratory animals were used. The time period of application of bands lasted from 2 to 15 hr; a bandaging of both hind limbs for 3 to 4 hr leads in the great majority of cases to death within 24 hr. The heavy shock state is accompanied by a decrease in the body temperature, hemoconcentration, hypotension, and edema in extremities used.[37] Just as in the crush syndrome, the exact mechanisms of the shock is not known.

E. Burn and Scald Models

Similarly as in the crush syndrome and tourniquette shock, these models exhibit a decrease in the body temperature, hemoconcentration, redistribution of body fluids, finally leading to a circulatory failure. Scalding is carried out by immersion into a water bath with a temperature of 70 to 90°C. Different body surface areas are exposed to the scalding bath. Mostly, 20% of the surface area are used in rats for 60 sec;[38] similar parameters have been also used in dogs.[39]

Walker and Mason[40] developed a standard burn model which uses a template to delineate the amount of exposed skin. Wolfe and Miller,[41] working with guinea pigs, used in-dwelling venous and arterial catheters for monitoring pressure changes during scalding of 50 to 70% of the body surface area. This technique was recently modified and precisely quantified.[2]

CONCLUSIONS

Many models of circulatory shock strongly resemble clinical states: hypovolemia (hemorrhage, severe dehydration, etc.), septic, and endotoxic shock. Modeling of cardiogenic forms of shock involves more difficulties. The majority of traumatic shock models are incompatible with todays ethics governing the experimentator-animal relationship: here, it is crucial to carefully consider the real importance of each particular experiment and opt for traumatic and painless models.

REFERENCES

1. **Hinshaw, L. B.,** Cardiovascular dysfunction in shock: an overview with emphasis on septic shock, in *Circulatory Shock: Basic and Clinical Implications,* Janssen, H. F. and Barnes, C. D., Eds., Academic Press, New York, 1985, 1.
2. **Cook. J. A., Olanoff, L. S., Wise, W. C., Tempel, G. E., Reines, H. D., and Halushka, P. V.,** Role of eicosanoids in endotoxic and septic shock, in *Circulatory Shock: Basic and Clinical Implications,* Janssen, H. F. and Barnes, C. D., Eds., Academic Press, New York, 1985.
3. **Wiggers, C. J.,** *Physiology of Shock,* Oxford University Press, New York, 1950.
4. **Altura, B. M., Lefer, A. M., and Schumer, W.,** Eds., *Handbook of Shock and Trauma,* Vol. 1, Raven Press, New York, 1983.
5. **Feola, M. and Stanton, M.,** Experimental cardiogenic shock, in *Handbook of Shock and Trauma,* Vol. 1, Altura, B. M., Lefer, A. M., and Schumer, W., Eds., Raven Press, New York, 1983, 437.
6. **Koo, A., Tse, T. F., and Yu, D. Y. C.,** Hepatic microvascular effects of terbutaline in experimental cardiogenic shock in rats, *Clin. Exp. Pharmacol. Physiol.,* 6, 495, 1979.
7. **Salazar, A. E.,** Experimental myocardial infarction. Induction of coronary thrombosis in intact closed-chest dogs, *Circ. Res.,* 9, 1351, 1961.
8. **Murphy, S. D., Charette, E. J. P., and Lynn, R. B.,** Experimental coronary artery thrombosis for production of cardiogenic shock, *Can. J. Surg.,* 13, 189, 1970.
9. **Lamson, P. and DeTurk, W.,** Studies on shock induced by hemorrhage, *J. Pharmacol. Exp. Ther.,* 83, 250, 1945.
10. **Share, L.,** Control of plasma ADH titres in hemorrhage: role of atrial and arterial receptors, *Am. J. Physiol.,* 215, 1384, 1968.
11. **Zimpfer, M., Manders, W. T., Barger, A. C., and Vatner, S. F.,** Pentobarbital alters compensatory neural and humoral mechanisms in response to hemorrhage, *Am. J. Physiol.,* 243, H713, 1982.
12. **Gagnon, J., Moore, J., Butkus, D., Verma, P., Reid, A., and Lake, C. R.,** Renal ischemia: effect of combined hemorrhage (H) and supra-renal aortic constriction (AC), *Fed. Proc.,* 41, 1009, 1982.
13. **Abel, F. L. and Murphy, Q. R.,** Mesenteric, renal, and iliac vascular resistance in dogs after hemorrhage, *Am. J. Physiol.,* 202, 978, 1962.
14. **McNeill, J. R., Stark, R. D., and Greenway, C. V.,** Intestinal vasoconstriction after hemorrhage: roles of vasopressin and angiotensin, *Am. J. Physiol.,* 219, 1342, 1970.
15. **Hinshaw, L. B.,** Comparison of responses of canine and primate species to bacteria and bacterial endotoxin, in *Shock in Low and High Flow States,* Forscher, B. K., et al., Eds., Excerpta Medica, Amsterdam, 1972, 245.
16. **Postel, J., Schloerb, P. R., and Furtado, D.,** Pathophysiologic alterations during bacterial infusion for the study of bacteremic shock, *Surg. Gynecol. Obstet.,* 141, 683, 1975.
17. **Weil, M. H., MacLean, L. D., Visscher, M. B., and Spink, W. W.,** Studies on the circulatory changes in the dog produced by endotoxin from gramnegative microorganisms, *J. Clin. Invest.,* 35, 1191, 1956.
18. **Weil, M. H.,** Current understanding of mechanisms and treatment of circulatory shock caused by bacterial infections, *Ann. Clin. Res.,* 9, 181, 1977.
19. **Richardson, J. D., Fry, D. E., Arsdall, L. V., and Flint, L. M., Jr.,** Delayed pulmonary clearance of gram-negative bacteria: the role of intraperitoneal sepsis, *J. Surg. Res.,* 26, 499, 1979.
20. **Stone, A. M., Stein, T., LaFortune, J., and Wise, L.,** Effect of steroids on the renovascular changes of sepsis, *J. Surg. Res.,* 26, 565, 1979.
21. **Nichols, R. L., Smith, J. W., and Balthazar, E. R.,** Peritonitis and intraabdominal abscess: an experimental model for the evaluation of human disease, *J. Surg. Res.,* 25, 129, 1978.
22. **Miller, H. I. and Whidden, S. J.,** Burns and trauma, in *Handbook of Shock and Trauma,* Vol. 1, Altura, B. M., Lefer, A. M., and Schumer, W., Eds., Raven Press, New York, 1983, 413.
23. **Best, C. H. and Solandt, P. Y.,** Studies in experimental shock, *Can. Med. Assoc. J.,* 43, 206, 1940.
24. **Cuthbertson, D.,** The metabolic changes consequent on injury, *Z. Tierphysiol.,* 12, 259, 1957.
25. **Cleghorn, R. A.,** Studies of shock produced by muscle trauma. III. The effect of serum, isinglass, glucose, certain salts, and adrenal cortical hormones on survival, *Can. J. Res. Med., Sci.,* 25, 86, 1947.
26. **Nobel, R. L. and Collip, J. B.,** A quantitative method for the production of experimental traumatic shock without hemorrhage in unanaesthetized animals, *J. Exp. Physiol.,* 31, 187, 1942.
27. **Noble, R. L.,** The development of resistance by rats and guinea pigs to amounts of trauma usually fatal, *Am. J. Physiol.,* 138, 346, 1943.
28. **Manning, J. W., Jr. and Hampton, J. K., Jr.,** Utilization of oxygen by normal and trauma-resistant rats following trauma and exposure to hypoxia, *Am. J. Physiol.,* 188, 99, 1957.
29. **Vigaš, M. and Németh, S.,** . The effect of pentobarbital anaesthesia on the metabolic reaction and resistance of rats traumatized in the Nobel-Collip-drum, *Physiol. Bohemoslov.,* 21, 149, 1972.
30. **Duncan, G. W. and Blalock, A.,** Uniform production of experimental shock by crush injury: possible relationship to clinical crush syndrome, *Ann. Surg.,* 115, 684, 1942.

31. **Swingle, W. W., Kleinberg, W., and Hays, H. W.**, Study of gelatin and saline as plasma substitutes, *Am. J. Physiol.*, 141, 329, 1944.
32. **Root, G. T. and Mann, F. C.**, Experimental study of shock with special reference to its effect on the capillary bed, *Surgery*, 12, 861, 1942.
33. **Koletsky, S. and Klein, D. E.**, Arterial pressure in tourniquet shock, *Proc. Soc. Exp. Biol. Med.*, 91, 486, 1956.
34. **Shipley, E. G., Meyer, R. K., and McShan, W. H.**, Shock produced by application of tourniquets to hind limbs of rats, *Proc. Soc. Exp. Biol. Med.*, 60, 340, 1945.
35. **Green, H. D., Dworkin, R. M., Antos, R. J., and Bergeron, G. A.**, Ischemic compression shock, with an analysis of local fluid loss, *Am. J. Physiol.*, 142, 494, 1944.
36. **Hamilton, J. J. and Haist, R. E.**, Studies on experimental shock in dogs, *Can. J. Res. Med. Sci.*, 23, 89, 1945.
37. **Stoner, H. B.**, Critical analysis of traumatic shock models, *Fed. Proc.*, 20 (suppl.9), 38, 1961.
38. **Loew, D. and Meng, K.**, Acute renal failure in experimental shock due to scalding, *Kidney Int.*, 10 (suppl.6) 81, 1976.
39. **Kirkebø, A., Hangan, A., and Tyssebotn, I.**, Blood flow heterogeneity in the renal cortex during burn shock in dogs, *Acta Physiol. Scand.*, 123, 205, 1985.
40. **Walker, H. L. and Mason, A. D.**, A standard animal burn, *J. Trauma*, 8, 1049, 1968.
41. **Wolfe, R. R. and Miller, H. I.**, Burn shock in untreated and saline-treated guinea pigs. Development of a model, *J. Surg. Res.*, 21, 269, 1976.
42. **Adams, H. R., Baxter, C. R., Keesler, D. D., Parker, J. L., and Senning, R. C.**, Hemodynamics during burn shock in guinea pigs, *Circ. Shock*, 7, 192, 1980.

Index

INDEX

Table 4
OCCURRENCE OF
LIPOFUSCIN IN TISSUES

Neurons
Neuroglia
Retinal pigment epithelium
Heart
Skeletal muscle
Liver
Reproductive organs
  Testes
  Prostate gland
  Epididymus
Endocrine glands
  Thyroid
  Adrenal gland
Salivary glands and digestive tract
Urinary system
Connective tissue

Note: For detailed information, see Reference 50 and the references mentioned therein.

substantia nigra and locus coeruleus.[41] If the melanin component of the neuromelanin granule is oxidatively solubilized, the lipofuscin component, as revealed by its fluorescence in the near-ultraviolet can be histochemically characterized and further subdivided into lipoprotein-like and unbound lipid components.

B. Histochemical Methods

Sudan stains (Sudan III) were introduced early in the history of lipopigments. Nile blue has been also used extensively, but today the periodic acid-Schiff (PAS) technique is preferred. Peracetic and performic acid oxidations in conjugation with the Schiff reagent also constitute effective stains for the pigment.

Of the other stains that have been used effectively in visualizing lipofuscin are neutral red, Sudan black B, the Schmore reaction (reduction of ferric ferricyanide to Prussian blue), aldehyde-fuchsin, and the Ziehe-Neelsen (acid-fast-carbol fuchsin) method. There are additional stains such as methylene blue, toluidine blue, thionine, gallocyanin, and cresyl violet which all appear to stain lipofuscin in paraffin sections. Lipofuscin also stains with silver nitrate and iodine, but these methods offer no advantage over the Sudan stains. Also, there are reports that they give a less distinct picture of the pigment than other methods, particularly the PAS technique. The pigment granules can be also visualized in semithin (0.5 to 4 μm) sections with toluidine blue.[42]

The staining properties of lipopigments (lipofuscin) are, of course, the result of certain chemical characteristics of the pigment. For example, the positive reaction of lipofuscin to the Prussian blue test appears to be due to the reducing groups, perhaps carbonyls, present in the pigment; similarly, the PAS procedure is due to the formation of indolyl sulphonic acids in the pigment. The selection of stains applicable here is quite large and the reader is directed to histochemistry textbooks when seeking more details. As a typical example, we present here the summary of histochemical properties of the adrenal lipopigment, which may well be applicable for studying the pigment accumulation in other tissues as well (Table 5).

Table 5
SUMMARY OF HISTOCHEMICAL PROPERTIES OF ADRENAL PIGMENT

Test	Method	Finding
Lipids	Sudan black	$+++$
Choline-containing phospholipids	Acid haematein	$++$
Plasmalogen phospholipids	Plasmal reaction	$+$
Phospholipids after acetone	OTAN	$++$
Phosphoglycerides	Gold hydroxamate	$\pm$
Sphingomyelin	NaOH-OTAN	$++$
Sulfatide and phosphoglycerides	Nile blue sulfate	$++$
Sulfatide	Cresyl violet-HOAc	$+++$
Cholesterol	Perchloric acid-naphthoquinone	$+++$
Unsaturated lipids	Primary fluorescence	$++?$
Unsaturated lipids	Performic acid-Schiff, Br-Ag	$+$
Unsaturated lipids	OsO_4	$++$
Free fatty acids	Cu^{++}-rubeanic acid	$+$
Protein	Ninhydrin-Schiff	$+$
Protein	Hg^{++}-bromophenol blue	$+$
Acid fast groups	Long-Ziehl-Neelsen	$++$
Acid groups (RNA?)	Methyl green pyronin	$+$
Acid groups (RNA?)	Azure B	$+$
Acid groups (RNA?) after RNAase	Azure B	$\pm$
Basic groups	Fast green stainability (pH 4)	$\pm$
Vicinal poly-OH groups	PAS	$+++$
Vicinal poly-OH groups after diastase	PAS	$+++$
Fe^{+++}	$Fe(CN)_6$	$+$
Acid phosphatase	Azo-dye	$\pm$
Acid phosphatase	Gomori lead	$++$
$NADH_2$-tetrazolium reductase	Tetra NBT	$\pm$
β-Glucuronidase	Ferric hydroxyquinoline	$\pm$
N-acetyl-β-glucosaminidase	Naphthol AS-BI	$++$
Peroxidase	3-3′ DAB medium	$+$
Peroxides (?)	Indophenol-HCl	$+$

C. Lipopigment Isolation Methods

Methods for lipopigment isolation were described for brain tissue, heart, and liver. The pioneering work in this respect was done by Siakotos and co-workers[43] and their procedures are briefly summarized in the following paragraphs.

In the isolation of brain lipofuscin, frozen brain samples are thawed overnight at 4°C, and the tissue mass is cut into sections of about 2 to 4-cm cubes and ground and blended in solution No. 1 (0.4 M sucrose, 0.05 M Tris base) for 20 to 30 sec. at a low speed (14,000 rpm). The proportion of the brain to the fluid is maintained at 0.5 kg/1ℓ of the solution (No. 1). The suspended brain tissue is passed through a Parr cell disruption bomb. At this stage the pH of the mixture is between 7.0 and 8.5. Then the mixture is centrifuged at 27,000 g for 30 min. The supernatant fluid and brown pigment on the upper surface of the centrifuge bottle are pooled, and the pellet is collected without dilution. Then the centrifuge bottles are rinsed with minimum solvent, and the rinsings resuspended with the pellet (by homogenization, five to six passes, 1200 rpm) and centrifuged at 170,000 g. The small quantity of pigment and supernatant fluid is collected and pooled with the supernatant obtained in the first step. The residual pellet arising from the second centrifugation is discarded. Combined supernatants obtained in the first and the second extraction are con-

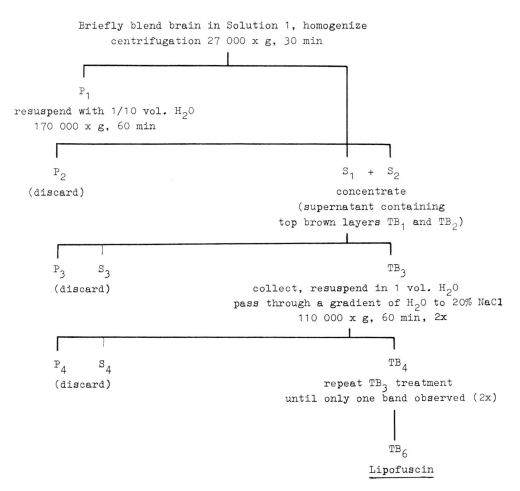

FIGURE 3. Flow scheme of brain lipofuscin isolation. (From Siakotos, A. N. and Munkres, K. D., in *Age Pigments*, Sohal, R. S., Ed., Elsevier, Amsterdam, 1981, 181. With permission).

centrated by repeated centrifugation (170,000 *g* for 60 min), and the floating brown pigment is collected by aspiration. The aspired material (TB$_3$) is resuspended in solution No. 1, homogenized, and centrifuged again. This process is continued until the total volume of the supernatants is decreased to 200 to 600 mℓ. The resulting pellets (P to P$_3$) are discarded and so is the supernatant fluid from the last centrifugation step (S$_3$). The aspired pigment enriched layer (TB$_3$) is resuspended in distilled water (equal volume of the aspired material) and layered on a linear gradient of distilled water to 20% NaCl and centrifuged at 110,000 *g* for 60 min; this step is repeated twice. Each time the top brown layer is aspired, resuspended in equal volume of distilled water, layered on the NaCl gradient, and centrifuged repeatedly until only a single brown band is observed on top of the centrifugation tube. After the first two centrifugation steps, the subsequent ones are done at 285,000 *g* for 60 min. The lipofuscin preparation is finally examined by phase contrast and fluorescence microscopy and judged to be pure when free of contaminating nonfluorescent substances. An alternative procedure to this just described was published also by Siakotos and Munkies;[43] however, the principle, i.e., fraction centrifugation of the pigment material, remains unaltered (Figure 3).

Two other tissues have been used for lipofuscin pigments besides the brain, namely the heart and liver. An outline of the isolation of heart lipofuscin is presented in Figure 4. The

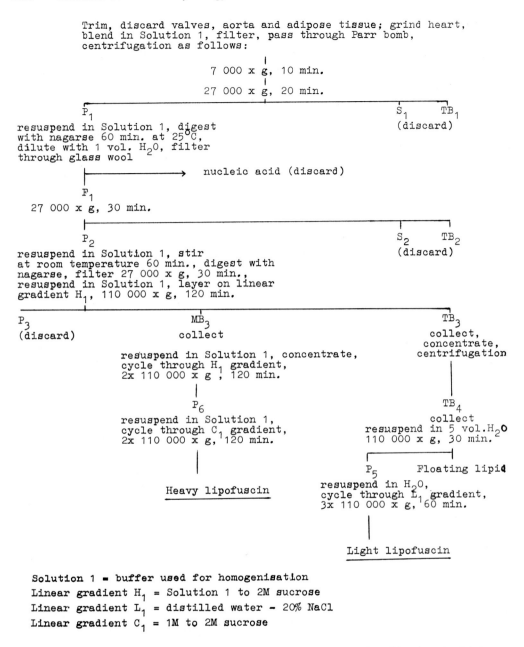

Trim, discard valves, aorta and adipose tissue; grind heart,
blend in Solution 1, filter, pass through Parr bomb,
centrifugation as follows:

7 000 x g, 10 min.

27 000 x g, 20 min.

P₁   S₁   TB₁
resuspend in Solution 1, digest   (discard)
with nagarse 60 min. at 25°C,
dilute with 1 vol. H₂O, filter
through glass wool

⟶ nucleic acid (discard)

F₁
27 000 x g, 30 min.

P₂   S₂   TB₂
resuspend in Solution 1, stir   (discard)
at room temperature 60 min., digest with
nagarse, filter 27 000 x g, 30 min.,
resuspend in Solution 1, layer on linear
gradient H₁, 110 000 x g, 120 min.

P₃   MB₃   TB₃
(discard)   collect   collect,
                            concentrate,
      resuspend in Solution 1, concentrate,   centrifugation
      cycle through H₁ gradient,
      2x 110 000 x g , 120 min.

           P₆   TB₄
      resuspend in Solution 1,   collect
      cycle through C₁ gradient,   resuspend in 5 vol. H₂O
      2x 110 000 x g, 120 min.   110 000 x g, 30 min.

                                P₅   Floating lipid
       Heavy lipofuscin   resuspend in H₂O,
                             cycle through L₁ gradient,
                             3x 110 000 x g, 60 min.

                            Light lipofuscin

Solution 1 = buffer used for homogenisation
Linear gradient H₁ = Solution 1 to 2M sucrose
Linear gradient L₁ = distilled water - 20% NaCl
Linear gradient C₁ = 1M to 2M sucrose

FIGURE 4. Flow scheme for heart lipofuscin isolation. (From Siakotos, A. N. and Munkres, K. D., in *Age Pigments*, Sohal, R. S., Ed., Elsevier, Amsterdam, 1981, 181. With permission.)

procedure has been originally worked out also by Siakotos et al.[43,44] Frozen heart samples are thawed, adipose tissue, aortae, valves, etc. are removed, and the remaining tissue is cut into 2 to 5-cm cubes. Next the material is grinded and blended in a 0.4 *M* sucrose-0.05 *M* Tris base. The pH of the resulting solution is adjusted to 8.0 and the resulting mixture is diluted to 750 mℓ/200 g of the heart tissue. The homogenate is next filtered through a 20-mesh sieve to remove connective tissue and cellular debris and disrupted in a Parr cell bomb. The pressure-treated homogenate is centrifuged at 7000 *g* for 10 min to break the foam that arises after pressure treatment. The defoamed homogenate is mixed and centrifuged again at 27,000 g for an additional 20 min. The floating fat layer (TB_1) and the supernatant (S_1) are discarded, and the resulting pellet (P_1) is resuspended in 500 mℓ of the original homogenization solution to which 25 mg of nagarse per adult heart was added. The mixture is digested with stirring for 60 min at room temperature (25°C); during this period pH is kept constant at 7.5 to 8.0. The resulting digest is diluted with 1 volume of distilled water, filtered through a 20-mesh sieve, and purified by two subsequent filtrations through glass wool. The resulting filtrate (F_1) is then centrifuged at 27,000 g for 30 min. In this stage a pellet (P_2), supernatant (S_2), and a floating lipid layer (TB_2) are obtained. S_2 and TB_2 are discarded, and the pellet P_2 is resuspended in the homogenizing solvent and homogenized thoroughly (at 1200 rpm, twice); next the material is redigested with nagarse under identical conditions as described above. The filtered digest is centrifuged at 270,000 g for 30 min to obtain the third pellet. This is resuspended in the homogenization solution, layered on a linear sucrose gradient (1.0 to 2.0 M sucrose), and centrifuged for 120 min at 110,000 g. After this centrifugation, a diffuse top-pigmented band is observed which is collected by aspiration, reconcentrated by adding one quarter of the aspired volume of 2 M sucrose, and centrifuged at 110,000 g for 15 min again. This yields a highly concentrated floating layer of pigmented material (TB_4); next, this is resuspended in 5 volumes of distilled water, centrifuged at 110,000 g for 30 min to remove the ''floating lipid'' fraction; the pellet from this step (P_5) is collected, resuspended in distilled water, and cycled three times through a distilled water-20% NaCl gradient. In each of these steps the floating layer of the brown pigment is collected, finally yielding a lipofuscin preparation.

As indicated in the flow sheet of this preparation there is a middle brown pigment layer (MB_3) present during the homogenizing solution-2 M sucrose gradient separation. This band is collected by aspiration, diluted with 1 volume of distilled water and concentrated as a pellet by centrifugation at 27,000 g for 30 min. The material can be recycled twice more through the same gradient (after being homogenized first) to finally yield the pellet P_6. This is resuspended in the homogenization solvent and cycled twice through the 1 to 2 M sucrose gradient containing 25% CsCl. The middle brown band is collected, yielding the fraction of the so-called heavy lipofuscin.

It is evident that the procedure for heart lipofuscin differs from that for brain lipofuscin in the need to disrupt the heart tissue by nonspecific proteolysis (nargarse). In the preparation scheme for liver lipofuscin, one has to take care about discarding the fraction of nucleic acids as well. The scheme is evident from Figure 5.

Basically, frozen liver samples are cut into cubes, grinded, and homogenized in 0.4 M sucrose-0.05 M Tris base (500 g of liver per 1 ℓ of homogenization solution). The mixture is filtered as in the preceding case, passed through a cell disruption bomb, and the resulting homogenate is centrifuged at 7000 g for 10 min to break the arising foam. Then centrifugation is continued at 27,000 g for 30 min yielding pellet P_1 and supernatant S_1. The pellet is collected and resuspended in the starting homogenization solution and 100 mg nagarse per 1 kg of the starting material is added. The mixture is digested 1 hr at room temperature, and filtered twice through a 20-mesh sieve and glass wool to remove the precipitated nucleic acids. The filtrate (F_1) is then concentrated by centrifugation at 27,000 g for 30 min, after which the supernatant (S_2) is collected and combined with the supernatant from the first step (S_1), while pellet P_2 is resuspended in an equal volume of 25% NaCl, homogenized, stirred at room temperature and layered over 0 to 2 M sucrose gradient (homogenization solution, centrifugation 110,000 g for 120 min). During this latter step pellet P_3 results, a middle band of brown pigment (MB_3), and a top band of brown pigment (TB_3). The MB_3 zone and P_4 are collected, resuspended in 1 volume 20% NaCl and recycled through the above sucrose gradient, until the top band is no longer visible. This final middle band is collected and desinged as heavy liver lipofuscin. The top band TB_4 is pooled with the top band TB_3, concentrated by adding one quarter of the pooled volume of 2 M sucrose, and centrifuged

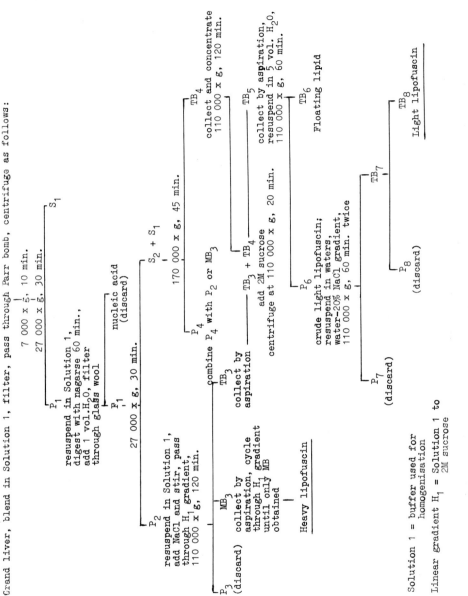

FIGURE 5.   Flow scheme for liver lipofuscin isolation. (From Siakotos, A. N. and Munkres, K. D., in *Age Pigments*, Sohal, R. S., Ed., Elsevier, Amsterdam, 1981, 181. With Permission.)

to obtain the concentrated TB_5 band. This band is centrifuged at 110,000 g for 20 min, resuspended in 5 volumes of water, and centrifuged at 110,000 g for 60 min to remove the "floating lipid" particles (TB_6). The arising light lipofuscin pellet (P_6) is resuspended in distilled water and cycled twice through a linear water-20% NaCl gradient (110,000 g for 60 min). The top bands arising in each step are collected, and the middle bands are cycled until no more top band is observed. The final isolate (TB_7) is collected by aspiration, resuspended in distilled water, and purified by centrifugation at 110,000 g for 60 min to yield TB_8.

D. Occurrence of Age Pigments (Lipofuscin) in Tissues

Although lipofuscin accumulation has been widely viewed as an age-associated process, it is well known that in some tissues its accumulation occurs in early life. This, however, refers to neurons and the brain.[41,67] As neuroscience methods are not involved in this monograph, let us limit ourselves to the fact that the accumulation and regional distribution of lipofuscin in the central nervous system is differential. More recently it was shown that lipofuscin or lipofuscin-like pigments occur also in the retinal pigment epithelium.[47,68]

However, there is also a distinct extraneural accumulation of lipofuscin. This occurs in the heart, skeletal muscle, liver, reproductive organs (testes, prostate gland, epididimus), and some endocrine glands as thyroid, adrenals, salivary glands, digestive tract, and in the urinary system and connective tissue.[70]

E. Ways of Modifying Lipofuscin Accumulation

In principle there are two ways that can be used to alter lipofuscin accumulation, namely, by the administration of certain drugs or by dietary regime.

By far the most studied drug in this respect is centrophenoxine (meclophenoxate, lucidril). In 1966 Nandy and Bourne[49] were the first to show that administration of centrophenoxine intraperitoneally or intramuscularly in guinea pigs (80 mg/kg/day) for 4 to 12 weeks resulted in a decrease in the amount of lipofuscin in the brain stem reticular formation. The same dosage was shown to be effective with rats and mice as well. In the latter species, experiments were carried out in the course of 2 to 11 months starting with mice 30 days old. From a large body of published data (for a review see Reference 50), it is evident that administration of centrophenoxine decreases the formation of the pigment during developmental stages as well as reduces its concentration in older animals. Particularly relevant to the aims of the present volume in this respect are the results of Spoerri et al.,[51] who studied the effect of this drug administration on lipofuscin content in cardiac muscle.

Another way of studying intracellular pigment accumulation is offered by tissue cultures. Spoerri and Glees[52] were the first to study the effect of centrophenoxine under these conditions. They reported that daily application of centrophenoxine to the cultures for 2 to 3 weeks resulted in fading out and vacuolation of the pigment soon after initial drug application. The blocking effect of centrophenoxine upon lipofuscin accumulation was also demonstrated with cultured neuroblastoma cells. The percentage of cells exhibiting pigment increased from 25% at 5 days of culture to 60% after 25 days of culture. Pigment formation was increased by treatment for 9 days with either papaverine (1.1×10^{-5} M) or prostaglandin E (5.6×10^{-5} M).[57] Pigment formation in papaverine-treated cells was considerably decreased by application of 1.0 or 3.4×10^{-4} M centrophenoxine for 90 days.

Though the results obtained with this drug are generally positive in terms of blocked lipofuscin accumulation or even its disappearance from already pigmented cells, the negative results of Tomonaga[53] in the guinea pig spinal cord and Riga and Riga[54] in the granule cell layer represent a warning regarding premature generalizations.

Another drug capable of decreasing the deposition of lipofuscin was shown to be potassium orotate (25 mg/kg, Wistar rats). The period needed for obtaining distinct results was over

50 days. Instead of potassium orotate, it is also possible to use the magnesium salt in a dosage of 100 mg/kg body mass. Kavain, applied in the same dosage, also showed a lowering effect upon lipofuscin accumulation.[55]

Since it was demonstrated in the past that during centrophenoxine treatment, lactic dehydrogenase, cytochrome oxidase, and succinic dehydrogenase activities were decreased while the activity of glucose-6-phosphatase was increased, it was proposed that pigment accumulation may be reduced by possible shunting of glucose metabolism toward the pentose cycle.[56] Whether this is so or not remains to be elucidated; it is, however, known that a decrease in lipofuscin in the central nervous system can be observed after treating rats with SH-containing substances.[42] It has already been mentioned that the administration of papaverine or prostaglandin E lead to enhanced accumulation of lipofuscin in cultured neuroblastoma cells.[57] Enhanced production of lipofuscin in nervous tissues was reported also following administration of ACTH[58] and perchlorperazine in rats[59] and tetanus toxin in mice.[60]

Regarding the experimental modifiability of lipofuscin occurrence, the following approaches are applicable in practice. Animals kept on a vitamin E-deficient diet[61] or, alternatively, on a diet producing physiological antioxidant deficiency[62] (32 weeks) show a greater accumulation of the fluorescent pigment in comparison to age-matching controls. On the other hand, the effects of vitamin E-supplemented diet on organic solvent-soluble lipofuscin in mice was studied by Csallany et al;[63] of the organs studied (liver, spleen, kidney, brain, uterus, lung, and heart), only the liver showed lower organic solvent-soluble pigment concentrations after vitamin E treatment.[69] Vitamin E supplementation showed a positive effect in terms of decreasing pigment concentration by 12 months of age.

The greater formation of fluorescent pigment can also be observed in rats (*Rattus rattus*) fed a diet high in polyunsaturated fatty acids and deficient concomitantly in vitamin E, selenium, sulfur-containing amino acids, and chromium.[62] If such a diet is supplemented with methionine and chromium, a significant reduction in the amount of the fluorescent pigment in the retinal pigment epithelium occurs. When the diet is supplemented with all the four types of nutrients, the amount of the fluorescent pigment drops down to very low levels. Similar effects to those observed in the retinal pigment epithelium can be seen in testes, renal tubule epithelial cells, lamina propria of the small intestine, and in the heart.[50] However, it was reported that the autofluorescence observed in the retinal pigment epithelium by the antioxidant deficiency is more pronounced than in other tissues. Also, no increase in the yellow fluorescent pigment was observed under these conditions in the liver, skeletal muscle, cerebrum, and hypothalamus. In the aorta or the lung, such a pigment was not detected at all. It is also worth mentioning that there are literary data available in which a vitamin E-deficient diet is combined with vitamin A-deficient or vitamin A-adequate diet. These studies, aimed to elucidate the role of both vitamins for the structure of the retinal pigment epithelium, indicated a slight increase of the fluorescent pigment in vitamin E-free as compared to vitamin E-adequate animals. (Sprague-Dawley rats, females, 5 months of the dietary regime).[64] There are some indications that in addition to lipofuscin, another fluorescent component is involved.

A larger number of lipofuscin bodies can be found in animals fed a low-protein diet (squirrel monkeys, 2% protein diet, 15 weeks).[65] If the animals are placed on a 25% protein-containing diet, a decrease in the lipofuscin content with an accompanying decrease in acid phosphatase and esterase activity is observed in neuronal perikarya; on the other hand, lipofuscin content increases in perineuronal glial cells. (From this observation the speculation about the role of glial cells in lipofuscin removal is derived.) The increase of lipofuscin in the motor cortex and cerebellum occurs not only in the experimental animals, but also in fetuses if mothers are kept on the low-protein diet regime.[66]

IV. OTHER PIGMENTS

There are at least two additional pigments that are worth mentioning. First is the so called ceroid (autofluorescent lipopigment) that can be obtained from postmortem brain samples of patients with neuronal ceroid-lipofuscinosis or from English setters affected with canine ceroid-lipofuscinosis. The flow-sheet for the preparation of this pigment is analogous to the procedures described for brain, heart, and liver lipofuscin (Figure 6).

The other pigment is called neuromelanin. This is a granular, naturally yellow to deep brown intraneuronal pigment which accumulates with age and which occurs in certain nuclei and brain regions as well as in dorsal root ganglia of the spinal cord.[45] It is found only in mammals, in the primates, carnivores, horse, and giraffe. The pigment is rich in sulfur and occurs in lysozomes of catecholamine-producing neurons. Its biochemical, histochemical, and electron-microscopic properties characterize it as melanized lipofuscin. The interesting thing about this pigment is that, contrary to lipofuscin, the content of neuromelanin declines with age.[46]

A detailed description of the current knowledge about these two categories of pigments are clearly beyond the scope of this chapter, and the reader is referred to more specialized reviews.[47,48]

V. DEPOSITS CAUSED BY DYSFUNCTION OF LYSOSOMES

The accumulation of lipofuscin is considered to be the consequence of the inability of lysosomal enzymes to degrade this compound. This, however, is only a single example of an accumulatory process resulting from lysosomal dysfunction. It has to be kept in mind that lysosomes represent the major site of intracellular digestion in all cell types, and this process is effected by a complex system of hydrolases (over 40 of which have been described up to now).

After the breakdown of intra- and extracellular components in the secondary lysosomes, the diffusible hydrolytic products can return to the cytoplasm and undigested residues are excreted by the exocytosis or are retained in the cell as residual bodies. Despite the apparent capacity of lysosomes to degrade a wide variety of macromolecules, lysosomes containing undigested materials occur in many cell types. Whether such undigested bodies are released from the cells is unknown. On the other hand, progressive accumulation of undigestible material due to lysosomal dysfunction is well known, e.g., in lysosomal storage diseases or in suitable animal models. The entities stored can be divided into two categories: they can either represent compounds that are normally present in the living body (e.g., glycosaminoglycans, glycolipids, phospholipids, cholesterolesters, triglycerides, glycogen, etc.) or materials which are in fact pathological (pigments, posttranslationally modified proteins).[71] Methods needed for studying the former category of stored compounds are mostly biochemical procedures, the description of which is beyond the scope of this volume. To elucidate the metabolic pathways of these depository processes, several animal models are available; these include cats with G_{M1} gangliosidosis,[72] cairn and West Highland terriers with Krabbe disease,[73] and Angus cattle with mannosidosis.[74] An important example in which a disturbance in the rate of exocytosis may play a role in a lysosomal disease occurs in the beige mutant mouse. This mutant is characterized by the inability to secrete lysosomal hydrolases from the kidney after androgenic stimulation.[75]

The second category of the stored materials is represented by compounds which the organism is unable to degrade because of structural modification. Thus lipofuscin is considered to be undigestible due to the peroxidation of lipids by free radicals and subsequent cross-linking of the macromolecule.[76,77]

Briefly blend whole brain in Solution 2, homogenize, dilute to 0,2 kg/l with Solution 2, centrifugation as follows:

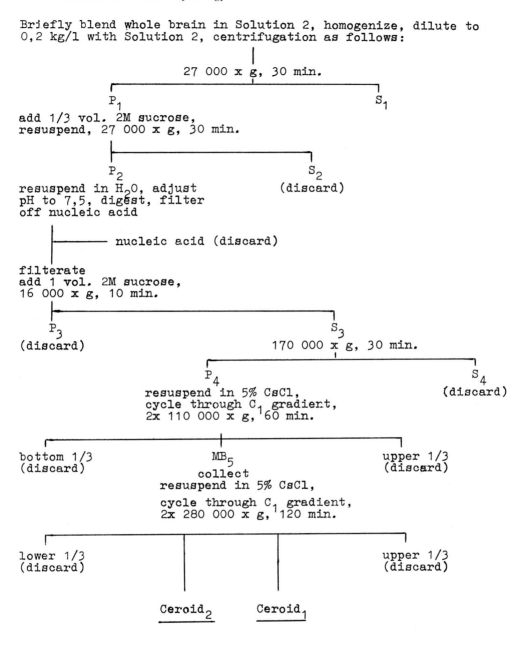

Solution 1 = 0,8M sucrose, 0,05M Tris base

Linear gradient C_1 = distilled water to 40% caesium chloride

FIGURE 6.   Flow scheme for ceroid isolation. (From Siakotos, A. N. and Munkres, K. D., in *Age Pigments*, Sohal, R. S., Ed., Elsevier, Amsterdam, 1981, 181. With Permission.)

Accumulation of such substances is best studied in cultured cells irradiated with ultraviolet light.[78] Recent studies using a fluorescence-activated cell sorter have established that when fibroblasts are cultured in vitro, the proliferative rate is inversely related to the rate of accumulation of fluorescent compounds.[79] It is also possible to follow the accumulation of the fluorescent pigment during the vitro aging of the cell culture.[78] There are reports in the literature that aside from the fluorescent bodies, high-molecular-weight glycoproteins are also accumulated in aged cells.[80]

VI. AMYLOID

A specific category of extracellular glycoprotein accumulation is amyloidosis. This process is characterized by accumulation of translucent acidophilic material. Deposition of amyloid in tissues can be demonstrated by positive birefringence in unstained tissues or in Congo Red stained specimens. In the electron microscope, amyloid-laden tissues reveal a large extracellular accumulation of fine fibers.[81]

Amyloid deposits consist of specific amyloid proteins to which other macromolecular compounds can be bound, e.g., fibronectin. Precursors of amyloid-specific proteins can be detected in serum even under normal circumstances. Description of the amyloid protein complex and characterization of its individual components is beyond the scope of this chapter, and readers interested in this area are referred to Reference.[82]

One component of amyloid is the so-called amyloid component P, which is immunologically identical with α-globulin. Aside from this, a plasma amyloid component was identified which is a glycoprotein with a characteristic pentagonal ultrastructure composed of five identical subunits. In terms of antigenicity, molecular mass, ultrastructure, and partial amino acid sequence it is undistinguishable from the amyloid P component. It was also proven that the plasma amyloid component P has about 60% homology with the so called C-reactive protein, a typical acute-phase reactant.

For isolation of the plasma amyloid P component and amyloid P component, several methods are available. Selective binding of the amyloid P component to fibronectin and C_4-binding protein can be used for this purpose.[83] A procedure yielding high quantities of plasma amyloid P component was described by Ohkubo et al.[84]

Briefly, the procedure runs as follows. Plasma is mixed with barium chloride (20 g/ℓ), stirred, and centrifuged at 1000 g for 30 min. The precipitate is suspended in 1 ℓ of sodium acetate (5 mmol/ℓ, pH 7.4), homogenized, and centrifuged. Sodium acetate extraction is repeated three times. The precipitate is finally resuspended in sodium acetate (200 mℓ/ℓ, pH 7.4) and stirred for 30 min before final centrifugation at 10,000 g. The supernatant is then brought to 30% saturation by ammonium sulfate and stirred. Next, the arising precipitate is removed by centrifugation at 10,000 g for 30 min. The supernatant is brought to 60% saturation, stirred, and centrifuged again. The precipitate is dissolved in Tris-HCl (20 mmol/ℓ, pH 8.0 containing 150 mmol/ℓ NaCl and 5 mmol/ℓ benzamidine) and dialyzed overnight against Tris-HCl buffer (20 mmol/ℓ, pH 8.0, containing 150 mmol/ℓ NaCl and 1 mmol/ℓ benzamidine). The dialysate is applied to a zinc-chalate-Sepharose® column equilibrated previously with Tris-HCl (20 mmol/ℓ, pH 8.0, containing 150 mmol/ℓ NaCl and 1 mmol/ℓ benzamidine). Plasma amyloid protein is eluted with a linear gradient of 1-histidine (0 to 30 mmol/ℓ in the equilibration buffer). Fractions containing plasma amyloid protein are pooled and dialyzed against sodium phosphate buffer (20 mmol/ℓ, pH 7.5). Further purification of the plasma amyloid protein is done by Red-Sepharose® column chromatography. The precipitate arising after sodium phosphate dialyses is removed by centrifugation and the supernatant is applied to a Red-Sepharose® column equilibrated with 20 mmol/ℓ sodium phosphate buffer (pH 7.5) and eluted with a linear KCl gradient (0 to 400 mmol/ℓ). For the determination of serum amyloid P component, it is possible to use the ELISA technique, western blotting, or radioimmunoassay.[85-87] Most of the studies regarding amyloid were done on necroptic material, as amyloid depositions occur in chronic inflammatory diseases. Following such depositions is possible also in animal models of chronic inflammation. Recently there appeared a possibility of studying experimental amyloidosis in CBA/J mice showing a high incidence of spontaneous amyloid formation,[88]

VII. DEPOSITS OF HEAVY METALS

Of the numerous inorganic compounds present in the body of an animal, there is a limited number of heavy metals which are subject to accumulatory processes. Basically these heavy metals are present in the form of salts or they are bound to proteins (enzymes). They may also be present as chelates (particularly if they are of exogenous origin). Accumulatory effects can be observed only with those metals which are relatively nontoxic and which undergo interactions with long-lived proteins.[89]

Even some metals with high toxicity can be deposited in the body provided that they are administered in the chelate form. Under in vitro conditions chelates are rather unreactive towards protein structures. In vivo, however, one may expect degradation of the chelating agent and liberation of the metal atom, after which the protein metal interaction may be observed. It has been demonstrated in the past that the nature of the chelating agent is of minor importance. Practically all chelating agents are capable of transporting metals within the body to tissues in which they can be deposited. Whether deposition of the metals occurs or not depends on the stability constant of the chelate. Of the long-lived proteins that are capable of interactions with heavy metals, collagen represents the best investigated model. As long as collagen represents a group of closely related proteins, it is necessary to keep in mind that the following considerations are related mainly to collagen types I and III. Monomeric collagen molecules contain a considerable number of polar groups and atoms with free electron pairs which are capable of binding metal ions. These binding sites permit a large number of combinations for different types of metal binding. Heavy metal atoms are bound to the collagen structure either by mono- or bitopic binding. The former case is much more frequent than the latter and it can be visualized in electron micrographs as the cross-striation pattern.[90] Bipolar intermolecular binding results in the formation of cross-links, which can be detected by different methods (e.g., chromatography and electrophoresis). Typically, when doses in units of milligrams per 100 g body weight per week are administered, alterations of protein properties can be seen after several weeks. The first change observed is the alteration in the mechanochemical properties (shrinkage temperature, swelling ability, solubility).[91] The reactivity of heavy metal complexes is influenced by the following factors: (1) complex stability, (2) the nature of the bond, (3) the size of the central atom, (4) the charge of the central atom, (5) the electronic structure of the central atom, and (6) steric relations, both of the central atom and the ligand.

There is no specific method available for the determination of the binding site of a metal atom in the protein structure. The following two general ways are the most common for obtaining such information: (1) studying the ion exchange reactions of complexes arising between different ligands and (2) following the mild proteolysis of protein-metal complexes without breaking the metal-protein bond. From the resulting peptide mixture, peptides containing the metal are isolated by routine separation techniques, and sequence studies may then be applied. It has to be emphasized that in some situations metals can be deposited as colloids (gold).

REFERENCES

1. **Harding, J. J.,** Nonenzymatic covalent posttranslational modification of proteins in vivo, *Adv. Protein Chem.,* 37, 247, 1985.
2. **Wold, F. and Moldave, K., Eds.,** *Methods in Enzymology,* Vol. 106A, Academic Press, New York, 1984.
3. **Sohal, R. S.,** *Age Pigments,* Elsevier, Amsterdam, 1981.
4. **Vlassara, H., Brownlee, M., and Cerami, A.,** Nonenzymatic glycosylation: role in the pathogenesis of diabetic complications, *Clin. Chem.,* 32, 337, 1986.

5. **Stevens, V. J., Fantl, W. J., Newman, C. B., Sims, R. V., Cerami, A., and Peterson, C. M.,** Acetaldehyde adducts with hemoglobin, *J. Clin. Invest.,* 67, 361, 1981.

6. **Pangor, S., Ulrich, P. C., Bencsath, F. A., and Cerami, A.,** Aging of proteins: isolation and identification of a fluorescent chromophore from the reaction of polypeptides with glucose, *Proc. Natl. Acad. Sci. U.S.A.,* 81, 2684, 1984.

7. **Dixon, H. B. F.,** A reaction of glucose with peptides, *Biochem. J.,* 129, 203, 1972.

8. **Koening, R. J., Blobstein, S. H., and Cerami, A.,** Structure of carbohydrate of hemoglobin A$_{Ic}$, *J. Biol. Chem.,* 252, 2992, 1977.

9. **Stevens, V. J., Rouser, C. A., Monnier, V. M., and Cerami, A.,** Diabetic cataract formation: Potential role of glycosylation of lens crystallins, (nonenzymatic glycosylation/sulphydryl oxidation), *Proc. Natl. Acad. Sci. U.S.A.,* 75, 2918, 1978.

10. **Brownlee, M. and Cerami, A.,** The biochemistry of the complications of diabetes mellitus, *Annu. Rev. Biochem.,* 50, 385, 1981.

11. **Wieland, O. H.,** Proteins modification by non-enzymatic glycosylation: possible role in the development of diabetic complications, *Mol. Cell. Endocrinol.,* 29, 125, 1983.

12. **Schleicher, E., Scheller, L., and Wieland, O. H.,** Quantitation of lysine-bound glucose of normal and diabetic erythrocyte membranes by HPLC analysis of furosine [ε-N(L-furosylmethyl)-L-lysine], *Biochem. Biophys. Res. Commun.,* 99, 1011, 1981.

13. **Gaines, K. C., Salhany, J. M., Tuma, D. J., and Sorrell, M. F.,** Reaction of acetaldehyde with human erythrocyte membrane proteins, *FEBS Lett.,* 75, M15, 1977.

14. **Ting, H. H. and Crabbe, M. J. C.,** Schiff base products in vivo, *Biochem. Soc. Trans.,* 10, 412, 1982.

15. **Eyre, D. R., Paz, M. A., and Gallop, P. M.,** Cross-linking in collagen and elastin, *Annu. Rev. Biochem.,* 53, 717, 1974.

16. **Harding, J. J. and Crabbe, M. J. C.,** in *The Eye,* Vol. 1b, Davson, H., Ed., Academic Press, New York, 1984, 207.

18. **Ink, S. L., Mehansho, H., and Henderson, L. M.,** The binding of pyridoxal to hemoglobin, *J. Biol. Chem.,* 257, 4753, 1982.

19. **Bucala, R., Fishman, J., and Cerami, A.,** Formation of covalent adducts between cortisol and 16 α-hydroxyestrone and protein: possible role in the pathogenesis of cortisol toxicity and systemic lupus erythematosus, (Heyns rearrangement/steroid-protein adducts/albumin/high-pressure liquid chromatography), *Proc. Natl. Acad. Sci. U.S.A.,* 79, 3320, 1982.

20. **Manabe, S., Bucala, R., and Cerami, A.,** Nonenzymatic addition Glucocorticoids to lens proteins in steroid-induced cataracts, *J. Clin. Invest.,* 74, 1803, 1984.

21. **Jain, S. K. and Hochstein, P.,** Polymerization of membrane components in aging red blood cells, *Biochem. Biophys. Res. Commun.,* 92, 247, 1980.

22. **Benedetti, A., Camporti, M., Fulceri, R., and Esterbauer, H.,** Cytotoxic aldehydes originating from the peroxidation of liver microsomal lipids. Identification of 4,5-dihydroxy-decenal, *Biochim. Biophys. Acta,* 792, 172, 1984.

23. **Andersen, S. O.,** Hardening of insect cuticle, *Annu. Rev. Entomol.,* 24, 29, 1979.

24. **Mason, H. S. and Peterson, B. W.,** Melanoproteins. I. Reactions between enzyme generated quinones and amino acids, *Biochem. Biophys. Acta,* 111, 134, 1965.

25. **Lipke, H. and Henzel, W.,** Arylated polypeptides of sarcophagid cuticle, *Insect Biochem.,* 11, 445, 1981.

26. **Jerina, D. M., Daly, J. W., Witkop, B., Zaltzman-Nirenberg, P., and Udenfriend, S.,** 1,2-Naphthalene oxide as an intermediate in the microsomal hydroxylation of naphthalene, *Biochemistry,* 9, 147, 1970.

27. **Jollow, D. I., Mitchell, J. R., Porter, W. Z., David, D. C., Gillette, J. R., and Brodie, B. B.,** Acetaminophen-induced hepatic necrosis. II. Role of covalent binding "in vivo", *J. Pharm. Exp. Ther.,* 187, 195, 1973.

28. **Nakagawa, Y., Hiraga, K., and Suga, T.,** On the mechanism of covalent binding of butylated hydroxytoluene to microsomal protein, *Biochem. Pharmacol.,* 32, 1417, 1983.

29. **Maggs, J. L., Grabowski, P. S., and Park, B. K.,** Drug-protein conjugates. V. Sex-linked differences in the metabolism and irreversible binding of 17 α-ethinylestradiol in the rat, *Biochem. Pharmacol.,* 32, 2793, 1983.

30. **Brownlee, M., Pangor, S., and Cerami, A.,** Covalent attachment of soluble proteins by nonenzymatically glycosylated collagen, *J. Exp. Med.,* 158, 1739, 1983.

31. **Higgins, P. J. and Bunn, H. F.,** Kinetic analysis of the nonenzymatic glycosylation of hemoglobin, *J. Biol. Chem.,* 256, 5204, 1981.

32. **Elleder, M.,** Chemical characterization of age pigments, in *Age Pigments,* Sohal, R. S., Ed., Elsevier, Amsterdam, 1981, 204.

33. **Chio, K. S. and Tappel, A. L.,** Synthesis and characterization of the fluorescent products derived from malonaldehyde and amino acids, *Biochemistry,* 7, 2821, 1969.

34. **Billard, C. J. and Tappel, A. L.,** Fluorescent products from reaction of peroxidizing polyunsaturated fatty acids with phosphatidyl ethanolamine and phenylalanine, *Lipids,* 8, 183, 1973.
35. **Shimasaki, H., Privett, O. S., and Hara, I.,** Studies of the fluorescent products of lipid oxidation in aqueous emulsion with glycine and on the surface on silica gel, *JAOCS,* 54, 119, 1977.
36. **Buttekus, H. and Bose, R. J.,** Amine malonaldehyde condensation products and their relative color contribution in the thiobarbituric acid test, *J. Food Sci.,* 49, 440, 1972.
37. **Essner, E. and Novikoff, A. B.,** Human hepatocellular pigments and lysosomes, *J. Ultrastruct. Res.,* 3, 374, 1960.
38. **Sosa, J. M.,** Aging of neurofibrils, *J. Gerontol.,* 7, 191, 1952.
39. **Novikoff, A. B.,** Lysosomes and related particles, in *The Cell,* Vol. II, Brachet, J. and Mirsky, A. E., Eds., Academic Press, New York, 1961, 423.
40. **Shauklin, W. M., Issidorides, M., and Nassar, T. K.,** Neurosecretion in human cerebellum, *J. Comp. Neurol.,* 107, 315, 1957.
41. **Barden, H.,** The biology and chemistry of neuromelanin, in *Age Pigments,* Sohal, R. S., Ed., Elsevier, Amsterdam, 1981, 156.
42. **Brizzee, K. R. and Ordy, J. M.,** Cellular features, regional accumulation and prospects of modification of age pigments in mammals, in *Age Pigments,* Sohal, R. S., Ed., Elsevier, Amsterdam, 1981, 102.
43. **Siakotos, A. N. and Munkres, K. D.,** Purification and properties of age pigments, in *Age Pigments,* Sohal, R. S., Ed., Elsevier, Amsterdam, 1981, 181.
44. **Siakotos, A. N. and Koppang, N.,** Procedure for the isolation of lipopigments from brain, heart and liver and their properties: a review, *Mech. Age Dev.,* 2, 177, 1973.
45. **van Woert, M. H. and Ambani, L. M.,** Biochemistry of neuromelanin, *Adv. Neurol.,* 5, 215, 1974.
46. **Mann, D. M. A., Yates, P. O., and Barton, C. M.,** Variations in melanin content with age in the human substantia nigra, *Biochem. Exp. Biol.,* 13, 137, 1977.
47. **Barden, H.,** The biology and chemistry of neuromelanin, in *Age Pigments,* Sohal, R. S., Ed., Elsevier, Amsterdam, 1981, 156.
48. **Barden, H.,** Further histochemical studies characterizing the lipofuscin component of human neuromelanin, *J. Neuropathol. Exp. Neurol.,* 37, 437, 1978.
49. **Nandy, K. and Bourne, G. H.,** Effect of centrophenoxine on the lipofuscin pigments in neurones of senile guinea pigs, *Nature,* 210, 313, 1966.
50. **Zucherman, B. M. and Geist, M. A.,** Effect of nutrition and chemical agents on lipofuscin formation, in *Age Pigments,* Sohal, R. S., Ed., Elsevier, Amsterdam, 1981, 283.
51. **Spoerri, P. E., Glees, P., and Ghazzawi, E. E.,** Accumulation of lipofuscin in the myocardium of senile guinea pigs: dissolution and removal of lipofuscin following dimethylaminoethyl chlorophenoxyacetate administration. An electron microscope study, *Mech. Age, Dev.,* 3, 311, 1974.
52. **Spoerri, P. E. and Gees, P.,** The effect of dimethylaminoethyl p-chlorophenoxyacetate on spinal ganglia neurons and satellite cells in culture. Mitochondrial changes in the aging neurons. An electron microscope study, *Mech. Age. Dev.,* 3, 131, 1974.
53. **Tomonaga, M., Izumiyama, N., and Kameyama, M.,** On the effect of centrophenoxine on the lipofuscin in the nerve cells of the aged guinea pigs, *Nippon Ronev Igakkai Zasshi,* 15, 355, 1978.
54. **Riga, S. and Riga, D.,** Effects of centrophenoxine on the lipofuscin pigments in the nervous systems of old rats, *Brain Res.,* 72, 265, 1974.
55. **Varkonyi, T., Domokos, H., Maurer, M., Zoltan, O. T., Scillik, B., and Foldi, M.,** Die Wirkung von DL-Kavain und Magnesium-orotat auf die feinstrukturellen neuropathologischen Veränderungen der experimentallen lymphogenen Enzephalopathie, *Z. Gerontol.,* 3, 254, 1970.
56. **Nandy, K.,** Further studies on the centrophenoxine on the lipofuscin pigment in the neurons of senile guinea pigs, *J. Gerontol.,* 23, 82, 1968.
57. **Schneider, F. H. and Nandy, K.,** Effect of entrophenoxine on lipofuscin formation in neuroblastoma cells in culture, *J. Gerontol.,* 32, 132, 1977.
58. **Sulkin, N. M. and Srivanij, P.,** The experimental production of senile pigments in the nerve cells of young rats, *J. Gerontol.,* 15, 2, 1960.
59. **Roizan, L.,** Some basic principles of molecular physiology. III. Ultracellular organelles as structural metabolic and pathogenic gradients, *J. Neuropathol. Exp. Neurol.,* 23, 209, 1964.
60. **Zacks, S. I. and Sheff, M. F.,** Studies on tetanus toxin. I. Formation of intramitochondrial dense granules in mice acutely poisoned with tetanus toxin, *J. Neuropathol. Exp. Neurol.,* 23, 306, 1964.
61. **Reddy, K., Fletcher, B., Tappel, A., and Tappel, A. L.,** Measurement and spectral characteristics of fluorescent pigments in tissues of rats as a function of dietary polyunsaturated fats and vitamin E, *J. Nutr.,* 103, 908, 1973.
62. **Katz, M. L., Stone, W. L., and Dratz, E. A.,** Fluorescent pigment accumulation in retinal pigment epithelium of antioxidant-deficient rats, *Invest. Ophthal. Vis. Sci.,* 17, 1049, 1978.
63. **Csallany, A. S., Ayaz, K. L., and Su, L. C.,** Effect of vitamin E and aging on tissue lipofuscin pigment concentration in mice, *J. Nutr.,* 107, 1792, 1972.